Carbon Nanodots

Mahendra Rai • Sushama Talegaonkar
József Prokisch
Editors

Carbon Nanodots

Synthesis and Applications in Health, Food and Agriculture

Springer

Editors
Mahendra Rai
Department of Chemistry
Federal University of Piauí (UFPI)
Teresina, Piauí, Brazil

József Prokisch
Institute of Animal Science,
Biotechnology, and Nature Conservation
University of Debrecen, Faculty of
Agricultural and Food Sciences and
Environmental Management
Debrecen, Hungary

Sushama Talegaonkar
Department of Pharmaceutics, School of
Pharmaceutical Sciences
Delhi Pharmaceutical Sciences and Research
University
New Delhi, Delhi, India

ISBN 978-981-95-4704-3 ISBN 978-981-95-4705-0 (eBook)
https://doi.org/10.1007/978-981-95-4705-0

This Springer imprint is published by the registered company Springer Nature Singapore Pte Ltd.
The registered company address is: 152 Beach Road, #21-01/04 Gateway East, Singapore 189721, Singapore

Preface

The rapid progress of nanoscience has ushered in new frontiers where materials at the nanoscale are transforming health, food, and agriculture. Among these, carbon nanodots (CNDs) have emerged as one of the most dynamic and sustainable innovations of recent years.

Carbon nanodots (CNDs), a novel class of carbon-based nanomaterials, have quickly attracted global attention due to their remarkable properties, bright photoluminescence, water solubility, chemical stability, and eco-friendly nature. In an era that increasingly values green technologies and sustainable innovations, CNDs are emerging as versatile tools with applications spanning human health, food safety, and agriculture.

The present volume, *Carbon Nanodots: Synthesis and Applications in Health, Food and Agriculture*, offers a comprehensive and up-to-date overview of this exciting field. Bringing together leading experts, the book examines the synthesis of CNDs through conventional, green, and AI-assisted approaches, alongside advanced characterization techniques. It further highlights their practical applications, ranging from disease diagnostics and targeted drug delivery to food preservation, smart packaging, and sustainable agriculture.

Organized into four sections, the volume begins with fundamental concepts and synthesis strategies. The second section explores biomedical frontiers, including applications in diagnostics, infectious disease management, and cancer therapy. The third section turns to food science and agriculture, showcasing innovations that improve processing, preservation, and crop health. The final section addresses critical issues of toxicity, environmental safety, and the challenges of scaling up production for real-world deployment.

This book is designed for a wide readership, including researchers, graduate students, industry professionals, and policymakers working in nanoscience, biotechnology, agritech, and food systems. We hope it will serve as both a reference and a springboard for interdisciplinary research, inspiring new directions in the sustainable use of carbon nanodots.

We extend our sincere gratitude to all contributing authors for their valuable chapters and insights, and to the reviewers and editorial team for their thoughtful feedback. A special note of appreciation is due to Vijayalakshmi Dayalan,

Production Editor, whose meticulous coordination and professionalism ensured the timely completion of this volume.

We trust this book will deepen scientific understanding of carbon nanodots and stimulate innovative applications with meaningful impact on health, food, and agriculture.

Teresina, Piauí, Brazil — Mahendra Rai
New Delhi, India — Sushama Talegaonkar
Debrecen, Hungary — József Prokisch

Contents

Editors and Contributors

About the Editors

Mahendra Rai is a UGC–Basic Science Research Faculty Fellow and former Head of the Department of Biotechnology at Sant Gadge Baba Amravati University, India. He is currently a Senior Visiting Professor at the Federal University of Piauí, Teresina, Brazil. His research expertise spans microbial biotechnology and nanobiotechnology, with a particular focus on the green synthesis of metal nanoparticles and carbon dots using plants and fungi for antimicrobial applications. He has published over 450 research papers and authored/edited 75 books. His research impact is reflected in an h-index of 93. He is the recipient of several prestigious honors, including the Father T.A. Mathias Award and the Medini Award (Government of India). He has also served as a visiting professor/scientist in several countries, including Brazil, Poland, Switzerland, and Italy. Recognized among the top 2% of global scientists in nanoscience and nanotechnology, he was recently honored with the 2025 Biology and Biochemistry India Leader Award by Research.com, where he was ranked 7th nationally in Biology and Biochemistry.

Sushama Talegaonkar is working as an Associate Professor in the Department of Pharmaceutics, School of Pharmaceutical Sciences, DPSRU, New Delhi. She has approximately 24 years of experience in teaching and research. She has been ranked among the top 2% of world scientists according to the global list compiled by Stanford University for the last five consecutive years, since 2020.

Dr. Talegaonkar has recently been granted two Indian patents and has filed three Indian patents that are currently under examination. She has published more than 210 research papers in high-impact-factor, internationally reputed journals and 30 papers in national journals, with an h-index of 63, an i10-index of 167, and more than 16,000 Google Scholar citations. She has travelled to Canada, United Kingdom, and Singapore to present her research papers. Her research paper titled "Development and bioavailability assessment of ramipril nanoemulsion formulation" has received the most cited research paper award on behalf of reputed publisher and the editorial board of the *European Journal of Pharmaceutics and*

Biopharmaceutics. She is also the recipient of UGC Research Award 2014–16. She is actively involved in developing a wide variety of smart and functionalized nano drug delivery systems for targeted drug delivery to treat challenging diseases like cancer (breast and triple negative breast cancer), rheumatoid arthritis, osteoporosis, etc. She is also a member of the Board of Governors, Hyderabad.

József Prokisch is a chemist specializing in food innovation and product development, with over 40 food products and several patents to his name. As a professor at the University of Debrecen, he teaches physical chemistry, automation, toxicology, biosensors, and functional food development. He has authored over 400 scientific publications and 6 patents and is the founder of multiple companies. His research focuses on nanotechnology, specifically the production and effects of selenium, tellurium, and carbon nanoparticles. Recently, he has been working on introducing sensors and automation in food technology. He actively participates in EU and Hungarian research projects and has collaborated with companies in Finland. He also spent two years as a postdoctoral researcher at the Institute of Reference Materials and Measurements (BCR-IRMM) in Belgium.

Contributors

D. Akila Department of Agricultural Microbiology, Tamil Nadu Agricultural University, Coimbatore, Tamil Nadu, India
Department of Food Process Engineering, Tamil Nadu Agricultural University, Coimbatore, Tamil Nadu, India

Graciela Dolores Avila-Quezada Facultad de Ciencias Agrotecnológicas, Universidad Autónoma de Chihuahua, Chihuahua, Chihuahua, Mexico

M. Balakrishnan Department of Agricultural Microbiology, Tamil Nadu Agricultural University, Coimbatore, Tamil Nadu, India
Department of Food Process Engineering, Tamil Nadu Agricultural University, Coimbatore, Tamil Nadu, India

Santan Barthwal ICFRE-Forest Research Institute, P.O. New Forest, Dehradun, Uttarakhand, India

Áron Béni Institute of Agricultural Chemistry and Soil Science, Faculty of Agricultural and Food Sciences and Environmental Management, University of Debrecen, Debrecen, Hungary

Rahul Bhagat Department of Biotechnology, Government Institute of Science, Chhatrapati Sambhajinagar, Maharashtra, India

Mohit Bhatt Department of Physics, School of Advanced Engineering, UPES, Dehradun, India

José Milton Elias de Matos Department of Chemistry, Federal University of Piaui (UFPI), Teresina, Brazil

C. Devadharshini Department of Agricultural Microbiology, Tamil Nadu Agricultural University, Coimbatore, Tamil Nadu, India
Department of Food Process Engineering, Tamil Nadu Agricultural University, Coimbatore, Tamil Nadu, India

Mohini Diwan School of Pharmaceutical Sciences, Delhi Pharmaceutical Sciences and Research University, New Delhi, India
Department of Pharmaceutics, School of Pharmaceutical Sciences, Delhi Pharmaceutical Sciences & Research University (DPSRU), Government of NCT of Delhi, Delhi, India

Ali Eslamifar Department of Clinical Research, Pasteur Institute of Iran, Tehran, Iran

Aniket K. Gade Department of Microbiology, Faculty of Biological and Veterinary Sciences, Nicolaus Copernicus University in Toruń, Toruń, Poland
Department of Biological Sciences and Biotechnology, Institute of Chemical Technology, Mumbai, Maharashtra, India

Kajal Gautam Department of Chemistry, School of Advanced Engineering, UPES, Dehradun, India

Mohammadhassan Gholami-Shabani Department of Mycology, Pasteur Institute of Iran, Tehran, Iran
Department of Labor Environment, Health, Safety, Environment and Energy, Ministry of Industry, Mine and Trade, Tehran, Iran

Arun Ghuge Department of Forensic Biology, Government Institute of Forensic Science, Nagpur, Maharashtra, India

K. M. Gijith Mohan Drug Delivery and Nanomedicine Research Laboratory, Department of Pharmaceutics, National Institute of Pharmaceutical Education and Research (NIPER)-Raebareli, Lucknow, India

Honey Goel Department of Pharmaceutics, University Institute of Pharmaceutical Sciences and Research, Baba Farid University of Health Sciences (BFUHS), Faridkot, Punjab, India

Patrycja Golińska Department of Microbiology, Faculty of Biological and Veterinary Sciences, Nicolaus Copernicus University in Toruń, Toruń, Poland

Patrycja Grabowska Faculty of Chemistry, Nicolaus Copernicus University in Torun, Torun, Poland

Indarchand Gupta Department of Biotechnology, Government Institute of Science, Chhatrapati Sambhajinagar, Maharashtra, India

Anna Ilnicka Faculty of Chemistry, Nicolaus Copernicus University in Torun, Torun, Poland

Avinash Ingle Biotechnology Centre, Department of Agricultural Botany, Dr. Panjabrao Deshmukh Agricultural University, Akola, Maharashtra, India

Keerti Jain Drug Delivery and Nanomedicine Research Laboratory, Department of Pharmaceutics, National Institute of Pharmaceutical Education and Research (NIPER)-Raebareli, Lucknow, India

Fatemehsadat Jamzivar Department of Mycology, Pasteur Institute of Iran, Tehran, Iran

Kiran Jeet Electron Microscopy and Nanoscience Lab, Punjab Agricultural University, Ludhiana, Punjab, India

Vandita Kakkar University Institute of Pharmaceutical Sciences, Panjab University, Chandigarh, India

Anu Kalia Electron Microscopy and Nanoscience Lab, Punjab Agricultural University, Ludhiana, Punjab, India

Gurpreet Kaur School of Pharmaceutical Sciences, Delhi Pharmaceutical Sciences and Research University, New Delhi, India

Parina Kumari University Institute of Pharmaceutical Sciences, Panjab University, Chandigarh, India
MM College of Pharmacy, Maharishi Markandeshwar University (MMDU), Mullana, Haryana, India

Parveen Kumari University Institute of Pharmaceutical Sciences, Panjab University, Chandigarh, India

Shiva Mahmoudi-Gahroui Department of Mycology, Pasteur Institute of Iran, Tehran, Iran

Maria Letizia Manca Department of Life and Environmental Sciences, University of Cagliari, University Campus, Monserrato, CA, Italy

Namish Manchanda Centre for Pharmaceutical Nanotechnology, Department of Pharmaceutics, National Institute of Pharmaceutical Education & Research (NIPER), Mohali, Punjab, India

Monalisa Mishra Neural Developmental Biology Lab, Department of Life Science, NIT Rourkela, Rourkela, India, Odisha

Arjun Muthu Doctoral School of Nutrition and Food Science, University of Debrecen, Debrecen, Hungary
Institute of Agricultural Chemistry and Soil Science, Faculty of Agricultural and Food Sciences and Environmental Management, University of Debrecen, Debrecen, Hungary

Rajashri R. Naik Faculty of Allied Medical Sciences, Al-Ahliyya Amman University, Amman, Jordan

Jigar Nandha Department of Regulatory Affairs, Alembic Pharmaceuticals Limited, Vadodara, Gujarat, India

Katayoon Nematpour Department of Labor Environment, Health, Safety, Environment and Energy, Ministry of Industry, Mine and Trade, Tehran, Iran

Duyen H. H. Nguyen Institute of Animal Science, Faculty of Agricultural and Food Sciences and Environmental Management, Biotechnology and Nature Conservation, University of Debrecen, Debrecen, Hungary
Institute of Life Sciences, Vietnam Academy of Science and Technology, Ho Chi Minh City, Vietnam
Doctoral School of Nutrition and Food Science, University of Debrecen, Debrecen, Hungary

Harris Panakkal Department of Biotechnology, Government Institute of Science, Chhatrapati Sambhajinagar, Maharashtra, India

Ravi Parashar University Institute of Pharmacy, Pt. Ravishankar Shukla University, Raipur, Chhattisgarh, India

Mrunali Patel Ramanbhai Patel College of Pharmacy, Charotar University of Science and Technology, Changa, Petlad, Anand, Gujarat, India

Rashmin Patel Ramanbhai Patel College of Pharmacy, Charotar University of Science and Technology, Changa, Petlad, Anand, Gujarat, India

József Prokisch Institute of Animal Science, Faculty of Agricultural and Food Sciences and Environmental Management, Biotechnology and Nature Conservation, University of Debrecen, Debrecen, Hungary

Mahendra Rai Department of Biotechnology, SGB Amravati University, Amravati, Maharashtra, India
Department of Chemistry, Federal University of Piauí – UFPI, Teresina, Piauí, Brazil
Institute of Animal Science, Faculty of Agricultural and Food Sciences and Environmental Management, Biotechnology and Nature Conservation, University of Debrecen, Debrecen, Hungary
Department of Chemistry, Federal University of Piaui (UFPI), Teresina, Brazil
Nanotechnology Laboratory, Department of Biotechnology, Sant Gadge Baba Amravati University, Amravati, Maharashtra, India

A. Ramalakshmi Department of Agricultural Microbiology, Tamil Nadu Agricultural University, Coimbatore, Tamil Nadu, India
Department of Food Process Engineering, Tamil Nadu Agricultural University, Coimbatore, Tamil Nadu, India

Mehdi Razzaghi-Abyaneh Department of Mycology, Pasteur Institute of Iran, Tehran, Iran

Surajita Sahu Neural Developmental Biology Lab, Department of Life Science, NIT Rourkela, Rourkela, India, Odisha

Reihaneh Seiad-Ahmadnezhad Department of Mycology, Pasteur Institute of Iran, Tehran, Iran

Ashok K. Shakya Michael Sayegh Faculty of Pharmacy, Aqaba University of Technology, Aqaba, Jordan

Masoomeh Shams-Ghahfarokhi Department of Mycology, Faculty of Medical Sciences, Tarbiat Modares University, Tehran, Iran

Sudhir S. Shende Department of Microbiology, Faculty of Biological and Veterinary Sciences, Nicolaus Copernicus University in Toruń, Toruń, Poland

Hukum Singh ICFRE-Forest Research Institute, P.O. New Forest, Dehradun, Uttarakhand, India

Malgorzata Skorupska Faculty of Chemistry, Nicolaus Copernicus University in Torun, Torun, Poland

Sareh Sohrabi Department of Medical Biotechnology, School of Advanced Medical Sciences and Technologies, Hamadan University of Medical Sciences, Hamadan, Iran
Samen Pharmaceutical Company, Mashhad, Iran

Preeti K. Suresh University Institute of Pharmacy, Pt. Ravishankar Shukla University, Raipur, Chhattisgarh, India

Mariusz Szkoda Faculty of Chemistry, Department of Chemistry and Technology of Functional Materials, Gdańsk University of Technology, Gdańsk, Poland
Advanced Materials Center, Gdańsk University of Technology, Gdańsk, Poland

Sushama Talegaonkar Department of Pharmaceutics, School of Pharmaceutical Sciences, Delhi Pharmaceutical Sciences and Research University (DPSRU), Government of NCT of Delhi, Delhi, India

Part I

Introduction and Synthesis of Carbon Dots

Carbon Nanodots in Health, Food, and Agriculture: An Overview

1

Mohini Diwan, Gurpreet Kaur, Mahendra Rai, and Sushama Talegaonkar

Abstract

Carbon nanodots (CNDs) are photoluminescent, carbon-based nanomaterials with unique features like high fluorescence, water solubility, surface tunability, and low toxicity. These properties have positioned them as multifunctional tools in biomedicine, food safety, environmental monitoring, and sustainable agriculture. In the biomedical field, CNDs support bioimaging, targeted drug delivery, and cancer therapy via photothermal and photodynamic approaches. In food safety, they function as sensitive sensors for detecting heavy metals, pesticides, and microbial contaminants, and are also used in smart packaging. Environmentally, CNDs help monitor pollutants in soil and water due to their chemical responsiveness. In agriculture, they enhance seed germination, nutrient uptake, stress tolerance, and crop yield, while supporting early disease detection, nanofertilization, and pest control.

Despite these advancements, challenges such as inconsistent synthesis, potential nanotoxicity, environmental persistence, and regulatory gaps must be addressed. Issues like unpredictable fluorescence in complex food matrices and impacts on soil microbiota raise concerns. Continued research is focused on refining CND synthesis, improving biocompatibility, and integrating CNDs

M. Diwan · G. Kaur
School of Pharmaceutical Sciences, Delhi Pharmaceutical Sciences and Research University, New Delhi, India

M. Rai
Department of Biotechnology, SGB Amravati University, Amravati, Maharashtra, India

Department of Chemistry, Federal University of Piauí – UFPI, Teresina, Piauí, Brazil

S. Talegaonkar (✉)
Department of Pharmaceutics, School of Pharmaceutical Sciences, Delhi Pharmaceutical Sciences and Research University (DPSRU), Government of NCT of Delhi, Delhi, India
e-mail: stalegaonkar@dpsru.edu.in

M. Rai et al. (eds.), *Carbon Nanodots*, https://doi.org/10.1007/978-981-95-4705-0_1

with precision technologies. This chapter underscores their broad applicability while stressing the importance of risk assessment and standardized methodologies to ensure their safe and scalable use in scientific and industrial settings.

Keywords

Carbon nanodots · Biosensors · Theranostics · Food safety · Fluorescence sensing · Environmental sensing · Nanofertilizers · Photoluminescence

1.1 Introduction

Carbon nanodots (CNDs) are a distinct class of carbonaceous nanomaterials characterized by their quasi-spherical morphology and nanometric size, typically less than 10 nm (Goryacheva et al., 2017; Sciortino et al., 2018; Đorđević et al., 2019; Qureshi et al., 2024). Carbon nanodots were first observed incidentally by Xu et al. (2004). During the purification of single-walled carbon nanotubes (SWCNTs) using gel electrophoresis (Khan et al., 2021), CNDs have since garnered significant attention owing to their unique optical behavior, chemical stability, and biocompatibility (Goryacheva et al., 2017; Sciortino et al., 2018). Structurally, CNDs can feature either amorphous or crystalline carbon cores and are rich in surface functional groups such as hydroxyl, carboxyl, and amino groups (Sciortino et al., 2018; Đorđević et al., 2019; Papaioannou et al., 2018).

The terms "carbon nanoclusters," "C-dots," or "carbon dots" (CDs) have been used synonymously with carbon nanodots in literature to refer to quasi-spherical, amorphous nanodots that lack a quantum confinement as opposed to the crystalline "carbon quantum dots" (CQDs) or "graphene quantum dots" (GQDs) (Cayuela et al., 2016).

Surface passivation is a deliberate modification of the outermost layer of carbon nanodots to control their interaction with the environment and tailor their properties for specific applications (Ross et al., 2020). It is an integral step in all synthesis techniques that involves the binding of a layer of functional groups to the carbonaceous core, which is further detrimental to the structure and function of CNDs (Sciortino et al., 2018).

These nanomaterials exhibit unique characteristics such as tunable optical properties, high photostability, and biocompatibility (Mocci et al., 2022). CNDs can be synthesized through various methods, including top-down and bottom-up approaches, often using inexpensive and readily available starting materials (Kang et al., 2023). Their applications span multiple fields, including bioimaging, biosensing, drug delivery, photodynamic therapy, and photocatalysis (Zaidi et al., 2022). The growing interest in CNDs has led to extensive research on their fundamental properties, synthesis techniques, and potential applications (Kang et al., 2023). The choice of synthesis pathway, along with the starting materials and reaction environment, strongly influences their morphology, surface chemistry, and resulting physicochemical properties (Sciortino et al., 2018; Xiao & Sun, 2018).

This chapter aims to provide a comprehensive overview of the synthesis, properties, and versatile applications of carbon nanodots (CNDs) across the health, food, and agriculture sectors. It highlights their photoluminescent behavior, biocompatibility, and functional versatility, while also addressing current challenges and future prospects to guide their safe and effective integration into biomedical diagnostics, food safety, smart packaging, sustainable agriculture, and environmental monitoring.

1.2 Unique Characteristics of CNDs

1.2.1 Optical Properties (Fluorescence, Photoluminescence)

One of the main properties of CNDs is their bright and tunable photoluminescence (PL), which spans from the ultraviolet to near-infrared regions (Papaioannou et al., 2018; Liu, 2020; Aditya Khatokar et al., 2021; Roy & Mamun, 2022; Azami et al., 2023). Many CNDs exhibit excitation-wavelength-dependent PL behavior, an attribute often attributed to size effects, surface defects, and heterogeneous surface states (Papaioannou et al., 2018; Liu, 2020; Roy & Mamun, 2022). The effect of solvent polarity and interaction on surface-related light emission in glucose-derived CNDs can be seen in Fig. 1.1, showing changes in the PL map above and below 320 nm (Papaioannou et al., 2018).

Mechanisms proposed to explain CND luminescence include quantum confinement, emission from surface energy traps, and contributions from surface-bound molecular fluorophores (Papaioannou et al., 2018; Roy & Mamun, 2022). Heteroatom doping (e.g., nitrogen, sulfur) and surface passivation have further been shown to enhance the quantum yield and control the emission properties of CNDs (Roy & Mamun, 2022; Azami et al., 2023). Additionally, certain CNDs demonstrate upconversion photoluminescence, wherein longer-wavelength excitation results in shorter-wavelength emission, adding to their potential in bioimaging and optoelectronics (Papaioannou et al., 2018; Roy & Mamun, 2022).

1.2.2 Biocompatibility and Low Toxicity

A key advantage of CNDs over traditional semiconductor quantum dots is their intrinsic biocompatibility and low toxicity. Composed mainly of carbon and devoid of heavy metals, CNDs have exhibited excellent cellular uptake, minimal generation of reactive oxygen species, and negligible cytotoxicity even at relatively high concentrations. Surface functionalization, such as PEGylation or glycosylation, has been shown to further enhance their biological compatibility and circulation time in vivo (Mao et al., 2016; Arcudi et al., 2019; Chan et al., 2021).

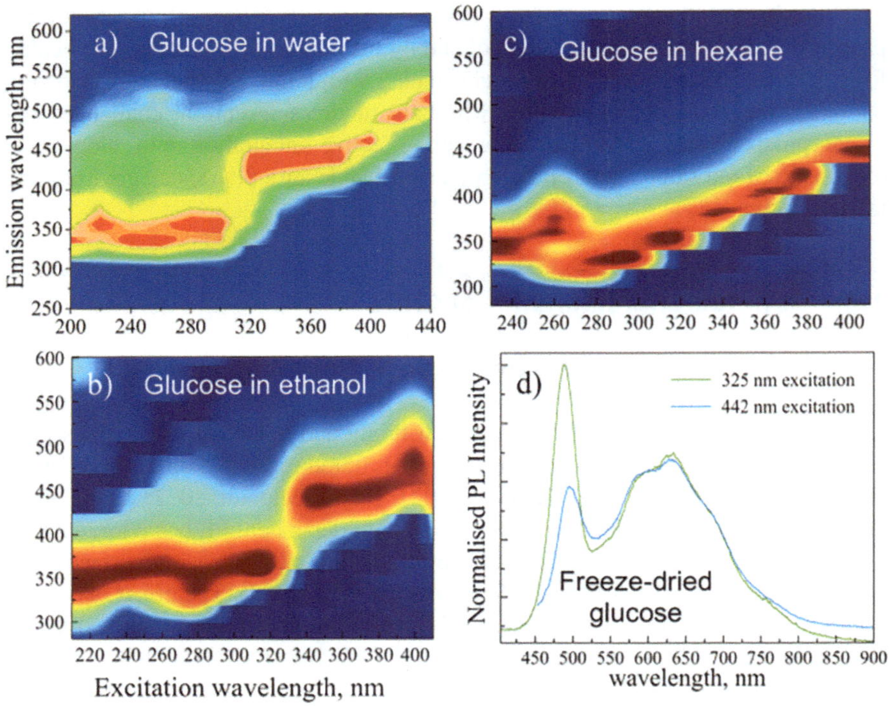

Fig. 1.1 PL maps glucose CNDs in water (**a**, relative polarity 1), ethanol (**b**, relative polarity 0.654), and hexane (**c**, relative polarity 0.009). Normalized PL data of freeze-dried glucose sample with two excitation wavelengths (325 and 442 nm) are also shown (**d**) (Papaioannou et al., 2018). (This work is licensed under a Creative Commons Attribution 4.0 International License (http://creativecommons.org/licenses/by/4.0/))

1.2.3 Surface Functionalization and Versatility

Surface engineering is central to the versatility of CNDs. The presence of abundant functional groups on their surface enables facile conjugation with a wide variety of molecules, ranging from synthetic polymers to biomolecules (Xiao & Sun, 2018; Arcudi et al., 2019; Đorđević et al., 2019; Swift et al., 2019). Surface modification allows for the tuning of solubility (hydrophilic or hydrophobic behavior), optical properties, and target specificity (Xiao & Sun, 2018; Arcudi et al., 2019). Moreover, surface passivation can drastically alter the electronic structure of CNDs, shifting absorption profiles and modifying fluorescence characteristics. Recent studies have also shown that the design of surface ligands (such as glycans) can significantly impact biological recognition and electronic behavior (Arcudi et al., 2019; Swift et al., 2019).

1.3 Scope of Applications in Health, Food, and Agriculture

Due to their advantageous optical, chemical, and biological properties, carbon nanodots (CNDs) have shown great potential for applications in a wide range of fields. In healthcare, CNDs are being explored as fluorescent probes for cellular and molecular imaging, vehicles for drug and gene delivery, and agents for photothermal and photodynamic therapies (Mao et al., 2016; Xiao & Sun, 2018; Chan et al., 2021). Their ability to emit strong fluorescence under excitation while maintaining low toxicity makes them ideal for noninvasive bioimaging.

In the food sector, the sensitivity of CNDs to environmental changes is leveraged for detecting food spoilage, monitoring contaminants like heavy metals and pesticides, and ensuring food quality control (Goryacheva et al., 2017; Xiao & Sun, 2018). Their use in fluorescent sensing platforms offers real-time, rapid, and low-cost analysis.

Within agriculture, CNDs have been utilized as nanocarriers for the controlled delivery of fertilizers and pesticides, improving efficacy and minimizing environmental impact (Chan et al., 2021). Additionally, their use in soil and plant health monitoring, based on fluorescence response to biochemical signals, offers a sustainable pathway for precision farming technologies (Fig. 1.2).

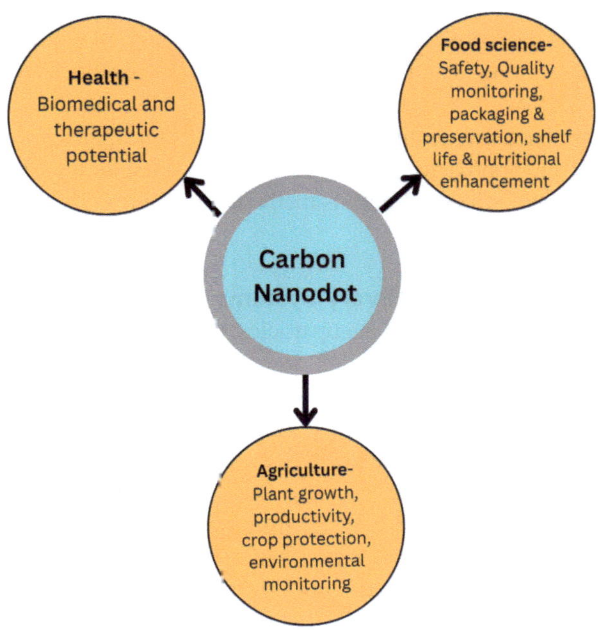

Fig. 1.2 Applications of carbon nanodots

1.3.1 Carbon Nanodots in Health

1.3.1.1 Biomedical Applications

Carbon nanodots (CNDs) have emerged as promising nanomaterials for biomedical applications due to their unique properties, including biocompatibility, photoluminescence, and ease of synthesis and functionalization (Khan et al., 2021; Ding et al., 2024). These nanoparticles have shown potential in various biomedical fields, such as bioimaging, biosensing, and therapy (Koutsogiannis et al., 2020; Khan et al., 2021). CNDs have been utilized for fluorescence imaging, magnetic resonance imaging, and photoacoustic imaging. They have also demonstrated promise in treating diseases like cancer, inflammation, and Alzheimer's (Ding et al., 2024). The versatility of CNDs has led to their application in cancer theranostics and as sensing probes for metal cations (Koutsogiannis et al., 2020). Furthermore, CNDs have shown potential in photocatalytic applications (Sivasankarapillai et al., 2020). Despite their promising features, further research is needed to overcome current obstacles and facilitate the clinical translation of CND-based technologies (Ding et al., 2024).

1.3.1.2 Imaging and Diagnostics (Cellular Imaging, Biosensors)

CNDs serve as effective fluorescent probes for cellular and in vivo imaging. Figure 1.4 shows NIR fluorescence images of mouse bodies after intravenous injections of carbon nanodots (Bao et al., 2018). Their potential in biosensing has been extensively explored, with applications in enzymatic, antigen–antibody, and nucleic acid detection (Koutsogiannis et al., 2020). CNDs also show promise in theranostics, combining diagnostic and therapeutic capabilities in a single platform (Boakye-Yiadom et al., 2019). Furthermore, CNDs have been utilized in drug delivery, antibacterial treatments, and cancer therapies, including photothermal and photodynamic approaches (Wang et al., 2022). Despite their advantages, challenges remain in standardizing synthesis methods and improving targeting specificity (Chung et al., 2020).

1.3.1.3 Drug Delivery Systems

These carbon-based nanoparticles, typically less than 10 nm in size, offer advantages such as biocompatibility, low toxicity, and tunable physicochemical properties (Kaurav et al., 2023; Bartkowski et al., 2024). CNDs have shown potential in cancer treatment, ocular diseases, infectious diseases, and brain disorders (Kaurav et al., 2023). Their unique optical properties make them suitable for bioimaging and theranostics (Khan et al., 2021; Bartkowski et al., 2024). CNDs can be functionalized for targeted drug delivery, improving efficacy and reducing side effects in chemotherapy, gene therapy, and antibiotic therapy (Debnath & Srivastava, 2021; Kaurav et al., 2023). Recent research has focused on CND-based systems for delivering anticancer agents like doxorubicin. Despite their promise, challenges remain in optimizing CND-based drug delivery systems for clinical applications, necessitating further preclinical toxicity studies (Debnath & Srivastava, 2021; Bartkowski et al., 2024).

1.3.1.4 Therapeutic Potential

Antibacterial and Antioxidant Properties

Carbon nanodots (CNDs) can also serve as antibacterial agents due to their unique properties and low toxicity (Lagos et al., 2023). These nanomaterials exhibit antibacterial activity through various mechanisms, including cell wall damage, biofilm disruption, and reactive oxygen species (ROS) generation (Wu et al., 2021; Zhao et al., 2023b). CNDs can be derived from different carbon sources, with those from natural sources inheriting beneficial properties. Their antibacterial efficacy can be enhanced through photoactivation, enabling photodynamic and photothermal therapies (Wu et al., 2021; Lagos et al., 2023). CNDs have shown effectiveness against both Gram-positive and Gram-negative bacteria, with mechanisms involving direct cell interaction, DNA binding, and gene expression modulation (Varghese & Balachandran, 2021). Importantly, CNDs may have a lower likelihood of inducing bacterial resistance compared to conventional antibiotics, making them attractive for clinical applications (Wu et al., 2021). Future research focuses on optimizing CND preparation methods and exploring their potential against other pathogens (Lagos et al., 2023; Zhao et al., 2023b).

As carbon nanodots (CNDs) are nanomaterials with unique properties, including excellent photoluminescence, biocompatibility, and easy surface modification (Khan et al., 2021), these nanoparticles exhibit both antioxidant and oxidant behaviors, depending on their structure and composition (Innocenzi & Stagi, 2023). CNDs have shown promising therapeutic potential in various biomedical applications, such as biosensing, bioimaging, and drug delivery (Zaidi et al., 2022). Their small size allows for intimate interactions with biological species, enabling them to cross impermeable membranes like the blood–brain barrier (Truskewycz et al., 2022). Recent research has focused on understanding the functional tailoring of CNDs through dopant and surface molecule modifications, which has provided insights into their antioxidant mechanisms (Khan et al., 2021). As nanomaterial platforms become more prevalent in clinical treatments, the commercialization of CND therapeutics is anticipated due to their multifunctional attributes and documented biocompatibility (Truskewycz et al., 2022).

Photothermal and Photodynamic Therapy

Carbon nanodots (CNDs) have shown potential for photothermal therapy (PTT) in cancer treatment. These nanoparticles exhibit high photothermal conversion efficiency, ranging from 35% for dopamine CNDs to 59% for sulfur and nitrogen co-doped CNDs (Li et al., 2016; Bao et al., 2018). CNDs demonstrate excellent photostability, biocompatibility, and tumor-targeting capabilities through passive accumulation after intravenous administration (Bao et al., 2018; Zhao et al., 2021). Notably, nitrogen-doped CNDs show strong near-infrared absorption, enabling their use in both photoacoustic imaging and PTT (Lee et al., 2016). These nanoparticles can effectively ablate cancer cells under laser irradiation while exhibiting minimal dark cytotoxicity (Li et al., 2016; Zhao et al., 2021). Furthermore, CNDs possess unique biodistribution properties, including renal clearance, enhancing their potential for clinical applications (Lee et al., 2016; Bao et al., 2018).

Carbon nanodots (CNDs) have also emerged as agents for photodynamic therapy (PDT) in cancer treatment and antimicrobial applications. These nanoparticles, typically 2–10 nm in size, offer advantages such as excellent photoactivity, biocompatibility, and low cost (Jovanović et al., 2023; Song et al., 2023). CNDs generate reactive oxygen species (ROS) under light exposure, causing damage to cancer cells and pathogens (Knoblauch & Geddes, 2020; Song et al., 2023). Near-infrared CNDs have shown potential for deep tissue penetration and superior imaging performance in cancer PDT (Song et al., 2023). Heteroatom and metal-doped CNDs have been developed to enhance their physicochemical properties and biological applications in image-guided PDT (Sekar et al., 2022). As intrinsic photosensitizers, CNDs offer an inexpensive and "green" alternative to traditional small molecule photosensitizers in antimicrobial PDT (Knoblauch & Geddes, 2020).

Figure 1.3 shows red-emitting carbon nanodots (RCNDs) prepared by Ji et al. (2020) and their subsequent functionalization with a folic acid ligand, resulting in RCNDs-TEG-FA, which demonstrated significantly improved water solubility and retained strong fluorescence even after 24 hours of incubation. Notably, these modified nanodots showed efficient intracellular reactive oxygen species (ROS) generation when exposed to a low-power 660 nm laser (0.1 W/cm^2), highlighting their promising nature for photodynamic therapy (PDT).

Tissue Engineering Applications

Carbon nanodots (CNDs) are also being tested as materials for tissue engineering and regenerative medicine applications due to their unique properties. CNDs exhibit excellent biocompatibility, low cytotoxicity, and biodegradability (Vedhanayagam et al., 2021; Farshidfar et al., 2023). Their tunable fluorescence and high photostability make them suitable for image-guided tissue engineering and live cell imaging (Jin et al., 2018; Zheng et al., 2022). CNDs can be incorporated into scaffolds, such as electrospun nanofiber mats, enhancing cell proliferation and guiding cell growth along the fibers. The presence of functional groups on CND surfaces plays a crucial role in supporting cellular activities (Jin et al., 2018; Farshidfar et al., 2023). Despite their potential, challenges remain, including scaffold inconsistency and the need for noninvasive monitoring techniques post-implantation (Vedhanayagam et al., 2021; Farshidfar et al., 2023). Ongoing research focuses on developing eco-friendly synthesis methods and exploring CND-based nanosystems for various tissue engineering applications (Zheng et al., 2022; Farshidfar et al., 2023).

1.3.2 Carbon Nanodots in Food Science

1.3.2.1 Food Safety and Quality Monitoring

Carbon nanodots offer promising applications in food safety and quality monitoring. They can be incorporated into sensors for detecting contaminants, spoilage, and

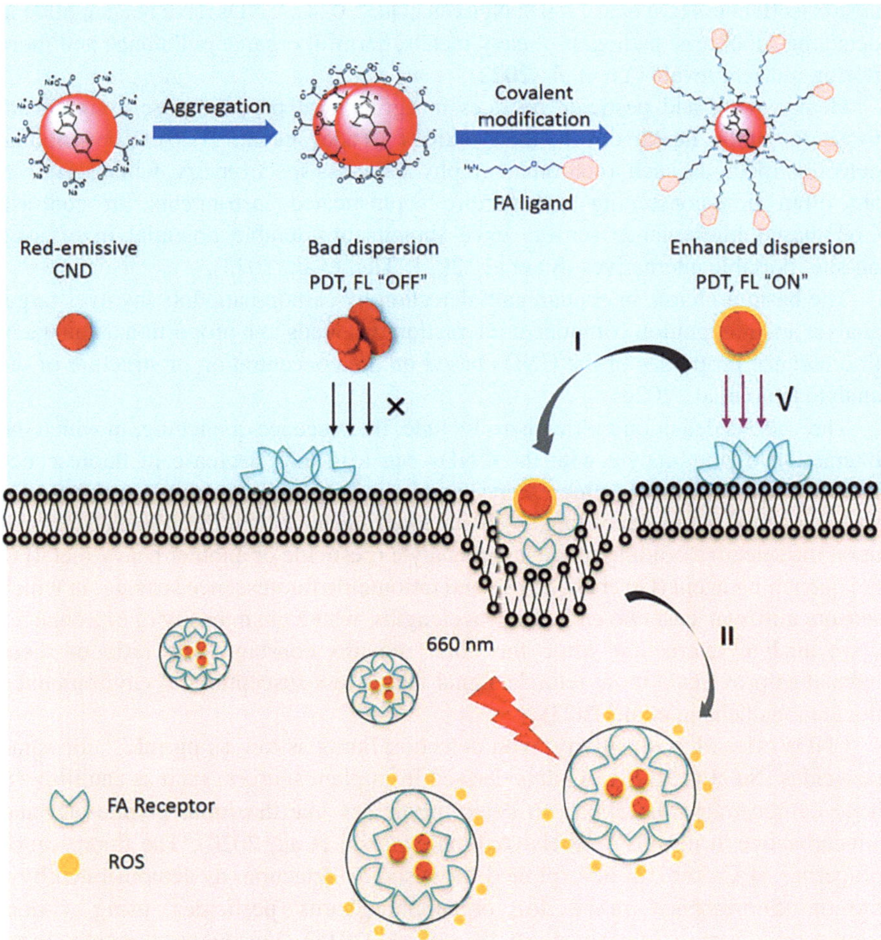

Fig. 1.3 Schematic illustration of the covalent functionalization of RCNDs and their targeted intracellular production of ROS through (I) FR-mediated endocytosis and (II) light irradiation. FL stands for fluorescence (Ji et al., 2020). (This work is licensed under a Creative Commons Attribution 4.0 International License (http://creativecommons.org/licenses/by/4.0/))

adulterants in food (Mohammadi & Jafari, 2020; Mary Isabella Sonali et al., 2022). These nanosensors provide faster, more accurate detection of microbes, toxins, and chemical species compared to traditional methods. Applications include intelligent packaging systems for assessing freshness and extending shelf life (Ghosh et al., 2022).

Detection of Contaminants (Pesticides, Heavy Metals)

Carbon nanodots (CNDs) have emerged as promising nanomaterials for detecting contaminants in food due to their unique optical properties, biocompatibility, and cost-effectiveness (Qu et al., 2018; Nguyen et al., 2024). CNDs have been applied in detecting foodborne pathogens, heavy metals, harmful organic pollutants, and monitoring nutrient levels (Yu et al., 2022).

Heavy metal and pesticide residues in agricultural produce pose a significant threat to public health due to their toxicity (Yadav et al., 2020). Conventional detection methods, such as chromatography and mass spectrometry, while sensitive, are often time-consuming and require sophisticated instruments. In contrast, CND-based fluorescence sensors have shown considerable potential in offering on-site, portable alternatives (Su et al., 2021; Tian et al., 2023).

The basic principle in contaminant detection by carbon nanodots involves target analyte and recognition component interaction that leads to a proportional change in fluorescence properties of the CNDs based on the concentration or structure of the analyte (Luo et al., 2020).

The various detection mechanisms include: fluorescence quenching, in which the interaction of the analyte with the CNDs can lead to a decrease in fluorescence intensity (Qu et al., 2018); fluorescence recovery ("Off-on" sensing) in which CNDs with quenched fluorescence (e.g., by metal ions) can have their fluorescence restored upon the selective binding of the target analyte (pesticide or another heavy metal) to the quenching agent (Qu et al., 2018); and ratiometric fluorescence sensing, in which sensors utilize at least two emission wavelengths, where the intensity of one changes upon analyte interaction while the other remains constant. The ratio of these intensities provides a more reliable signal that is less susceptible to environmental fluctuations (Zhang et al., 2023).

CNDs offer high sensitivity, with detection limits as low as pg mL^{-1} for some pesticides (Su et al., 2021). C-dots derived from plant sources, such as cauliflower, have demonstrated the ability to detect pesticides like diazinon, glyphosate, and amicarbazone at ng mL^{-1} levels (Ashrafi Tafreshi et al., 2020). The fluorescence properties of C-dots can be exploited for pesticide detection, as demonstrated by a turn-on fluorescence probe for organophosphorus pesticides using C-dots synthesized from waste paper ash (Lin et al., 2018). Various recognition units, including enzymes, antibodies, aptamers, and molecularly imprinted polymers, have been employed in conjunction with carbon nanomaterials for pesticide detection (Su et al., 2021). These innovative approaches offer potential for improving food safety and environmental monitoring (Mishra et al., 2022).

Heavy metals such as mercury, cadmium, copper, iron, chromium, cobalt, silver, etc. can be detected by CNDs (Soni et al., 2024). CNDs can be used in fluorometric sensing, where the presence of metal ions causes either fluorescence quenching or enhancement (Yadav & Lahariya, 2024). Additionally, CNDs have been employed in electrochemical sensors for heavy metal detection, offering high sensitivity down to parts per trillion levels (Zhou et al., 2025). These sensors exploit the high surface area, electrical conductivity, and electrocatalytic activity of carbon nanomaterials to preconcentrate metal ions and amplify signals at electrode interfaces (Song et al., 2024). Aluminum ions were detected using a carbon dot fluorescence-based colorimetric sensor with a smartphone, with a detection limit of 5.55 μM (Wei et al., 2021). Mercury ions were detected using sulfur-doped carbon dots (SCDs) in which abnormal fluorescence enhancement was the basis of detection, with a detection limit of 5.16 nm (Bavya et al., 2025). For lead ion detection, nitrogen and sulfur co-doped carbon dots (NSCDs) were immobilized using chitosan-modified filter paper, which showed a linear relationship between mean fluorescence intensity and target (lead) with a detection limit of 4.50 μmol L^{-1} (Aboobakri et al., 2024) (Table 1.1).

Monitoring Freshness and Spoilage

Carbon nanodots have the potential to revolutionize food safety monitoring, particularly in detecting adulterants, spoilage, and ensuring freshness. They offer unique properties such as high selectivity, biocompatibility, and real-time analysis capabilities (Shruti et al., 2024). CNDs can monitor freshness and spoilage by various mechanisms:

Detection of Volatile Amines Food spoilage is often associated with the production of volatile organic compounds, particularly biogenic amines like ammonia and trimethylamine. Functionalized carbon nanodots can be designed to interact specifically with these amines, leading to a change in their optical properties (e.g., fluorescence intensity or color). This change can then serve as an indicator of spoilage (Szczepankowska et al., 2023).

pH Monitoring Carbon nanodots exhibit pH-dependent optical properties, primarily their fluorescence. This sensitivity arises from the protonation and deprotonation of functional groups present on their surface. For example, some CNDs may exhibit stronger fluorescence at acidic pH and weaker fluorescence at basic pH, or vice versa. The wavelength of emitted light can also shift (Ehtesabi et al., 2020; Liu, 2020).

Detection of Pathogens Some carbon nanodot-based sensors can be designed to detect specific foodborne pathogens, providing an early warning of potential contamination and spoilage (Mei et al., 2022).

These nanomaterials can detect various food quality indicators through mechanisms such as fluorescence quenching and recovery (Qu et al., 2018), fluorescent resonance energy transfer, photoinduced electron transfer, and internal filtering effects (Tian et al., 2023). Recent applications include histamine-responsive CDs for visual monitoring of food spoilage (Qin et al., 2022) and pH-responsive CD-based strips for assessing shrimp freshness (Ezati et al., 2023). Compared to traditional analytical methods, CD-based sensors offer rapid, sensitive, and potentially portable solutions for food analysis (Luo et al., 2020).

1.3.2.2 Food Packaging and Preservation

Antimicrobial Packaging

Carbon nanodots (CNDs) have emerged as promising materials for antimicrobial packaging in food science. These nanoparticles can be synthesized from environmentally friendly sources, including microorganisms and food by-products (Moradi et al., 2023). CNDs exhibit excellent antibacterial and antioxidant properties, making

Table 1.1 A few recent examples of contaminant detection by carbon nanodots

S.No.	Target contaminant	Carbon nanodot type and its synthesis	Key findings	Reference
1.	Fe^{3+}	Multicolor CDs derived from food waste via hydrothermal carbonization	Selective detection of Fe^{3+} ions; luminescence response range: 1–50 μM	Zhou et al. (2018)
2.	Hg^{2+} and Cu^{2+}	Nitrogen-doped CDs synthesized from urea and EDTA via the pyrolytic method	Fluorescence ON–OFF–ON sensor for Hg^{2+} and Cu^{2+} detection with a detection limit of 6.2 and 2.304 nM, respectively, in aqueous solutions, 0.1 μM in a filter paper device and 50 μM in a microfluidic device	Patir and Gogoi (2019)
3.	Al^{3+}	Blue-emission carbon dots (CDs) synthesized in a one-pot solvothermal method using naringin as a precursor	Ratiometric fluorescence sensing for the analysis of Al^{3+} with a detection limit of 113.8 nM. Use of smartphones and RGB analysis	Wei et al. (2021)
4.	Hg^{2+}, Pb^{2+}, Fe^{3+}, and Cu^{2+}	CDs derived from green tea residue via hydrothermal synthesis	Simultaneous detection of Pb^{2+}, Fe^{3+}, Cu^{2+}, and Hg^{2+} in food samples. The quantitative range of Hg^{2+}, Pb^{2+}, Fe^{3+} and Cu^{2+} are 0.0401–16.0 $\mu g \cdot g^{-1}$, 1.04–14.5 $\mu g \cdot g^{-1}$, 0.0558–55.8 $\mu g \cdot g^{-1}$, 0.635–31.8 $\mu g \cdot g^{-1}$	Zhang et al. (2024)
5.	Diazinon, glyphosate, and amicarbazone	Carbon nanodots prepared via the hydrothermal synthesis technique from cauliflower	Detection through fluorescence quenching, estimated limit of detection for diazinon, amicarbazone, and glyphosate was 0.25, 0.5, and 2 ng mL^{-1}, respectively	Ashrafi Tafreshi et al. (2020)
6.	Thiophanate methyl (TM)	Copper-modified dual-emission CDs synthesized via the hydrothermal method	Dual-emission CDs exhibit fluorescence quenching upon TM interaction; detection limits: 0.75 μM (pear), 0.78 μM (apple)	Yue et al. (2022)
7.	Parathion	Nitrogen-doped fluorescent CDs synthesized via the hydrothermal method using rutinum and o-phenylenediamine	Detection limits: 0.11 ng/mL; linear ranges: 0–7.5 μg/L and 12.5–62.5 μg/L; successful application in Chinese cabbage, cantaloupe, and cowpea	Jiang et al. (2023)

them effective against pathogenic and spoilage bacteria in food (Cui et al., 2024). When incorporated into packaging materials, CNDs enhance mechanical strength, UV barrier properties, and water resistance (Zhao et al., 2023a).

CNDs can be incorporated in packaging materials in several ways, such as direct incorporation into polymer matrices, surface coating, electrospinning, immobilization on packaging surfaces, etc. (Gautam et al., 2022; Al Ragib et al., 2023; Deepika et al., 2023; Manikandan & Min, 2023). Studies have demonstrated the integration of CDs into various biopolymers, such as carrageenan and cellulose nanofibers, resulting in films with improved tensile strength, hydrophobicity, and gas barrier properties (Wagh et al., 2023; Sangeetha et al., 2024).

They can also be used to develop active and intelligent packaging systems that extend food shelf life and monitor freshness (Moradi et al., 2023; Zhao et al., 2023a). For instance, CNDs that exhibit pH-dependent fluorescence or color changes can be used to detect the spoilage of meat or fish (Fu et al., 2022). A carrageenan-CD composite film successfully tracked milk spoilage through color changes, while a cellulose nanofiber-CD-anthocyanin film monitored the freshness of various meats (Wagh et al., 2023; Sangeetha et al., 2024).

Shelf-Life Enhancement

CNDs can improve shelf life and food safety by acting as antioxidants, antimicrobials, photoluminescent agents, and UV-light blockers. CNDs, especially when photoactivated by light, can generate reactive oxygen species (ROS) such as superoxide radicals and hydroxyl radicals. These ROS can damage bacterial cell membranes, proteins, and DNA, leading to cell death (Saita & Kawasaki, 2023). CNDs can act as photosensitizers. Upon irradiation with visible light, they can transfer energy to surrounding oxygen molecules, generating cytotoxic singlet oxygen that effectively kills bacteria (Hutton et al., 2016). Furthermore, CNDs have shown efficacy in combating drug-resistant bacteria and biofilm infections in food preservation applications (Cui et al., 2024). Despite their potential, further research is needed to optimize CND production parameters and assess their long-term safety and degradability in food systems (Zhao et al., 2023a; Cui et al., 2024).

1.3.2.3 Potential Role in Nutrition Enhancement

Carbon nanodots and other nanoparticles show promising applications in food science and nutrition. They can enhance food safety by detecting pathogens and contaminants, improve food packaging to extend shelf life, and increase the bioavailability of micronutrients like vitamins and iron (Singh et al., 2017; Arshad et al., 2021; Mary Isabella Sonali et al., 2022). Specifically, nitrogen-phosphorus-doped carbon nanodots (NPCDs) demonstrate antimicrobial properties against foodborne pathogens and potential for food packaging applications. NPCDs also show anticancer effects and can be used for cellular bio-imaging (Bajpai et al., 2020). While nanotechnology offers numerous benefits for nutritional enhancement and food safety, concerns remain about the potential toxicity of nanoparticles in food systems. Further research is needed to fully understand the physiological effects and safety of nanomaterials in food applications (Arshad et al., 2021; Mary Isabella Sonali et al., 2022).

1.3.3 Carbon Nanodots in Agriculture

1.3.3.1 Plant Growth and Productivity

Carbon nanodots (CNDs) have been shown to enhance agricultural productivity as these carbon-based nanoparticles exhibit unique properties, including photoluminescence, biocompatibility, and low toxicity (Li et al., 2023). CNDs have shown potential in improving crop growth, photosynthesis, nutrient absorption, and stress resistance (Chowmasundaram et al., 2023; Li et al., 2023). They can enhance plant tolerance to various biotic and abiotic stresses, including heat, drought, toxic chemicals, and pathogens. The surface functionalities of CNDs can be tuned to regulate plant physiological processes (Zia et al., 2024). However, challenges remain in scaling up CND production for agricultural applications, and their environmental impact needs further assessment (Peng et al., 2024). Despite these challenges, CNDs offer a promising toolbox for sustainable agriculture, potentially revolutionizing crop production systems to address global food security concerns (Chowmasundaram et al., 2023; Zia et al., 2024) (Fig. 1.4).

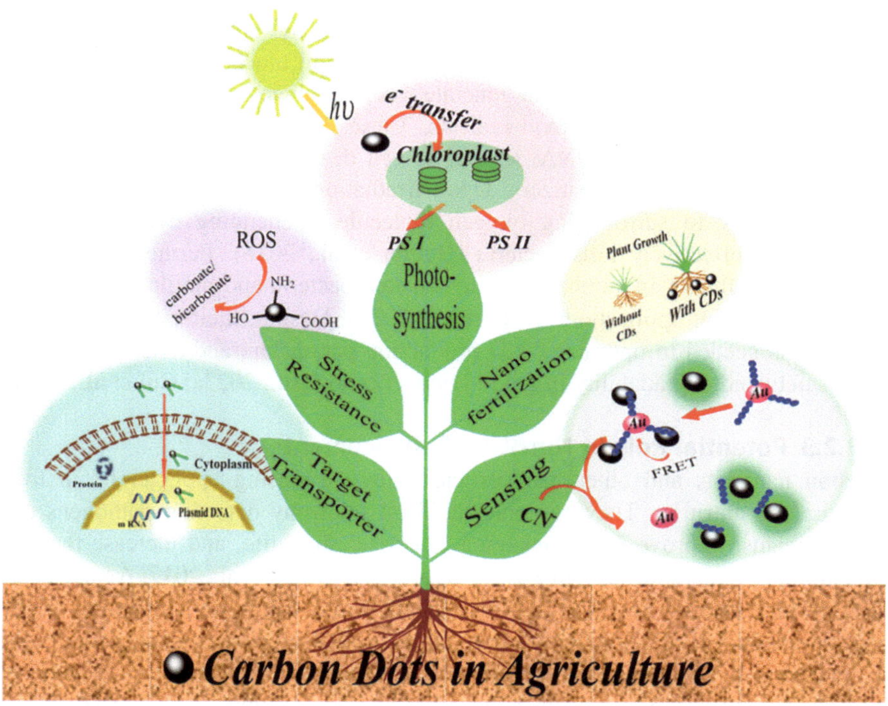

Fig. 1.4 Applications of CNDs in agriculture (Chaudhary et al., 2024). (This work is licensed under a Creative Commons Attribution 4.0 International License (http://creativecommons.org/licenses/by/4.0/))

Improved Seed Germination and Nutrient Uptake

Carbon nanodots have shown promising applications in agriculture, particularly for improving seed germination and plant growth under stress conditions. These carbon-based nanomaterials can enhance germination rates, stress tolerance, and nutrient uptake in crops. Water-soluble carbon nanoparticles have been found to promote seed germination and post-germination growth in lettuce under salinity stress (Baz et al., 2020; Li et al., 2023). Carbon nanomaterials used in seed priming can regulate gene pathways and metabolism, leading to improved germination and stress resistance (Fonseca et al., 2024). Additionally, carbon nanotubes have demonstrated positive effects on seed germination and plant growth, although the exact mechanisms remain unclear (Patel et al., 2020). These nanomaterials offer potential for sustainable agricultural practices, as they can be produced from biowaste, making them cost-effective and environmentally friendly. However, the effectiveness of carbon nanomaterials varies depending on concentration and crop species (Fonseca et al., 2024).

Enhanced Photosynthesis and Stress Resistance

Nanomaterials, particularly carbon nanoparticles, show promise in enhancing crop productivity and stress tolerance, addressing global food security challenges (Swift et al., 2019; Zhao et al., 2020). These materials can augment photosynthesis, improve nutrient uptake, and bolster plant defense mechanisms against abiotic stressors like drought, salinity, and heavy metal toxicity (Nawaz et al., 2023; Cao et al., 2024). CNPs' effectiveness depends on their physicochemical properties, application method, and concentration. They can stimulate seed germination, promote root and shoot growth, and increase biomass and grain yield (Zhao et al., 2020). Nanobiotechnology offers a sustainable approach to crop stress management by strengthening physical barriers, improving photosynthesis, and activating defense-related genes (Nawaz et al., 2023). However, further research is needed to optimize NP dosages and application techniques and assess potential environmental impacts. Integration with precision agriculture technologies could enhance the targeted application of nanofertilizers and nanopesticides (Swift et al., 2019; Cao et al., 2024).

1.3.3.2 Sustainable Crop Protection

Carbon nanodots (CNDs) can also be used as tools for sustainable crop protection and enhancement. These carbon-based nanoparticles, typically smaller than 50 nm, exhibit unique properties including photoluminescence, biocompatibility, and low toxicity (Li et al., 2023). CNDs have shown potential in improving plant tolerance to various biotic and abiotic stresses, such as heat, drought, toxic chemicals, and pathogens (Zia et al., 2024). Their versatile surface chemistry allows for tuning of physiological processes in plants, enhancing growth, photosynthesis, and nutrient absorption (Li et al., 2023). CNDs also demonstrate applications in smart packaging, food safety detection, and crop preservation post-harvest. Their light-harvesting and conversion capabilities make them particularly useful for improving crop photosynthesis (Guo et al., 2022). While CNDs show great promise in agriculture, further research is needed to fully explore their potential for innovation in technology-based farming practices (Chaudhary et al., 2024).

Nanofertilizers and Nanopesticides

First, carbon nanodots (CNDs) as nanofertilizers can be a great boon for agriculture. These small (<10 nm), fluorescent nanoparticles can enhance crop yields through various mechanisms. CNDs promote seed germination, nutrient absorption, and plant growth due to their small size and hydrophilic surface groups. They can also act as carriers for the targeted delivery of agrochemicals and genetic material (Okey-Onyesolu et al., 2021; Guo et al., 2022). The luminescence and electron transfer properties of CNDs may enhance photosynthesis, while their antioxidant properties improve crop stress resistance (Guo et al., 2022). Nitrogen-doped CDs have shown significant potential as fertilizers, increasing mung bean sprout yield and enabling in vivo monitoring of plant growth (Wang et al., 2019). However, further research is needed to optimize CNDs concentrations for different crops, as high levels may cause phytotoxicity (Aacharya & Chhipa, 2019).

As pesticides, Carbon nanodots (CNDs) offer advantages such as controlled release, enhanced efficacy, and reduced environmental impact compared to traditional pesticides (Okey-Onyesolu et al., 2021; Chandel et al., 2022). These nanoparticles exhibit excellent optical properties, including high quantum yields and photostability, making them suitable for developing ultrasensitive fluorescent sensors for pesticide detection. Plant-based CNDs have demonstrated detection limits as low as 0.25 ng/mL for certain pesticides in real samples, showcasing their potential for food safety and environmental monitoring (Ashrafi Tafreshi et al., 2020). Furthermore, CND-based optical sensors, including colorimetric and photoluminescent types, have been developed for pesticide recognition in wastewater, incorporating various recognition elements such as nanoparticles, enzymes, and aptamers (Rasheed, 2023). These advancements highlight the potential of CNDs as sustainable and environmentally friendly tools for pesticide management in agriculture.

Disease Detection and Management

Carbon nanodots (CNDs) are emerging nanomaterials with unique properties that make them promising for biomedical applications, particularly in disease detection and management. Their excellent photoluminescence, biocompatibility, and easy surface modification enable their use in bioimaging, biosensing, and therapeutics. CNDs have shown potential in advanced diagnostics, including in vitro/in vivo bioimaging and thermoelectric biosensing (Khan et al., 2021; Kapat et al., 2023). Their therapeutic applications extend to drug delivery, photothermal therapy, and photodynamic therapy (Zaidi et al., 2022). CNDs have also demonstrated utility in tissue regeneration and as antioxidants (Khan et al., 2021; Kapat et al., 2023). Recent advancements in CND synthesis and functionalization have improved their efficacy in cancer nanomedicine, offering reduced side effects and increased tumor accumulation compared to traditional therapies (Bayda et al., 2021). Despite these promising developments, further research is needed to address toxicity concerns and facilitate bench-to-bedside translation (Kapat et al., 2023). Nanosensors are emerging as powerful tools for early disease detection, offering high sensitivity and rapid on-site screening of biomarkers. These sensors exploit unique properties of nanomaterials, such as large surface area and reactive sites, to achieve exceptional sensitivity in analyte recognition (Munawar et al., 2019; Leong et al., 2022). Various approaches have been developed, including optical, electrical, and mechanical transducers (Munawar et al., 2019; Arndt et al., 2020). Machine learning algorithms are increasingly integrated with nanosensors to enhance their design, fabrication, and predictive capabilities (Leong et al., 2022). A promising application of nanosensors is in exhaled breath analysis, which offers noninvasive, accurate, and potentially inexpensive disease diagnosis. This approach could detect metabolic products associated with various diseases, potentially serving as a self-test kit and guiding patients to appropriate specialists (Lagopati et al., 2023).

Nanosensors also enable sensitive and selective sensing of reactive oxygen species (ROS). Carbon nanodots (CNDs) have emerged as versatile nanoprobes for detecting hydrogen peroxide and glucose, with chemiluminescent CDs achieving high quantum yields and low detection limits. Near-infrared chemiluminescent CDs have been successfully applied for in vivo imaging of inflammatory hydrogen peroxide (Shen et al., 2020). A novel electrochemiluminescence platform combining MoS2 nanosheets and CDs has demonstrated high sensitivity in detecting the breast cancer biomarker HER2 in human serum samples (Gutiérrez-Gálvez et al., 2023). Additionally, paper-based sensors functionalized with fluorescent CDs have shown potential for optical detection of various nanomaterials through different photoluminescence quenching mechanisms (Papadopoulou et al., 2021). All these advancements in nanosensor technology offer promising applications in disease diagnostics, treatment, and nanomaterial detection.

1.3.3.3 Environmental Monitoring (Soil and Water Quality)

Carbon nanodots (CNDs) can play a role in environmental monitoring due to their unique properties and versatile applications. CNDs exhibit excellent biocompatibility, stable optical properties, and low cost, making them ideal for detecting various pollutants in complex environmental matrices (Zhu et al., 2022). These nanomaterials have been successfully employed in sensing metals, dyes, pesticides, and antibiotics (Hebbar et al., 2023). CNDs offer advantages such as bright luminescence, tunable emission, and low cytotoxicity, enabling their use in biomedical applications like bioimaging and targeted drug delivery (Manisha et al., 2019). To enhance CND performance in environmental monitoring, strategies like heteroatom-doping and surface passivation have been explored (González-González et al., 2022). However, challenges remain in understanding the structure–property relationships and dynamic luminescence processes of CNDs (Zhu et al., 2022). Ongoing research focuses on improving sensitivity, selectivity, and detection efficiency to address the limitations of conventional detection methods (González-González et al., 2022) (Table 1.2).

Table 1.2 Summary of applications of carbon nanodots in agriculture

Application area	Key benefits	Mechanisms of action	Challenges	References
Plant growth and productivity	Enhanced photosynthesis, nutrient uptake, and biomass yield	Photoluminescence improves light absorption; regulation of plant hormones and gene expression	Crop-specific responses; scalability of production; unknown long-term effects	Chowmasundaram et al. (2023), Li et al. (2023), Peng et al. (2024), and Zia et al. (2024)
Seed germination	Improved germination rate under stress (e.g., salinity, drought)	Activation of metabolic and stress-response pathways; water retention enhancement	Dose sensitivity; variability in performance across species	Baz et al. (2020), Patel et al. (2020), Li et al. (2023), and Fonseca et al. (2024)
Stress resistance	Increased tolerance to heat, drought, heavy metals, and pathogens	Antioxidant activity, activation of defense genes, regulation of ROS levels	Lack of clarity on mechanism; potential toxicity at high concentrations	Swift et al. (2019), Zhao et al. (2020), Nawaz et al. (2023), and Cao et al. (2024)
Nanofertilizers	Targeted nutrient delivery, higher efficiency, lower input requirements	Small size and hydrophilic nature facilitate uptake; enhanced photosynthesis	Phytotoxicity at high concentrations; optimization needed for specific crops	Aacharya and Chhipa (2019), Wang et al. (2019), Okey-Onyesolu et al. (2021), and Guo et al. (2022)
Nanopesticides	Controlled release, enhanced effectiveness, reduced environmental contamination	Surface-functionalized CNDs bind or degrade pesticide molecules; photothermal action against pathogens	Regulatory barriers; lack of standardized application protocols	Ashrafi Tafreshi et al. (2020), Okey-Onyesolu et al. (2021), Chandel et al. (2022), and Rasheed (2023)
Disease detection	Early detection of plant diseases and foodborne pathogens	Fluorescence and biosensing for pathogen markers; in vivo imaging	Integration into field-ready diagnostic tools	Shen et al. (2020), Khan et al. (2021), Papadopoulou et al. (2021), Zaidi et al. (2022), and Kapat et al. (2023)
Environmental monitoring	Detection of soil/water contaminants (metals, pesticides, antibiotics)	Luminescent sensing; high surface area for pollutant interaction	Interference from complex matrices; need for improved sensitivity and specificity	Manisha et al. (2019), González-González et al. (2022), Zhu et al. (2022), and Hebbar et al. (2023)

1.4 Challenges and Future Prospects

Carbon nanomaterials, including carbon nanodots (CNDs), have shown promise in diverse applications such as biomedical diagnostics, therapeutics, tissue engineering, food safety, and agriculture due to their unique properties, including biocompatibility, excellent photoluminescence, and easy surface modification (Sivasankarapillai et al., 2020; Zhao et al., 2020; Shahazi et al., 2023; Singh et al., 2023). Their advantages include simple synthesis, low cost, high resistance to photobleaching, and non-blinking fluorescent behavior (Shi et al., 2019; Luo et al., 2020). Applications extend to detecting food contaminants like metal ions, pesticides, and foodborne pathogens, regulating plant physiology and seed germination, and serving roles in bioimaging, biosensing, and cancer nanotheranostics through photodynamic and photothermal therapies (Qu et al., 2018; Jia et al., 2020; Khan et al., 2021; Roy & Mamun, 2022; Fonseca et al., 2024).

However, several challenges persist in their widespread adoption. Toxicity concerns and biocompatibility issues remain significant hurdles in biomedical and food applications, necessitating further research to ensure safety for human use and to minimize health risks (Holmannova et al., 2022; Shahazi et al., 2023; Nguyen et al., 2024). In agriculture, potential nanotoxicity to soil microbes and plants, environmental risks, and the need for further investigation into the transport and translocation of CNDs within plant organs present obstacles to their sustainable use. Furthermore, the carbon nanodot–microbe–plant nexus in agroecosystems and ecological risk assessments at broader levels require attention (Prokisch et al., 2024). Additional limitations include the complex composition of food samples causing interferences, unexplained diversity in fluorescent properties, low quantum yields, and the lack of uniformity in CND structure and properties (Shi et al., 2019; Sharma & Soni, 2024).

Despite these challenges, ongoing preclinical studies and recent advancements continue to support the potential of CNDs in areas such as cancer therapy, drug delivery, and improving crop productivity (Jia et al., 2020; Holmannova et al., 2022; Singh et al., 2023). Future research will likely focus on optimizing synthesis methods, enhancing CND properties, developing more specific and sensitive sensing probes, and expanding their applications across domains while ensuring rigorous regulatory assessments for clinical and commercial translation (Sivasankarapillai et al., 2020; Khan et al., 2021; Nguyen et al., 2024).

1.5 Conclusion

Carbon nanodots (CNDs) have emerged as versatile nanomaterials with significant potential in a wide range of fields, owing to their unique photoluminescence, biocompatibility, low toxicity, and tunable surface chemistry. In biomedical applications, CNDs are being explored for diagnostics, imaging, and targeted drug delivery due to their excellent fluorescence properties and functionalization capabilities. In food science, CNDs contribute to improved safety and quality monitoring through rapid, sensitive detection of contaminants like pesticides, heavy metals, and foodborne pathogens. Their integration into smart packaging systems also represents a breakthrough in shelf-life extension and spoilage prevention.

In the field of agriculture, CNDs enhance seed germination, nutrient uptake, stress resistance, and photosynthesis, offering a sustainable approach to improving crop productivity. Their role as nanofertilizers and nanopesticides further underscores their utility in precision agriculture, while also aiding in disease detection and management. Additionally, CNDs show promise in environmental monitoring, especially for assessing soil and water quality, where their luminescent and sensing capabilities are valuable for detecting pollutants like heavy metals and organic contaminants.

Despite these promising applications, several challenges hinder the full-scale implementation of CNDs. These include unresolved concerns about toxicity, environmental accumulation, and long-term impacts on human and ecological health. Inconsistencies in synthesis methods, variability in physicochemical properties, and a lack of standardized safety and regulatory frameworks also need to be addressed. Future research must focus on green synthesis, improving quantum yields, enhancing target specificity, and establishing robust safety protocols to enable responsible and effective deployment.

Overall, carbon nanodots represent a transformative nanotechnology platform with the potential to significantly advance healthcare, food safety, agriculture, and environmental sustainability. With continued innovation and rigorous assessment, they could play a key role in addressing global challenges related to health, food security, and environmental protection.

References

Aacharya, R., & Chhipa, H. (2019). Nanocarbon fertilizers: Implications of carbon nanomaterials in sustainable agriculture production. In *Carbon nanomaterials for agri-food and environmental applications*. https://doi.org/10.1016/B978-0-12-819786-8.00015-3

Aboobakri, E., Heidari, T., & Jahani, M. (2024). Determination of lead(II) in food samples using a functionalized paper-based fluorescent sensor modified by carbon dots. *Luminescence, 39*(2), e4690. https://doi.org/10.1002/bio.4690

Aditya Khatokar, J., Vinay, N., Sanjay, B., Bhargava, S., Sudhir Bale, A., Kolekar, T. R., Singh, S., Umarani, S., & Huddar, S. A. (2021). Carbon nanodots: Chemiluminescence, fluorescence and photoluminescence properties. *Materials Today, 43*, 3928–3931. https://doi.org/10.1016/j.matpr.2021.02.582

Al Ragib, A., Al Amin, A., Alanazi, Y. M., Kormoker, T., Uddin, M., Siddique, M. A. B., & Barai, H. R. (2023). Multifunctional carbon dots in nanomaterial surface modification: A descriptive review. *Carbon Research, 2*(1). https://doi.org/10.1007/s44246-023-00069-x

Arcudi, F., Đorđević, L., & Prato, M. (2019). Design, synthesis, and functionalization strategies of tailored carbon nanodots. *Accounts of Chemical Research, 52*(8), 2070–2079. https://doi.org/10.1021/acs.accounts.9b00249

Arndt, N., Tran, H. D. N., Zhang, R., Xu, Z. P., & Ta, H. T. (2020). Different approaches to develop nanosensors for diagnosis of diseases. *Advanced Science, 7*(24). https://doi.org/10.1002/advs.202001476

Arshad, R., Gulshad, L., Haq, I. U., Farooq, M. A., Al-Farga, A., Siddique, R., Manzoor, M. F., & Karrar, E. (2021). Nanotechnology: A novel tool to enhance the bioavailability of micronutrients. *Food Science & Nutrition, 9*(6), 3354–3361. https://doi.org/10.1002/fsn3.2311

Ashrafi Tafreshi, F., Fatahi, Z., Ghasemi, S. F., Taherian, A., & Esfandiari, N. (2020). Ultrasensitive fluorescent detection of pesticides in real sample by using green carbon dots. *PLoS One, 15*(3), e0230646. https://doi.org/10.1371/journal.pone.0230646

Azami, M., Wei, J., Valizadehderakhshan, M., Jayapalan, A., Ayodele, O. O., & Nowlin, K. (2023). Effect of doping heteroatoms on the optical behaviors and radical scavenging properties of carbon nanodots. *The Journal of Physical Chemistry C, Nanomaterials and Interfaces, 127*(15), 7360–7370. https://doi.org/10.1021/acs.jpcc.3c00953

Bajpai, V. K., Khan, I., Shukla, S., Kumar, P., Chen, L., Anand, S. R., Tripathi, K. M., Bhati, A., Kang, S.-M., Lee, H., Kwak, C. H., Huang, M., Sonkar, S. K., Huh, Y. S., & Han, Y.-K. (2020). N,P-doped carbon nanodots for food-matrix decontamination, anticancer potential, and cellular bio-imaging applications. *Journal of Biomedical Nanotechnology, 16*(3), 283–303. https://doi.org/10.1166/jbn.2020.2899

Bao, X., Yuan, Y., Chen, J., Zhang, B., Li, D., Zhou, D., Jing, P., Xu, G., Wang, Y., Holá, K., Shen, D., Wu, C., Song, L., Liu, C., Zbořil, R., & Qu, S. (2018). In vivo theranostics with near-infrared-emitting carbon dots—Highly efficient photothermal therapy based on passive targeting after intravenous administration. *Light: Science and Applications, 7*(1). https://doi.org/10.1038/s41377-018-0090-1

Bartkowski, M., Zhou, Y., Nabil Amin Mustafa, M., Eustace, A. J., & Giordani, S. (2024). CARBON DOTS: Bioimaging and anticancer drug delivery. *Chemistry – A European Journal, 30*(19). https://doi.org/10.1002/chem.202303982

Bavya, V., Rajan, T. P. D., & Suresh, K. I. (2025). Design of fluorescence enhancing sensor for mercury detection via bamboo cellulose-derived carbon dots. *Langmuir, 41*(2), 1333–1343. https://doi.org/10.1021/acs.langmuir.4c03942

Bayda, S., Amadio, E., Cailotto, S., Frión-Herrera, Y., Perosa, A., & Rizzolio, F. (2021). Carbon dots for cancer nanomedicine: A bright future. *Nanoscale Advances, 3*(18). https://doi.org/10.1039/d1na00036e

Baz, H., Creech, M., Chen, J., Gong, H., Bradford, K., & Huo, H. (2020). Water-soluble carbon nanoparticles improve seed germination and post-germination growth of lettuce under salinity stress. *Agronomy, 10*(8). https://doi.org/10.3390/agronomy10081192

Boakye-Yiadom, K. O., Kesse, S., Opoku-Damoah, Y., Filli, M. S., Aquib, M., Joelle, M. M. B., Farooq, M. A., Mavlyanova, R., Raza, F., Bavi, R., & Wang, B. (2019). Carbon dots: Applications in bioimaging and theranostics. *International Journal of Pharmaceutics, 564*(April), 308–317. https://doi.org/10.1016/j.ijpharm.2019.04.055

Cao, Y., Turk, K., Bibi, N., Ghafoor, A., Ahmed, N., Azmat, M., Ahmed, R., Ghani, M. I., & Ahanger, M. A. (2024). Nanoparticles as catalysts of agricultural revolution: Enhancing crop tolerance to abiotic stress: A review. *Frontiers in Plant Science, 15*(January), 1–23. https://doi.org/10.3389/fpls.2024.1510482

Cayuela, A., Soriano, M. L., Carrillo-Carrión, C., & Valcárcel, M. (2016). Semiconductor and carbon-based fluorescent nanodots: The need for consistency. *Chemical Communications (Cambridge), 52*(7), 1311–1326. https://doi.org/10.1039/c5cc07754k

Chan, M., Chen, B., Ngo, L. T., Huang, W., Li, C., Liu, R., & Hsiao, M. (2021). Natural carbon nanodots: Toxicity assessment and theranostic biological application. *Pharmaceutics, 13*. https://doi.org/10.3390/pharmaceutics13111874

Chandel, M., Kaur, K., Sahu, B. K., Sharma, S., Panneerselvam, R., & Shanmugam, V. (2022). Promise of nano-carbon to the next generation sustainable agriculture. *Carbon, 188*. https://doi.org/10.1016/j.carbon.2021.11.060

Chaudhary, M., Singh, P., Singh, G. P., & Rathi, B. (2024). Structural features of carbon dots and their agricultural potential. *ACS Omega, 9*(4). https://doi.org/10.1021/acsomega.3c04638

Chowmasundaram, Y. A. P., Tan, T. L., Nulit, R., Jusoh, M., & Rashid, S. A. (2023). Recent developments, applications and challenges for carbon quantum dots as a photosynthesis enhancer in agriculture. *RSC Advances, 13*(36). https://doi.org/10.1039/d3ra01217d

Chung, Y. J., Kim, J., & Park, C. B. (2020). Photonic carbon dots as an emerging nanoagent for biomedical and healthcare applications. *ACS Nano, 14*(6). https://doi.org/10.1021/acsnano.0c02114

Cui, F., Ning, Y., Wang, D., Li, J., Li, X., & Li, T. (2024). Carbon dot-based therapeutics for combating drug-resistant bacteria and biofilm infections in food preservation. *Critical Reviews in Food Science and Nutrition, 64*(2), 203–219. https://doi.org/10.1080/10408398.2022.2105801

Debnath, S. K., & Srivastava, R. (2021). Drug delivery with carbon-based nanomaterials as versatile nanocarriers: Progress and prospects. *Frontiers in Nanotechnology, 3*. https://doi.org/10.3389/fnano.2021.644564

Deepika, Kumar, L., & Gaikwad, K. K. (2023). Carbon dots for food packaging applications. *Sustainable Food Technology, 1*, 185–199. https://doi.org/10.1039/D2FB00020B

Ding, H., Xiao, T., Ren, F., Qiu, Y., Shen, Z., Chen, X., Mijowska, E., & Chen, H. (2024, March). Carbon-based nanodots for biomedical applications and clinical transformation prospects. *BMEMat*, 1–19. https://doi.org/10.1002/bmm2.12085

Đorđević, L., Arcudi, F., & Prato, M. (2019). Preparation, functionalization and characterization of engineered carbon nanodots. *Nature Protocols, 14*, 2931–2953. https://doi.org/10.1038/s41596-019-0207-x

Ehtesabi, H., Hallaji, Z., Najafi Nobar, S., & Bagheri, Z. (2020). Carbon dots with pH-responsive fluorescence: A review on synthesis and cell biological applications. *Mikrochimica Acta, 187*(2), 150. https://doi.org/10.1007/s00604-019-4091-4

Ezati, P., Khan, A., Rhim, J.-W., Kim, J. T., & Molaei, R. (2023). pH-Responsive strips integrated with resazurin and carbon dots for monitoring shrimp freshness. *Colloids and Surfaces B: Biointerfaces, 221*(113013), 113013. https://doi.org/10.1016/j.colsurfb.2022.113013

Farshidfar, N., Fooladi, S., Nematollahi, M. H., & Iravani, S. (2023). Carbon dots with tissue engineering and regenerative medicine applications. *RSC Advances, 13*(21). https://doi.org/10.1039/d3ra02336b

Fonseca, J. D. d. S., Wojciechowska, E., Kulesza, J., & Barros, B. S. (2024). Carbon nanomaterials in seed priming: Current possibilities. *ACS Omega, 9*(45), 44891–44906. https://doi.org/10.1021/acsomega.4c07230

Fu, B., Liu, Q., Liu, M., Chen, X., Lin, H., Zheng, Z., Zhu, J., Dai, C., Dong, X., & Yang, D.-P. (2022). Carbon dots enhanced gelatin/chitosan bio-nanocomposite packaging film for perishable foods. *Chinese Chemical Letters, 33*(10), 4577–4582. https://doi.org/10.1016/j.cclet.2022.03.048

Gautam, B., Huang, M.-R., Ali, S. A., Yan, A.-L., Yu, H.-H., & Chen, J.-T. (2022). Smart thermoresponsive electrospun nanofibers with on-demand release of carbon quantum dots for cellular uptake. *ACS Applied Materials & Interfaces, 14*(35), 40322–40330. https://doi.org/10.1021/acsami.2c10810

Ghosh, T., Raj, G. V. S. B., & Dash, K. K. (2022). A comprehensive review on nanotechnology based sensors for monitoring quality and shelf life of food products. *Measurement: Food, 7*(100049), 100049. https://doi.org/10.1016/j.meafoo.2022.100049

González-González, R. B., Morales-Murillo, M. B., Martínez-Prado, M. A., Melchor-Martínez, E. M., Ahmed, I., Bilal, M., Parra-Saldívar, R., & Iqbal, H. M. N. (2022). Carbon dots-based nanomaterials for fluorescent sensing of toxic elements in environmental samples: Strategies for enhanced performance. *Chemosphere, 300*. https://doi.org/10.1016/j.chemosphere.2022.134515

Goryacheva, I. Y., Sapelkin, A. V., & Sukhorukov, G. B. (2017). Carbon nanodots: Mechanisms of photoluminescence and principles of application. *TrAC Trends in Analytical Chemistry, 90*, 27–37. https://doi.org/10.1016/j.trac.2017.02.012

Guo, B., Liu, G., Wei, H., Qiu, J., Zhuang, J., Zhang, X., Zheng, M., Li, W., Zhang, H., Hu, C., Lei, B., & Liu, Y. (2022). The role of fluorescent carbon dots in crops: Mechanism and applications. *SmartMat, 3*(2). https://doi.org/10.1002/smm2.1111

Gutiérrez-Gálvez, L., Sulleiro, M. V., Gutiérrez-Sánchez, C., García-Nieto, D., Luna, M., Pérez, E. M., García-Mendiola, T., & Lorenzo, E. (2023). MoS2-carbon nanodots as a new electrochemiluminescence platform for breast cancer biomarker detection. *Biosensors, 13*(3). https://doi.org/10.3390/bios13030348

Hebbar, A., Selvaraj, R., Vinayagam, R., Varadavenkatesan, T., Kumar, P. S., Duc, P. A., & Rangasamy, G. (2023). A critical review on the environmental applications of carbon dots. *Chemosphere, 313*. https://doi.org/10.1016/j.chemosphere.2022.137308

Holmannova, D., Borsky, P., Svadlakova, T., Borska, L., & Fiala, Z. (2022). Carbon nanoparticles and their biomedical applications. *Applied Sciences (Switzerland), 12*(15). https://doi.org/10.3390/app12157865

Hutton, G. A. M., Reuillard, B., Martindale, B. C. M., Caputo, C. A., Lockwood, C. W. J., Butt, J. N., & Reisner, E. (2016). Carbon dots as versatile photosensitizers for solar-driven catalysis with redox enzymes. *Journal of the American Chemical Society, 138*(51), 16722–16730. https://doi.org/10.1021/jacs.6b10146

Innocenzi, P., & Stagi, L. (2023). Carbon dots as oxidant-antioxidant nanomaterials, understanding the structure-properties relationship. A critical review. *Nano Today, 50*. https://doi.org/10.1016/j.nantod.2023.101837

Ji, D. K., Reina, G., Guo, S., Eredia, M., Samorì, P., Ménard-Moyon, C., & Bianco, A. (2020). Controlled functionalization of carbon nanodots for targeted intracellular production of reactive oxygen species. *Nanoscale Horizons, 5*(8). https://doi.org/10.1039/d0nh00300j

Jia, Q., Zhao, Z., Liang, K., Nan, F., Li, Y., Wang, J., Ge, J., & Wang, P. (2020). Recent advances and prospects of carbon dots in cancer nanotheranostics. *Materials Chemistry Frontiers, 4*(2). https://doi.org/10.1039/c9qm00667b

Jiang, L., Yuan, L., Gao, S., Xiang, Y., Song, F., Ma, W., Wan, J., Ji, X., & Tu, Y. (2023). Facile hydrothermal synthesis of N-doped fluorescent carbon dots for selective detection of insecticide parathion. *Analytical Methods, 15*(19), 2376–2381. https://doi.org/10.1039/d3ay00124e

Jin, L., Ren, K., Xu, Q., Hong, T., Wu, S., Zhang, Y., & Wang, Z. (2018). Multifunctional carbon dots for live cell staining and tissue engineering applications. *Polymer Composites, 39*(1). https://doi.org/10.1002/pc.23903

Jovanović, S., Marković, Z., Budimir, M., Prekodravac, J., Zmejkoski, D., Kepić, D., Bonasera, A., & Marković, B. T. (2023). Lights and dots toward therapy—Carbon-based quantum dots as new agents for photodynamic therapy. *Pharmaceutics, 15*(4). https://doi.org/10.3390/pharmaceutics15041170

Kang, Z., Yang, B., & Prato, M. (2023). Carbon nanodots: Nanolights illuminating a bright future. *Small, 19*(31). https://doi.org/10.1002/smll.202304703

Kapat, K., Semwal, N., Chillarge, A., & Aswani, A. (2023). Multifunctional carbon nanodot-based advanced diagnostics and therapeutics. *Advanced Therapeutics, 6*(11). https://doi.org/10.1002/adtp.202300189

Kaurav, H., Verma, D., Bansal, A., Kapoor, D. N., & Sheth, S. (2023). Progress in drug delivery and diagnostic applications of carbon dots: A systematic review. *Frontiers in Chemistry, 11*. https://doi.org/10.3389/fchem.2023.1227843

Khan, S., Dunphy, A., Anike, M. S., Belperain, S., Patel, K., Chiu, N. H. L., & Jia, Z. (2021). Recent advances in carbon nanodots: A promising nanomaterial for biomedical applications. *International Journal of Molecular Sciences, 22*(13). MDPI. https://doi.org/10.3390/ijms22136786

Knoblauch, R., & Geddes, C. D. (2020). Carbon nanodots in photodynamic antimicrobial therapy: A review. *Materials, 13*(18). https://doi.org/10.3390/ma13184004

Koutsogiannis, P., Thomou, E., Stamatis, H., Gournis, D., & Rudolf, P. (2020). Advances in fluorescent carbon dots for biomedical applications. *Advances in Physics: X, 5*(1). https://doi.org/10.1080/23746149.2020.1758592

Lagopati, N., Valamvanos, T. F., Proutsou, V., Karachalios, K., Pippa, N., Gatou, M. A., Vagena, I. A., Cela, S., Pavlatou, E. A., Gazouli, M., & Efstathopoulos, E. (2023). The role of nanosensors in breath analysis for early and non-invasive disease diagnosis. *Chemosensors, 11*(6). https://doi.org/10.3390/chemosensors11060317

Lagos, K. J., García, D., Cuadrado, C. F., de Souza, L. M., Mezzacappo, N. F., da Silva, A. P., Inada, N., Bagnato, V., & Romero, M. P. (2023). Carbon dots: Types, preparation, and their boosted antibacterial activity by photoactivation. Current status and future perspectives. *Wiley Interdisciplinary Reviews: Nanomedicine and Nanobiotechnology, 15*(4). https://doi.org/10.1002/wnan.1887

Lee, C., Kwon, W., Beack, S., Lee, D., Park, Y., Kim, H., Hahn, S. K., Rhee, S. W., & Kim, C. (2016). Biodegradable nitrogen-doped carbon nanodots for non-invasive photoacoustic imaging and photothermal therapy. *Theranostics, 6*(12). https://doi.org/10.7150/thno.16923

Leong, Y. X., Tan, E. X., Leong, S. X., Lin Koh, C. S., Thanh Nguyen, L. B., Ting Chen, J. R., Xia, K., & Ling, X. Y. (2022). Where nanosensors meet machine learning: Prospects and challenges in detecting disease X. *ACS Nano, 16*(9). https://doi.org/10.1021/acsnano.2c05731

Li, Y., Zhang, X., Zheng, M., Liu, S., & Xie, Z. (2016). Dopamine carbon nanodots as effective photothermal agents for cancer therapy. *RSC Advances, 6*(59). https://doi.org/10.1039/c6ra02932a

Li, G., Xu, J., & Xu, K. (2023). Physiological functions of carbon dots and their applications in agriculture: A review. *Nanomaterials, 13*(19). https://doi.org/10.3390/nano13192684

Lin, B., Yan, Y., Guo, M., Cao, Y., Yu, Y., Zhang, T., Huang, Y., & Wu, D. (2018). Modification-free carbon dots as turn-on fluorescence probe for detection of organophosphorus pesticides. *Food Chemistry, 245*, 1176–1182. https://doi.org/10.1016/j.foodchem.2017.11.038

Liu, M. (2020). Optical properties of carbon dots: A review. *Nanoarchitectonics, 1*, 1–12. https://doi.org/10.37256/nat.112020124.1-12

Luo, X., Han, Y., Chen, X., Tang, W., Yue, T., & Li, Z. (2020). Carbon dots derived fluorescent nanosensors as versatile tools for food quality and safety assessment: A review. *Trends in Food Science & Technology, 95*. https://doi.org/10.1016/j.tifs.2019.11.017

Manikandan, V., & Min, S. C. (2023). Biofabrication of carbon quantum dots and their food packaging applications: A review. *Food Science and Biotechnology, 32*(9), 1159–1171. https://doi.org/10.1007/s10068-023-01309-x

Manisha, H., Priya Swetha, P. D., Shim, Y. B., & Prasad, K. S. (2019). Revisiting fluorescent carbon nanodots for environmental, biomedical applications and puzzle about fluorophore impurities. *Nano-Structures & Nano-Objects, 20*. https://doi.org/10.1016/j.nanoso.2019.100391

Mao, Q.-X., Han, L., Shu, Y., Chen, X.-W., & Wang, J.-H. (2016). Improving the biocompatibility of carbon nanodots for cell imaging. *Talanta, 161*, 54–61. https://doi.org/10.1016/j.talanta.2016.08.031

Mary Isabella Sonali, J., Subhashree, S., Senthil Kumar, P., & Veena Gayathri, K. (2022). New analytical strategies amplified with carbon-based nanomaterial for sensing food pollutants. *Chemosphere, 295*(133847), 133847. https://doi.org/10.1016/j.chemosphere.2022.133847

Mei, Y., He, C., Zeng, W., Luo, Y., Liu, C., Yang, M., Kuang, Y., Lin, X., & Huang, Q. (2022). Electrochemical biosensors for foodborne pathogens detection based on carbon nanomaterials: Recent advances and challenges. *Food and Bioprocess Technology, 15*(3), 498–513. https://doi.org/10.1007/s11947-022-02759-7

Mishra, S., Mishra, S., Patel, S. S., Singh, S. P., Kumar, P., Khan, M. A., Awasthi, H., & Singh, S. (2022). Carbon nanomaterials for the detection of pesticide residues in food: A review. *Environmental Pollution, 310*(119804), 119804. https://doi.org/10.1016/j.envpol.2022.119804

Mocci, F., de Villiers Engelbrecht, L., Olla, C., Cappai, A., Casula, M. F., Melis, C., Stagi, L., Laaksonen, A., & Carbonaro, C. M. (2022). Carbon nanodots from an in silico perspective. *Chemical Reviews, 122*(16), 13709–13799. https://doi.org/10.1021/acs.chemrev.1c00864

Mohammadi, Z., & Jafari, S. M. (2020). Detection of food spoilage and adulteration by novel nanomaterial-based sensors. *Advances in Colloid and Interface Science, 286*(102297), 102297. https://doi.org/10.1016/j.cis.2020.102297

Moradi, M., Molaei, R., Kousheh, S. A., Guimarães, J. T., & McClements, D. J. (2023). Carbon dots synthesized from microorganisms and food by-products: Active and smart food packaging applications. *Critical Reviews in Food Science and Nutrition, 63*(14), 1943–1959. https://doi.org/10.1080/10408398.2021.2015283

Munawar, A., Ong, Y., Schirhagl, R., Tahir, M. A., Khan, W. S., & Bajwa, S. Z. (2019). Nanosensors for diagnosis with optical, electric and mechanical transducers. *RSC Advances, 9*(12). https://doi.org/10.1039/c8ra10144b

Nawaz, A., Rehman, H. u., Usman, M., Wakeel, A., Shahid, M. S., Alam, S., Sanaullah, M., Atiq, M., & Farooq, M. (2023). Nanobiotechnology in crop stress management: an overview of novel applications. *Discover Nano, 18*(1). https://doi.org/10.1186/s11671-023-03845-1

Nguyen, D. H. H., El-Ramady, H., & Prokisch, J. (2024). Food safety aspects of carbon dots: A review. *Environmental Chemistry Letters, 23*(1), 337–360. https://doi.org/10.1007/s10311-024-01779-3

Okey-Onyesolu, C. F., Hassanisaadi, M., Bilal, M., Barani, M., Rahdar, A., Iqbal, J., & Kyzas, G. Z. (2021). Nanomaterials as nanofertilizers and nanopesticides: An overview. *ChemistrySelect, 6*(33). https://doi.org/10.1002/slct.202102379

Papadopoulou, E. L., Biffi, G., Senthamizhan, A., Martín-García, B., Carzino, R., Krahne, R., & Athanassiou, A. (2021). Paper sensors based on fluorescence changes of carbon nanodots for optical detection of nanomaterials. *Sustainability (Switzerland), 13*(21). https://doi.org/10.3390/su132111896

Papaioannou, N., Marinović, A., Yoshizawa, N., Goode, A., Fay, M., Khlobystov, A., Titirici, M., & Sapelkin, A. (2018). Structure and solvents effects on the optical properties of sugar-derived carbon nanodots. *Scientific Reports, 8*. https://doi.org/10.1038/s41598-018-25012-8

Patel, D. K., Kim, H. B., Dutta, S. D., Ganguly, K., & Lim, K. T. (2020). Carbon nanotubes-based nanomaterials and their agricultural and biotechnological applications. *Materials, 13*(7). https://doi.org/10.3390/ma13071679

Patir, K., & Gogoi, S. K. (2019). Nitrogen-doped carbon dots as fluorescence ON-OFF-ON sensor for parallel detection of copper(ii) and mercury(ii) ions in solutions as well as in filter paper-based microfluidic device. *Nanoscale Advances, 1*(2), 592–601. https://doi.org/10.1039/c8na00080h

Peng, W., Li, P., Zia, M., Bhuiyan, S. A., Liu, Y., & Chen, D. (2024). A comparative review on carbon nanomaterials and their impact on plant growth with the lens of green chemistry principles. *Global Challenges, 2400008*, 1–16. https://doi.org/10.1002/gch2.202400008

Prokisch, J., Nguyen, D. H. H., Muthu, A., Ferroudj, A., Singh, A., Agrawal, S., Rajput, V. D., Ghazaryan, K., El-Ramady, H., & Rai, M. (2024). Carbon nanodot–microbe–plant nexus in agroecosystem and antimicrobial applications. *Nanomaterials, 14*(15). https://doi.org/10.3390/nano14151249

Qin, Y., Huang, P., & Wu, F.-Y. (2022). Histamine-responsive dye-incorporated carbon dots for visual monitoring of food spoilage. *Sensors and Actuators. B, Chemical, 365*(131911), 131911. https://doi.org/10.1016/j.snb.2022.131911

Qu, J. H., Wei, Q., & Sun, D. W. (2018). Carbon dots: Principles and their applications in food quality and safety detection. *Critical Reviews in Food Science and Nutrition, 58*(14), 2466–2475. Taylor and Francis. https://doi.org/10.1080/10408398.2018.1437712

Qureshi, Z. A., Dabash, H., Ponnamma, D., & Abbas, M. K. G. (2024). Carbon dots as versatile nanomaterials in sensing and imaging: Efficiency and beyond. *Heliyon, 10*(11), e31634. https://doi.org/10.1016/j.heliyon.2024.e31634

Rasheed, T. (2023). Carbon dots as robust class of sustainable and environment friendlier nano/optical sensors for pesticide recognition from wastewater. *TrAC Trends in Analytical Chemistry, 160*. https://doi.org/10.1016/j.trac.2023.116957

Ross, S., Wu, R.-S., Wei, S.-C., Ross, G. M., & Chang, H.-T. (2020). The analytical and biomedical applications of carbon dots and their future theranostic potential: A review. *Journal of Food and Drug Analysis, 28*(4), 677–695. https://doi.org/10.38212/2224-6614.1154

Roy, T., & Mamun, S. M. A. A. (2022). Carbon nanomaterials towards transformation in agriculture: A review. *International Journal of Research Publication and Reviews*. https://doi.org/10.55248/gengpi.2022.3.5.32

Saita, S., & Kawasaki, H. (2023). Carbon nanodots with a controlled N structure by a solvothermal method for generation of reactive oxygen species under visible light. *Luminescence, 38*(2), 127–135. https://doi.org/10.1002/bio.4428

Sangeetha, U. K., Sudhakaran, N., Parvathy, P. A., Abraham, M., Das, S., De, S., & Sahoo, S. K. (2024). Coconut husk-lignin derived carbon dots incorporated carrageenan based functional film for intelligent food packaging. *International Journal of Biological Macromolecules, 266*(Pt 1), 131005. https://doi.org/10.1016/j.ijbiomac.2024.131005

Sciortino, A., Cannizzo, A., & Messina, F. (2018). Carbon nanodots: A review—from the current understanding of the fundamental photophysics to the full control of the optical response. *C++*. https://doi.org/10.3390/C4040067

Sekar, R., Basavegowda, N., Jena, S., Jayakodi, S., Elumalai, P., Chaitanyakumar, A., Somu, P., & Baek, K. H. (2022). Recent developments in heteroatom/metal-doped carbon dot-based image-guided photodynamic therapy for cancer. *Pharmaceutics, 14*(9). https://doi.org/10.3390/pharmaceutics14091869

Shahazi, R., Majumdar, S., Saddam, A. I., Mondal, J., Rahman, M. M., & Alam, M. M. (2023). Carbon nanomaterials for biomedical applications: A comprehensive review. *Nano Carbons, 1*(1), 448. https://doi.org/10.59400/n-c.v1i1.448

Sharma, V., & Soni, V. (2024). Advancement in carbon nanoparticle synthesis and their application: A comprehensive review. *American Journal of Nano Research and Applications, 12*(1), 1–14. https://doi.org/10.11648/j.nano.20241201.11

Shen, C. L., Zheng, G. S., Wu, M. Y., Wei, J. Y., Lou, Q., Ye, Y. L., Liu, Z. Y., Zang, J. H., Dong, L., & Shan, C. X. (2020). Chemiluminescent carbon nanodots as sensors for hydrogen peroxide and glucose. *Nanophotonics, 9*(11). https://doi.org/10.1515/nanoph-2020-0233

Shi, X., Wei, W., Fu, Z., Gao, W., Zhang, C., Zhao, Q., Deng, F., & Lu, X. (2019). Review on carbon dots in food safety applications. *Talanta, 194*. https://doi.org/10.1016/j.talanta.2018.11.005

Shruti, A., Bage, N., & Kar, P. (2024). Nanomaterials based sensors for analysis of food safety. *Food Chemistry, 433*(137284), 137284. https://doi.org/10.1016/j.foodchem.2023.137284

Singh, T., Shukla, S., Kumar, P., Wahla, V., & Bajpai, V. K. (2017). Application of nanotechnology in food science: Perception and overview. *Frontiers in Microbiology, 8*, 1501. https://doi.org/10.3389/fmicb.2017.01501

Singh, J., Nayak, P., Singh, G., Khandai, M., Sarangi, R. R., & Kar, M. K. (2023). Carbon nanostructures as therapeutic cargoes: Recent developments and challenges. *C-Journal of Carbon Research, 9*(1). https://doi.org/10.3390/c9010003

Sivasankarapillai, V. S., Vishnu Kirthi, A., Akksadha, M., Indu, S., Dhiviya Dharshini, U., Pushpamalar, J., & Karthik, L. (2020). Recent advancements in the applications of carbon nanodots: Exploring the rising star of nanotechnology. *Nanoscale Advances, 2*(5), 1760–1773. Royal Society of Chemistry. https://doi.org/10.1039/c9na00794f

Song, J., Gao, X., Yang, M., Hao, W., & Ji, D. K. (2023). Recent advances of photoactive near-infrared carbon dots in cancer photodynamic therapy. *Pharmaceutics, 15*(3). https://doi.org/10.3390/pharmaceutics15030760

Song, H., Huo, M., Zhou, M., Chang, H., Li, J., Zhang, Q., Fang, Y., Wang, H., & Zhang, D. (2024). Carbon nanomaterials-based electrochemical sensors for heavy metal detection. *Critical Reviews in Analytical Chemistry, 54*(7), 1987–2006. https://doi.org/10.1080/10408347.2022.2151832

Soni, H., Jain, V., Ballal, S., Ariffin, I. A., Chahar, M., Saini, S., Bhattu, M., Singh, H., Bechelany, M., & Singh, J. (2024). From structure to sensing: Molecular mechanistic insights into plant-derived carbon dots for heavy metal ion detection. *Nanomaterials (Basel), 14*(21). https://doi.org/10.3390/nano14211766

Su, D., Li, H., Yan, X., Lin, Y., & Lu, G. (2021). Biosensors based on fluorescence carbon nanomaterials for detection of pesticides. *Trends in Analytical Chemistry, 134*(116126), 116126. https://doi.org/10.1016/j.trac.2020.116126

Swift, T. A., Oliver, T. A. A., Carmen Galan, M., & Whitney, H. M. (2019). Functional nanomaterials to augment photosynthesis: Evidence and considerations for their responsible use in agricultural applications. *Interface Focus, 9*(1). https://doi.org/10.1098/rsfs.2018.0048

Szczepankowska, J., Khachatryan, G., Khachatryan, K., & Krystyjan, M. (2023). Carbon dots-types, obtaining and application in biotechnology and food technology. *International Journal of Molecular Sciences, 24*(19), 14984. https://doi.org/10.3390/ijms241914984

Tian, J., An, M., Zhao, X., Wang, Y., & Hasan, M. (2023). Advances in fluorescent sensing carbon dots: An account of food analysis. *ACS Omega, 8*(10), 9031–9039. https://doi.org/10.1021/acsomega.2c07986

Truskewycz, A., Yin, H., Halberg, N., Lai, D. T. H., Ball, A. S., Truong, V. K., Rybicka, A. M., & Cole, I. (2022). Carbon dot therapeutic platforms: Administration, distribution, metabolism, excretion, toxicity, and therapeutic potential. *Small, 18*(16). https://doi.org/10.1002/smll.202106342

Varghese, M., & Balachandran, M. (2021). Antibacterial efficiency of carbon dots against Gram-positive and Gram-negative bacteria: A review. *Journal of Environmental Chemical Engineering, 9*(6). https://doi.org/10.1016/j.jece.2021.106821

Vedhanayagam, M., Raja, I. S., Molkenova, A., Atabaev, T. S., Sreeram, K. J., & Han, D. W. (2021). Carbon dots-mediated fluorescent scaffolds: Recent trends in image-guided tissue engineering applications. *International Journal of Molecular Sciences, 22*(10). https://doi.org/10.3390/ijms22105378

Wagh, R. V., Khan, A., Priyadarshi, R., Ezati, P., & Rhim, J.-W. (2023). Cellulose nanofiber-based multifunctional films integrated with carbon dots and anthocyanins from Brassica oleracea for active and intelligent food packaging applications. *International Journal of Biological Macromolecules, 233*(123567), 123567. https://doi.org/10.1016/j.ijbiomac.2023.123567

Wang, A., Kang, F., Wang, Z., Shao, Q., Li, Z., Zhu, G., Lu, J., & Li, Y. Y. (2019). Facile synthesis of nitrogen-rich carbon dots as fertilizers for mung bean sprouts. *Advanced Sustainable Systems, 3*(3). https://doi.org/10.1002/adsu.201800132

Wei, W., Huang, J., Gao, W., Lu, X., & Shi, X. (2021). Carbon dots fluorescence-based colorimetric sensor for sensitive detection of aluminum ions with a smartphone. *Chemosensors (Basel), 9*(2), 25. https://doi.org/10.3390/chemosensors9020025

Wu, Y., Li, C., van der Mei, H. C., Busscher, H. J., & Ren, Y. (2021). Carbon quantum dots derived from different carbon sources for antibacterial applications. *Antibiotics, 10*(6). https://doi.org/10.3390/antibiotics10060623

Xiao, L., & Sun, H. (2018). Novel properties and applications of carbon nanodots. *Nanoscale Horizons, 3*(6). https://doi.org/10.1039/c8nh00106e

Xu, X., Ray, R., Gu, Y., Ploehn, H. J., Gearheart, L., Raker, K., & Scrivens, W. A. (2004). Electrophoretic analysis and purification of fluorescent single-walled carbon nanotube fragments. *Journal of the American Chemical Society, 126*(40), 12736–12737. https://doi.org/10.1021/ja040082h

Yadav, R., & Lahariya, V. (2024). Fluorometric sensing of metal ions by carbon nano dots – A review. *Current Materials Science, 18*(4), 413–428. https://doi.org/10.2174/0126661454270676231127043124

Yadav, H., Kumar, R., & Sankhla, M. S. (2020). Residues of pesticides and heavy metals in crops resulting in toxic effects on living organism. *Journal of Seybold Report.* https://www.researchgate.net/profile/Mahipal-Singh-Sankhla/publication/343291021_Residues_of_Pesticides_and_Heavy_Metals_in_Crops_Resulting_in_Toxic_Effects_on_Living_Organism/links/5f2195cc92851cd302c5fd81/Residues-of-Pesticides-and-Heavy-Metals-in-Crops

Yu, Y., Zhang, L., Gao, X., Feng, Y., Wang, H., Lei, C., Yan, Y., & Liu, S. (2022). Research progress in the synthesis of carbon dots and their application in food analysis. *Biosensors (Basel), 12*(12), 1158. https://doi.org/10.3390/bios12121158

Yue, X., Zhu, C., Gu, R., Hu, J., Xu, Y., Ye, S., & Zhu, J. (2022). Copper-modified double-emission carbon dots for rapid detection of thiophanate methyl in food. *Foods, 11*(21), 3336. https://doi.org/10.3390/foods11213336

Zaidi, Z., Maiti, N., Ali, M. I., Sharma, G., Moin, S., Padhy, H., Balaji, G. L., & Sundramurthy, V. P. (2022). Fabrication, characteristics, and therapeutic applications of carbon-based nanodots. *Journal of Nanomaterials, 2022*. https://doi.org/10.1155/2022/8031495

Zhang, J., Chen, H., Xu, K., Deng, D., Zhang, Q., & Luo, L. (2023). Current progress of ratiometric fluorescence sensors based on carbon dots in foodborne contaminant detection. *Biosensors (Basel), 13*(2). https://doi.org/10.3390/bios13020233

Zhang, L., Cai, Z., Liu, Y., Fan, Y., & She, Y. (2024). Fluorescent enhanced endogenous carbon dots derived from green tea residue for multiplex detection of heavy metal ions in food. *Frontiers in Sustainable Food Systems, 8*. https://doi.org/10.3389/fsufs.2024.1431792

Zhao, Q., Song, W., Zhao, B., & Yang, B. (2020). Spectroscopic studies of the optical properties of carbon dots: Recent advances and future prospects. *Materials Chemistry Frontiers, 4*(2). https://doi.org/10.1039/c9qm00592g

Zhao, S., Yan, L., Cao, M., Huang, L., Yang, K., Wu, S., Lan, M., Niu, G., & Zhang, W. (2021). Near-infrared light-triggered lysosome-targetable carbon dots for photothermal therapy of cancer. *ACS Applied Materials and Interfaces, 13*(45). https://doi.org/10.1021/acsami.1c15926

Zhao, L., Zhang, M., Mujumdar, A. S., & Wang, H. (2023a). Application of carbon dots in food preservation: A critical review for packaging enhancers and food preservatives. *Critical Reviews in Food Science and Nutrition, 63*(24), 6738–6756. https://doi.org/10.1080/10408398.2022.2039896

Zhao, W. B., Liu, K. K., Wang, Y., Li, F. K., Guo, R., Song, S. Y., & Shan, C. X. (2023b). Antibacterial carbon dots: Mechanisms, design, and applications. *Advanced Healthcare Materials, 12*(23). https://doi.org/10.1002/adhm.202300324

Zheng, S., Tian, Y., Ouyang, J., Shen, Y., Wang, X., & Luan, J. (2022). Carbon nanomaterials for drug delivery and tissue engineering. *Frontiers in Chemistry, 10*. https://doi.org/10.3389/fchem.2022.990362

Zhou, Y., Liu, Y., Li, Y., He, Z., Xu, Q., Chen, Y., Street, J., Guo, H., & Nelles, M. (2018). Multicolor carbon nanodots from food waste and their heavy metal ion detection application. *RSC Advances, 8*(42), 23657–23662. https://doi.org/10.1039/c8ra03272f

Zhou, Q., Fu, L., & Zhu, J. (2025). Electrochemical sensors go nano: Carbon nanomaterials for ultrasensitive heavy metal analysis. *Current Nanoscience, 21*(4), 596–612. https://doi.org/10.2174/0115734137281774231214054405

Zhu, P., Zhang, Z., Zhang, T., & Chen, W. (2022). Construction of carbon-dot-based fluorescent sensors and their applications in environmental pollution analysis. *Scientia Sinica Chimica, 52*(7). https://doi.org/10.1360/SSC-2022-0096

Zia, M. U., Sambasivam, P. T., Chen, D., Bhuiyan, S. A., Ford, R., & Li, Q. (2024). A carbon dot toolbox for managing biotic and abiotic stresses in crop production systems. *EcoMat, 6*(5), 1–26. https://doi.org/10.1002/eom2.12451

Synthesis Methods of Carbon Nanodots 2

Anna Ilnicka, Malgorzata Skorupska, Patrycja Grabowska, and Mariusz Szkoda

Abstract

This chapter provides a comprehensive overview of the various methods employed to synthesize carbon dots (CDs), a novel class of fluorescent materials with extensive applications in biomedicine and optoelectronics. It discusses top-down and bottom-up strategies, highlighting their benefits regarding raw material availability, scalability, and procedural simplicity. Additionally, the chapter emphasizes the unique photoluminescent characteristics of CDs, particularly their tunable emission and quantum confinement effects. Applications in cell imaging, photodynamic therapy, drug delivery, and optoelectronic devices such as light-emitting diodes (LEDs) are explored. Factors influencing the properties of carbon dots, including precursor choice, reaction conditions, doping, and post-synthesis treatments, are summarized. The chapter concludes by presenting an outlook on future developments in carbon dot synthesis and their potential advancements in biomedical imaging, sensors, and other emerging technologies.

Keywords

Carbon nanodots · Laser ablation · Plasma discharge · Hydrothermal method · Pyrolysis · Electrochemical method

A. Ilnicka (✉) · M. Skorupska · P. Grabowska
Faculty of Chemistry, Nicolaus Copernicus University in Torun, Torun, Poland
e-mail: ailnicka@umk.pl

M. Szkoda
Faculty of Chemistry, Department of Chemistry and Technology of Functional Materials, Gdańsk University of Technology, Gdańsk, Poland

Advanced Materials Center, Gdańsk University of Technology, Gdańsk, Poland

M. Rai et al. (eds.), *Carbon Nanodots*, https://doi.org/10.1007/978-981-95-4705-0_2

2.1 Introduction

Carbon nanodots (CNDs) or carbon dots (CDs) are a class of carbon-based nanoparticles typically below 10 nm in size. Over the last decade, they have attracted remarkable attention thanks to their unique physicochemical properties, which include strong and tunable photoluminescence, biocompatibility, low toxicity, and high chemical stability. Furthermore, they can be synthesized using relatively simple and cost-effective methods, often from abundant raw materials such as biomass or common organic compounds (Baker & Baker, 2010; Chauhan et al., 2022; Liu et al., 2007). One of the most intriguing aspects of carbon nanodots is their luminescent behavior, which can often be observed under UV irradiation (Li et al., 2019; Song et al., 2022). This luminescence can be modulated, both in intensity and emission wavelength, by controlling factors such as particle size, surface functional groups, and chemical dopants (e.g., nitrogen, sulfur, phosphorus). As a result, CNDs have found utility in a wide range of applications (Baker & Baker, 2010; Gong et al., 2022; Kottam & Smrithi, 2021; Vasin et al., 2021):

Biological Imaging. Their bright, stable fluorescence and low cytotoxicity make them excellent fluorescent probes for live cell imaging, tissue labeling, and potential use in theranostics (combining therapy and diagnostics).

Chemical and Biological Sensing. CNDs can be functionalized with specific molecules or ligands to detect ions, small molecules, or biomolecules through changes in fluorescence intensity or wavelength.

Catalysis. By their large surface area, unique electronic structure, and potential for heteroatom doping, CNDs can serve as catalysts or co-catalysts in organic transformations, photocatalysis, or electrochemical reactions.

Photovoltaics and Optoelectronics. Their ability to absorb photons and re-emit light efficiently has driven research into employing CNDs in solar cells, light-emitting diodes (LEDs), and other optoelectronic devices.

Environmental Applications. In addition to their potential use in the photocatalytic degradation of pollutants, CNDs—particularly those derived from biomass—offer eco-friendly routes and may help in developing sustainable, "green" nanotechnologies.

Beyond these immediate applications, the mechanisms underlying CND photoluminescence and the role of surface/edge states remain active research areas. Studies have highlighted that fluorescence can arise from multiple sources: quantum confinement effects (similar to semiconductor quantum dots), emissive defects, surface functional groups, and molecular fluorophores that form during the synthesis process (Cadranel et al., 2019; Hang et al., 2022; Soni et al., 2021). This scientifically intriguing complexity drives further investigation into the fundamental chemistry and physics of carbon nanodots. Interestingly, the synthetic diversity of CNDs sets them apart from many other nanomaterials. Researchers have developed numerous strategies to obtain CNDs, ranging from "top-down" approaches (breaking down bulk carbon sources, such as graphite, carbon nanotubes, or other carbon allotropes) to "bottom-up" methods (assembling carbon-containing small molecules through thermal decomposition, hydro/solvothermal reactions, or polymerization pathways) (Baker & Baker, 2010; Banger et al., 2023; Dhamodharan et al., 2022; Kottam & Smrithi, 2021). Many of these synthetic routes are simple and can be performed in standard laboratory glassware or commonly available reactors, lowering barriers to entry for researchers from a variety of disciplines. From a historical standpoint, carbon dots were initially recognized around the early 2000s as a by-product of other carbon nanostructure syntheses, for instance, the production of carbon nanotubes (Qi et al., 2016; Xu et al., 2004). Their distinctive optical properties, combined with an inherently lower toxicity profile than traditional semiconductor quantum dots (e.g., CdSe and PbS, which contain heavy metals), quickly propelled CNDs to the forefront of nanomaterial research. Indeed, carbon nanodots now represent a safer, more environmentally benign alternative in many applications that require fluorescence. Given these compelling features (broad-spectrum photoluminescence, straightforward and potentially scalable syntheses, and favorable biocompatibility), CNDs continue to gain traction in multidisciplinary fields, including materials science, chemistry, biology, engineering, and medicine. Researchers remain focused on optimizing their synthesis, improving luminescence efficiencies, fine-tuning surface modifications, and integrating them into complex devices and sensors (Banger et al., 2023; Li et al., 2012; Xiao & Sun, 2018). The main goal of this chapter is to present an overview of the most prominent methods for synthesizing CNDs, offering insights into both the historical context and the latest developments in this rapidly evolving field. A concise historical perspective on the discovery and early investigations of carbon nanodots is provided. The principal synthetic techniques are outlined, including physical approaches such as laser ablation and plasma discharge, chemical and electrochemical routes, and "one-pot" and green protocols. Critical factors that influence the shape, size, composition, and photoluminescent properties of carbon nanodots are discussed. The chapter concludes with a summary of key findings and highlights prospective directions for future research, focusing on the development and large-scale production of carbon nanodots.

2.2 Historical Perspective and Overview of Synthetic Approaches

The concept of carbon nanostructures can be traced back to the 1980s, marking the beginning of significant advancements in nanomaterial science. A key development occurred in 1985 with the discovery of fullerenes by Kroto et al. (1985). This breakthrough introduced a novel class of carbon molecules with unique properties, contributing to the development of nanomaterials. In 1991, Iijima (1991) further

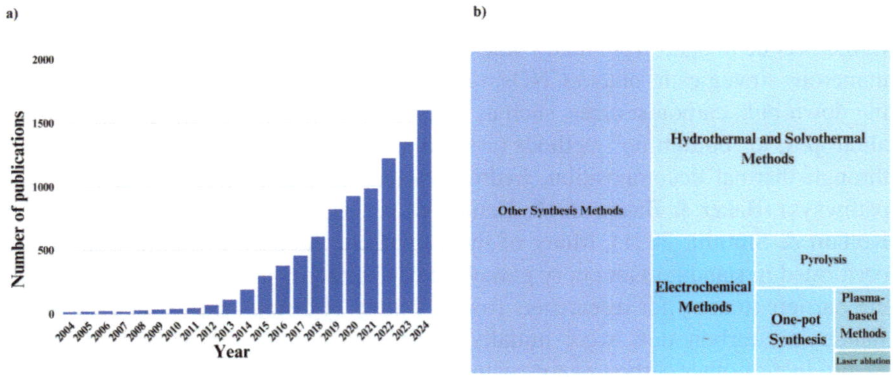

Fig. 2.1 (**a**) Annual number of publications related to carbon dots, based on data from the Scopus database; (**b**) distribution of publications on different synthesis methods for carbon dots in 2024, according to the Scopus database

expanded the field by reporting and describing carbon nanotubes, whose exceptional optical and physicochemical properties created new avenues for research and usage (Shoukat & Khan, 2021). These discoveries laid the foundation for the extensive exploration of carbon-based nanomaterials, which continue to drive innovation across various scientific disciplines.

Carbon nanodots (CNDs) were first identified and described as "fluorescent nanoparticles" by Xu et al. in 2004 during the purification of single-walled carbon nanotubes using the electrophoretic method. The term "quantum carbon nanodots" was first used by Sun et al. in 2006. The carbon dots, around 5 nm in diameter, were produced via laser ablation of a mixture of graphite powder and cement. Since these initial discoveries, interest in the synthesis of carbon nanodots grew slowly; however, the field has been expanding significantly from 2012 onward, with a growing number of research articles dedicated to their synthesis (Fig. 2.1a). Figure 2.1b presents the distribution of publications on different synthesis methods for carbon dots in 2024.

The most frequently reported synthetic approaches for carbon dots (CDs) are the hydrothermal and solvothermal methods, widely preferred due to their simplicity, versatility, and ease of control over reaction conditions. Significant attention has been dedicated to alternative synthesis techniques as well, notably electrochemical methods, pyrolysis, and one-pot syntheses, which offer benefits such as reduced complexity, shorter reaction times, or compatibility with various precursor materials. In contrast, methods like plasma-based synthesis and laser ablation are represented to a lesser extent in the literature, likely due to their higher technical demands, specialized equipment, or limited scalability. Collectively, these data reflect the diversity and ongoing innovation in the field, highlighting researchers' continuous exploration of novel synthetic strategies aimed at improving carbon dot properties, scalability, and applicability across emerging biomedical and technological applications.

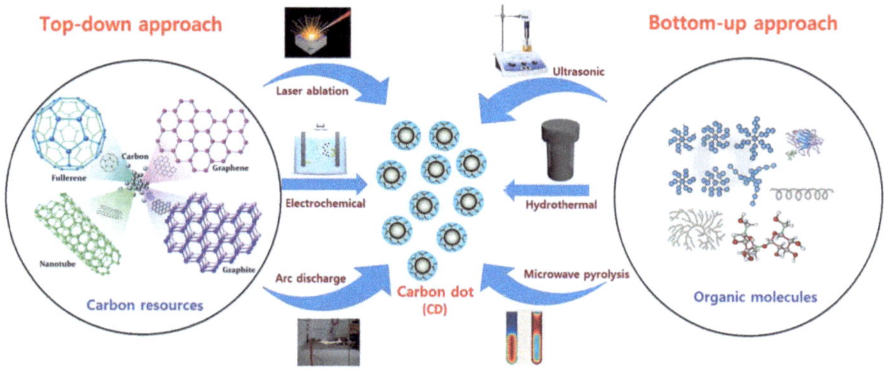

Fig. 2.2 Schematic illustrations of various methods for preparation of CDs; reproduced with permission from the *Journal of Materials Chemistry A*, 2017 (De & Karak, 2017)

Methods of synthesizing CNDs have been explored extensively, leading to the development of a variety of methods that can be broadly classified into two main categories: top-down and bottom-up approaches (Fig. 2.2). Each synthesis strategy confers distinct advantages that are critical in tailoring material properties to specific application needs, thereby enhancing the potential of CNDs in a diverse array of fields, including bioimaging, biosensing, and energy conversion technologies. Understanding and optimizing these synthesis methods remains vital for maximizing the performance and versatility of carbon nanodots.

Top-down strategies involve breaking down larger carbon structures, e.g., carbon powder, graphite, graphene, and carbon nanotubes (Kaczmarek et al., 2021; Shi et al., 2022), and include several methods for producing carbon dots. These involve laser ablation (Cui et al., 2020; Li et al., 2022), electrochemical methods (Li et al., 2024; Wu et al., 2021), and plasma-based methods (Choi et al., 2020; Ma et al., 2020; Qin et al., 2021). Laser ablation involves directing a pulse of laser energy onto a solid surface, leading to the generation of CNDs (Banger et al., 2023). Laser ablation is a highly effective method that can precisely control the size and properties of the carbon nanodots (Yogesh et al., 2021b), but it often requires high energy input and expensive equipment. Electrochemical synthesis of carbon nanodots offers a simple, environmentally friendly, and efficient method that allows for controlling the properties of CNDs by manipulating electrochemical parameters. This technique holds great potential in sustainable chemistry, enabling the synthesis of CNDs with applications in various fields (Banger et al., 2023). Plasma-based methods provide an efficient, rapid, and environmentally friendly approach without the need for harsh chemicals or high temperatures (Nevar et al., 2021). In this process, a plasma environment generates reactive species that break down carbon-containing precursors, leading to the formation of nanometer-sized carbon dots (Treepet et al., 2025; Tyutrin et al., 2025). The physical and chemical characteristics of the resulting carbon nanodots, such as size, surface functionalization, and optical properties, can be precisely controlled by adjusting the plasma parameters (Ma et al., 2020), making this technique highly versatile for various applications.

Bottom-up approaches involve constructing CNDs from small precursor molecules using various methods. The most commonly used techniques include hydrothermal and solvothermal methods (Guo et al., 2020; Yang et al., 2023; Zheng et al., 2024), pyrolysis (Lad et al., 2024; Lu et al., 2022; Sawalha et al., 2022a,b), and chemical syntheses such as one-pot synthesis (Kim et al., 2022; Zhang et al., 2025a, b). Hydrothermal and solvothermal synthesis methods use water or organic solvents at high temperatures and pressures to produce CNDs from precursors. These techniques are valued for their simplicity and efficiency. Additionally, they grant precise control over the resulting material's composition, allowing for tailored characteristics (Ndlwana et al., 2021). Pyrolysis is a widely employed thermal decomposition method of synthesizing carbon nanodots (CNDs). In this process, precursor materials are heated to high temperatures under an inert gas atmosphere, resulting in the carbonization and subsequent formation of CNDs. This technique allows for exact control over critical material properties, such as particle size, morphology, crystallinity, and photoluminescence, by carefully adjusting reaction parameters like temperature, duration, and atmosphere composition. However, the pyrolysis method typically requires elevated temperatures, contributing to increased energy demands and higher operational costs, potentially limiting scalability and environmental sustainability. In contrast, the one-pot synthesis strategy integrates all reaction stages into a single reaction vessel, significantly streamlining the synthetic procedure. This approach offers distinct advantages, including lower production costs, improved efficiency, reduced reaction time, and simplified experimental setups. By minimizing multiple handling and purification steps, one-pot synthesis methods are particularly attractive for scalable, eco-friendly production of CNDs (Yao et al., 2020).

Choosing the specific method of synthesizing CNDs is primarily guided by their intended application and desired production scale, ranging from small-scale laboratory research to large-scale industrial manufacturing. Critical factors influencing this decision include precursor availability, reproducibility, scalability, and the ability to precisely control the resulting materials' physicochemical properties, such as size distribution, surface chemistry, optical performance, and biocompatibility. Achieving an optimal balance between efficiency, cost-effectiveness, and tailored material characteristics is crucial for maximizing the potential of CNDs in diverse technological and biomedical fields. Consequently, the following subsections will comprehensively explore how specific synthesis approaches affect key parameters of CNDs, highlighting their suitability for targeted applications such as bioimaging, sensing, drug delivery, and energy conversion, and addressing their feasibility for practical implementation.

2.3 Physical Methods

2.3.1 Laser Ablation

Laser ablation is one of the pioneering top-down methods for producing carbon nanodots (CNDs). It involves bombarding carbon-based samples, such as graphite (Kaczmarek et al., 2021; Narasimhan et al., 2017; Russo et al., 2014; Tarasenka et al., 2017), graphene oxide (GO) (Małolepszy et al., 2018; Novoa et al., 2019; Russo et al., 2016), or carbon nano-onions (Calabro et al., 2018) with a laser beam while the sample is immersed in a liquid (e.g., water, ethanol, or surfactant solutions) (Tian & Yin, 2019). As the laser beam ablates the surface, micropores and nanopores are formed and ejected carbon fragments disperse into the solution to form quantum-sized carbon dots (Naskar et al., 2022; Reyes et al., 2016a). Advantages of laser ablation include high product purity with minimal or no need for additional chemical reagents (Banger et al., 2023). However, a small fraction of diamond-like structures and some amorphous carbon were identified alongside graphene quantum dots in some cases (Kang et al., 2020a,b,c,d; Novoa et al., 2019; Yogesh et al., 2021a). Broad tunability is possible through control over laser power, pulse frequency, laser type (femtosecond vs. nanosecond), and wavelength (Puliyasseri et al., 2024; Xu et al., 2017). Drawbacks include the high cost of laser equipment and relatively low production yields (Chai et al., 2022). Graphene quantum dots (GQDs) are a class of graphene-based materials classified as zero-dimensional (0D) graphitic nanomaterials, converted from two-dimensional (2D) graphene. Different strategies to incorporate heteroatoms (such as N, B, S, P, Si, and O) into GQDs, along with a thorough examination of the resulting dopant configurations, are essential for enhancing their practical performance. By precisely tuning the type and distribution of heteroatom groups, researchers can tailor the optical, electronic, and catalytic properties of GQDs to satisfy a wide range of technologically relevant applications (Algethami & Abdelhamid, 2024; Kang et al. 2020a,b,c,d; Qian et al., 2014; Sohal et al., 2021).

The conditions under which the laser ablation process is conducted strongly influence the final product's properties. Using pulses at 1064, 532, and 355 nm with varying irradiation times during laser ablation in acetone affects the carbon dot structure. Using a short-wavelength UV laser source with 355 nm laser irradiation can produce carbon quantum dot-like structures, whereas switching to a longer wavelength, 1064 nm in particular, leads to agglomeration (Reyes et al., 2016b). Dual-beam pulsed ablation is another technique involving lasers, proposed as an ultrafast method and used to fabricate homogeneous CQDs with a high quantum yield (Cui et al., 2020). The two-step irradiation applied to the graphite target led to highly efficient production of carbon nanodots. However, ablation in PEG was only effective with a 1064-nm laser pulse; ablation with a 355-nm laser resulted in an unexpectedly low nanoparticle production rate (Małolepszy et al., 2018). The luminescence of CDs can be enhanced by increasing the laser irradiation time from 30 minutes to 3 hours (Torrisi et al. 2025a, b).

The choice of medium is very important, and nanoparticles can be stabilized through the application of a steric or electrostatic protective layer. Ionic liquids (ILs) are good media for producing and stabilizing metallic nanoparticles via laser ablation (Castro et al., 2016). The synthesis of GQDs through femtosecond laser ablation of highly oriented pyrolytic graphite in a water medium is a scalable and environmentally friendly method (Russo et al., 2014).

In the laser ablation process, elevated local pressure can influence the properties of the final product, while intense heat enables the target material to break down into micro- or nanoparticles cleanly, without leaving residues, thus minimizing the risk of contamination or impurity introduction (Cui et al., 2020). The interaction of the laser with the interface between graphite and the surrounding liquid induces extreme conditions, with temperatures reaching up to 4000 K and pressures as high as 10 GPa, leading to vaporization and ejection of surface material (Nishtha et al., 2025). Other key factors influencing the properties of carbon dots (CDs) synthesized via pulsed laser ablation in liquid (PLAL) include various laser parameters and experimental settings. Elements such as the laser's wavelength, pulse length, repetition frequency, ablation time, and energy delivered per pulse all play a significant role in determining the final characteristics of the CDs (Cui et al., 2020; Gonçalves et al., 2010; Kaczmarek et al., 2021; Kang et al. 2020a,b,c,d; Małolepszy et al., 2018; Narasimhan et al., 2017; Nishtha et al., 2025; Novoa et al., 2019; Russo et al., 2016; Santiago et al., 2016; Torrisi et al. 2025a, b; Yogesh et al., 2021a; Zhang et al. 2025a, b). Changing this distance between the material surface and the laser focus also affects the concentration and distribution of energy, which can determine the efficiency of the ablation process and the type of nanoparticles produced (Cutroneo et al., 2024; Sattarahmady et al., 2017; Torrisi et al. 2025a, b). A longer time of exposure to laser radiation may result in higher ablation efficiency and more uniform CDs (Li et al., 2012; Sun et al., 2006; Tang et al., 2013; Zhu et al., 2015). However, exposure that is too long may lead to excessive degradation of the material and changes in its properties.

2.3.2 Plasma Discharge Methods

Plasma-based methods utilize highly ionized gas (plasma) to interact with a carbon source, eroding its surface to form nanoparticles. Common approaches include arc discharge and microwave discharge (Huang et al., 2015; Mohammadzaheri et al., 2022; Qu et al., 2012; Selvaraj et al., 2025; Wang et al., 2012; Wang et al., 2015). These methods can rapidly yield large quantities of CNDs under suitable conditions, but they require specialized equipment and precise control over parameters such as pressure, gas composition, and energy input.

Generating an arc discharge between two graphite electrodes in water can be a means of obtaining a variety of composite materials, like TiO_2 nanostructures coupled with CDs (Biazar et al., 2018) or CQDs and a small amount of graphene oxide (GO) sheets (Chao-Mujica et al., 2021). In general, different compositions and their heteroatom doping give composites of CQDs significant and versatile application potential in areas such as photocatalysis (Biazar et al., 2018; Dey et al., 2014) and electrocatalysis, optronics (Hernandez-Tabares et al., 2021), and biomedicine, where CQDs can play the role of theragnostic agents. To sustain the current during the arc discharge process, the distance between electrodes is precisely regulated with an automated micropositioning system (Chao-Mujica et al., 2021). Additionally, arc stability can be improved through the use of an electrical stabilization device. An increase in the discharge current and voltage leads to a rise in temperature, which in turn promotes the evaporation of the electrode material. This process has a significant impact on the morphology and size of the resulting carbon dots (Biazar et al., 2018; Chao-Mujica et al., 2021; Hernandez-Tabares et al., 2021; Yang et al., 2019). Furthermore, longer discharge times not only typically yield a higher amount of product but can also impact the particle size distribution of carbon nanodots (Javed & O'Carroll, 2021; Kaczmarek et al., 2024; Papaioannou et al., 2019). The parameters of the reaction atmosphere are another crucial element of the arc discharge process. The type of background gas influences arc stability, cooling rate, and the possibility of doping, some commonly used gases being helium, argon, nitrogen, or their mixtures (El-Zein et al., 2016; Wang et al. 2020a, b). Also, higher gas pressure can result in smaller particles due to faster plasma cooling, while lower pressure may lead to more CD aggregation (Chauhan et al., 2022).

Applying microwave discharge parameters, such as gas pressure, discharge time, background gas type, microwave power, temperature, frequency, exposure time, and precursor concentration, it is possible to control the size, morphology, and properties of the carbon dots (de Medeiros et al., 2019; Ma et al., 2020). The influence of these parameters on the CDs is crucial, as changes in the synthesis conditions can lead to carbon dots with different sizes, morphologies, optical (e.g., photoluminescent) properties, and chemical features. This allows for tailoring the synthesized CDs to various applications, such as photonics, bio-optics, or chemical sensing (Ke et al., 2018; Liu et al., 2014; Ma et al., 2019a, b; Wang et al., 2015). Carbon dots have also been found to be useful in electrochemical applications, where their incorporation into electrode materials leads to increased interlayer spacing and enhanced ion transport efficiency (Dong et al., 2021).

2.4 Chemical Methods

2.4.1 Hydrothermal and Solvothermal Methods

The hydrothermal method is one of the most commonly used and, at the same time, most accessible techniques for obtaining nanomaterials, including carbon nanodots. It has gained widespread recognition due to its simple equipment setup, relatively low operational costs, and the ability to process natural and readily available raw materials such as carbohydrates or plant biomass (Kang et al. 2020a, b, c, d). With the growing demand for sustainable and environmentally friendly production methods, hydrothermal synthesis of CNDs stands out as an eco-friendly alternative

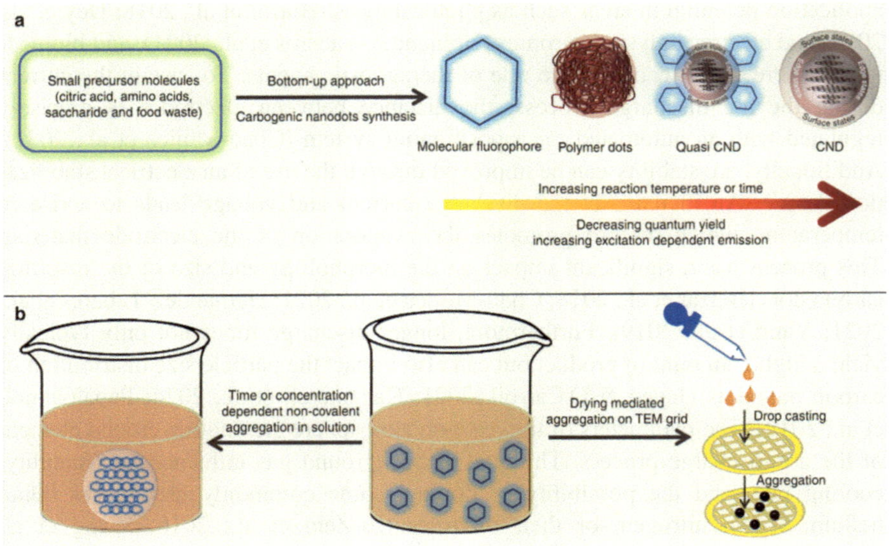

Fig. 2.3 Bottom-up approach to the synthesis of carbon nanodots from small molecular precursors: (**a**) a schematic of CND synthesis by creating a mixture of molecular fluorophores, polymer dots, quasi-CNDs, and CNDs; (**b**) aggregation of fluorophores on TEM grids during the drying process, resulting in the formation of aggregated structures; reproduced with permission from *Nature Communications*, 2019 (Verma N.C. et al., 2019)

to more complex and costly techniques. The hydrothermal process is classified as a bottom-up method, which involves building nanostructures from small molecular units (Fig. 2.3).

In the case of CNDs, carbon precursors are subjected to high temperatures and pressures in a tightly sealed reactor, typically a stainless steel autoclave lined with Teflon. These conditions allow the initiation of a series of chemical reactions that lead to the formation of nanoparticles. The typical temperature range used in this method is between 120 °C and 250 °C, and the reaction time may vary, from several minutes to several hours, depending on the material used and the desired properties of the final product. Unlike the solvothermal method, which utilizes organic solvents, the hydrothermal method employs water as the reaction medium (Huo et al., 2023). This is a significant advantage, as it eliminates the need for toxic and more expensive solvents, thereby increasing the safety of the process and reducing its environmental impact. Additionally, water can act as a reagent participating in condensation or hydrolysis reactions, supporting the transformation of complex organic structures into forms more susceptible to carbonization.

One of the major advantages of the hydrothermal method is its exceptional versatility regarding the types of carbon precursors that can be used. A wide range of organic materials can serve as starting substances for CND synthesis, from simple molecules such as citric acid and urea (Strauss et al., 2020), sucrose (Papaioannou et al., 2018), and resorcinol (Barik & Mohapatra, 2022) or amino acids, to plant waste and by-products of the food industry, such as leaves (Raveendran & Renuka, 2022; Sawalha et al., 2023), fruit peels and juices (Aouadi et al., 2024; Aslan & Eskalen, 2021; Li et al., 2018), nut shells, and other forms of biomass waste (Sabet & Mahdavi, 2019). The scientific literature increasingly reports examples of unconventional carbon sources. For instance, Ana and Camacho (2019) proposed an approach to obtain carbon quantum dots using cellulose nanocrystals extracted from *Cladophora rupestris* as the carbon precursor, achieving a yield of $54.6 \pm 2.71\%$. They noted that the color change of the solution indicated successful conversion of natural precursors (polysaccharides) into carbon quantum dots dispersed in the supernatant and hydrochar (Zhou et al., 2018), i.e., aggregated black particles, which is very typical for the hydrothermal method. Their research demonstrated that as the fiber size decreases, the cellulose (polysaccharide) chain becomes more exposed, which facilitates fragmentation of the polymer into smaller sugar monomers.

The mechanism of CND formation has been described many times. One proposed mechanism was presented in the study by Quang et al. (2022), in which orange juice was used as the carbon precursor. The carbon source underwent successive stages of dehydration, condensation, polymerization, followed by aromatization and passivation at elevated temperatures in a sealed Teflon-lined vessel (He et al., 2017). To obtain CNDs, the orange juice was heated to 120 °C for 150 minutes. Under these conditions, the concentration of aromatic carbon clusters increased up to a critical point at which supersaturation and rapid nucleation occurred, leading to the formation of quantum dots. It can therefore be assumed that the precursors of carbon nanodots, as small organic molecules, are capable of forming intermediate chain-like compounds through condensation reactions. These intermediates then undergo polymerization or aggregation into polymer-like CDs (carbon clusters) through covalent, non-covalent, or other types of interactions. The final stage of CND formation is based on high-temperature carbonization, resulting in the development of a carbonaceous core (Ding et al., 2018; Su et al., 2017; Tian et al., 2017). The use of renewable raw materials makes the hydrothermal synthesis of carbon nanodots significantly more environmentally friendly compared to traditional approaches that rely on nonrenewable chemical precursors. This strategy aligns with the principles of green chemistry by reducing the emission of harmful substances and minimizing the environmental impact of the process, further emphasizing the attractiveness of this method in the context of sustainable development.

The quality and properties of the resulting carbon dots (CDs) are strongly influenced by various synthesis parameters, such as temperature, reaction time, solution pH, reagent concentration, and internal pressure within the reactor (He et al., 2023). The ability to precisely adjust these parameters allows for control over the size, structure, oxidation level, and optical and surface properties of the nanomaterial, making the method highly versatile and adaptable for a wide range of applications.

In the case of solvothermal synthesis, the choice of solvent plays a crucial role as well. Solvents are essential in regulating the dehydration and carbonization processes of the precursors, which directly determine the morphology and properties of the final nanomaterial (Ding et al., 2018). A carefully selected reaction medium can influence the kinetics and mechanisms of carbon dot formation, which in turn affects the morphology and distribution of sp^2-conjugated domains of varying sizes within the CD structure. The presence of these domains is key to the luminescent properties of the material, as they affect charge carrier trapping and the mechanisms of light emission (Tian et al., 2017). Nevertheless, despite the wide availability of carbon sources and the relatively simple and well-controlled synthesis procedures, both the hydrothermal and solvothermal methods share a significant drawback. The main limitation of these techniques lies in the time-consuming and multi-step purification processes required for the final product. These steps not only prolong the total production time but also considerably reduce the overall efficiency of the process. Moreover, the need for additional chemical reagents and specialized laboratory equipment during separation and purification can increase operational costs and negatively impact the environmental friendliness of the entire procedure.

2.4.2 Pyrolysis

Pyrolysis is a thermal decomposition process of organic materials conducted at elevated temperatures, typically in an inert atmosphere. This method is frequently employed in the synthesis of carbon nanodots (CNDs). To control the size and tailor the properties of the resulting nanomaterials, various carbon precursors similar to those used in hydrothermal methods are utilized, including sucrose (Sawalha et al., 2022a, b), organic compounds (Stachowska et al., 2021), biowaste (Sawalha et al., 2020), and biomass (Lad et al., 2024; Sharma et al., 2024). Modifying synthesis parameters such as temperature, reaction time, and the type of atmospheric environment significantly influences the structural and morphological features of the produced CNDs (Krysmann et al., 2012; Zhang et al., 2022). Sawalha et al. (2022a, b) investigated the effect of carbonization temperature on the optical properties of CNDs synthesized from olive solid waste. Their findings indicated that temperature plays a crucial role in determining the production efficiency of CNDs. The highest yield was observed at 300 °C, which was attributed to the presence of oxidized functional groups on the surface of the resulting nanodots. In contrast, increasing the carbonization temperature to 600 °C led to a significant reduction in degradation efficiency, which they ascribed to enhanced electron–hole recombination within the CND structure at higher temperatures. Figure 2.4 illustrates the synthesis pathway of CNDs from olive solid waste via pyrolysis, followed by surface functionalization using various organic compounds such as ethylenediamine and dopamine.

Mild pyrolysis, typically performed at relatively low temperatures of about 200–300 °C, is a widely adopted strategy for synthesizing carbon nanodots from organic precursors with low melting points. Under these controlled conditions, the precursors readily condense into nanostructured carbon, allowing for precise manipulation of particle size and properties. By preventing excessive carbonization, mild pyrolysis also helps preserve functional groups, which can be critical for tuning the optical, electronic, and surface characteristics of the resulting nanomaterials.

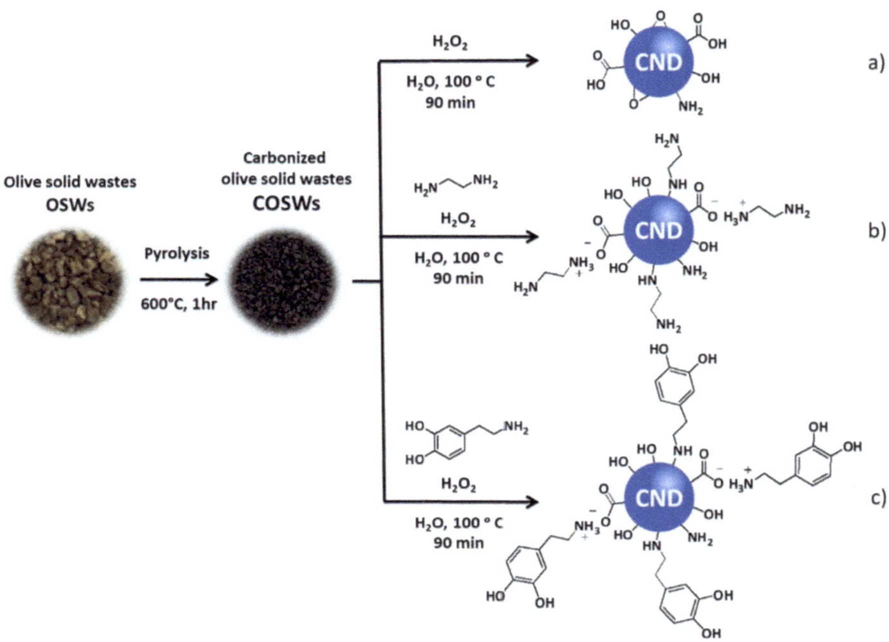

Fig. 2.4 Synthesis procedure for (**a**) carbon nanodots (CNDs), (**b**) CNDs functionalized with ethylenediamine (EDA), and (**c**) CNDs functionalized with dopamine (Dopa); reproduced with permission from *Carbon*, 2020 (Sawalha et al., 2020)

Krysmann et al. (Krysmann et al., 2012) proposed a mechanism for the formation of carbonaceous nanoparticles from low-molecular-weight compounds, highlighting their impact on photoluminescence (PL) properties. They demonstrated that distinct photoluminescent species correspond to three different stages of the pyrolytic process. In their study, citric acid (CA) and ethanolamine (EA) were subjected to pyrolysis within a temperature range of 180 °C to 400 °C. At lower temperatures, the process led to the formation of CNP precursors with strong PL, attributed to the presence of organic fluorophores. As the temperature increased, the pyrolysis progressed, and the precursors gradually transformed into spherical carbon nanodots (CNPs). At higher temperatures, the PL originated primarily from carbonaceous structures forming the CNP cores.

The pyrolysis process begins with the careful selection of carbon-rich raw materials. The procedure itself typically occurs at between 300 °C and 800 °C under oxygen-free conditions, which is essential to prevent combustion. Characteristics of the pyrolysis products are strongly influenced by feedstock composition, reaction temperature, and residence time, all of which must be precisely controlled. In the context of biomass pyrolysis, this thermal conversion process leads to the formation of low-molecular-weight substances. Zhang et al. (2022) reported a straightforward synthesis of carbon nanodots via pyrolytic decomposition of rice straw. Based on thermogravimetric (TG) analysis, they divided the pyrolysis process into four stages: initial drying below 100 °C, preheating between 110 °C and 220 °C, release of volatile compounds between 220 °C and 600 °C, and final carbonization above 600 °C. Some reports also indicate that pyrolysis can be conducted in the presence of oxygen. For instance, Ma et al. (2019a, b) used citric acid (CA) and ethylenediamine (EDA) as precursors; EDA, due to its simple structure containing two amine groups, served as a small-molecule nitrogen dopant. The reagents were mixed at room temperature and heated in a crucible up to 170 °C for 20 minutes. Their study identified three key parameters influencing the PL properties of the resulting carbon dots: reaction temperature, nitrogen concentration, and the nitrogen-to-oxygen (N:O) ratio. An increase in temperature was found to reduce PL intensity, likely due to excessive carbonization. The optimal nitrogen doping level and N:O ratio were essential for tuning optical properties and enhancing quantum yield (QY). Moreover, a short reaction time of 20 minutes facilitated efficient material formation, likely due to direct reagent interaction. Prolonged thermal treatment diminished PL intensity, presumably as a result of further carbonization. These findings suggest that polymerization, crosslinking, and carbonization steps occur most efficiently within a brief time frame, yielding materials with desirable characteristics. Success in this process depends on the deliberate selection of precursors, precise control of experimental conditions, and the ability to tailor the properties of the resulting nanostructures while ensuring adherence to safety standards and allowing for process adjustments when necessary (Otten et al., 2022).

2.5 Electrochemical Methods

During the electrophoretic purification of single-walled carbon nanotubes (SWCNTs), Xu et al. (2004) unintentionally discovered carbon nanodots (CNDs) exhibiting photoluminescent properties. Since this initial discovery, numerous approaches have been developed for synthesizing photoluminescent CNDs, such as wet chemical oxidation, laser ablation, pyrolytic methods, microwave-assisted synthesis, and electrochemical techniques (Baker & Baker, 2010; Hola et al., 2014; Li et al., 2013; Zheng et al., 2015). Among these methods, electrochemical synthesis stands out due to its rapid processing time and moderate experimental conditions. Particularly noteworthy is the ease with which electrochemical parameters, including applied voltage and current density, can be adjusted to precisely control the size, chemical composition, and photoluminescent characteristics of CNDs. This subchapter introduces electrochemical methods for synthesizing carbon nanodots (CNDs), describing the mechanisms by which carbon precursors are converted into these nanomaterials. Key subtypes of electrochemical approaches are discussed, highlighting their application in fine-tuning the size and chemical composition of CNDs. This fine-tuning directly influences the properties and stability of the resulting nanodots, offering significant control over their performance in various applications.

2.5.1 Overview and Rationale for Electrochemical Synthesis of CNDs

Electrochemical synthesis methods have become increasingly popular due to their inherent advantages in producing carbon nanodots (CNDs) with controllable size, surface chemistry, and photoluminescent properties (Baker, 2010; Michalet et al., 2005). Electrochemical reactions leverage specific redox potentials applied directly at electrode surfaces, facilitating the precise generation of highly reactive intermediates, such as hydroxyl radicals (·OH) in aqueous media (Lu et al., 2013). These radicals are particularly effective at breaking carbon–carbon bonds within bulk carbon substrates, transforming macroscopic materials into nano-sized fragments with desirable luminescent characteristics. An additional advantage of electrochemical synthesis involves ionic intercalation into structured carbon materials, including graphite and carbon nanotubes (Lu et al., 2013). This intercalation effectively weakens interlayer interactions, facilitating exfoliation and the subsequent release of photoluminescent carbon fragments characteristic of CNDs. Moreover, the versatility of electrochemical approaches extends to bottom-up synthesis strategies, where small organic precursors or monomers undergo redox-driven polymerization (Yan et al., 2010). Under carefully controlled oxidation or reduction conditions, these molecules polymerize into cross-linked oligomers, ultimately yielding fluorescent carbon cores upon further electrochemical processing. Electrochemical methods for the synthesis of CNDs can generally be classified into two main categories: top-down and bottom-up approaches. Top-down electrochemical approaches primarily involve the oxidative "etching" of bulk carbon substrates, such as graphite rods, multi-walled carbon nanotubes (MWCNTs), carbon fibers, or even biomass-derived carbon materials (Lamba et al., 2024). Figure 2.5 schematically illustrates the preparation of carbon nanodots using this top-down electrochemical method, showing the key steps involved in the controlled breakdown of bulk carbon substrates into nanoscale carbon dots.

During these processes, electro-generated radical species, particularly hydroxyl radicals, incrementally break down large carbon structures into smaller, oxygen-enriched, and photoluminescent fragments. Research has indicated that tuning electrochemical parameters, such as applied potential, electrolyte composition, supporting ions, and pH, can directly influence both the size distribution and surface functional groups of the resultant CNDs (Bao et al., 2011; Zhao et al., 2008). Conversely, bottom-up electrochemical methods initiate synthesis from simple

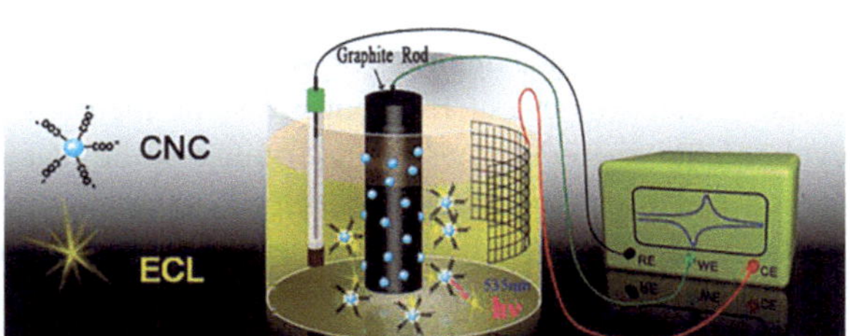

Fig. 2.5 Electrochemical production of C-dots from a graphite rod; reproduced with permission from the *Journal of the American Chemical Society*, 2009 (Zheng et al., 2009).

organic molecules, including ethanol or other low-molecular-weight organic compounds. These precursor molecules undergo electrochemically induced polymerization, where the intermediates cross-link and progressively form stable, photoluminescent carbon cores. Such bottom-up approaches allow for precise molecular-level control over the CNDs' final characteristics, making them particularly attractive for applications requiring highly specific functional properties. Both top-down and bottom-up electrochemical methods are advantageous due to their straightforward nature, minimal environmental impact, and lack of reliance on harsh chemical reagents beyond a sacrificial electrode and suitable electrolytes. Additionally, these methods typically operate under mild temperature and pressure conditions, which significantly reduces energy consumption and operational complexity. These features highlight electrochemical synthesis as a promising strategy for scaling up the production of CNDs in an environmentally friendly, economically feasible, and industrially adaptable manner (Qi et al., 2016).

2.5.2 Electrochemical Etching as a Top-Down Method for the Synthesis of CNDs

Among top-down methods for synthesizing carbon nanodots (CNDs), electrochemical etching stands out as one of the most commonly employed approaches. In this method, carbon nanodots are generated by selectively "etching" or detaching smaller, fluorescent carbon fragments from larger carbon-based substrates (Long et al., 2012; L. Zheng et al., 2009; J. Zhou et al., 2007). Electrochemical etching involves applying an electric potential or current to conductive carbonaceous materials, such as graphite rods, carbon fibers, activated carbon, or other naturally abundant and inexpensive carbon sources. These materials serve as sacrificial electrodes, gradually disintegrating to yield nano-sized carbon fragments with distinct photoluminescent properties. One significant advantage of electrochemical etching is the ease with which the resulting CNDs' size, chemical composition, and optical characteristics can be precisely controlled (Ullah et al., 2024). By modifying key experimental parameters, including the electrolyte's pH, concentration, and chemical composition, as well as the electrolysis technique employed (e.g., potentiostatic, galvanostatic, or cyclic voltammetry), researchers can tailor the synthesis process to produce CNDs optimized for specific applications. For instance, varying the electrolyte pH can significantly impact the surface charge, functional groups, and stability of the resulting CNDs, while changing the electrolyte concentration and composition influences the efficiency of carbon fragmentation and the resultant fluorescence intensity (Roy et al., 2015; Zuo et al., 2016). Additionally, the mode of electrolysis (such as potentiostatic methods, galvanostatic approaches, or cyclic voltammetry) allows precise tuning of the reaction kinetics and the extent of carbon electrode decomposition, directly affecting the yield and uniformity of the synthesized CNDs. Furthermore, electrochemical etching offers practical benefits, such as simplicity, cost-effectiveness, and minimal environmental impact, due to the use of abundant and inexpensive materials combined with mild operating conditions.

This approach has been successfully demonstrated in numerous studies, emphasizing its effectiveness in scaling up production and developing sustainable industrial processes for the fabrication of CNDs.

2.5.2.1 Role of Radical Intermediates and Ionic Intercalation

The role of radical intermediates and ionic intercalation in electrochemical top-down synthesis of carbon nanodots (CNDs) is crucial for understanding and optimizing their production. Early research by Zhou et al. (2007) demonstrated the generation of blue-emitting carbon nanodots by cycling the electrochemical potential of multi-walled carbon nanotubes (MWCNTs) within an organic electrolyte solution containing tetrabutylammonium perchlorate (TBAP). In this context, TBAP acts as an intercalation agent, disrupting the integrity of the nanotube walls and facilitating their gradual fragmentation into photoluminescent CNDs (Long et al., 2012). In aqueous systems, a similar phenomenon occurs through the electrochemical oxidation of water molecules at the anode surface, producing highly reactive hydroxyl radicals (·OH). These radicals act as powerful oxidative agents capable of efficiently cleaving carbon–carbon bonds within bulk graphite structures or carbon nanotube walls. Zhao et al. provided compelling evidence for this mechanism by applying anodic potentials to graphite rods immersed in a sodium phosphate buffer solution (Zhao et al., 2008). They obtained stable dispersions of photoluminescent CNDs, approximately 2–3 nm in size, attributed to the synergistic effects of radical-mediated oxidation and potential phosphate ion intercalation into graphite layers. The precise control of these radical-induced oxidation processes and the extent of ionic intercalation can be effectively managed by adjusting electrochemical conditions, including applied potential, current density, and electrolyte composition. Bao et al. highlighted the influence of applied potential during the electrochemical oxidation of carbon fibers, reporting a clear relationship between increasingly positive potentials and a reduction in the average size of resulting nanodots (Bao et al., 2011). Higher potentials enhance radical activity, promoting extensive carbon lattice cleavage, thus reducing the particle size. Furthermore, the study illustrated how manipulating these electrochemical parameters could alter surface chemistry, specifically controlling the ratio and type of functional groups (such as oxygen- or nitrogen-containing groups) attached to CND surfaces. Such surface functionalization directly impacts the optical properties of CNDs, including luminescence wavelength and quantum yield. Overall, the interplay between radical intermediates and ionic intercalation provides a versatile and precise method of fine-tuning the structural and optical characteristics of electrochemically synthesized CNDs (Chattopadhyay & Bag, 2024; Mahto et al., 2024). This insight is essential for optimizing synthesis protocols aimed at specific applications ranging from bioimaging and sensing to optoelectronic devices.

2.5.2.2 Influence of pH and Electrolyte Composition

The electrolyte composition and pH significantly influence the electrochemical synthesis of carbon nanodots (CNDs), particularly in top-down approaches. Alkaline conditions, in particular, have proven highly beneficial due to their ability to enhance the generation of hydroxyl radicals (·OH) at the anode surface (Baker et al., 2005). These radicals accelerate the fragmentation of bulk carbon substrates, efficiently converting larger carbon structures into nanoscale, luminescent fragments. For example, research conducted by Li et al. highlighted the effectiveness of an alkaline electrolyte containing sodium hydroxide (NaOH) and ethanol, demonstrating that this combination facilitated rapid electrochemical oxidation of graphite electrodes (Li et al., 2010). This approach yielded highly photoluminescent CNDs with superior fluorescence intensity and stability. Conversely, attempts to replicate these results in strongly acidic environments failed to produce significant quantities of fluorescent carbon materials, underscoring the critical role of alkaline conditions in promoting radical-mediated carbon cleavage. Moreover, the choice of electrolyte components extends beyond pH to include specific additives capable of influencing the intercalation and doping processes during electrochemical synthesis. Ionic liquids, such as 1-butyl-3-methylimidazolium tetrafluoroborate ([BMIM][BF4]), have been particularly notable due to their ability to penetrate and intercalate within graphite layers under an applied electric bias (Lu et al., 2011). This penetration induces structural expansion and exfoliation of graphite sheets into functionalized graphene nanosheets or carbon quantum dots. Such ionic liquid-mediated synthesis allows precise tuning of the doping levels, surface oxygenation, and other functional group modifications, thereby enabling the controlled emission properties of the resulting CNDs. Lu et al. demonstrated that adjusting the composition and water content of the ionic liquid electrolyte could effectively modulate the optical characteristics of synthesized CNDs (Lu et al., 2009). Specifically, variations in electrolyte formulation made it feasible to shift the photoluminescence emission from shorter wavelengths (blue) to longer wavelengths (green or even yellow), providing a versatile tool for fabricating CNDs tailored to specific optical applications.

2.5.3 Bottom-Up Electrochemical Approaches

While top-down electrochemical etching techniques have gained significant attention, bottom-up electrochemical synthesis methods have also emerged as valuable alternatives for preparing carbon nanodots (CNDs) (Deng et al., 2014; Li et al., 2011). These bottom-up strategies focus on the direct electrochemical conversion of small molecular precursors, such as alcohols (e.g., ethanol, methanol) or organic acids, into fluorescent carbon dots via oxidative or reductive processes carried out within an electrochemical cell.

The formation mechanism of bottom-up electrochemically synthesized CNDs typically involves several sequential stages (Ming et al., 2012; Nguyen et al., 2023; Rigodanza et al., 2018):

(a) *Initial Decomposition of Molecular Precursors*. Under an applied electrical current, small organic molecules undergo oxidation or reduction, forming highly reactive intermediate species, such as free radicals or carbocation intermediates.

These reactive species serve as building blocks for subsequent polymerization reactions.

(b) *Polymerization and Intermediate Formation.* Reactive intermediates generated in the initial step readily combine, initiating polymerization or cyclization reactions. This results in the formation of oligomeric or polymeric intermediates that gradually increase in molecular weight and complexity, laying the groundwork for further carbonization.

(c) *Carbonization and Core Formation.* Subsequent dehydration and cross-linking reactions between intermediate polymeric chains lead to the progressive development of stable, condensed carbon cores. These cores typically exhibit sizes in the nanoscale range, often a few nanometers in diameter, and are characterized by structural stability and inherent photoluminescence.

(d) *Surface Oxidation and Functionalization.* During or immediately after carbonization, reactive radicals and oxygen-based species (such as hydroxyl radicals (·OH) or superoxide anions ($O_2\cdot^-$)) generated at the electrode interface further oxidize or passivate the surfaces of the newly formed carbon cores. This step results in the introduction of various surface functional groups, including carboxyl, hydroxyl, and carbonyl functionalities, which greatly enhance the colloidal stability, dispersibility, and optical properties of the resulting CNDs.

Unlike top-down methods, bottom-up electrochemical syntheses tend to be highly sensitive to the specific conditions used, particularly regarding the applied voltage, electrolyte composition, and local electrochemical environment (Nguyen et al., 2023; Shi et al., 2023). The precise control of these parameters is crucial, as the successful synthesis of stable and highly luminescent carbon dots relies on the careful regulation of intermediate radical populations and polymerization kinetics. Deng et al. noted, for instance, that increasing the applied potential to higher oxidation states could, paradoxically, yield larger carbon dots due to enhanced radical formation, facilitating more extensive cross-linking and cluster growth (Deng et al., 2014). This intricate interplay between radical-mediated polymerization, carbonization, and subsequent surface functionalization highlights the delicate balance required to obtain carbon nanodots with controlled sizes and desirable luminescent properties. As such, bottom-up electrochemical approaches offer valuable versatility and precise tunability for the tailored synthesis of carbon nanodots suited to various advanced applications, including bioimaging, sensing, optoelectronics, and energy technologies.

Electrochemical synthesis methods, whether utilizing top-down oxidative fragmentation or bottom-up radical-induced polymerization, represent versatile and highly effective strategies for producing carbon nanodots (CNDs). These methods uniquely combine simplicity, scalability, and precise tunability, allowing researchers and industry professionals to finely control the structural and optical characteristics of the resulting CNDs (Banger et al., 2023; Dhamodharan et al., 2022; Rocco et al., 2023). In top-down electrochemical methods, the controlled oxidation of bulk carbonaceous electrodes, typically mediated by hydroxyl radicals or ionic intercalation, efficiently converts larger carbon substrates into uniformly sized, stable, and highly fluorescent nanoparticles. This method facilitates precise tuning of particle size and surface chemistry by simply adjusting reaction parameters, such as applied voltage, electrolyte composition, and solution conditions. Conversely, bottom-up electrochemical approaches involve the controlled polymerization of small molecular precursors, driven by electrochemically induced radical species generated via oxidation or reduction processes. These radicals initiate and propagate polymerization reactions, resulting in nanoscale carbon cores following subsequent carbonization and surface passivation steps. This strategy provides excellent flexibility in precisely engineering CNDs' size, surface functionality, and photoluminescent properties, by carefully managing electrochemical reaction conditions and precursor selection (Rocco et al., 2023). Collectively, electrochemical methods offer a platform for sustainable, environmentally benign production of high-quality carbon nanodots at scalable levels suitable for practical industrial applications. Beyond their practical advantages, electrochemical synthesis also provides invaluable insights into the fundamental relationships between CNDs' structure, size, surface chemistry, and photophysical properties. Thus, electrochemical synthesis emerges as a robust and advantageous approach, not only facilitating the practical manufacturing of high-quality CNDs but also enhancing the foundational knowledge essential for developing next-generation carbon-based nanomaterials.

2.6 Factors Influencing CND Properties

In recent years, carbon nanodots have attracted significant attention due to their unique chemical, physical, and photophysical properties, positioning them as highly promising materials for a variety of applications (Hassanvand et al., 2021; Omer et al., 2022; Prajapati et al., 2024; Zaidi et al., 2022). Methods for obtaining these nanomaterials have evolved significantly, leading to a deeper understanding and development of the various synthesis methods discussed here. In addition to the synthesis methods themselves, the selection of an appropriate precursor, the optimization of reaction conditions, heteroatom doping, and post-synthesis processes are also important in view of the desired properties of carbon nanodots (Chauhan et al., 2022; Kottam & Smrithi, 2021; Moniruzzaman & Kim, 2023; Li et al., 2021).

A key aspect in the production of carbon nanodots is the selection of an appropriate carbon precursor. Ahlawat et al. (2024) synthesized carbon nanodots via pyrolysis using citric acid, urea, and polyethyleneimine as precursors. The study demonstrates that different precursor combinations result in differences in absorption peaks, fluorescence emission, and particle size. Yang et al. (2021) prepared nitrogen-functionalized carbon nanodots using a one-pot hydrothermal method with various isomers of phenylenediamine as precursors; isomer type influences morphology, optical properties, and quantum yield. The selection of an appropriate precursor plays a crucial role in determining the properties of carbon nanodots, and different precursor combinations lead to variations in absorption peaks, fluorescence emission, particle size, and quantum yield.

Doping with heteroatoms such as nitrogen (Wang et al., 2023), sulfur (Cillari et al., 2025), phosphorus (Mathew et al., 2020), boron (Fu et al., 2024), silicon (Gong et al., 2022), and oxygen (Du et al., 2024) significantly impacts the properties of carbon nanodots. Dhanush et al. (Dhanush & Sethuraman, 2021) proposed a hydrothermal route for synthesizing N-, B-, and S-doped carbon nanodots from *Syzygium cumini* fruit extract. The doping of heteroatoms modifies the structural and optical properties of the biogenic carbon nanodots, making them more efficient for catalytic applications. Among the dopants, nitrogen (N) most notably enhanced the catalytic performance of carbon nanodots in the reduction of Alizarine yellow R azo dye. Azami et al. (2023) introduced a one-step closed-vessel microwave method by which they synthesized phosphorus- and boron-doped carbon nanodots. Phosphorus and boron doping in carbon nanodots significantly affects their optoelectronic properties and antioxidant activity, modifying both core structure and surface functionality. Heteroatom doping in carbon nanodots is a promising strategy for modifying their physicochemical properties, opening up new possibilities for adaptation to suitable applications.

Choosing the right synthesis method and precisely adjusting its parameters are key to determining the properties of the resulting carbon nanodots. Wang et al. (Wang et al., 2020a, b) developed a one-step solvothermal synthesis to produce full-color fluorescent carbon quantum dots from o-phenylenediamine and selected acid reagents. The type of acid reagent significantly influenced particle size and optical properties of the nanodots, allowing for controlled fluorescence emission, ranging from blue to red and even white light. Sawalha et al. (Sawalha et al., 2022a, b) presented a study focusing on the optimization of carbon nanodot synthesis by combining pyrolysis and chemical oxidation with hydrogen peroxide. The effects of pyrolysis temperature, equivalents of H_2O_2, and oxidation reaction time on production yield were studied. Changes in these parameters affected both the fluorescence properties and the quantum yield of the carbon nanowires. The choice of synthesis method and their optimization are essential to control the properties of CNDs, including their size, fluorescence, and quantum yield. By carefully tuning the synthesis parameters, it is possible to tailor the optical properties of carbon nanodots to specific applications.

Post-synthesis processing is another crucial step in the production of carbon nanotubes. The selection of the appropriate method significantly impacts the purity of the final product and its properties (Bartolomei & Prato, 2023; Hu et al., 2024; Moniruzzaman & Kim, 2023; Ullal et al., 2024). Various purification techniques are commonly used after synthesis, including dialysis (Otten et al., 2022), electrophoresis (Cardo et al., 2023), centrifugation (Sam et al., 2021), chromatography (Michaud et al., 2021), and filtration (Ru et al., 2021). The choice of a specific method depends on the type of impurities present and the desired properties of the final material for specific applications.

2.7 Conclusions and Perspectives

Methods of synthesizing carbon nanodots (CNDs) exhibit substantial diversity, both in terms of the nature of the carbon precursors employed and the specific reaction parameters, such as temperature, pressure, reaction duration, and the presence of dopants or catalysts. This variability allows the production of CNDs with highly tunable structural, chemical, and optical properties. The selection of a particular synthetic route is dictated by multiple considerations, including the intended field of application (e.g., bioimaging, photocatalysis, optoelectronics), economic and scalability constraints (laboratory-scale versus industrial-scale production), and the required uniformity in particle size distribution, crystallinity, and surface functionality. A critical aspect of CND synthesis involves achieving precise control over key physicochemical parameters, such as photoluminescence behavior (including emission wavelength, quantum yield, and lifetime), aqueous dispersibility, surface charge, environmental stability, and biocompatibility. These properties are of paramount importance in determining the efficacy and safety of CNDs, particularly in biomedical and environmental contexts. Contemporary trends in CND synthesis increasingly emphasize the development of environmentally benign and economically viable protocols. These approaches frequently utilize renewable raw materials or organic waste streams as carbon sources, aligning with the principles of green chemistry and sustainable nanotechnology. In parallel, significant efforts are being directed toward the refinement of synthesis protocols to allow for reproducible and highly controlled production, minimizing batch-to-batch variability and the need for extensive post-synthetic processing.

As the application potential of CNDs continues to expand across a broad range of technological domains, including medical diagnostics, environmental sensing, energy conversion and storage, and advanced electronic devices, the necessity for standardized, scalable, and cost-effective manufacturing strategies becomes increasingly evident. Progress in this area will be essential to facilitate the transition from laboratory-scale investigations to industrial-scale implementation and to support the integration of CNDs into commercially viable technologies.

References

Ahlawat, A., Dhiman, T. K., Solanki, P. R., & Rana, P. S. (2024). Facile synthesis of carbon dots via pyrolysis and their application in photocatalytic degradation of rhodamine B (RhB). *Environmental Science and Pollution Research, 31*(34), 46752–46759.

Algethami, F. K., & Abdelhamid, H. N. (2024). Heteroatoms-doped carbon dots as dual probes for heavy metal detection. *Talanta, 273*, 125893.

Ana, J. C. S., & Camacho, D. H. (2019). Influence of precursor size in the hydrothermal synthesis of cellulose-based carbon nanodots and its application towards solar cell sensitization. *Materials Chemistry and Physics, 228*, 187–193.

Aouadi, A., Saoud, D. H., Bouafia, A., Mohammed, H. A., Gamal, H. G., Achouri, A., et al. (2024). Unveiling the antioxidant power: synthesis and characterization of lemon and orange peel-derived carbon quantum dots with exceptional free radical scavenging activity. *Biomass Conversion and Biorefinery*, 1–14.

Aslan, M., & Eskalen, H. (2021). A study of carbon nanodots (carbon quantum dots) synthesized from tangerine juice using one-step hydrothermal method. *Fullerenes, Nanotubes, and Carbon Nanostructures, 29*(12), 1026–1033.

Azami, M., Wei, J., Valizadehderakhshan, M., Jayapalan, A., Ayodele, O. O., & Nowlin, K. (2023). Effect of doping heteroatoms on the optical behaviors and radical scavenging properties of carbon nanodots. *The Journal of Physical Chemistry C, 127*(15), 7360–7370.

Baker, M. (2010). Nanotechnology imaging probes: smaller and more stable. *Nature Methods, 7*(12), 957–962.

Baker, S. N., & Baker, G. A. (2010). Luminescent carbon nanodots: emergent nanolights. *Angewandte Chemie (International Ed. in English), 49*(38), 6726–6744.

Baker, G. A., Baker, S. N., Pandey, S., & Bright, F. V. (2005). An analytical view of ionic liquids. *Analyst, 130*(6), 800–808.

Banger, A., Gautam, S., Jadoun, S., Jangid, N. K., Srivastava, A., Pulidindi, I. N., et al. (2023). Synthetic methods and applications of carbon nanodots. *Catalysts, 13*(5), 858.

Bao, L., Zhang, Z. L., Tian, Z. Q., Zhang, L., Liu, C., Lin, Y., et al. (2011). Electrochemical tuning of luminescent carbon nanodots: From preparation to luminescence mechanism. *Advanced Materials, 23*(48), 5801–5806.

Barik, B., & Mohapatra, S. (2022). Selective and sensitive fluorescence turn-on detection of bilirubin using resorcinol-sucrose derived carbon dot. *Analytical Biochemistry, 654*, 114813.

Bartolomei, B., & Prato, M. (2023). The importance of the purification step and the characterization of the products in the synthesis of carbon nanodots. *Small, 19*(31), 2206714.

Biazar, N., Poursalehi, R., & Delavari, H. (2018). Optical and structural properties of carbon dots/TiO2 nanostructures prepared via DC arc discharge in liquid. *AIP Conference Proceedings, 1920*(1). https://doi.org/10.1063/1.5018965

Cadranel, A., Margraf, J. T., Strauss, V., Clark, T., & Guldi, D. M. (2019). Carbon nanodots for charge-transfer processes. *Accounts of Chemical Research, 52*(4), 955–963.

Calabro, R. L., Yang, D.-S., & Kim, D. Y. (2018). Liquid-phase laser ablation synthesis of graphene quantum dots from carbon nano-onions: Comparison with chemical oxidation. *Colloid and Interface Science, 527*, 132–140.

Cardo, L., Martínez-Parra, L., Cesco, M., Echeverría-Beistegui, B. M., Martínez-Moro, M., Herrero-Álvarez, N., et al. (2023). Luminescent carbon nanodots doped with gadolinium (III): Purification criteria, chemical and biological characterization of a new dual fluorescence/MR imaging agent. *Small, 19*(31), 2206442.

Castro, H. P., Souza, V. S., Scholten, J. D., Dias, J. H., Fernandes, J. A., Rodembusch, F. S., et al. (2016). Synthesis and characterisation of fluorescent carbon nanodots produced in ionic liquids by laser ablation. *Chemistry, 22*(1), 138–143.

Chai, Y., Feng, Y., Zhang, K., & Li, J. (2022). Preparation of fluorescent carbon dots composites and their potential applications in biomedicine and drug delivery—A review. *Pharmaceutics, 14*(11), 2482.

Chao-Mujica, F. J., Garcia-Hernández, L., Camacho-López, S., Camacho-López, M., Camacho-López, M. A., Reyes Contreras, D., et al. (2021). Carbon quantum dots by submerged arc discharge in water: Synthesis, characterization, and mechanism of formation. *Journal of Applied Physics, 129*(16). https://doi.org/10.1063/5.0040322

Chattopadhyay, J., & Bag, B. (2024). Carbon quantum dot electrocatalysts in hydrogen evolution reaction. *Energy Technology, 12*(1), 2300913.

Chauhan, D. S., Quraishi, M., & Verma, C. (2022). Carbon nanodots: Recent advances in synthesis and applications. *Carbon Letters, 32*(7), 1603–1629.

Choi, D., Yeom, H., You, K., Kim, J., Seong, D., Yoon, E., & Lee, H.-C. (2020). Generation of carbon nanowhiskers, nanotips, and nanodots by controlling plasma environment: Ion energy and radical effects. *Carbon, 162*, 423–430.

Cillari, R., Acúrcio, R. C., Barateiro, A., Florindo, H. F., Mauro, N., & Cavallaro, G. (2025). Harnessing sulfur-doped carbon nanodots conjugated with IDO inhibitors act as a dual-mode breast cancer immunotherapy. *Journal of Controlled Release, 381*, 113575.

Cui, L., Ren, X., Wang, J., & Sun, M. (2020). Synthesis of homogeneous carbon quantum dots by ultrafast dual-beam pulsed laser ablation for bioimaging. *Materials Today Nano, 12*, 100091.

Cutroneo, M., Silipigni, L., Malinsky, P., Slepicka, P., Franco, D., & Torrisi, L. (2024). Polyvinylalcohol composite filled with carbon dots produced by laser ablation in liquids. *Polymers, 16*(10), 1390.

De Bibekananda, Karak Niranjan (2017). Recent progress in carbon dot–metal based nanohybrids for photochemical and electrochemical applications *Journal of Materials Chemistry A* 5(5) 1826–1859.

de Medeiros, T. V., Manioudakis, J., Noun, F., Macairan, J.-R., Victoria, F., & Naccache, R. (2019). Microwave-assisted synthesis of carbon dots and their applications. *Journal of Materials Chemistry C, 7*(24), 7175–7195.

Deng, J., Lu, Q., Mi, N., Li, H., Liu, M., Xu, M., et al. (2014). Electrochemical synthesis of carbon nanodots directly from alcohols. *Chemistry–A European Journal, 20*(17), 4993–4999.

Dey, S., Govindaraj, A., Biswas, K., & Rao, C. N. R. (2014). Luminescence properties of boron and nitrogen doped graphene quantum dots prepared from arc-discharge-generated doped graphene samples. *Chemical Physics Letters, 595-596*, 203–208. https://doi.org/10.1016/j.cplett.2014.02.012

Dhamodharan, D., Byun, H.-S., Shree, M. V., Veeman, D., Natrayan, L., & Stalin, B. (2022). Carbon nanodots: Synthesis, mechanisms for bio-electrical applications. *Journal of Industrial and Engineering Chemistry, 110*, 68–83.

Dhanush, C., & Sethuraman, M. (2021). Independent hydrothermal synthesis of the undoped, nitrogen, boron and sulphur doped biogenic carbon nanodots and their potential application in the catalytic chemo-reduction of Alizarine yellow R azo dye. *Spectrochimica Acta Part A: Molecular and Biomolecular Spectroscopy, 260*, 119920.

Ding, H., Wei, J. S., Zhang, P., Zhou, Z. Y., Gao, Q. Y., & Xiong, H. M. (2018). Solvent-controlled synthesis of highly luminescent carbon dots with a wide color gamut and narrowed emission peak widths. *Small, 14*(22), 1800612.

Dong, S., Song, Y., Fang, Y., Zhu, K., Ye, K., Gao, Y., et al. (2021). Microwave-assisted synthesis of carbon dots modified graphene for full carbon-based potassium ion capacitors. *Carbon, 178*, 1–9. https://doi.org/10.1016/j.carbon.2021.02.094

Du, C., Lv, Y., Yu, H., Zhang, Y., Zhu, H., Dong, W., et al. (2024). In situ synthesis of oxygen-doped carbon quantum dots embedded in MIL-53 (Fe) for efficient degradation of oxytetracycline. *Environmental Science and Pollution Research, 31*(18), 26686–26698.

El-Zein, A., Talaat, M., El-Aragi, G., & El-Amawy, A. (2016). Electrical characteristics of nonthermal gliding arc discharge reactor in argon and nitrogen gases. *IEEE Transactions on Plasma Science, 44*(7), 1155–1159.

Fu, Q., Sun, S., Lu, K., Li, N., & Dong, Z. (2024). Boron-doped carbon dots: Doping strategies, performance effects, and applications. *Chinese Chemical Letters, 35*(7), 109136.

Gonçalves, H., Jorge, P. A., Fernandes, J., & da Silva, J. C. E. (2010). Hg (II) sensing based on functionalized carbon dots obtained by direct laser ablation. *Sensors and Actuators B: Chemical, 145*(2), 702–707.

Gong, X., Zheng, S., Zhao, X., & Vomiero, A. (2022). Engineering high-emissive silicon-doped carbon nanodots towards efficient large-area luminescent solar concentrators. *Nano Energy, 101*, 107617.

Guo, F., Bao, L., Wang, H., Larson, S. L., Ballard, J. H., Knotek-Smith, H. M., et al. (2020). A simple method for the synthesis of biochar nanodots using hydrothermal reactor. *MethodsX, 7*, 101022.

Hang, N. T. N., Van Canh, N., Hoa, N. H., Du, P. D., Le, P. H., & Nguyen, V. (2022). Co-assembled hybrid of carbon nanodots and molecular fluorophores for efficient solar-driven water evaporation. *Carbon, 199*, 462–468.

Hassanvand, Z., Jalali, F., Nazari, M., Parnianchi, F., & Santoro, C. (2021). Carbon nanodots in electrochemical sensors and biosensors: A review. *ChemElectroChem, 8*(1), 15–35.

He, G., Shu, M., Yang, Z., Ma, Y., Huang, D., Xu, S., et al. (2017). Microwave formation and photoluminescence mechanisms of multi-states nitrogen doped carbon dots. *Applied Surface Science, 422*, 257–265.

He, Z., Sun, Y., Zhang, C., Zhang, J., Liu, S., Zhang, K., & Lan, M. (2023). Recent advances of solvent-engineered carbon dots: A review. *Carbon, 204*, 76–93.

Hernandez-Tabares, L., Darias-Gonzalez, J. G., Chao-Mujica, F. J., Ledo-Pereda, L. M., Antuch, M., Carrillo-Barroso, E., et al. (2021). Stabilization methods in the submerged arc discharge synthesis of carbon nanostructures. *Journal of Nanomaterials, 2021*(1), 6550809. https://doi.org/10.1155/2021/6550809

Hola, K., Zhang, Y., Wang, Y., Giannelis, E. P., Zboril, R., & Rogach, A. L. (2014). Carbon dots—Emerging light emitters for bioimaging, cancer therapy and optoelectronics. *Nano Today, 9*(5), 590–603.

Hu, Y., Seivert, O., Tang, Y., Karahan, H. E., & Bianco, A. (2024). Carbon dot synthesis and purification: Trends, challenges and recommendations. *Angewandte Chemie International Edition, 63*(48), e202412341.

Huang, X., Li, Y., Zhong, X., Rider, A. E., & Ostrikov, K. (2015). Fast microplasma synthesis of blue luminescent carbon quantum dots at ambient conditions. *Plasma Processes and Polymers, 12*(1), 59–65. https://doi.org/10.1002/ppap.201400133

Huo, Y., Xiu, S., Meng, L.-Y., & Quan, B. (2023). Solvothermal synthesis and applications of micro/nano carbons: A review. *Chemical Engineering Journal, 451*, 138572.

Iijima, S. (1991). Helical microtubules of graphitic carbon. *Nature, 354*(6348), 56–58.

Javed, N., & O'Carroll, D. M. (2021). Long-term effects of impurities on the particle size and optical emission of carbon dots. *Nanoscale Advances, 3*(1), 182–189. https://doi.org/10.1039/D0NA00479K

Kaczmarek, A., Hoffman, J., Morgiel, J., Mościcki, T., Stobiński, L., Szymański, Z., & Małolepszy, A. (2021). Luminescent carbon dots synthesized by the laser ablation of graphite in polyethylenimine and ethylenediamine. *Materials, 14*(4), 729.

Kaczmarek, A., Wisniewska, A., Mościcki, T., & Hoffman, J. (2024). The luminescence of laser-produced carbon nanodots: The effect of aggregation in PEI solution. *Materials, 17*(7), 1573.

Kang, C., Huang, Y., Yang, H., Yan, X. F., & Chen, Z. P. (2020a). A review of carbon dots produced from biomass wastes. *Nanomaterials, 10*(11), 2316.

Kang, S., Jeong, Y. K., Jung, K. H., Son, Y., Kim, W. R., Ryu, J. H., & Kim, K. M. (2020b). One-step synthesis of sulfur-incorporated graphene quantum dots using pulsed laser ablation for enhancing optical properties. *Optics Express, 28*(15), 21659–21667.

Kang, S., Jeong, Y. K., Ryu, J. H., Son, Y., Kim, W. R., Lee, B., et al. (2020c). Pulsed laser ablation based synthetic route for nitrogen-doped graphene quantum dots using graphite flakes. *Applied Surface Science, 506*, 144998.

Kang, S., Jung, K. H., Mhin, S., Son, Y., Lee, K., Kim, W. R., et al. (2020d). Fundamental understanding of the formation mechanism for graphene quantum dots fabricated by pulsed laser fragmentation in liquid: Experimental and theoretical insight. *Small, 16*(38), 2003538. https://doi.org/10.1002/smll.202003538

Ke, C.-B., Lu, T.-L., & Chen, J.-L. (2018). Capacitively coupled plasma discharge of ionic liquid solutions to synthesize carbon dots as fluorescent sensors. *Nanomaterials, 8*(6), 372.

Kim, K. W., Kwon, Y. M., Kim, S. Y., & Kim, J. Y. H. (2022). One-pot synthesis of UV-protective carbon nanodots from sea cauliflower (Leathesia difformis). *Electronic Journal of Biotechnology, 56*, 22–30.

Kottam, N., & Smrithi, S. (2021). Luminescent carbon nanodots: Current prospects on synthesis, properties and sensing applications. *Methods and Applications in Fluorescence, 9*(1), 012001.

Kroto, H. W., Heath, J. R., O'Brien, S. C., Curl, R. F., & Smalley, R. E. (1985). C60: Buckminsterfullerene. *Nature, 318*(6042), 162–163.

Krysmann, M. J., Kelarakis, A., Dallas, P., & Giannelis, E. P. (2012). Formation mechanism of carbogenic nanoparticles with dual photoluminescence emission. *Journal of the American Chemical Society, 134*(2), 747–750.

Lad, U. M., Chunekar, N. P., Dave, D. J., Desai, B. N., Suthar, D. H., & Modi, C. K. (2024). Luminous insights: Harnessing carbon nanodots from black seed powder via pyrolysis for bioimaging and antifungal investigations. *Journal of Fluorescence, 34*(6), 2895–2906.

Lamba, R., Yukta, Y., Mondal, J., Kumar, R., Pani, B., & Singh, B. (2024). Carbon dots: Synthesis, characterizations, and recent advancements in biomedical, optoelectronics, sensing, and catalysis applications. *ACS Applied Bio Materials, 7*(4), 2086–2127.

Li, H., He, X., Kang, Z., Huang, H., Liu, Y., Liu, J., et al. (2010). Water-soluble fluorescent carbon quantum dots and photocatalyst design. *Angewandte Chemie International Edition, 49*(26), 4430–4434.

Li, H., Ming, H., Liu, Y., Yu, H., He, X., Huang, H., et al. (2011). Fluorescent carbon nanoparticles: Electrochemical synthesis and their pH sensitive photoluminescence properties. *New Journal of Chemistry, 35*(11), 2666–2670.

Li, H., Kang, Z., Liu, Y., & Lee, S.-T. (2012). Carbon nanodots: Synthesis, properties and applications. *Journal of Materials Chemistry*, *22*(46), 24230–24253.

Li, L., Wu, G., Yang, G., Peng, J., Zhao, J., & Zhu, J.-J. (2013). Focusing on luminescent graphene quantum dots: Current status and future perspectives. *Nanoscale, 5*(10), 4015–4039.

Li, J., Zhang, L., Li, P., Zhang, Y., & Dong, C. (2018). One step hydrothermal synthesis of carbon nanodots to realize the fluorescence detection of picric acid in real samples. *Sensors and Actuators B: Chemical, 258*, 580–588.

Li, X., Yan, L., Si, J., Xu, H., & Xu, Y. (2019). Tuning the photoluminescence property of carbon dots by ultraviolet light irradiation. *RSC Advances, 9*(22), 12732–12736.

Li, S., Li, L., Tu, H., Zhang, H., Silvester, D. S., Banks, C. E., et al. (2021). The development of carbon dots: From the perspective of materials chemistry. *Materials Today, 51*, 188–207.

Li, H., Zhang, Y., Ding, J., Wu, T., Cai, S., Zhang, W., et al. (2022). Synthesis of carbon quantum dots for application of alleviating amyloid-β mediated neurotoxicity. *Colloids and Surfaces B: Biointerfaces, 212*, 112373.

Li, Y., Liang, J., Li, J., Liao, L., Guo, J., & Jia, J. (2024). Tuning the carboxyl content of carbon quantum dots by an electrochemical exfoliation method to enhance Fenton-like reactions for contaminants removal in a wide pH range. *Journal of Environmental Chemical Engineering, 12*(2), 112008.

Liu, H., Ye, T., & Mao, C. (2007). Fluorescent carbon nanoparticles derived from candle soot. *Angewandte Chemie, 119*(34), 6593–6595.

Liu, Y., Xiao, N., Gong, N., Wang, H., Shi, X., Gu, W., & Ye, L. (2014). One-step microwave-assisted polyol synthesis of green luminescent carbon dots as optical nanoprobes. *Carbon, 68*, 258–264. https://doi.org/10.1016/j.carbon.2013.10.086

Long, Y.-M., Zhou, C.-H., Zhang, Z.-L., Tian, Z.-Q., Bao, L., Lin, Y., & Pang, D.-W. (2012). Shifting and non-shifting fluorescence emitted by carbon nanodots. *Journal of Materials Chemistry, 22*(13), 5917–5920.

Lu, J., Yang, J.-X., Wang, J., Lim, A., Wang, S., & Loh, K. P. (2009). One-pot synthesis of fluorescent carbon nanoribbons, nanoparticles, and graphene by the exfoliation of graphite in ionic liquids. *ACS Nano, 3*(8), 2367–2375.

Lu, J., Yang, J., Wang, J., Lim, A., Wang, S., & Loh, K. J. C. (2011). ACS Nano 2009, *3*, 2367–2375

Lu, J., Yang, J., Wang, J., Lim, A., Wang, S., Loh, K. J. L., Poh, C. K., Zhan, D., Lai, L., Lim, S. H., Wang, L., Liu, X., Gopal Sahoo, N., Li, C., Shen, Z., & Lin, J. (2013). *ACS Nano 2009, 3*, 2367–2375; (b) J. *2*, 377–386.

Lu, W., Guo, Y., Zhang, J., Yue, Y., Fan, L., Li, F., et al. (2022). A high catalytic activity nanozyme based on cobalt-doped carbon dots for biosensor and anticancer cell effect. *ACS Applied Materials & Interfaces, 14*(51), 57206–57214.

Ma, C. A., Yin, C., Fan, Y., Yang, X., & Zhou, X. (2019a). Highly efficient synthesis of N-doped carbon dots with excellent stability through pyrolysis method. *Journal of Materials Science, 54*(13), 9372–9384.

Ma, X., Li, S., Hessel, V., Lin, L., Meskers, S., Gallucci, F. (2019b). Synthesis of luminescent carbon quantum dots by microplasma process. *Chemical Engineering and Processing-Process Intensification, 140*, 29–35.

Ma, X., Li, S., Hessel, V., Lin, L., Meskers, S., & Gallucci, F. (2020). Synthesis of N-doped carbon dots via a microplasma process. *Chemical Engineering Science, 220*, 115648.

Mahto, B., Mahanty, B., Hait, S., Hussain, S. J. M. S., & B, E. (2024). A review of coal-based carbon and graphene quantum dots: Synthesis, properties, and applications. *Materials Science and Engineering: B, 304*, 117386.

Małolepszy, A., Błonski, S., Chrzanowska-Giżyńska, J., Wojasiński, M., Płocinski, T., Stobinski, L., & Szymanski, Z. J. A. P. A. (2018). Fluorescent carbon and graphene oxide nanoparticles synthesized by the laser ablation in liquid. *Applied Physics A, 124*, 1–7.

Mathew, R. M., Zachariah, E. S., Jose, J., Thomas, T., John, J., Titus, T., et al. (2020). Synthesis, characterization and evaluation of tunable thermal diffusivity of phosphorus-doped carbon nanodot. *Applied Physics A, 126*, 1–9.

Michalet, X., Pinaud, F. F., Bentolila, L. A., Tsay, J. M., Doose, S., Li, J. J., et al. (2005). Quantum dots for live cells, in vivo imaging, and diagnostics. Science, *307*(5709), 538–544.

Michaud, V., Pracht, J., Schilfarth, F., Damm, C., Platzer, B., Haines, P., et al. (2021). Well-separated water-soluble carbon dots via gradient chromatography. *Nanoscale, 13*(30), 13116–13128.

Ming Hai, Ma Zheng, Liu Yang, Pan Keming, Yu Hang, Wang Fang, Kang Zhenhui (2012). Large scale electrochemical synthesis of high quality carbon nanodots and their photocatalytic property *Dalton Transactions* 41(31) 9526–9531.

Mohammadzaheri, M., Siahpoush, V., & Asgari, A. (2022). Characterization of N-doped carbon quantum dots synthesized by DBD-based cold atmospheric pressure plasma jet. *Plasma Processes and Polymers, 19*(12), 2200089. https://doi.org/10.1002/ppap.202200089

Moniruzzaman, M., & Kim, J. (2023). Synthesis and post-synthesis strategies for polychromatic carbon dots toward unique and tunable multicolor photoluminescence and associated emission mechanism. *Nanoscale, 15*(34), 13858–13885.

Narasimhan, A. K., Santra, T. S., Rao, M. R., & Krishnamurthi, G. (2017). Oxygenated graphene quantum dots (GQDs) synthesized using laser ablation for long-term real-time tracking and imaging. *RSC Advances, 7*(85), 53822–53829.

Naskar, N., Wagner, M., Räder, H. J., Qi, H., Kaiser, U., Weil, T., & Chakrabortty, S. (2022). Molecular insights of carbon nanodots formation and their two-photon emission properties. *Advanced Photonics Research, 3*(2), 2100092. https://doi.org/10.1002/adpr.202100092

Ndlwana, L., Raleie, N., Dimpe, K. M., Ogutu, H. F., Oseghe, E. O., Motsa, M. M., et al. (2021). Sustainable hydrothermal and solvothermal synthesis of advanced carbon materials in multidimensional applications: A review. *Materials, 14*(17), 5094.

Nevar, A., Tarasenka, N., Nedelko, M., & Tarasenko, N. (2021). Carbon nanodots with tunable luminescence properties synthesized by electrical discharge in octane. *Carbon Letters, 31*(1), 39–46.

Nguyen, D.-K., Le, T.-S., Le, Q.-T., & Mai, X.-D. (2023). The roles of intermediate fluorophores on the optical properties of bottom-up synthesized carbon nanodots. *HPU2 Journal of Science: Natural Sciences and Technology, 2*(2), 68–82.

Nishtha, Awasthi, D., & Sarma, A. K. J. C. (2025). Recent advances in biomass-derived quantum dots, synthesis, and their applications. *ChemistrySelect, 10*(6), e202404611.

Novoa-De León, I., Johny, J., Vázquez-Rodríguez, S., García-Gómez, N., Carranza-Bernal, S., Mendivil, I., et al. (2019). Tuning the luminescence of nitrogen-doped graphene quantum dots synthesized by pulsed laser ablation in liquid and their use as a selective photoluminescence on–off–on probe for ascorbic acid detection. *Carbon, 150*, 455–464.

Omer, K. M., Al-Hashimi, B., Mohammadi, S., Salimi, A., Salih, Y. M., Hassan, A. Q., et al. (2022). Carbon nanodots as sensitive and selective nanomaterials in pharmaceutical analysis. *Journal of Materials Science, 57*(30), 14217–14245.

Otten, M., Hildebrandt, M., Kühnemuth, R., & Karg, M. (2022). Pyrolysis and solvothermal synthesis for carbon dots: Role of purification and molecular fluorophores. *Langmuir, 38*(19), 6148–6157.

Papaioannou, N., Marinovic, A., Yoshizawa, N., Goode, A. E., Fay, M., Khlobystov, A., et al. (2018). Structure and solvents effects on the optical properties of sugar-derived carbon nanodots. *Scientific Reports, 8*(1), 6559.

Papaioannou, N., Titirici, M.-M., & Sapelkin, A. (2019). Investigating the effect of reaction time on carbon dot formation, structure, and optical properties. *ACS Omega, 4*(26), 21658–21665. https://doi.org/10.1021/acsomega.9b01798

Prajapati, B. G., Pandey, V., Sharma, S., Patel, S., Shah, D. P., & Kapoor, D. U. (2024). Carbon nanodots: An illuminating paradigm in production, characterization, and oncological targeting methodologies—A review. *BioNanoScience*, 1–20.

Puliyasseri, R., Haribabu, J., Ramirez-Tagle, R., & Sastikumar, D. (2024). Synthesis of nitrogen-doped graphene oxide by nanosecond pulsed laser ablation of graphene in liquid for bioimaging applications. *Diamond and Related Materials, 143*, 110909. https://doi.org/10.1016/j.diamond.2024.110909

Qi, B.-P., Bao, L., Zhang, Z.-L., & Pang, D.-W. (2016). Electrochemical methods to study photoluminescent carbon nanodots: Preparation, photoluminescence mechanism and sensing. *ACS Applied Materials & Interfaces, 8*(42), 28372–28382.

Qian, Z., Shan, X., Chai, L., Ma, J., Chen, J., & Feng, H. (2014). Si-doped carbon quantum dots: A facile and general preparation strategy, bioimaging application, and multifunctional sensor. *ACS Applied Materials & Interfaces, 6*(9), 6797–6805. https://doi.org/10.1021/am500403n

Qin, X., Liu, J., Zhang, Q., Chen, W., Zhong, X., & He, J. (2021). Synthesis of yellow-fluorescent carbon nano-dots by microplasma for imaging and photocatalytic inactivation of cancer cells. *Nanoscale Research Letters, 16*, 1–9.

Qu, S., Wang, X., Lu, Q., Liu, X., & Wang, L. (2012). A biocompatible fluorescent ink based on water-soluble luminescent carbon nanodots. *Angewandte Chemie-International Edition, 51*(49), 12215.

Quang, N. K., Hieu, N. N., Bao, V. V. Q., Phuoc, V. T., Ngoc, L. X. D., Doc, L. Q., et al. (2022). Hydrothermal synthesis of carbon nanodots from waste wine cork and their use in biocompatible fluorescence imaging. *New Carbon Materials, 37*(3), 595–602.

Raveendran, P. V., & Renuka, N. (2022). Hydrothermal synthesis of biomass-derived carbon nanodots: Characterization and applications. *Materials Chemistry and Physics, 288*, 126236.

Reyes, D., Camacho, M., Camacho, M., Mayorga, M., Weathers, D., Salamo, G., et al. (2016a). Laser ablated carbon nanodots for light emission. *Nanoscale Research Letters, 11*(1), 424. https://doi.org/10.1186/s11671-016-1638-8

Reyes, D., Camacho, M., Camacho, M., Mayorga, M., Weathers, D., Salamo, G., et al. (2016b). Laser ablated carbon nanodots for light emission. *Nanoscale Research Letters, 11*, 1–11.

Rigodanza, F., Đorđević, L., Arcudi, F., & Prato, M. (2018). Customizing the electrochemical properties of carbon nanodots by using quinones in bottom-up synthesis. *Angewandte Chemie International Edition, 57*(18), 5062–5067.

Rocco, D., Moldoveanu, V. G., Feroci, M., Bortolami, M., & Vetica, F. (2023). Electrochemical synthesis of carbon quantum dots. *ChemElectroChem, 10*(3), e202201104.

Roy, P., Chen, P.-C., Periasamy, A. P., Chen, Y.-N., & Chang, H.-T. (2015). Photoluminescent carbon nanodots: Synthesis, physicochemical properties and analytical applications. *Materials Today, 18*(8), 447–458.

Ru, Y., Sui, L., Song, H., Liu, X., Tang, Z., Zang, S. Q., et al. (2021). Rational design of multicolor-emitting chiral carbonized polymer dots for full-color and white circularly polarized luminescence. *Angewandte Chemie, 133*(25), 14210–14218.

Russo, P., Hu, A., Compagnini, G., Duley, W. W., & Zhou, N. Y. (2014). Femtosecond laser ablation of highly oriented pyrolytic graphite: A green route for large-scale production of porous graphene and graphene quantum dots. *Nanoscale, 6*(4), 2381–2389.

Russo, P., Liang, R., Jabari, E., Marzbanrad, E., Toyserkani, E., & Zhou, Y. N. (2016). Single-step synthesis of graphene quantum dots by femtosecond laser ablation of graphene oxide dispersions. *Nanoscale, 8*(16), 8863–8877.

Sabet, M., & Mahdavi, K. (2019). Green synthesis of high photoluminescence nitrogen-doped carbon quantum dots from grass via a simple hydrothermal method for removing organic and inorganic water pollutions. *Applied Surface Science, 463*, 283–291.

Sam, S., Malinga, S. P., & Mabuba, N. (2021). Carbon nanodots embedded on a polyethersulfone membrane for cadmium (II) removal from water. *Membranes, 11*(2), 114.

Santiago, S., Lin, T., Yuan, C., Shen, J., Huang, H., & Lin, C. (2016). Origin of tunable photoluminescence from graphene quantum dots synthesized via pulsed laser ablation. *Physical Chemistry Chemical Physics, 18*(32), 22599–22605.

Sattarahmady, N., Rezaie-Yazdi, M., Tondro, G., & Akbari, N. J. (2017). Bactericidal laser ablation of carbon dots: An in vitro study on wild-type and antibiotic-resistant Staphylococcus aureus. *Journal of Photochemistry and Photobiology B: Biology, 166*, 323–332.

Sawalha, S., Silvestri, A., Criado, A., Bettini, S., Prato, M., & Valli, L. (2020). Tailoring the sensing abilities of carbon nanodots obtained from olive solid wastes. *Carbon, 167*, 696–708.

Sawalha, S., Assali, M., Nasasrah, A., Salman, M., Nasasrah, M., Jitan, M., et al. (2022a). Optical properties and photoactivity of carbon nanodots synthesized from olive solid wastes at different carbonization temperatures. *RSC Advances, 12*(8), 4490–4500.

Sawalha, S., Assali, M., Raddad, M., Ghneem, T., Sawalhi, T., Almasri, M., et al. (2022b). Broad-spectrum antibacterial activity of synthesized carbon nanodots from d-glucose. *ACS Applied Bio Materials, 5*(10), 4860–4872.

Sawalha, S., Abdallah, S., Barham, A., Badawi, H., Barham, Z., Ghareeb, A., et al. (2023). Green synthesis of fluorescent carbon nanodots from sage leaves for selective anticancer activity on 2D liver cancer cells and 3D multicellular tumor spheroids. *Nanoscale Advances, 5*(21), 5974–5982.

Selvaraj, Y., Govindaraj, R., Carrasco, M. F., Natarajan, M., Velauthapillai, D., & Devaraj, K. (2025). Optimization of carbon dot deposition on polycarbonate substrates via non-thermal plasma: Exploring its superhydrophobic properties and pH sensing capabilities. *Surfaces and Interfaces, 58*, 105759.

Sharma, S., Kumar, R., Kumar, K., & Thakur, N. (2024). Sustainable applications of biowaste-derived carbon dots in eco-friendly technological advancements: A review. *Materials Science and Engineering B, 305*, 117414.

Shi, W., Han, Q., Wu, J., Ji, C., Zhou, Y., Li, S., et al. (2022). Synthesis mechanisms, structural models, and photothermal therapy applications of top-down carbon dots from carbon powder, graphite, graphene, and carbon nanotubes. *International Journal of Molecular Sciences, 23*(3), 1456.

Shi, L., Wang, B., & Lu, S. (2023). Efficient bottom-up synthesis of graphene quantum dots at an atomically precise level. *Matter, 6*(3), 728–760.

Shoukat, R., & Khan, M. I. (2021). Carbon nanotubes: A review on properties, synthesis methods and applications in micro and nanotechnology. *Microsystem Technologies*, 1–10.

Sohal, N., Maity, B., & Basu, S. (2021). Recent advances in heteroatom-doped graphene quantum dots for sensing applications, *RSC Advances, 11*(41), 25586–25615.

Song, S.-Y., Liu, K.-K., Cao, Q., Mao, X., Zhao, W.-B., & Wang, Y. (2022). Ultraviolet phosphorescent carbon nanodots. *Light: Science & Applications. 11*(1), 146.

Soni, N., Singh, S., Sharma, S., Batra, G., Kaushik, K., Rao, C., et al. (2021). Absorption and emission of light in red emissive carbon nanodots. *Chemical science, 12*(10), 3615–3626.

Stachowska, J. D., Murphy, A., Mellor, C., Fernandes, D., Gibbons, E. N., Krysmann, M. J., et al. (2021). A rich gallery of carbon dots based photoluminescent suspensions and powders derived by citric acid/urea. *Scientific Reports, 11*(1), 10554.

Strauss, V., Wang, H., Delacroix, S., Ledendecker, M., & Wessig, P. (2020). Carbon nanodots revised: The thermal citric acid/urea reaction. *Chemical Science, 11*(31), 8256–8266.

Su, Y., Zhang, M., Zhou, N., Shao, M., Chi, C., Yuan, P., & Zhao, C. (2017). Preparation of fluorescent N, P-doped carbon dots derived from adenosine 5′-monophosphate for use in multicolor bioimaging of adenocarcinomic human alveolar basal epithelial cells. *Microchimica Acta, 184*, 699–706.

Sun, Y.-P., Zhou, B., Lin, Y., Wang, W., Fernando, K. S., Pathak, P., et al. (2006). Quantum-sized carbon dots for bright and colorful photoluminescence. *Journal of the American Chemical Society, 128*(24), 7756–7757.

Tang, J., Kong, B., Wu, H., Xu, M., Wang, Y., Wang, Y., et al. (2013). Carbon nanodots featuring efficient FRET for real-time monitoring of drug delivery and two-photon imaging. *Advanced Materials, 25*(45), 6569–6574.

Tarasenka, N., Stupak, A., Tarasenko, N., Chakrabarti, S., & Mariotti, D. J. C. (2017). Structure and optical properties of carbon nanoparticles generated by laser treatment of graphite in liquids. *ChemPhysChem, 18*(9), 1074–1083.

Tian, X.-T., & Yin, X.-B. (2019). Carbon dots, unconventional preparation strategies, and applications beyond photoluminescence. *Small, 15*(48), 1901803. https://doi.org/10.1002/smll.201901803

Tian, Z., Zhang, X., Li, D., Zhou, D., Jing, P., Shen, D., et al. (2017). Full-color inorganic carbon dot phosphors for white-light-emitting diodes. *Advanced Optical Materials, 5*(19), 1700416.

Torrisi, L., Cutroneo, M., Silipigni, L., Torrisi, A., Signorile, A., Manno, D. (2025a). Luminescent carbon dots structure by charcoal laser ablation in biocompatible liquid. *Fullerenes, Nanotubes and Carbon Nanostructures*, 33(8), 1–11.

Torrisi, L., Torrisi, A., & Cutroneo, M. J. F. (2025b). Luminescence enhancement of carbon dots synthesized by intense CW laser at 450 nm irradiating biocompatible solutions. *Fullerenes, Nanotubes, and Carbon Nanostructures, 33*(2), 111–122.

Treepet, S., Duangmanee, T., Chokradjaroen, C., Kim, K., Saito, N., & Watthanaphanit, A. (2025). Solution plasma synthesis of nitrogen-doped carbon dots from glucosamines: Comparative fluorescence modulation for dopamine detection. *Carbon, 231*, 119705. https://doi.org/10.1016/j.carbon.2024.119705

Tyutrin, A., Rakevich, A., & Martynovich, E. (2025). Microplasma-assisted synthesis and regulation the luminescence performance of glucose-based carbon dots by sodium borohydride. *Diamond and Related Materials, 112168*.

Ullah, M., Awan, U. A., Ali, H., Wahab, A., Khan, S. U., Naeem, M., et al. (2024). Carbon dots: New rising stars in the carbon family for diagnosis and biomedical applications. *Journal of Nanotheranostics, 6*(1), 1.

Ullal, N., Mehta, R., & Sunil, D. (2024). Separation and purification of fluorescent carbon dots–an unmet challenge. *Analyst, 149*(6), 1680–1700.

Vasin, A., Kysil, D., Rusavsky, A., Isaieva, O., Zaderko, A., Nazarov, A., & Lysenko, V. (2021). Synthesis and luminescent properties of carbon nanodots dispersed in nanostructured silicas. *Nanomaterials (Basel), 11*(12), 3267.

Verma Navneet C., Yadav Aditya, Nandi Chayan K. (2019). Paving the path to the future of carbogenic nanodots *Nature Communications* 10(1) 239.

Wang, J., Wang, C.-F., & Chen, S. (2012). Amphiphilic egg-derived carbon dots: Rapid plasma fabrication, pyrolysis process, and multicolor printing patterns. *Angewandte Chemie (International Ed. in English), 51*(37), 9297–9301. https://doi.org/10.1002/anie.201204381

Wang, Z., Lu, Y., Yuan, H., Ren, Z., Xu, C., & Chen, J. (2015). Microplasma-assisted rapid synthesis of luminescent nitrogen-doped carbon dots and their application in pH sensing and uranium detection. *Nanoscale, 7*(48), 20743–20748. https://doi.org/10.1039/C5NR05804J

Wang, C., Li, D., Lu, Z., Song, M., & Xia, W. (2020a). Synthesis of carbon nanoparticles in a non-thermal plasma process. *Chemical Engineering Science, 227*, 115921. https://doi.org/10.1016/j.ces.2020.115921

Wang, L., Li, W., Yin, L., Liu, Y., Guo, H., Lai, J., et al. (2020b). Full-color fluorescent carbon quantum dots. *Science Advances, 6*(40), eabb6772.

Wang, H., Wang, Q., Wang, Q., Dong, W., Liu, Y., Hu, Q., et al. (2023). Metal-free nitrogen-doped carbon nanodots as an artificial nanozyme for enhanced antibacterial activity. *Journal of Cleaner Production, 411*, 137337.

Wu, Y., Cao, M., Zhao, Q., Wu, X., Guo, F., Tang, L., et al. (2021). Novel high-hydrophilic carbon dots from petroleum coke for boosting injection pressure reduction and enhancing oil recovery. *Carbon, 184*, 186–194.

Xiao, L., & Sun, H. (2018). Novel properties and applications of carbon nanodots. *Nanoscale Horizons, 3*(6), 565–597.

Xu, X., Ray, R., Gu, Y., Ploehn, H. J., Gearheart, L., Raker, K., & Scrivens, W. A. (2004). Electrophoretic analysis and purification of fluorescent single-walled carbon nanotube fragments. *Journal of the American Chemical Society, 126*(40), 12736–12737.

Xu, H., Yan, L., Nguyen, V., Yu, Y., & Xu, Y. J. (2017). One-step synthesis of nitrogen-doped carbon nanodots for ratiometric pH sensing by femtosecond laser ablation method. *Applied Surface Science, 414*, 238–243.

Yan, X., Cui, X., & Li, L. (2010). Synthesis of large, stable colloidal graphene quantum dots with tunable size. *Journal of the American Chemical Society, 132*(17), 5944–5945.

Yang, J.-S., Pai, D. Z., & Chiang, W.-H. (2019). Microplasma-enhanced synthesis of colloidal graphene quantum dots at ambient conditions. *Carbon,153*, 315–319.

Yang, P.-C., Ting, Y.-X., Gu, S., Gandomi, Y. A., & Hsieh, C.-T. (2021). Fluorescent nitrogen-doped carbon nanodots synthesized through a hydrothermal method with different isomers. *Journal of the Taiwan Institute of Chemical Engineers, 123*, 302–309.

Yang, P.-C., Panda, P. K., Li, C.-H., Ting, Y.-X., Ashraf Gandomi, Y., & Hsieh, C.-T. (2023). Hydrothermal synthesis of functionalized carbon nanodots and their clusters as ionic probe for high sensitivity and selectivity for sulfate anions with excellent detection level. *Polymers, 15*(12), 2655.

Yao, N., Wang, H., Zhang, L., Yue, D., & Tian, M. (2020). One-pot solvothermal synthesis of silane-functionalized carbon nanodots as compatibilizers for the immiscible TPU/MVQ blends. *Applied Surface Science, 530*, 147124.

Yogesh, G. K., Shukla, S., Sastikumar, D., & Koinkar, P. (2021a). Progress in pulsed laser ablation in liquid (PLAL) technique for the synthesis of carbon nanomaterials: A review. *Applied Physics A, 127*(11), 810. https://doi.org/10.1007/s00339-021-04951-6

Yogesh, G. K., Shukla, S., Sastikumar, D., & Koinkar, P. (2021b). Progress in pulsed laser ablation in liquid (PLAL) technique for the synthesis of carbon nanomaterials: A review. *Applied Physics A, 127*, 1–40.

Zaidi, Z., Maiti, N., Ali, M. I., Sharma, G., Moin, S., Padhy, H., et al. (2022). Fabrication, characteristics, and therapeutic applications of carbon-based nanodots. *Journal of Nanomaterials, 2022*(1), 8031495.

Zhang, X., Zhang, K., Yao, X., Wu, Z., Lv, C., & Li, Q. (2022). Facile synthesis of blue fluorescent carbon nanodots based on the pyrolysis of straw for iron (III) detection and cellular imaging. *International Journal of Environmental Analytical Chemistry, 102*(16), 4768–4783.

Zhang, L., Zhao, C., Wang, S., Zhao, M., Liu, M., Zhou, H., et al. (2025a). Synthesis of biomass-derived N-doped carbon nanodots via a one-pot hydrothermal approach for enhanced lithium extraction in flow-electrode capacitive deionization. *Journal of Environmental Chemical Engineering, 115677*.

Zhang, Z., Wu, C., Hu, J., Li, C., Liu, Y., Lei, B., & Zheng, M. J. A. A. B. M. (2025b). Recent advances of carbon dots: Synthesis, plants applications, prospects, and challenges. *ACS Applied Bio Materials, 8*(2), 935–961.

Zhao, Q.-L., Zhang, Z.-L., Huang, B.-H., Peng, J., Zhang, M., & Pang, D.-W. (2008). Facile preparation of low cytotoxicity fluorescent carbon nanocrystals by electrooxidation of graphite. *Chemical Communications, 41*, 5116–5118.

Zheng, L., Chi, Y., Dong, Y., Lin, J., & Wang, B. J. J. (2009). Electrochemiluminescence of water-soluble carbon nanocrystals released electrochemically from graphite. *Journal of the American Chemical Society, 131*(13), 4564–4565.

Zheng, X. T., Ananthanarayanan, A., Luo, K. Q., & Chen, P. J. (2015). Glowing graphene quantum dots and carbon dots: Properties, syntheses, and biological applications. *Small, 11*(14), 1620–1636.

Zheng, G.-S., Shen, C.-L., Deng, Y., Liu, K.-K., Zang, J.-H., Dong, L., et al. (2024). Rational design multi-color-emissive chemiluminescent carbon nanodots in a single solvothermal reaction. *Nano Research, 17*(6), 4651–4660.

Zhou, J., Booker, C., Li, R., Zhou, X., Sham, T.-K., Sun, X., & Ding, Z. J. (2007). An electrochemical avenue to blue luminescent nanocrystals from multiwalled carbon nanotubes (MWCNTs). *Journal of the American Chemical Society, 129*(4), 744–745.

Zhou, Y., Liu, Y., Li, Y., He, Z., Xu, Q., Chen, Y., et al. (2018). Multicolor carbon nanodots from food waste and their heavy metal ion detection application. *RSC Advances, 8*(42), 23657–23662.

Zhu, S., Song, Y., Zhao, X., Shao, J., Zhang, J., & Yang, B. (2015). The photoluminescence mechanism in carbon dots (graphene quantum dots, carbon nanodots, and polymer dots): Current state and future perspective. *Nano Research, 8*, 355–381.

Zuo, P., Lu, X., Sun, Z., Guo, Y., & He, H. (2016). A review on syntheses, properties, characterization and bioanalytical applications of fluorescent carbon dots. *Microchimica Acta183*, 519–542.

Green Synthesis of Carbon Nanodots

3

Surajita Sahu and Monalisa Mishra

Abstract

Carbon dots (CDs) are carbon-based nanoparticles that have emerged as nanomaterials, possessing physicochemical, optical, and biocompatible properties. Carbon dots are a family of fluorescent nanoscale materials that behave like quantum dots, usually between 1 and 10 nm in diameter. Due to their fluorescence, lower toxicity, water-solubility, tunable emission across a broad range of wavelengths, and ease of modification properties, they hold great potential for applications in bioimaging, biosensing, medicine, drug delivery, electrocatalysis, and photocatalysis. Recently, green synthesis procedures have arisen as sustainable and environmentally acceptable alternatives to traditional chemical methods, utilizing natural precursors such as herbal extracts, biomass, agro-waste, and organic debris. These approaches have provided benefits like cost-efficiency and affordability, low toxicity, and a lesser impact on the environment. This chapter highlights the synthesis of green carbon dots, their synthesis methods, and beneficial characteristics resulting from animal-, plant-, and agro-waste-based resources and possible applications in biomedicine.

Keywords

Carbon dots · Natural precursors · Top-down · Bottom-up

S. Sahu · M. Mishra (✉)
Neural Developmental Biology Lab, Department of Life Science, NIT Rourkela, Rourkela, Odisha, India
e-mail: mishramo@nitrkl.ac.in

M. Rai et al. (eds.), *Carbon Nanodots*, https://doi.org/10.1007/978-981-95-4705-0_3

3.1 Introduction

Carbon is a crucial component in recent science and technology, assisting notably in multiple sectors of industry as well as society as a whole. Carbon-based nanomaterials have gained popularity in recent years due to their potential uses in a variety of sectors (Aksu & Güzdemir, 2025). These nanomaterials include carbon nanotubes, graphene, carbon quantum dots, carbon nanofibers, carbon dots, and other carbon-based derivatives with varying physicochemical properties (Patel et al., 2019). Among these, carbon dots (CDs) are an outstanding fluorescent nanomaterial that is chemically stable and biocompatible. CDs are usually less than 10 nm and constitute a family of zero-dimensional quasi-spherical carbon nanoparticles. These are mostly made up of a carbogenic backbone with surface functionalities incorporating hydrogen (H), oxygen (O), and nitrogen (N) (Dubey, 2023). The hydrophilic or hydrophobic properties of CDs are influenced by their surface functionalities, which allow them to disperse and stabilize in a variety of solvents. As a result, altering the CD's surface can offer a way to regulate its characteristics based on necessities (Zhao & Zhu, 2018). Adding heteroatoms, such as nitrogen and sulfur, to the system can also regulate the optical characteristics and electrical conductivity of CDs. These benefits show that CDs are a viable substitute for semiconductor QDs (Chen et al., 2022). These materials have garnered a lot of attention in a variety of fields, including biological applications, because of their exceptional mechanical, electrical, thermal, optical, and chemical qualities as well as their distinctive structural dimensions (Xu et al., 2019; Semeniuk et al., 2019). Furthermore, CDs can be easily dispersed in organic solvents and water media, have high quantum yield, and are nontoxic for living systems, hence emerging as sustainable fluorescent nanomaterials in various applications (Vithalani et al., 2020).

This chapter focuses on the synthesis of green CDs, highlighting eco-friendly and sustainable approaches over conventional chemical methods. We have discussed a variety of natural precursors, such as plant extracts, animal-derived materials, and agro-waste, underlining their benefits, which include low toxicity, affordability, and environmental friendliness. This chapter covers two basic fabrication strategies: top-down and bottom-up, with examples of hydrothermal treatment, microwave-assisted procedures, pyrolysis, and ultrasonication. In general, the chapter promotes the conversion of biological waste into valuable nanoparticles that aid in improvements in technology and sustainability.

3.2 Green Synthesis

3.2.1 Why is It Called Green Synthesis?

The 12 principles of green chemistry proposed by Paul Anastas and John Warner, as stated in the Pollution Prevention Act of 1990, aim to promote sustainability. They focus on limiting the hazardous substances of the chemical products and processes produced, following the pollution prevention hierarchy (Fig. 3.1). The goals of green

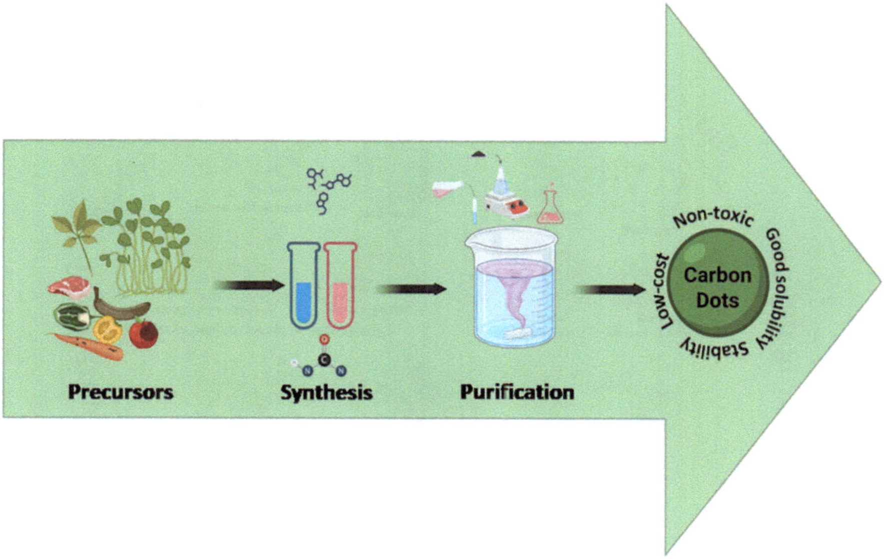

Fig. 3.1 Schematic representation of the synthesis of CDs from natural precursors

synthesis are to achieve sustainability at the molecular level. The term *green synthesis* is used because it aligns with the following principles of green chemistry: (1) prevention of waste rather than treatment after formation; (2) maximization of atom economy; (3) use of safer and less toxic chemical syntheses; (4) design of secure products and chemicals; (5) avoidance of hazardous solvents and auxiliary substances; (6) design of energy-efficient reactions; (7) use of renewable raw materials; (8) minimization of derivative formation; (9) reduction of waste through the use of catalysts; (10) development of chemicals that decompose into nontoxic compounds after use; (11) facilitation of real-time monitoring; and (12) use of inherently safer chemicals to minimize the risk of accidents (Beach et al., 2009). The three key features of the green synthesis framework are:

(a) All phases of the chemical life cycle, from production to disposal, are considered in green chemistry.
(b) Green chemistry designs chemical products and processes that aim to eliminate inherent hazards.
(c) It functions as a unified system to design products that permit sustainability.

3.3 Natural Precursors for Green Synthesis of C-Dots

Generally, there are three aspects that need to be considered for green synthesis: preferred solvent medium, a non-toxic reducing agent, and selecting eco-friendly materials for the stabilization of NPs (Cheviron et al., 2014). Traditional methods

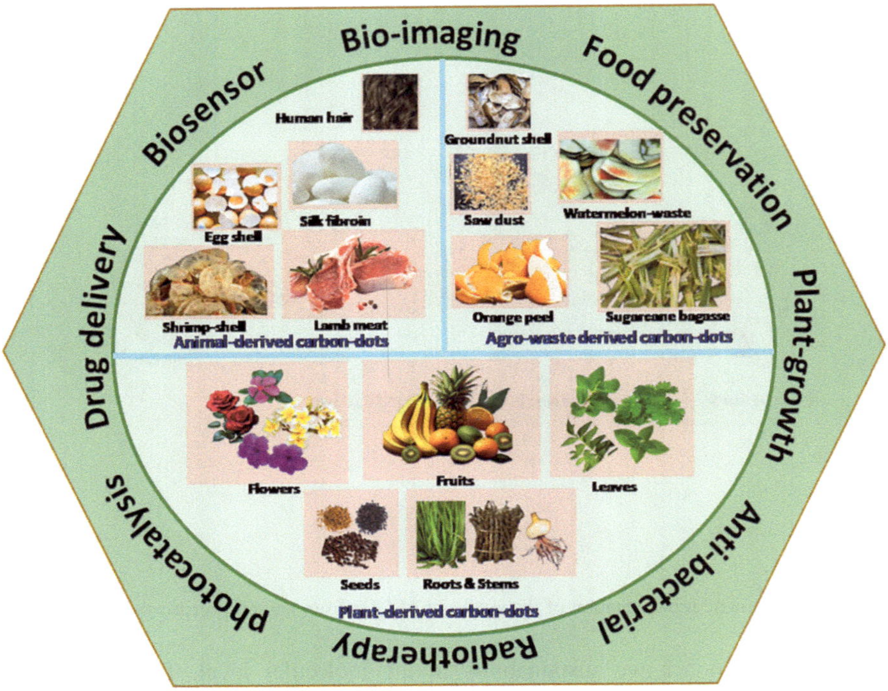

Fig. 3.2 CDs derived from animal, plant, and agro-waste sources, along with their applications in various fields

used to synthesize carbon-based NPs use energy-consuming processes and highly reactive substances, which are environmentally unfriendly. Therefore, to tackle this issue, green synthesis is introduced, which uses natural sources such as plants, animals, and agro-waste derivatives to synthesize NPs (Fig. 3.2).

3.3.1 Plant-Based Precursors

Plant-based CDs synthesis has gained attention due to its excellent fluorescent nature' efficacy for metal detection. They have been effectively used by researchers to detect elements including lead, chromium, mercury, iron, and arsenic in soil and water samples. Because of their remarkable durability, sensitivity, and selectivity, these CDs eliminate undesired precipitation during detection. A variety of plant sources have been utilized to yield CDs optimized for metal detection, which include turmeric leaves (*Curcuma longa*), tulsi leaves (*Ocimum tenuiflorum*), beetroot pulp (*Beta vulgaris*), papaya leaves (*Carica papaya*), baelpatra fruit (*Aegle marmelos*), kiwi fruit peels (*Actinidia deliciosa*), *Catharanthus roseus*, *Tinospora cordifolia* leaves, *Sophora japonica* leaves, and many more (Suseem, 2024). The advantages of plant-based materials over traditional materials are highlighted by their uniqueness in producing CDs. Their variable surface chemistry, huge surface area, tiny size, stability, and biocompatibility make them extremely sensitive and effective and provide sustainable sensing technology solutions. A few applications of plant-based CDs are discussed below:

1. CDs and folic acid-conjugated CDs, synthesized using citrus peels, provide better optical properties in MCF-7 breast cancer cells under different light excitations. Up to 48 h, CDs and folic acid-conjugated CDs maintain their intensity of photostability (Gudimella et al., 2021).
2. CDs at a lower concentration (20 μg/ml) act as a bioimaging agent as observed in SKBR3 breast cancer cells. They also penetrate to nucleus and crosses the cell membrane due to their small size via endocytosis (Fatahi et al., 2019).
3. In vivo detection of Fe^{3+} and H^{+} is accomplished via blue fluorescent CDs and green fluorescent CDs, respectively (Diao et al., 2018).
4. In vivo toxicological studies of CDs synthesized from *Ruellia simplex* show better bio-imaging potential with cysteamine modification with less toxicity studies in *Drosophila melanogaster* (Pansari et al., 2024).
5. Eco-friendly and biocompatible, the CDs made from kiwi and pear demonstrated a high fluorescence production and minimal toxicity in zebrafish embryos and human epithelial cancer cells (Suseem, 2024).
6. CDs with optical tuning from star jasmine (*Trachelospermum jasminoides*) leaves emitted the full visible spectrum at 466 nm, 521 nm, and 625 nm wavelengths. These green fluorophores can be used with LEDs because of their photostable emission characteristics. Furthermore, R2-CDs function to detect chloroform and water as dual-mode detectors both visually and fluorometrically (Kumari & Sahu, 2022).

3.3.2 Animal-Based Precursors

Biomass from humans and animals has demonstrated a great deal of promise for the environmentally friendly and sustainable synthesis of luminous CDs, which have numerous applications in bio-imaging and sensing. The biomass contains abundant proteins, carbohydrates, fats, minerals, and vitamins as sources for the synthesis of CDs. Researchers have used egg white for the synthesis of CDs as a greener approach. These CDs serve as biocompatible agents for cell imaging, as a fluorescent probe for the detection of Fe^{3+} metal ion, as a constituent of multifunctional microgels used in optical devices (Zhang et al., 2015). Another naturally occurring precursor, silk fibroin from *Bombyx mori*, is used for the production of doped CDs and carbon nanorods (Dubey, 2023; Sharma et al., 2024a, b). These carbon nanorods derived from silkworm powder are good in detecting Bi^{3+} metal ions from river samples and function as a potential radical scavenger (Sharma et al., 2024a, b). CDs derived from bones of pork, sheep, and bovine via a hydrothermal approach are naturally doped with nitrogen, sulfur, carbon, oxygen, and calcium. These bone-derived CDs are employed as sensing probes for the detection and discrimination of heavy metals like Cu^{2+}, Ag^{+}, Hg^{2+}, Pb^{2+}, and Fe^{3+} (Fu et al., 2022). One more animal-based Grass carp fish scale is used to synthesize CDs via a one-step pyrolytic approach that develops as a fluorescent probe with great biocompatibility and optical properties. The fluorescence property of grass carp fish-derived CDs is readily quenched by Ag^{+} ions due to electron transfer. But when glutathione is added, it binds with Ag^{+} and the fluorescence is restored, preventing the quenching of CDs. Because of this interaction between Ag^{+} ions and glutathione, these CDs exhibit selective fluorescence recovery. Thus, grass carp fish-derived CDs also detect the presence of glutathione (Zhang et al., 2024). Recently, elephant manure has been utilized for the synthesis of blue fluorescent CDs via a one-pot hydrothermal approach. These CDs exhibited excellent photostability, strong UV absorption, and resistance to photobleaching. They functioned as sensitive "turn-off" fluorescence sensors for Hg^{2+}, Fe^{2+}, and Fe^{3+}, performing well in drinking water and having low detection limits. The multifunctional potential of ECDs in sensing, agriculture, and biological applications was highlighted by their antioxidant activity as well as their ability to stimulate growth and the synthesis of fragrance compounds in fragrant rice (Seesuea et al., 2024). CDs developed from animal and human biomass offer researchers novel experimental applications and sustainably expand this field.

3.3.3 Agro-Waste-Derived Precursors

Developing CDs from waste to wealth includes products from agro-waste organic matter, such as leaves, stems, wheat waste, corn surplus, dry fruit shells, pulp, manure, and animal feathers. These wastes require biological, chemical, and physical processing to increase the yield and quality of carbon dots. This pretreatment allows these wastes to break down into simpler ones with altered physicochemical properties. Biomass from agro-waste has the potential to produce reliable raw materials in a sustainable approach with minimum processing. These waste materials contain a high amount of carbon, which is significant to prepare carbon-based NPs that are environment-friendly and cost-effective. Researchers have used pistachio shells (*Pistacia vera*) and groundnut shells (*Arachis hypogaea*) for the synthesis of carbon quantum dots that have biomedical applications. The outer protective layer of pistachio shell is abundant in bioactive components, which include polyphenols, unsaturated fatty acids, dietary fibers, essential oils, and tocopherols. These shells are also considered a good source of antioxidants. Carbon quantum dots that are made from waste of pistachio shells and groundnut shells are easily available and have potential for biosensing, bioimaging, drug delivery, and diagnosis (Manjubaashini et al., 2024). Another agro-waste, sugarcane bagasse, is abundant in cellulose and hemicellulose, which are sufficient to synthesize carbon-based nanomaterials. These agro-wastes are produced in huge quantities and are easily biodegradable, affordable, and recyclable. Utilizing sugarcane bagasse as a precursor for agricultural waste management provides the foundation to develop carbon-based quantum dots, which have noble optical qualities and a variety of uses (Baweja

& Jeet, 2019). Sugarcane bagasse is also used to develop a carbon dot sensor dipstick, which, when doped with nitrogen, gains the potential to detect tetracycline (Alfi et al., 2022). More such examples of carbon dots from banana and potato peels, rich in cellulose, are used in drug deliveries because of their biocompatibility. CDs synthesized from mango and carrot peels are rich in polyphenols and carotenoids and have antioxidant properties. Animal-derived waste, such as shrimp shells and crab shells, also helps in CDs synthesis. These shells have chitosan, a biopolymer that exhibits anti-microbial properties (Sharma et al., 2024a, b).

CDs synthesized from bio-waste are good at energy storage and conversion, help in making low-cost sensors, support smart farming, and improve device performance. Furthermore, they are excellent at cleaning up the environment, purifying water, and detecting contaminants to deal with pollution issues. Overall, they turn waste into useful tools supporting sustainability.

3.4 Top-Down and Bottom-Up Approaches for Green Synthesis

Synthesis of CDs from natural precursors is primarily divided into two types: "top-down" and "bottom-up" (Fig. 3.3). In the top-down approach, large carbon molecules are fragmented into smaller nanosized particles. Typical carbon sources include coal, candle soot, graphene oxide, natural graphite, and carbon fibers. These "top-down" methods involve techniques such as ultrasonic treatment, chemical oxidation, and hydrothermal or solvothermal processing (Ozyurt et al., 2023). In contrast, the "bottom-up" approach produces CDs from appropriate molecular precursors such as glucose, fructose, citric acid, and ascorbic acid in specific circumstances (Nazibudin et al., 2023). The bottom-up approach is accomplished using physicochemical and electrochemical techniques, such as heteroatom doping, altering functionalization, and nanohybrids when producing or processing CDs. One benefit of using a bottom-up approach is that it allows for easy surface alterations to produce CDs.

3.4.1 Top-Down Approaches

The following section discusses the most widespread top-down techniques:

3.4.1.1 Chemical Oxidation

A wide range of carbon-based materials, including GO (graphene oxide), coal, carbon nanotubes, carbon black, and activated carbon, have been utilized extensively to produce CDs using chemical oxidation or oxidative cutting (Egbedina et al., 2022). Several studies suggest adopting this technique because of its exceptional benefits, which include low cost, excellent purity, huge production yield, and easy size management. The majority of carbon-based materials have been dissolved with potent oxidizing acids. CDs made with this method are typically common in

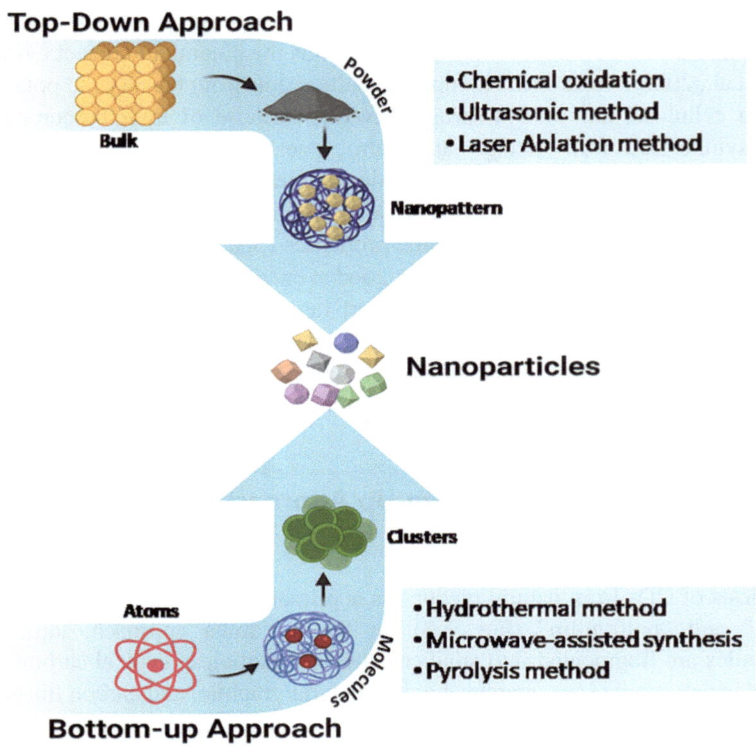

Fig. 3.3 Top-down and bottom-up approaches for the synthesis of CDs

functional groups, which allows for more comprehensive sensor applications (Jing et al., 2023). For example, the C60 cage is employed as a precursor for graphene oxide-based quantum dots (GOQDs) through chemical oxidation with H_2SO_4, $KMnO_4$, and H_2O_2 (Ozyurt et al., 2023). The produced GOQDs possess numerous oxygen sites and a diameter of about 2.5 nm. Compared to graphene oxide nanosheets, hydrophilic graphene oxide quantum dots (GOQDs) with a high concentration of oxygen-containing groups are more effective peroxidase mimics (Özyurt, 2022). An alternative study suggests GQDs produced without the need for any post-treatment procedures by employing H_2O_2 as an oxidizing agent and graphene nanosheets (GNs) and graphene oxide (GO) as precursors (Ozyurt et al., 2023). Peng and Travas-Sejdic (2009) employed concentrated H_3PO_2 to dehydrate carbohydrates and produce luminous CQDs. After the generated carbaceous materials were treated with HNO_3 to separate into individual CQDs, amine-terminated compounds were used in passivation procedures to produce luminescent CQDs (Peng & Travas-Sejdic, 2009). Similarly, studies also reported that H_2SO_4 and H_3PO_2 were applied as an oxidant to prepare CDs from natural sources such as tomato, muskmelon fruit, and green tea leaf residues (Hu et al., 2020; Jing et al., 2023). The limitation of chemical oxidation is that sometimes a strong oxidizing agent leads to the production of harmful material, which is neutralized by additional purification steps that increase the cost of the chemical oxidation method. Nevertheless, CDs made by chemical oxidation have excellent photoluminescent qualities as well as better water solubility.

3.4.1.2 Ultrasonic Approach

Ultrasonication is a simple and fast method of synthesizing CDs. In this method, a high-energy ultrasound wave is responsible for breaking large carbon molecules into simpler ones. Here, the sound waves form and collapse small bubbles that cause the breakdown of carbon material into CDs (Kumar et al., 2020; Ozyurt et al., 2023; Lakshmi et al., 2024). One example of ultrasonication-based synthesis involves the use of crab shells as a precursor. Prior to ultrasonication treatment, the obtained crab shells were dried and ground into a fine powder. After filtering and centrifuging the resultant solution, highly luminous CDs were obtained, yielding a 14.5% quantum yield (QY) (Dehvari et al., 2019). Another example of GQD production involves using GO with $KMnO_4$ as an oxidant under ultrasonication, achieving synthesis in a relatively short time (4 h). These green synthesis-made GQDs exhibit a high quantum yield of 27.8% (Zhu et al., 2015). CDs synthesized from cigarette ash via ultrasonication exhibit vivid green fluorescence when functionalized with polyethylene oxide (Huang et al., 2018; Bandi et al., 2018). The primary obstacle to CD preparation with ultrasonic treatment is unequal sample heating, which is mostly related to the local thermal effect of ultrasonic waves. Also, it requires a high energy cost. Therefore, further research is required to enhance CD production using ultrasonic therapy in light of these reported limitations.

3.4.1.3 Laser Ablation Method

Laser ablation is a top-down approach to break large carbon materials into smaller carbon quantum dots (CQDs) physically. This technique provides high-purity synthesis of quantum dots (Shorgar et al., 2023). The preparation of nanomaterials, particularly CQDs, has benefited from the use of pulsed laser ablation (PLA), a novel and promising technique, because of its simplicity and reduced processing time. The local high pressure and high temperature during the laser ablation process can immediately cut a target into residue-free nanoparticles, preventing the entrance of contaminants and removing impurities. The impact of laser parameters, including laser energy, laser pulse width, and ablation time, on CQDs has been documented in multiple studies. Furthermore, the photoluminescence emission of the resulting CQDs by PLA has thus been thoroughly investigated and primarily exhibits an excitation-dependent characteristic. Nevertheless, conventional PLA remains less effective, and the resulting CQDs are not uniform in size and shape, which hinders their potential use (Cui et al., 2020).

3.4.2 Bottom-Up Approaches

3.4.2.1 Hydrothermal Approach

A hydrothermal process uses high temperatures and pressures to break carbon precursors into CDs. It is a cost-effective and environmentally friendly approach to synthesize CDs (Huston et al., 2021). As discussed above, the sources of organic compounds are glucose, fructose, amines, and organic acids such as citric acid. Citric acid is one of the widely used sources for synthesizing CQDs because of the presence of carboxylic, hydroxylic, and carbonyl groups (Nazibudin et al., 2023). After hydrothermally synthesizing carbon quantum dots functionalized with polyamines by carbonizing citric acid with branched polyethyleneimine (BPEI), the quantum yield (QY) was determined at 42.5% (Yan et al., 2019). Hydrothermal procedures use lignin, silk, and gelatin sources to enhance CD yield up to 38%. This includes heteroatom doming of nitrogen, phosphorus, and sulfur elements, which reduces the band gaps and increases performance and quantum yield. For example, strawberry juice is used to produce carbon nanoparticles doped with nitrogen when treated at 180 °C for 12 h (Khan et al., 2024). A simple and sustainable hydrothermal process was used to create CDs utilizing cabbage as a carbon source. The cabbage was smashed and then autoclaved for 5 h at 140 °C. The solution was then filtered and centrifuged for 15 min at 12,000 rpm after cooling, producing a brown solution (Alam et al., 2015).

CDs produced via hydrothermal routes possess applications in biosensing, bioimaging, and other optical devices (Wang et al., 2022). Nevertheless, this approach is time consuming and typically results in lower yields (Ross et al., 2020).

3.4.2.2 Microwave-Assisted Method

This method also synthesizes green CDs in an eco-friendly manner. In this technique, organic substances are carbonized using microwave radiation (Khan et al., 2024). Another ecologically favorable source for producing CDs is amino acids. In one study, the carbon precursor in a microwave-assisted process was casein, a protein obtained from milk. A semi-solid combination was produced by heating the casein solution in the microwave. After centrifuging this mixture, the CD-containing clear supernatant was removed and diluted with distilled water (Bajpai et al., 2019). Studies have also demonstrated that combining urea as a carbon source with bovine serum albumin can produce CDs using microwave treatment (Liu et al., 2016). Microwave also uses sources like citric acid or glucose doped with nitrogen-rich elements like urea in water as a solvent. After the synthesis of CDs, they are purified using a dialysis and centrifugation process (Ahmad et al., 2025). The CDs synthesized using microwave-assisted methods have potential for biosensing, drug delivery, solar cells, and photocatalysis (de Medeiros et al., 2019). This approach uses minimal synthesis time, yet lacks large-scale production (Kaiming et al., 2021).

3.4.2.3 Pyrolysis Approach

It is a simple process that uses 300–800 °C temperature and carbonizes organic substances to manufacture CDs. This approach follows a thermochemical way of synthesis, which is performed without oxygen or using inert atmospheric conditions (Kang et al., 2020; Dhenadhayalan et al., 2020). The pyrolysis process often uses high concentrations of acid or alkali to break carbon precursors into nanoparticles (Zahid et al., 2018). Watermelon peel, orange peel, coffee grounds, and plant leaves are just a few examples of the various biomass materials that can be utilized as carbon sources for making C-dots through the pyrolysis process (Divyabharathi et al., 2025). CDs that are formed can be modified by altering the pyrolysis parameters such as time, temperature, and reaction systems' pH (Kang et al., 2020). The ideal temperature for the pyrolysis route for plant leaves is found to be 350 °C, especially pine needles (11.8% quantum yield), lotus leaves (15.3% quantum yield), and oriental leaves (16.4% quantum yield) (Ayisha Naziba et al., 2024). The subsequently formed C-dots exhibit high salinity, robust stability in solutions with a broad pH range, bright blue luminescence, and outstanding water solubility (Kang et al., 2020). For example, CDs synthesized via pyrolysis have the potential to image HeLa cells (Mehta et al., 2014). Additionally, the C-dots produced by pyrolysis techniques can also be applied to anticancer drug delivery due to their porosity (Mansi & Gaurav, 2023). This approach needs to operate at ideal temperatures. A low pyrolysis temperature could stop the carbon supply from completely carbonizing into C-dots. However, at extremely high temperatures, the carbon source is over-oxidized, and the surface structure of C-dots is damaged, resulting in a loss in optical performance (Kang et al., 2020).

3.4.3 Structural and Morphological Comparison Between Top-Down and Bottom-Up Approaches

Top-down and bottom-up approaches show both similarities and differences, which are observed via structural and morphological characteristics (Pillar-Little et al., 2018). Bottom-Up CDs naturally develop a rich surface functionalization, and their optical response is excellent without the need for additional processing. Bottom-up CDs are therefore frequently utilized in their as-synthesized form, with a subsequent purification step (Nagarajan et al., 2023). In top-down, post-synthesis functionalization processes, one usually needs to modify the surface groups to fulfill the required characteristics. According to AFM measurements, the top-down nanoparticles had a significantly greater average size of roughly 5 nm, whereas the bottom-up nanoparticles had an average size of roughly 2 nm. Nanoparticles made with bottom-up exhibit Raman G-band at 1598 cm^{-1} and D-band at 1356 cm^{-1}, while nanoparticles made with top-down exhibit slightly displaced G and D bands at 1608 cm^{-1} and 1357 cm^{-1}, indicating structural changes related to the fabrication process. Top-down CDs showed simpler ATR spectra compared to the bottom-up, even after functionalization (Bruno et al., 2021).

3.5 Conclusion

The traditional way of producing CDs using chemicals had a detrimental effect on the environment. To address this, researchers have developed CDs using "green synthesis," a more sustainable and eco-friendly approach. This chapter focuses on natural routes for producing CDs, utilizing natural raw materials such as plant and animal waste, and on top-down and bottom-up methodologies for CDs synthesis. It highlights how waste can be transformed into valuable materials and applied across various fields, including medicine, the food industry, chemical industry, and agriculture. We also discuss the development of CDs from a variety of organic and inorganic compounds, including citric acid, ascorbic acid, ammonium citrate, glucose, graphite, gelatin, and lignin. Their use in biomedical applications is made extremely convenient by their excellent biocompatibility, photoluminescent property, and low cytotoxicity. In the near future, the increasing use of CDs in a variety of environmental applications is going to persist and grow to reach significant heights.

References

Ahmad, S., Salma, U., Alam, M. Z., Mohasin, M., Parveen, H., Alimuddin, & Khan, S. A. (2025). Introduction to green carbon dots. In *Green carbon dots: Sustainable analytical approaches* (pp. 1–26). American Chemical Society.

Aksu, M., & Güzdemir, Ö. (2025). Food waste-derived carbon quantum dots and their applications in food technology: A critical review. *Food and Bioprocess Technology, 18*(8), 6753–6778.

Alam, A. M., Park, B. Y., Ghouri, Z. K., Park, M., & Kim, H. Y. (2015). Synthesis of carbon quantum dots from cabbage with down-and up-conversion photoluminescence properties: Excellent imaging agent for biomedical applications. *Green Chemistry, 17*(7), 3791–3797.

Alfi, A. A., Alamrani, N. A., Azher, O. A., Snari, R. M., Abumelha, H. M., Al-Ahmed, Z. A., & El-Metwaly, N. M. (2022). Development of carbon dots sensor dipstick from sugarcane bagasse agricultural waste toward all-cellulose-derived tetracycline sensor. *Journal of Materials Research and Technology, 19*, 4697–4707.

Ayisha Naziba, T., Praveen Kumar, D., Karthikeyan, S., Sriramajayam, S., Djanaguiraman, M., Sundaram, S., et al. (2024). Biomass derived biofluorescent carbon dots for energy applications: Current progress and prospects. *The Chemical Record, 24*(6), e202400030.

Bajpai, S. K., D'souza, A., & Suhail, B. (2019). Blue light-emitting carbon dots (CDs) from a milk protein and their interaction with Spinaciaoleracea leaf cells. *International Nano Letters, 9*, 203–212.

Bandi, R., Devulapalli, N. P., Dadigala, R., Gangapuram, B. R., & Guttena, V. (2018). Facile conversion of toxic cigarette butts to N, S-codoped carbon dots and their application in fluorescent film, security ink, bioimaging, sensing and logic gate operation. *ACS Omega, 3*(10), 13454–13466.

Baweja, H., & Jeet, K. (2019). Economical and green synthesis of graphene and carbon quantum dots from agricultural waste. *Materials Research Express, 6*(8), 0850g8.

Beach, E. S., Cui, Z., & Anastas, P. T. (2009). Green Chemistry: A design framework for sustainability. *Energy & Environmental Science, 2*(10), 1038–1049.

Bruno, F., Sciortino, A., Buscarino, G., Soriano, M. L., Ríos, Á., Cannas, M., et al. (2021). A comparative study of top-down and bottom-up carbon nanodots and their interaction with mercury ions. *Nanomaterials, 11*(5), 1265.

Chen, Z., Liu, Y., & Kang, Z. (2022). Diversity and tailorability of photoelectrochemical properties of carbon dots. *Accounts of Chemical Research, 55*(21), 3110–3124.

Cheviron, P., Gouanvé, F., & Espuche, E. (2014). Green synthesis of colloid silver nanoparticles and resulting biodegradable starch/silver nanocomposites. *Carbohydrate Polymers, 108*, 291–298.

Cui, L., Ren, X., Wang, J., & Sun, M. (2020). Synthesis of homogeneous carbon quantum dots by ultrafast dual-beam pulsed laser ablation for bioimaging. *Materials Today Nano, 12*, 100091.

de Medeiros, T. V., Manioudakis, J., Noun, F., Macairan, J. R., Victoria, F., & Naccache, R. (2019). Microwave-assisted synthesis of carbon dots and their applications. *Journal of Materials Chemistry C, 7*(24), 7175–7195.

Dehvari, K., Liu, K. Y., Tseng, P. J., Gedda, G., Girma, W. M., & Chang, J. Y. (2019). Sonochemical-assisted green synthesis of nitrogen-doped carbon dots from crab shell as targeted nanoprobes for cell imaging. *Journal of the Taiwan Institute of Chemical Engineers, 95*, 495–503.

Dhenadhayalan, N., Lin, K. C., & Saleh, T. A. (2020). Recent advances in functionalized carbon dots toward the design of efficient materials for sensing and catalysis applications. *Small, 16*(1), 1905767.

Diao, H., Li, T., Zhang, R., Kang, Y., Liu, W., Cui, Y., et al. (2018). Facile and green synthesis of fluorescent carbon dots with tunable emission for sensors and cells imaging. *Spectrochimica Acta Part A: Molecular and Biomolecular Spectroscopy, 200*, 226–234.

Divyabharathi, R., Kalidasan, B., SakthiSuriya Raj, J. S., Chinnasamy, S., Pandey, A. K., Giri, J., & Fatehmulla, A. (2025). Harnessing agro-wastes for sustainable nanomaterials: The promise of biofluorescent carbon dots. *Environmental Progress & Sustainable Energy, 44*(1), e14525.

Dubey, P. (2023). An overview on animal/human biomass-derived carbon dots for optical sensing and bioimaging applications. *RSC Advances, 13*(50), 35088–35126.

Egbedina, A. O., Bolade, O. P., Ewuzie, U., & Lima, E. C. (2022). Emerging trends in the application of carbon-based materials: A review. *Journal of Environmental Chemical Engineering, 10*(2), 107260.

Fatahi, Z., Esfandiari, N., Ehtesabi, H., Bagheri, Z., Tavana, H., Ranjbar, Z., & Latifi, H. (2019). Physicochemical and cytotoxicity analysis of green synthesis carbon dots for cell imaging. *EXCLI Journal, 18*, 454.

Flores-Contreras, E. A., González-González, R. B., Pablo Pizaña-Aranda, J. J., Parra-Arroyo, L., Rodríguez-Aguayo, A. A., Iñiguez-Moreno, M., et al. (2024). Agricultural waste as a sustainable source for nanoparticle synthesis and their antimicrobial properties for food preservation. *Frontiers in Nanotechnology, 6*, 1346069.

Fu, L., Liu, T., Yang, F., Wu, M., Yin, C., Chen, L., & Niu, N. (2022). A multi-channel array for metal ions discrimination with animal bones derived biomass carbon dots as sensing units. *Journal of Photochemistry and Photobiology, A: Chemistry, 424*, 113638.

Gudimella, K. K., Appidi, T., Wu, H. F., Battula, V., Jogdand, A., Rengan, A. K., & Gedda, G. (2021). Sand bath assisted green synthesis of carbon dots from citrus fruit peels for free radical scavenging and cell imaging. *Colloids and Surfaces B: Biointerfaces, 197*, 111362.

Hu, Z., Jiao, X. Y., & Xu, L. (2020). The N, S co-doped carbon dots with excellent luminescent properties from green tea leaf residue and its sensing of gefitinib. *Microchemical Journal, 154*, 104588.

Huang, H., Cui, Y., Liu, M., Chen, J., Wan, Q., Wen, Y., et al. (2018). A one-step ultrasonic irradiation assisted strategy for the preparation of polymer-functionalized carbon quantum dots and their biological imaging. *Journal of Colloid and Interface Science, 532*, 767–773.

Huston, M., DeBella, M., DiBella, M., & Gupta, A. (2021). Green synthesis of nanomaterials. *Nanomaterials, 11*(8), 2130.

Jing, H. H., Bardakci, F., Akgöl, S., Kusat, K., Adnan, M., Alam, M. J., et al. (2023). Green carbon dots: Synthesis, characterization, properties and biomedical applications. *Journal of Functional Biomaterials, 14*(1), 27.

Kaiming, K., Baoyou, L., Ju, H., Hongwei, R., Limin, W., & Gang, Y. (2021). Green preparation and application of carbon quantum dots. *IOP Conference Series: Earth and Environmental Science., 826*(1), 012036. IOP Publishing.

Kang, C., Huang, Y., Yang, H., Yan, X. F., & Chen, Z. P. (2020). A review of carbon dots produced from biomass wastes. *Nanomaterials, 10*(11), 2316.

Khan, A. U., Liu, Y., Wang, S., Ullah, M. W., Chen, Q., Zhang, D., et al. (2024). Advancements in the green synthesis of carbon dots for sustainable development. *Sustainable Materials and Technologies, 41*, e01004.

Kumar, R., Kumar, V. B., & Gedanken, A. (2020). Sonochemical synthesis of carbon dots, mechanism, effect of parameters, and catalytic, energy, biomedical and tissue engineering applications. *Ultrasonics Sonochemistry, 64*, 105009.

Kumari, R., & Sahu, S. K. (2022). A new insights into multicolor emissive carbon dots using Trachelospermum jasminoidesleaves for the application of WLEDs. *Colloids and Surfaces A: Physicochemical and Engineering Aspects, 647*, 128959.

Lakshmi, D. S., Ravindran, L., Mary, P. M., Rathika, G., Sreekala, M. S., & Munisamy, S. (2024). Valorization of crustacean shells: Preparation, characterization, and chemical modification of nano-chitin biopolymer membrane for application in water purification. *Waste and Biomass Valorization, 15*(11), 6067–6090.

Liu, X., Li, T., Hou, Y., Wu, Q., Yi, J., & Zhang, G. (2016). Microwave synthesis of carbon dots with multi-response using denatured proteins as carbon source. *RSC Advances, 6*(14), 11711–11718.

Manjubaashini, N., Bargavi, P., & Balakumar, S. (2024). Carbon quantum dots derived from agro waste biomass for pioneering bioanalysis and in vivo bioimaging. *Journal of Photochemistry and Photobiology A: Chemistry, 454*, 115702.

Mansi, M., & Gaurav, S. (2023). Synthesis and applications of carbon dots from waste biomass. In *Carbon dots in analytical chemistry* (pp. 319–328). Elsevier.

Mehta, V. N., Jha, S., Singhal, R. K., & Kailasa, S. K. (2014). Preparation of multicolor emitting carbon dots for HeLa cell imaging. *New Journal of Chemistry, 38*(12), 6152–6160.

Nagarajan, D., Gangadharan, D., & Venkatanarasimhan, S. (2023). Synthetic strategies toward developing carbon dots via top-down approach. In *Carbon dots in analytical chemistry* (pp. 1–13). Elsevier.

Nazibudin, N. A., Zainuddin, M. F., & Abdullah, C. A. C. (2023). Hydrothermal synthesis of carbon quantum dots: An updated review. *Journal of Advanced Research in Fluid Mechanics and Thermal Sciences, 101*, 192–206.

Özyurt, D. (2022). *Synthesis of carbon dots from low-cost precursors and study of concentration-dependent emission properties* [Doctoral dissertation, Swinburne].

Ozyurt, D., Al Kobaisi, M., Hocking, R. K., & Fox, B. (2023). Properties, synthesis, and applications of carbon dots: A review. *Carbon Trends, 12*, 100276.

Pansari, P., Sahu, S., Mishra, M., Gupta, P. K., & Durga, G. (2024). In-vivo toxicological study of cysteamine modified carbon dots derived from Ruellia simplex on fruit fly for potential bioimaging. *Biocatalysis and Agricultural Biotechnology, 61*, 103359.

Patel, K. D., Singh, R. K., & Kim, H. W. (2019). Carbon-based nanomaterials as an emerging platform for theranostics. *Materials Horizons, 6*(3), 434–469.

Peng, H., & Travas-Sejdic, J. (2009). Simple aqueous solution route to luminescent carbogenic dots from carbohydrates. *Chemistry of Materials, 21*(23), 5563–5565.

Pillar-Little, T. J., Wanninayake, N., Nease, L., Heidary, D. K., Glazer, E. C., & Kim, D. Y. (2018). Superior photodynamic effect of carbon quantum dots through both type I and type II pathways: Detailed comparison study of top-down-synthesized and bottom-up-synthesized carbon quantum dots. *Carbon, 140*, 616–623.

Ross, S., Wu, R. S., Wei, S. C., Ross, G. M., & Chang, H. T. (2020). The analytical and biomedical applications of carbon dots and their future theranostic potential: A review. *Journal of Food and Drug Analysis, 28*(4), 677.

Seesuea, C., Sansenya, S., Thangsunan, P., & Wechakorn, K. (2024). Green synthesis of elephant manure-derived carbon dots and multifunctional applications: Metal sensing, antioxidant, and rice plant promotion. *Sustainable Materials and Technologies, 39*, e00786.

Semeniuk, M., Yi, Z., Poursorkhabi, V., Tjong, J., Jaffer, S., Lu, Z. H., & Sain, M. (2019). Future perspectives and review on organic carbon dots in electronic applications. *ACS Nano, 13*(6), 6224–6255.

Sharma, S., Kumar, R., Kumar, K., & Thakur, N. (2024a). Sustainable applications of biowaste-derived carbon dots in eco-friendly technological advancements: A review. *Materials Science and Engineering: B, 305*, 117414.

Sharma, N., Sharma, A., Park, M., & Lee, H. J. (2024b). Silkworm-derived carbon nano rods (swCNR) for detection of bismuth ions (Bi3+) in aquatic medium and their antiradical properties. *Heliyon, 10*(13), e33572.

Shorgar, N., Bhati, I., & Jhalora, P. (2023). Laser ablation synthesis of quantum dots. In *Quantum dots* (pp. 53–75). Elsevier.

Suseem, S. R. (2024). Plant-based carbon dots are a sustainable alternative to conventional nanomaterials for biomedical and sensing applications. *Nano Express, 5*(1), 012002.

Vithalani, R., Patel, D., Modi, C. K., & Suthar, D. H. (2020). Glowing photoluminescene in carbon-based nanodots: Current state and future perspectives. *Journal of Materials Science, 55*(21), 8769–8792.

Wang, B., Cai, H., Waterhouse, G. I., Qu, X., Yang, B., & Lu, S. (2022). Carbon dots in bioimaging, biosensing and therapeutics: A comprehensive review. *Small Science, 2*(6), 2200012.

Xu, Q., Li, W., Ding, L., Yang, W., Xiao, H., & Ong, W. J. (2019). Function-driven engineering of 1D carbon nanotubes and 0D carbon dots: Mechanism, properties and applications. *Nanoscale, 11*(4), 1475–1504.

Yan, F., Sun, Z., Zhang, H., Sun, X., Jiang, Y., & Bai, Z. (2019). The fluorescence mechanism of carbon dots, and methods for tuning their emission color: A review. *Microchimica Acta, 186*, 1–37.

Zahid, M. U., Pervaiz, E., Hussain, A., Shahzad, M. I., & Niazi, M. B. K. (2018). Synthesis of carbon nanomaterials from different pyrolysis techniques: A review. *Materials Research Express, 5*(5), 052002.

Zhang, Z., Sun, W., & Wu, P. (2015). Highly photoluminescent carbon dots derived from egg white: Facile and green synthesis, photoluminescence properties, and multiple applications. *ACS Sustainable Chemistry & Engineering, 3*(7), 1412–1418.

Zhang, Y., Lei, C., Dong, P., Fu, P., Zhang, Y., & Hua, R. (2024). Green synthesis of carbon dots from fish scales for selective turn off–on detection of glutathione. *RSC Advances, 14*(5), 3578–3587.

Zhao, P., & Zhu, L. (2018). Dispersibility of carbon dots in aqueous and/or organic solvents. *Chemical Communications, 54*(43), 5401–5406.

Zhu, Y., Wang, G., Jiang, H., Chen, L., & Zhang, X. (2015). One-step ultrasonic synthesis of graphene quantum dots with high quantum yield and their application in sensing alkaline phosphatase. *Chemical Communications, 51*(5), 948–951.

AI-Assisted Synthesis of Carbon Dots and Their Applications

4

Kiran Jeet and Anu Kalia

Abstract

Due to their diverse physicochemical properties and varied beneficial characteristics such as high stability, superior biocompatibility, unique optical properties, economical and eco-friendly synthesis pathway, presence of numerous functional groups, and excellent electron mobility, carbon dots (CDs) have been a subject of great interest among researchers. Artificial intelligence (AI), including machine learning (ML), has emerged as a transformative force in the realm of materials science, offering a diverse array of models to decipher complex relationships between material properties and structures. Using the experimental data of the previously reported research, algorithms are designed to create the link between the input and the output parameters. Once the link is established, a little mutation with experimentally designed parameters may readily provide the required features of the nanomaterial. Among the many attributes, quantum yield (QY) is regarded as one of the most significant features from an application perspective. Currently, AI-based CDs are being synthesized to get the appropriate quantum yield and are subsequently used in many sectors.

Keywords

Carbon dots (CDs) · Artificial Intelligence (AI) · Machine learning (ML) · Quantum Yield (QY) · 0D (zero dimension)

K. Jeet (✉) · A. Kalia
Electron Microscopy and Nanoscience Lab, Punjab Agricultural University, Ludhiana, Punjab, India
e-mail: kiranjeet@pau.edu

M. Rai et al. (eds.), *Carbon Nanodots*, https://doi.org/10.1007/978-981-95-4705-0_4

4.1 Introduction

The element carbon, which is one of the most common in our environment and has the periodic table symbol "C," has a great variety of structural shapes and morphologies. Carbon is a vital building component of both life and materials. Carbon can exist in three different hybridization states, such as sp, sp^2, and sp^3, due to its ability to have different bonding options. A variety of bonding options leads to different geometries of carbon, ranging from linear to trigonal planar to tetrahedral and other shapes. The exceptional bonding properties and structural diversity of carbon are exemplified by its versatility, encompassing graphite, a notably soft substance typically linked to pencils, and diamond, the hardest allotrope, characterized by its intricate lattice structure. Besides diamond and graphite, other allotropes of carbon comprise fullerene (C_{60}), carbon nanotubes, graphene, nanofibers, graphene nanoribbons, graphene quantum dots, and carbon quantum dots.

Carbon-based quantum dots, also referred to as carbon dots (CDs), have dimensions smaller than 10 nm. They fall under the category of 0D (zero-dimensional) nanomaterials. Figure 4.1 indicates the different categories of carbon-based materials and nanomaterials. CDs are spherical nanoparticles featuring an amorphous to nanocrystalline core composed of sp^2 hybridized carbon, with graphene oxide sheets integrated in a manner akin to diamond-like sp^3 hybridized carbon (Lim et al., 2015). The core–shell architecture of carbon dots (CDs) has a central carbon core with chemical functional groups attached or modified on its surface. The surface generally contains many functional groups, such as amino, epoxy, carbonyl, aldehyde, hydroxyl, and carboxylic acid, whereas the carbon core structure is composed of sp^2 and sp^3 carbon atoms. Due to the presence of several surface groups that may interact with ions, chemical compounds, polymers, DNA, and proteins, functionalized CDs can be tailored to fulfil particular needs across various applications, hence modifying their characteristics.

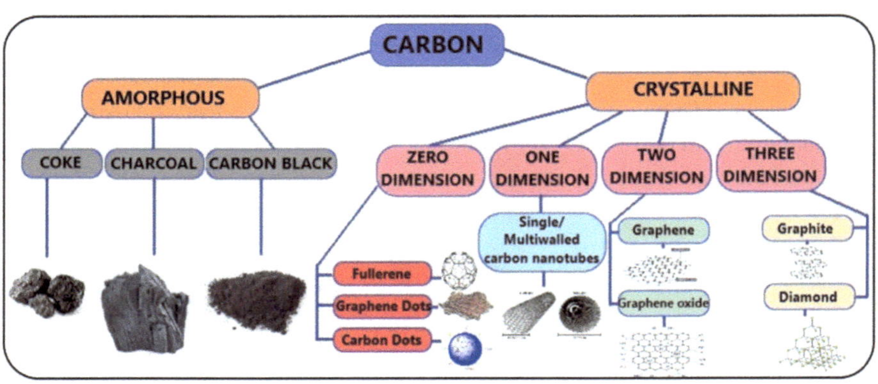

Fig. 4.1 Overview of carbon classification from amorphous materials to 0D, 1D, 2D, and 3D crystalline forms

CDs are a general term that refers to a diverse array of nanosized fluorescent carbon materials. These materials are classified based on their specific carbon core structure, surface groups, and properties. Specifically, they include graphene quantum dots, carbon quantum dots, carbon nanodots, and carbonized polymer dots (Xia et al., 2019; Singh et al. 2023a, b).

The initial report of carbon nanoparticles exhibiting fluorescence behavior was made in 2004, when they were inadvertently obtained during the purification of single-walled CNTs (Xu et al., 2004). In 2006, Sun and colleagues designated the term CDs for these fluorescent nanoparticles; nonetheless, their low quantum yield (QY) of around 10% was the primary factor hindering their acceptance (Sun et al., 2006). Subsequently, in 2013, Yang's team synthesized carbon dots (CDS) with an 80% quantum yield (QY) via a one-step synthesis technique with citric acid (CA) and ethylene diamine as precursors (Zhu et al., 2013).

CDs exhibit superior photostability, a higher quantum yield, lower toxicity levels, and exceptional biocompatibility compared to conventional nanomaterials such as cadmium/lead, rare earths, and metal oxides (Shen et al., 2012). Additionally, some CDs are derived from abundant, low-cost sources, which enable their use in widely varied fields.

The types and compositions of surface groups, the dimensions of π-conjugated domains, and the variability of oxygen/nitrogen concentration in carbon cores significantly affect the absorption properties of CDs. Typically, they exhibit significant absorption in the ultraviolet (UV) spectrum (200–400 nm), with a tail extending into the visible region, where the absorption bands are attributed to the $\pi-\pi^*$ transition of the C=C bond or the $n-\pi^*$ transition of the C=O/C=N bond (Xu et al., 2004; Jiang et al., 2013; Liu et al., 2018; Xu et al. 2017a, b; Huang et al., 2013; Yang et al., 2018). Some CDs that emit red or near-infrared light often include π-conjugated electrons in the sp^2 domains or the surface groups/polymer chains that are connected to them. This is what makes them absorb light in the 500–800 nm range (Liu et al. 2020a, b, c; Yuan et al., 2017; Qu et al., 2016).

The fluorescence characteristics of CDs are further categorized into down-conversion and up-conversion fluorescence, where the former is referred to as Stokes and the latter as anti-Stokes. The fluorescence origins of CDs primarily include surface states (Bao et al., 2011, 2015; Ding et al., 2016; Wang et al., 2014; Lu et al., 2014; Song et al., 2015; Liu et al. 2015a, b, 2017), quantum confinement effects (Jiang et al., 2015; Yan et al., 2013; Kim et al., 2012), edge defects (Cao et al., 2013; Liu et al., 2013; Song et al., 2014; Santiago et al., 2016), and multi-emissive centers (Dhenadhayalan et al., 2016; Chen et al., 2018; Ju et al., 2017). The surface characteristics, encompassing the extent of surface oxidation and the presence of surface functional groups, directly influence the fluorescence of carbon dots (CDs) (Liu et al., 2019; Xia et al., 2018). Surface oxidation, induced by oxygen-related groups, can generate the defect (Bao et al., 2011). Diverse defects can result in several emission sites on the surface of CDs, hence intensifying their fluorescence (Bao et al., 2011, 2015).

Defects arising from O-related functional groups at the periphery of carbon dots or from the overlap of two spectral bands with intrinsic and extrinsic states can be utilized to alter the fluorescence of carbon dots. The deficiency will enlarge, and the redshift of the fluorescence will become more pronounced as the surface oxygen level rises. Multiple fluorescent centers contribute to the adjustable fluorescence of carbon dots, arising from the hybridization of several tiny emissive molecules on the emissive carbon cores (Ju et al., 2017).

Recently, there has been a lot of interest in the phosphorescence characteristics of CDs, particularly long-lifetime room-temperature phosphorescence (RTP). Two criteria should be prioritized to accomplish RTP: the first aims to suppress nonradiative transitions by limiting vibration and rotation, while the second focuses on improving spin–orbit coupling by introducing transition metals to increase intersystem crossing efficiency (Tao et al., 2018; Jiang et al., 2018). Additionally, other luminescent properties of CDs, including electro-chemiluminescence (Chen et al., 2020; Jie et al., 2019) and chemiluminescence (Molaei 2019; Shen et al., 2020), have been successfully investigated and revealed. Different optical characteristics of CDs enable their extensive applications in varied fields. The characteristics of CDs are significantly influenced by their structure, which is in turn contingent upon the production method and the precursor employed. This chapter delineates the importance of machine learning in the synthesis of carbon dots (CDs), augmented by case studies and examples that illustrate how machine learning methodologies enhance the synthesis of CDs with desired functionalities, improve their properties, and broaden their applicability across various domains.

4.2 Production of CDs

4.2.1 Conventional Method

Both top-down and bottom-up strategies are used in the manufacturing of carbon quantum dots. The techniques used for the production of carbon dots (CD) under the top-down approach include laser ablation, arc discharge, ultrasonic synthesis, chemical exfoliation, and electrochemical methods. Laser ablation is a technique that uses high-energy laser pulses to irradiate the surface of a target substance. This method subjects the material to a thermodynamic condition that produces elevated temperatures and pressures, resulting in fast heating, melting, and evaporation into a plasma state. The resultant vapors then solidify upon exposure to cold, yielding nanoparticles (Li et al., 2010, 2022).

In the arc discharge method, a high-current arc is generated between two graphite electrodes, often with their tips immersed in deionized water, producing a plasma that enables carbon nanomaterial formation. During this process, the bubbles formed around the arc acted as a small reactor, isolated from the atmosphere by water. In the submerged arc discharge in water process, the product is partitioned into three phases, namely, floating material, suspended material, and precipitate. The suspended material in the water is then meticulously removed from the precipitate. This CQD suspension is employed in its original form or may be progressively concentrated or dried as needed (Kim et al., 2017).

Carbon quantum dots may be generated via the electrochemical oxidation of carbon source materials, with precise regulation of reaction parameters such as voltage and current (Hou et al., 2015). An appropriate carbon source is chosen, and an electrode is fabricated. Thereafter, the carbon source is solubilized in an electrolyte solution, and the electrode is submerged in the aforementioned solution. As a result, the carbon atoms and metal ions in the solution undergo an electrochemical process in the presence of an electric current or potential, leading to the synthesis of carbon quantum dots (Li et al., 2010, 2022).

The ultrasonic-assisted approach is recognized for its cost-effectiveness and ease of operation in the production of CDs. In the ultrasonic process, alternating high-pressure and low-pressure waves are produced, leading to the development and collapse of tiny bubbles in the liquid. Consequently, macroscopic carbon materials were fragmented into nanoscale carbon dots by intense hydrodynamic shear pressures generated by the cavitation of microscopic bubbles (Zhuo et al., 2012).

Bottom-up approaches encompass the application of chemical oxidation, microwave techniques, and hydrothermal processes. Chemical oxidation refers to the oxidative transformation of organic raw materials into carbon quantum dots by dissolving the carbon source in a chemical oxidant solution, often using a reaction process with nitric acid as the predominant oxidant. The chemical oxidation process necessitates no specialized equipment, and the carbon sources needed are varied, simple, and readily accessible, rendering this method an efficient and scalable synthesis technique (Ge et al., 2021; Lim et al., 2015).

The microwave technique utilizes microwave radiation to swiftly heat the carbon source, resulting in carbonization or pyrolysis events and the consequent production of carbon quantum dots. This technology has several advantages, including rapid reaction time, elevated yield, little energy use, and the elimination of organic solvents (Liu et al., 2017).

The hydrothermal approach utilizes a hydrothermal process to create carbon quantum dots by dissolving a carbon source in a combination of water and an organic solvent. A carbon source is solubilized in water or an organic solvent and subsequently subjected to treatment at elevated temperature and pressure within a high-pressure reactor. The reaction often lasts several hours, during which the size and shape of the carbon quantum dots may be controlled by modifying the water vapor pressure during the process (Yang et al., 2012).

While both bottom-up and top-down methodologies have been employed to synthesize CQDs, the former is predominantly favored for its cost-effectiveness and environmental sustainability (Chae et al., 2017). Notwithstanding its benefits, the bottom-up strategy exhibits inadequate control over the dimensions of CQDs, while the top-down solutions incur higher costs. In order to resolve this issue and enhance the synthesis of CDs, artificial neural network (ANN) and machine learning (ML) approaches were used to optimize the synthesis of CDs and their sensing capabilities.

4.2.2 AI- and ML-Assisted Synthesis of CDs

Artificial intelligence (AI) focuses on the creation of intelligent machines capable of executing perception, decision-making, and problem-solving activities. Machine learning (ML) is a subset of artificial intelligence (AI) that uses computational algorithms to derive mathematical models capable of executing certain tasks, such as prediction, dimensionality reduction, and grouping, directly from collected data (Russell & Norvig, 2020). This method is especially beneficial when the correlation between the variables in the experiment and the resultant results is unclear. The growing volume of accessible experimental data, coupled with pervasive computer resources, has led to the implementation of intricate ML models across several fields of research and industry (Liu et al. 2020a, b, c; Silver et al., 2017; Senior et al., 2020). Its efficacy in fields such as recognition, physics, chemistry, biology, and materials research has rendered it a flexible and potent tool. One of its key benefits is its ability to accelerate the development of novel materials by analyzing previously documented successful and unsuccessful experiments. ML provides a robust mechanism for the expedited formulation of effective protocols for nanoparticle synthesis, and possibly for the creation of novel nanoparticles with intricate features. Numerous studies have been published on the synthesis of various nanoparticle categories, such as metals, semiconductors, carbon-based materials, and polymers, using machine learning algorithms and an extensive approach to regulate the synthesis parameters. The review conducted by Tao et al. (2021) explores machine learning (ML) algorithms relevant to nanoparticle manufacturing, while also highlighting essential methods for acquiring comprehensive datasets (Fig. 4.2).

ML algorithms are primarily concerned with anticipating a relationship between experimental input circumstances to get the desired result. The input experimental conditions might include reaction duration, temperature, reactant concentration, precursor type, and so on. Similarly, the intended outcome or property may be the nanoparticle's size and shape, reaction yield, functionalization/capping agent on the nanoparticles, fluorescence quantum yield in the case of CDs, and so on. To infer a relationship between the input parameters and the nanoparticle characteristics, ML algorithms necessitate a dataset that includes synthetic recipes and the related outcomes from previous experiments.

The scientific community has been developing ML technologies for nanoparticle production. Machine learning techniques are predominantly employed for two functions in nanoparticle synthesis: predicting synthetic outcomes and planning experiments. Tree-based, instance-based, and deep learning techniques are applicable for predicting nanoparticle characteristics. Heuristic and Bayesian optimization techniques may be employed to design nanoparticle production experiments.

The fundamental component of tree-based algorithms is the decision tree, a diagram resembling a flowchart that associates possible experimental outcomes (e.g., nanoparticle size or shape) with a series of decisions concerning experimental parameters, such as designated temperature ranges or reagent concentration intervals (Breiman, 2001; Friedman, 2001, 2002; Criminisi et al., 2012). Deep learning approaches rely on the amalgamation of several nonlinear functions (LeCun et al., 2015). Given a set of conditions x and characteristics y to be predicted, a nonlinear

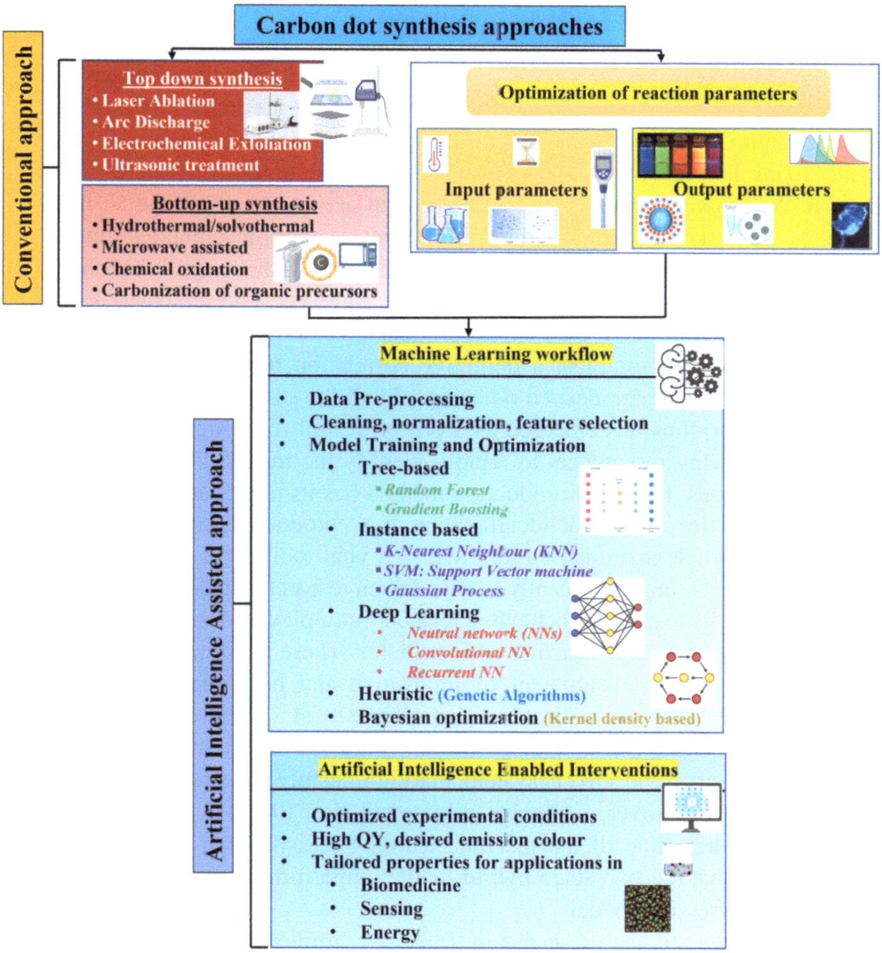

Fig. 4.2 Flowchart illustrating conventional (top-down and bottom-up) and AI-assisted approaches for carbon dot synthesis

function *f* may be established so that *y* is approximately equal to *f*(*x*). Heuristic methods for experimental design apply concepts from natural evolution to propose a new set of trials, based on the adaptation of circumstances used in previous research (Hansen, 2006; Waltraud & Arnold, 2008). Bayesian optimization strategies, in contrast to heuristic methods, utilize machine learning algorithms to model the relationships between experimental conditions and outcomes (Shahriari et al., 2016).

The distinguishing feature that renders any CD superior to others is its high fluorescence quantum yield (QY). The efficacy of high-QY CDs is contingent upon the optimization of many parameters, including reaction temperature, precursor mass, ramp rate, and reaction duration. Nonetheless, contemporary carbon dots

(CDs) documented in the literature frequently involve optimized preparation by regulating one reaction parameter while maintaining others constant, and consistently overlook the interrelation of reaction parameters during the synthesis process of CDs, which are required to be monitored throughout the analysis. Consequently, comprehending the essential elements of the manufacturing process is vital for synthesizing high quantum yield carbon dots, necessitating controlled evolution to achieve the inherent relationships derived from certain transformational methods. Extensive research has been conducted in recent years on machine learning-driven methodologies for synthesizing carbon dots. It is beyond the scope to encompass all studies; therefore, a limited number of studies have been presented in Table 4.1, illustrating the synthesis of CDs via experimental methods, followed by the application of ML tools to establish a definitive correlation between the diverse input parameters to achieve the desired outcomes. In all the studies discussed, the use of modelling algorithms has emerged as a powerful tool for optimization of the CD synthesis, allowing researchers to rapidly identify the most efficient and effective synthesis methods. Each CD production method has its merits and limitations. The primary difficulties include the identification of a reproducible method that provides a consistent particle size distribution. ML is a vital tool in the field of CD-based materials, contributing significantly to data-driven research and development. It may be used to anticipate material behaviors, evaluate complex information, and uncover the relationships between structure and attributes. These features significantly contribute to the quick development of innovative and high-quality CD materials intended for a variety of applications. Senanayke et al. (2022) employed machine learning techniques to ascertain the principal parameters, including reaction method, purification process, and solvent system, that significantly influence the CD emission characteristics. Moreover, a precise color prediction (94%) was attained by the optimal R-square value of the machine learning classification model. The study demonstrated that the two-stage hybrid model outperformed the single-stage ANN k-ensemble regression model.

Alongside the studies presented in Table 4.1, further noteworthy research has emerged recently, highlighting the application of machine learning in the synthesis of CDs for specific applications. Zirui Bian et al. (2024) documented the synthesis of CDs demonstrating antibacterial properties by the use of four distinct machine learning methodologies (KNN: K-Nearest Neighbor, RF: Random Forest, XGBoost: Extreme Gradient Boosting, and SVM: Support Vector Machine). Ensemble learning approaches have proven to be the most effective on the data. The correlation between CD characteristics and antibacterial efficacy was anticipated and elucidated, facilitating the advancement of high-performance nanoparticles and their prospective therapeutic use.

The lack of energy supply is one of the most critical factors affecting human sustenance on Earth. However, the energy is primarily derived from non-renewable resources, which are depleting at a greater pace. Furthermore, the burning of coal, natural gas, and other nonrenewable resources leads to the emission of greenhouse gases. Therefore, the renewable resources of power must be identified, such as hydrogen gas. Nguyen et al. (2022) documented the application of a Bayesian genetic algorithm (BGA) machine learning model to forecast the strategy for the

Table 4.1 Synthesis process, input parameters, and machine learning/artificial intelligence algorithms used for the generation of carbon dots

Sl. No.	Precursor	Synthesis method	Input parameters	ML/AI algorithm	Desired output	Reference
1.	MnO_2 + CDs	Solvothermal	CD ratio, reaction time	–	Supercapacitor capacity	Xu et al. (2017a, b)
2.	Fennel seeds	Hydrothermal	Emission spectra, time, pH	ML + PL analysis (principal component analysis (PCA), multiple curve resolution (MCR-ALS), non-negative matrix factorization with sparseness)	PL color classification	Dager et al. (2019)
3.	Tapioca flour	Pyrolysis	Temperature, mass, reaction time	ANN, RSM (Levenberg–Marquardt back propagation (LMBP) algorithm)	High QY and stability	Yahaya Pudza et al. (2019)
4.	N, S-Doped CDs	Doping + coating	Dopant ratio, light intensity	DSSC + ML	PCE in solar cells	Wang et al. (2019a, b)
5.	p-Dihydroxybenzene (p-Db) ethylenediamine (EDA)	Hydrothermal	Mass of precursors and the volume of alkaline catalysts	Regression ML model	QY 39.3%	Han et al. (2020)
6.	Citric acid + DMSO	Solvothermal	Temperature, ratio, solvents	ML-guided tuning	RGB emission	Chen et al. (2020)
7.	Arginine	Hydrothermal	Arginine %, pH, time	Regression	Gene delivery efficiency	Rezaei and Hashemi (2021)
8.	Citric acid + urea or ethylenediamine	Hydrothermal (bottom-up)	pH, temperature, molar ratio	Supervised ML/ANN	Emission wavelength prediction	Senanayake et al. (2022)

(continued)

Table 4.1 (continued)

Sl. No.	Precursor	Synthesis method	Input parameters	ML/AI algorithm	Desired output	Reference
9.	Citric acid + amine	Hydrothermal	Molar ratio, temperature	Regression	Red-emitting CDs	Luo et al. (2022)
10.	Quercetin	Hydrothermal	Concentration, heating time	Classification	Cancer targeting	Chandrasekaran et al. (2022)
11.	Ni + carbon source	Solvothermal	Catalyst load, temperature, pH	Bayesian genetic algorithm	HER activity	Nguyen et al. (2022)
12.	p-Phenylenediamine urea + citric acid	Hydrothermal and ultrasonic-assisted approach	Time, solvent, temperature, precursor combination, and precursor ratio	Extreme gradient boosting (XGBoost), random forest (RF), light gradient boosting machine (LGBM), ridge regression (Ridge), least absolute shrinkage and selection operator (LASSO), and support vector regression (SVR) *Best Model RF*	Multi-colored CD	Chen et al. (2023)
13.	Mopan Persimmon	Hydrothermal	Carbonization temperature, concentration	Regression	Fe^{3+} sensing	Ma et al. (2023)
14.	TF-conjugated Dox	Conjugation + hydrothermal	Ligand ratio, drug loading	In silico + regression	Cytotoxicity prediction	Mahani et al. (2021)
15.	2,7-Naphthalenediol, urea + EDA	Hydrothermal	Precursors, temperature	Regression + DL	Tunable full-color emission	Guo et al. (2024)
16.	Glucose + amine	Microwave	Temperature, pH, time	KNN, RF, SVM, XGB	Antibacterial activity	Bian et al. (2024)
17.	Plant biomass	Hydrothermal	pH, concentration, time	Supervised ML	Emission prediction	Xing et al. (2024)

18.	Biomass soot (Marigold flower Extract Corn starch)	Top-down	Carbonization temperature	Pattern recognition	Sweat fingerprint detection	Yadav et al. (2024)
19.	Citric acid + urea and around 22 other precursors	Hydrothermal	Combination of variable precursors	k-nearest neighbor (kNN), decision tree (DT), SVM, random forest (RF), Adaboot, and Bagging *Best Model: RF*	Predictable photoluminescence (PL)	Xing et al. (2024)
20.	Herb biomass	Microwave synthesis	Precursor blend, wavelength	Sensor array + ML	Drug metabolite classification	Li et al. (2025)
21.	2,2′-Dithiosalicylic acid (DTSA) + melamine	Hydrothermal	Ion concentration, pH	K-means clustering	Al^{3+}/F^{-} sensing	Zhang et al. (2025)

rational design and optimization of transient metal-incorporated carbon quantum dot (M@CQD)-based hydrogen evolution reaction (HER) electrocatalysts. The nickel-doped carbon dots on a three-dimensional reduced graphene oxide (3D rGO) conductor demonstrated superior performance under optimal conditions of catalyst loading, electrode type, temperature, and electrolyte pH for enhanced hydrogen evolution reaction (HER) properties regarding electronic structure. The outcomes were subsequently evaluated by density functional theory (DFT).

Jiajun Li et al. (2025) reported the green synthesis of N-doped carbon dots (NCDs) from biomass derived from three different herbs as the precursor. Three NCDs exhibited commendable blue fluorescence characteristics and sensing skills for the differentiation of warfarin (WAR) and its metabolites. The sensor array, in conjunction with the machine learning model, exhibited significant resistance to interference in the identification of the WAR, even with unfamiliar samples. The approach offers extensive opportunities for biomass-derived nanoparticles and presents an efficient and viable solution for pharmaceutical investigation by utilizing machine learning techniques.

Jun Bo Lu et al. (2022) presented an effective synthesis approach utilizing machine learning to aid researchers in the production of red fluorescent CDs. This technique can rapidly and effectively forecast the predetermined circumstances of CD synthesis.

Qiang Zhang et al. (2025) described a unique method that utilizes machine learning to improve the accuracy and efficiency of fluorescence-based detection for the sequential quantitative measurement of aluminum (Al^{3+}) and fluoride (F^{-}) ions in aqueous solutions. The suggested technique entails the manufacture of sulfur-functionalized carbon dots (C-dots) as fluorescent probes, exhibiting fluorescence amplification upon contact with Al^{3+} ions, with a detection limit of 4.2 nmol/L. Fluorescence fingerprints were retrieved utilizing a cross-platform computer vision library in Python, followed by data preprocessing. Subsequently, the fingerprint data undergoes cluster analysis utilizing the K-means model from machine learning, and the average Silhouette coefficient demonstrates exceptional model performance.

4.3 Application of CDs

4.3.1 Bioimaging

Bioimaging denotes a noninvasive method for real-time visualization of biological systems. Bioimaging strives to minimize interference with life processes. Moreover, it is often employed to acquire information regarding the examined specimens' three-dimensional structure externally, without medical intervention. Bioimaging broadly refers to techniques for visualizing living cells that have been preserved for research purposes. Bioimaging involves the examination of subcellular structures, whole cells, tissues, and multicellular organisms.

CDs are recognized as advantageous options for bioimaging owing to their enhanced adjustable photoluminescence (PL), reduced toxicity, hydrophilic characteristics, and excellent opto-chemical stability under UV irradiation. Most CDs predominantly produce blue emission with a narrow excitation spectrum, thus constraining bioimaging applications (Wang et al. 2019a, b; Marković et al., 2020).

Nanomaterials such as CDs can be coupled with proteins, peptides, antibodies, or other small molecules to monitor cancer cells, identify specific biomarkers, or facilitate medication delivery. Iron oxides utilized for MRI represent a significant instance of CDs employed in clinical applications. CDs have been documented as safe for bioimaging in multiple in vitro and in vivo contexts. Semiconductor quantum dots (SQDs), including those containing cadmium and selenium, represent older options for bioimaging but are hindered by several limitations, such as substantial size, cellular accumulation, toxicity from heavy metal synthesis, and blinking, which restricts their application at the single-particle level. However, CDs are much superior nanomaterials for cellular and bioimaging due to their enhanced biocompatibility, low toxicity, and nonblinking characteristics (Bozrova et al., 2018; Molaei, 2019).

Graphene quantum dots (GQDs), a subtype of CDs, were utilized for the first time in cancer cell imaging by Zhu et al. (2011), demonstrating stable fluorescence emission. Numerous CDs derived from natural precursors have been reported to undergo testing for cellular imaging properties in various breast cancer cell lines (T47D, MCF-7) and cervical cancer (HeLa) (Peng et al., 2012; Fan et al., 2014; Song et al., 2018). The examined CDs-graphene QDs, boron-doped GQDs, and linseed-derived CDs showed little toxicity, significant water solubility, resistance to photobleaching, and enhanced biocompatibility. The characteristic of possessing a diminutive size (under 10 nm) allows CDs to investigate minute biological entities (Yang et al., 2009). Cao et al. (2007) employed poly (propionyl ethylenimine-coethylenimine) for surface passivation and two-photon excitations to produce water-soluble carbon quantum dots (CQDs) with significant luminescence and a modest size (~5 nm). The CDs successfully tagged the cell membrane and cytoplasm but failed to penetrate the nucleus when incubated with MCF-7 cells, as indicated by fluorescence microscopy observations. Labeling these CQDs with the HIV-1-derived translocation protein TAT (Trans Activator of Transcription) may facilitate nuclear probing by surmounting cell membrane barriers and improving intracellular labeling efficiency (Cao et al., 2007).

In the realm of in vivo biological imaging, Yang et al. (2009) performed a comparative analysis of their synthesized PEG1500-functionalized CDs and commercial cadmium selenide/zinc sulfide quantum dots (QDs). The CDs displayed a high quantum yield of 20% and showed enhanced biocompatibility and luminous characteristics relative to CdSe/ZnS QDs. The environmentally friendly CDs, noted for reduced cytotoxicity and improved bioimaging efficacy, exhibited considerable promise for investigating diminutive biological structures with minimal injection volumes in in vivo applications.

Zheng et al. (2015) produced unique multicolor emissive carbon quantum dots with cancer-targeting capabilities by pyrolyzing D-glucose and L-aspartic acid in another investigation. These CDs could function as biological imaging agents for the precise diagnosis of noninvasive gliomas without the need for additional targeting molecules. They demonstrated remarkable biocompatibility and could easily penetrate the blood–brain barrier. In in vivo trials with mice, the CDs were administered, and after 15 min, fluorescence imaging revealed exceptional contrast, with the CDs mostly concentrating at the glioma site, demonstrating their significant selectivity.

Srivastava et al. (2012) synthesized iron oxide-doped carbon quantum dots (IO-CQDs) for multimodal (MRI/fluorescence) bioimaging. The IO-CQDs were mainly taken up by RAW 264.7 cells, with fluorescence predominantly observed in the cytoplasm. For in vivo imaging, the IO-CQDs were administered to rats via intravenous injection. Fluorescence signals from the IO-CQDs were detected in the spleen slide samples. The MRI results indicated increased signals in the cerebral blood vessels under both T1 and T2 sequences. The biocompatibility of CQDs enables the integration of additional imaging technologies with their fluorescence imaging for multimodal bioimaging.

4.3.2 Drug Delivery

Chemotherapies, although extensively employed for tumor and illness treatment, frequently exhibit imprecise targeting and may result in toxicity and drug resistance. Targeted medication administration has emerged as a viable method to improve drug bioavailability and efficacy while reducing unwanted effects. CDs are regarded as efficient dual nanocarriers that integrate bioimaging and targeted drug delivery functionalities while exhibiting low cytotoxicity.

Jing et al. (2011) engineered multifunctional core–shell drug capsules using coaxial electro-spraying. These capsules comprised olive oil as a reservoir for oil-soluble pharmaceuticals, magnetite (Fe_3O_4), CDs, and a dual-layer porous titania (TiO_2) shell for precise drug administration. CDs incorporated into these carriers emitted in situ fluorescent signals for monitoring medication release. The amino groups on the surface of CDs enabled targeted interactions with tumor cells, hence minimizing damage from drug leakage into healthy cells.

In a separate study conducted by Jia et al. (2016), CDs were produced via a microwave pyrolysis process and subsequently cross-linked with PEGylated oxidized alginate. The modified CDs were conjugated with doxorubicin (DOX) for the purpose of anticancer medication delivery. In vitro studies indicated that DOX CDs could release medicines in a pH-dependent way within the acidic microenvironment of malignancies. Their fluorescent characteristics facilitated imaging-guided drug distribution during tumor treatment, underscoring the significance of CDs in imaging-guided drug administration.

A formulation of CDs combined with doxorubicin, cisplatin, and docetaxel has been created, demonstrating significant tumor targeting, enhanced anti-tumor efficacy, and reduced adverse effects (Yang et al., 2016; Mahani et al., 2021; Sawpari et al., 2023).

Chandrasekaran et al. (2022) documented the application of nitrogen-doped carbon quantum dots (NDCQDs) as carriers for the delivery of quercetin (QC). In vitro investigations demonstrated a notable decrease in the survival rates of colon carcinoma cell lines, principally attributable to the enhanced absorption of QC in the NDCQDs–FA–QC combination and elevated cumulative drug concentrations within tumors.

CDs are also manufactured from graphene-based materials (GQD) that possess graphene-like properties and can be employed in biomedical applications. Graphene-based carbon quantum dots possess many carboxylic groups, facilitating their functionalization with active biomolecules (Kortel et al., 2020). This renders them advantageous for therapy development and drug delivery. Liu et al. (2015a, b) demonstrated the efficacy of graphene-based CDs as a low-cytotoxicity inhibitor for the aggregation of Aß peptides, which stick to the peptides' hydrophobic core. Functionalized GQDs have proven highly effective in identifying cancer receptors and selectively delivering chemotherapeutic agents, such as doxorubicin (DOX) or cisplatin, to the cell nucleus, thereby enhancing cytotoxicity, preventing unintended transport into normal tissues, and inhibiting drug resistance (Deng et al., 2018). Graphene-based CDs, extensively utilized in photodynamic therapy (PDT), have demonstrated considerable therapeutic advantages (Choi et al., 2014).

The management of brain tumors and neurodegenerative disorders is complicated by the impermeability of the blood–brain barrier. CDs have been utilized alongside tumor-penetrating medicines for the treatment of brain malignancies. RGERPPR penetrating peptides conjugated with CDs were employed to penetrate tumor vascular walls (Yang et al., 2013). Gao et al. (2018) synthesized CDs functionalized with RGERPPR and maleimide polyethylene glycol amino succinimide succinate, successfully facilitating glioma targeting and in vivo bioimaging.

4.3.3 Gene Therapy

Gene therapy represents a cutting-edge method for rectifying gene expression and addressing a range of diseases. In gene therapy, various types of nanoparticles are utilized to transport genes to targeted DNA locations. Gene therapy is considered a long-lasting and potentially curative treatment approach for various disorders (Dunbar et al., 2018). This technique utilizes a vector that demonstrates high efficiency in gene transfection, enabling the transfer of genetic components within cells. Recent reports indicate that CDs can serve as gene nanocarriers that are trackable through imaging (Ghosh et al., 2019; Han & Na, 2019; Pham et al., 2023). Consequently, among the diverse range of nanoparticles, CDs have garnered significant interest in the field of gene therapy as a nonviral vector.

Liu et al. (2012) utilized a straightforward microwave-assisted pyrolysis technique to produce CDs. The remaining branched PEI on CDs enhanced their fluorescence intensity and imparted a positive charge to the CDs. The positive charge facilitated the condensation of DNA on the CD surface through electrostatic interactions with PEI. As a result, DNA may be internalized by HepG2 cells, leading to elevated gene expression levels relative to control PEI.

Cheng et al. (2014) formulated a DNA coagulant utilizing polycationic-b-polysulfobetaine grafted carbon quantum dots, which impeded protein adsorption and preserved gene transfection efficacy at elevated serum concentrations. These CD-based nanomaterials demonstrated exceptional fluorescence emission within transfected cells and were very appropriate for therapeutic stealth applications.

Ghosh et al. (2019) synthesized CD-PAMAM conjugates (CDPs) by utilizing sweet lemon peels as a renewable source for CDs and conjugating them with many generations of polyamidoamine (PAMAM) dendrimers (Ghosh et al., 2019). Among the conjugates of CDP, CDP3 emerges as a promising gene transfer vector for triple-negative breast cancer (TNBC) gene therapy due to its superior gene complexation and protective capabilities, minimal toxicity, blood compatibility, and improved gene transfection efficacy, thereby presenting an intriguing theranostic approach for future treatment (Ghosh et al., 2019). Moreover, CDP3 exhibited highly specific detection of Cu (II) ions, potentially aiding in the assessment of the metastatic stage of TNBC, characterized by elevated concentrations of Cu (II) ions.

Rezaei and Hashemi (2021) engineered chitosan-derived carbon dots functionalized with arginine to perform as an effective pseudo-homogeneous gene transfer vehicle (Rezaei & Hashemi, 2021). The CQDs coated with poly-l-lysine (PLL), known as CD/PLL core–shell nanoparticles, demonstrate significant potential for gene delivery applications owing to their robust photoluminescence properties in both solid and solution states under physiological conditions, alongside their remarkable cellular uptake in human embryonic kidney −293T (HEK-293T) cells without causing toxicity (Hasanzadeh et al., 2021).

4.3.4 Combating Pathogenic Microbes

There has been an alarming rise in antimicrobial resistance (AMR) among human pathogenic microbes, largely due to widespread use of different types of antibiotics and other pharmaceutical products (Murray et al., 2022; Salam et al., 2023; Sharma et al., 2024). The pathogenic microorganisms exhibiting AMR for multiple different drugs or multidrug resistance are difficult to treat with the conventional arsenal of antimicrobial compounds, posing a significant threat to global health (IALP, 2025). Therefore, it is imperative to develop new and effective alternatives to conventional antibiotics (Ruiz-Pérez et al., 2024). Nanomaterials, including 0-, 1-, and 2-dimensional nanostructures, have showcased promising antimicrobial properties to combat pathogenic infections and diseases (Gao & Zhang, 2021; Konwar et al., 2022). These materials exhibit antimicrobial properties, which could be traced down to a few physical, biochemical, or molecular mechanisms (Gajic et al., 2025; León-Buitimea et al., 2020). Among the emerging nanoscale antimicrobial materials, fluorescent carbon dots have garnered considerable attention due to their unique properties, as discussed in Sects. 4.1 and 4.2 of this chapter. The antimicrobial potential of the C dots can be efficiently utilized for combating a wide range of pathogens.

4.3.4.1 Bacterial Pathogens

The CDs exhibit broad-spectrum antibacterial efficiency through diverse mechanisms for both Gram-positive and Gram-negative bacterial genera (Hussen et al., 2024). However, the extent of antibacterial efficacy and the effective concentrations vary among the two types of bacterial genera owing to differences in the cell wall structure and chemical composition (Wang et al., 2025). The presence of lipopolysaccharide-based outer membrane in Gram-negative bacteria confers a strong negative surface charge, enabling electrostatic binding of the positively charged carbon dots, which renders effective membrane disruption and allows the CDs to trespass through the barrier structures to reach inside the cell (Vadakkan et al., 2025). Furthermore, the thin peptidoglycan cell wall layer facilitates easier movement of the CDs from the external milieu inside the cell cytoplasm (Lin et al. 2019). A study on fluorescent CDs derived from glucosamine hydrochloride and m-phenylenediamine precursors revealed a significantly higher uptake of the CDs by Gram-negative bacterial cultures (*E. coli*, *Klebsiella pneumoniae*, and *Pseudomonas aeuroginosa*) as compared to *S. aureus* (Samphire et al., 2022). Likewise, *Artemisia argyi* leaf-derived CDs exhibited selective high antibacterial potential against *Escherichia coli*, *Pseudomonas aeruginosa*, and *Proteusbacillus vulgaris* (Gram-negative bacterial genera) (Wang et al., 2020). However, the fluorescent dye-tagged biomass-derived CDs (FITC-CDs) exhibited high selectivity for adsorption/internalization in Gram-positive bacteria even in the presence of co-existing Gram-negative and eukaryotic mammalian cells (Xu et al., 2025). Another similar study showcased the selective adsorption and antibacterial potential of D-cysteine-derived chiral CDs for Gram-positive bacteria (Song et al., 2024).

Enhanced membrane lipid peroxidation events occur in Gram-negative bacteria in comparison to Gram-positive bacteria in response to supplementation of CDs in the growth media. A recent study involving the sour whey-derived CDs showcased lower minimum inhibitory concentrations (MIC) of the CDs for Gram-negative *Salmonella typhimurium* in comparison to Gram-positive *Listeria monocytogenes* (Esmaeili Koutamehr et al., 2023). Unlike Gram-negative genera, the presence of teichoic acid units in the Gram-positive bacteria imparts a negative charge to the cell wall, favoring positively charged CDs to interact effectively. On evaluation, the CDs@hematite nanostructures revealed a sustainable antibacterial potential preferentially for Gram-positive bacteria (Moradlou et al., 2019). Furthermore, surface functionalization or doping of the CDs can further enhance the antibacterial action spectrum with lower FICI values on synergistic or additive interactions of the loaded functional groups or molecules. For instance, treatment of *E. coli* and *Staphylococcus aureus* with the quartanized CDs obtained through one-pot synthesis using dimethyl diallyl ammonium chloride and glucose as precursors significantly downregulated the genes involved in the citrate pathway in *E. coli* (Zhao et al., 2022). Likewise, an in vivo mouse model study on acute peritonitis caused by *E. coli* involved intraperitoneal administration of quartanized CDs with promising results at par with that of the gentamicin antibiotic control (Zhang et al., 2023). However, another study involving the evaluation of curcumin-loaded folic acid CD nanocomposite as a potent antibacterial agent revealed a MIC of 12.5 μg/mL against both Gram-positive *S. aureus* and Gram-negative *Pseudomonas aeuroginosa* (Serag et al., 2024). Zinc-doping of the plant biomass mix (Joshina)-derived CDs exhibited zones of inhibition representing equivalent antibacterial efficacy for *S. aureus* and *Serratia marcescens* (Gram-negative) (Tariq et al., 2024). These studies invariably showcased the unique potential of different types of CDs as putative antibacterial agents involving low mammalian cell cytotoxicity and amenability for multifunctional integration (e.g., loading of drug molecules, photothermal/photodynamic activation, or scaffold incorporation) (Table 4.2). Therefore, clinical applications of CDs could be foreseen for wound therapies, development of antibacterial coatings, and control of resistant infections caused by multidrug-resistant bacterial pathogens.

4.3.4.2 Fungal Pathogens

Recent research findings indicated a strong antifungal potential of carbon dots with plausible applications in biomedicine, agriculture, and preservation industries. A recent study on the evaluation of microwave-assisted synthesis of N-doped CDs adsorbed on cellulosic matrix revealed that these CDs-cellulosic matrices exhibited considerable antifungal potential against *Postia placenta* and *Trametes versicolor* fungal genera through adsorption, followed by penetration, and a significant decrease in enzyme activities, which led to hyphal death (Zhao et al., 2025). Similar to interactions of CDs in prokaryotic bacteria, the predominant mechanism of antifungal activity of CDs involves interaction of the CDs with the lipid membranes, leading to disruption and leakage of the cell contents (Table 4.2). Thus, the membrane lipid–CDs interactions are similarly affected by the surface charge and size dimensions of CDs, with positively charged CDs being more effective in facilitating electrostatic adsorption of CDs to the negatively charged fungal cell walls. Gold-conjugated CDs (Au@CDs) demonstrated size-dependent antifungal activity against *Candida albicans* (Priyadarshini et al., 2018). The Au@CDs got efficiently internalized, as confirmed through the fluorescence signals, which can allow for real-time tracking of cellular uptake (Priyadarshini et al., 2018). The positively charged surface of the functionalized CDs can also inhibit the formation of biofilm due to the prevention of cell adhesion like yeast-like fungal genera such as *Candida* (Sturabotti et al., 2024a). Furthermore, mammalian cell line (COS-7 fibroblast)-based studies revealed that the CDs remained nontoxic to the mammalian fibroblasts while effectively reducing the fungal load in both planktonic and biofilm states (Sturabotti et al., 2024a).

The photodynamic induction of ROS by CDs (especially S or N-doped CDs) on exposure to UV or visible wavelengths leads to oxidative stress within fungal cells, which impairs cellular function and contributes significantly to cell death by disruption of the enzymatic activity and chelation of the Fe^{3+} ions to prevent Fenton reactions (Chen et al., 2024). DNA degradation in the CDs-treated hyphal cells is the other mechanism of antifungal activity. Degradable vitamin C-derived CDs were observed to penetrate through the cell wall and membrane to target the nucleus of two test plant pathogenic fungal genera, *Rhizoctonia solani* and *Pyricularia grisea*

Table 4.2 Antimicrobial potential of different types of carbon dots

Antimicrobial property	CD type and size	Functionalization	Organism(s)	MIC (µg/mL)	Reference
Antibacterial	NPS-codoped CDs (~5 nm)	N, P, S doping	*Salmonella typhi*, *Pseudomonas aeruginosa*, *Bacillus subtilis*	5–100	Pant et al. (2023)
	Quaternized CDs (~5 nm)	Quaternization	*Escherichia coli*	Not specified	Zhang et al. (2023)
	P-doped CDs (~5 nm)	Phosphorus doping	*Escherichia coli*, *Staphylococcus aureus*	1.23–1.44 mg/mL	Chai et al. (2021)
	Various CDs (~5–10 nm)	Various functionalizations	Various bacteria	Not specified	Yu et al. (2023)
Antifungal	CD-Gu^{+}-AmB (~5 nm, amphotericin B conjugate)	Guanidinium + Amphotericin B	*Candida albicans* (planktonic and biofilm)	Potent antifungal and antibiofilm activity	Li et al. (2019)
	Ultrasmall positively-charged CDs	Melt-synthesis, positive charge	*Candida albicans*	Significant antifungal efficacy in mice	Chen et al. (2024)
	N-doped carbon quantum dots	Nitrogen doping	*Postia placenta*, *Trametes versicolor*	Strong inhibition of fungal decay via ROS and membrane disruption	Zhao et al. (2025)
	~>10 nm, three classes	Varied surface groups; importantly positive charge	*Candida albicans*	In vitro inhibition of adhesion/ biofilm and better larval survival	Sturabotti, Camilli, Georgian Moldoveanu, et al. (2024a)
	Similar bioactive CDs	Same as above	*Candida albicans*	Confirmed antifungal bioactivity in vitro	Sturabotti, Camilli, Leonelli, and Vetica (2024b)

(continued)

Table 4.2 (continued)

Antimicrobial property	CD type and size	Functionalization	Organism(s)	MIC (µg/mL)	Reference
Antiviral	Porphyrin-derived CDs (size not specified)	Porphyrin	SARS-CoV-2 (N-protein dimer inhibition)	Inhibits nucleocapsid dimerization at 4–10 µg/mL; 75% reduction in CPE at 4 µg/mL	Fibriani et al. (2023)
	Cur-CQDs (~5 nm)	Curcumin-derived	Enterovirus 71	EC_{50}: 0.2 µg/mL ($CC_{50} \approx$ 452 µg/mL)	Lin, Chang, et al. (2019c)
	Glycyrrhizic acid CDs (~2–10 nm)	Glycyrrhizic acid	Influenza A, PRRSV, SARS-CoV surrogate	Broad antiviral activity (exact EC_{50} not specified)	Tong et al. (2020)
	Boronic acid (e.g., 4-APBA) CDs (~2–10 nm)	Boronic acids	HSV-1, HCoV-229E	Significant viral inhibition (reduced titers)	Gharpure and Ankamwar (2021)
	Cur-CQDs derived from curcumin	Curcumin	Japanese encephalitis virus	Blocks viral entry in vitro; binds E protein	Chen et al. (2022)

(Li et al., 2018). The internalized CDs damaged the nuclear membrane and chromatin structure, besides downregulating the non-ribosomal peptide synthetase (NRPS) genes, which are essential for fungal virulence and survival. Confocal microscopy confirmed the nuclear localization of these CDs, highlighting their potential as gene-targeting antifungal agents (Li et al., 2018). With low toxicity and high biocompatibility, CDs offer a safe and effective alternative to conventional antifungals. However, further research is needed to optimize their surface chemistry, improve species-specific targeting, and validate long-term safety and efficacy in clinical and agricultural settings.

4.3.4.3 Viral Pathogens

Several research findings indicated the virus-neutralizing effect of the CDs (Table 4.2). COVID-19, a viral lung illness characterized by significant respiratory distress and fever, has resulted in substantial global fatalities recently (Zhou et al., 2021). Researchers are developing swift diagnostic methods for instances when immune cells generate mediators (e.g., interleukins, α-defensins) that trigger inflammation.

CDs are utilized for the diagnosis and treatment of the condition (Xue et al., 2022). The herbs are superior to medicinal plants as precursors of CDs in combating COVID-19 (Łoczechin et al., 2019; Naik et al., 2022). For instance, curcumin-derived CDs showed efficacy against enterovirus (EV-71), whereas garlic-derived CQDs were subsequently shown to reduce inflammatory cytokines and inhibit coronavirus attachment and penetration (Lin et al., 2019a, b, c; Kalkal et al., 2021). CDs were generated using the hydrothermal treatment of ethylenediamine and citric acid as carbon precursors, followed by modification with boronic acid ligands, demonstrating concentration-dependent inactivation of coronaviruses (Łoczechin et al., 2019).

CDs were produced using the hydrothermal treatment of ethylenediamine and citric acid as carbon precursors, followed by modification with boronic acid ligands, demonstrating concentration-dependent inactivation of coronaviruses (Łoczechin et al., 2019).

4.3.4.4 Mechanisms of Antimicrobial Action

Carbon dots display antimicrobial activity by targeting microbial cells through both physical and chemical pathways (Hussen et al., 2024). The nanoscale dimensions of CDs allow for easy movement across the membrane barriers (cell membrane and other unit membranes enclosing the cell organelles) (Liu et al. 2020a, b, c); moreover, a size-dependent antibacterial activity has been reported (Du et al., 2023; Sun et al., 2021). The process is initiated due to the high surface area of the CDs and surface functionalization with different molecules, which enable electrostatic interactions with the microbial cell membrane, leading to adsorption on the cell surface and disruption of the membrane (particularly bacterial cell membranes) (Salvi et al., 2024). The predominant mode of antimicrobial activity is through oxidative burst stress caused due to the generation of a variety of radical moieties or reactive oxygen species (ROS), particularly superoxide anions and hydroxyl radicals on exposure to UV or visible light wavelengths (Wang et al., 2025). This refers to the photodynamic properties of the CDs, which can be tailored by doping of the hetero-atoms such as nitrogen or sulfur (Miao et al., 2020). The generated ROS moieties are highly reactive and interact with different macromolecules and structures in the cell, such as membrane lipids, cytoskeletal proteins, enzymes, and genomic DNA (Tumilaar et al., 2024). The lipids in the unit membranes are highly reactive in nature and are the primary molecules to interact with the ROS generated outside the cell surface (Duché & Sanderson, 2024), leading to lipid peroxidation, i.e., formation of malondialdehyde and disruption of the lipid bilayer (Juan et al., 2021). This causes leakage of cellular contents and eventually leads to cell death. Once internalized inside the cell, CDs can interfere with a variety of metabolic pathways due to the formation of complexes with the biomolecules and cause enzyme deactivation, thereby inhibiting different cellular processes, including the vital DNA replication and RNA transcription (Li et al., 2024). The DNA fragmentation and chromosomal aberrations are the other manifestations of the interaction of the CDs with the nucleic acid in the cell. Their tunable surface chemistry and amenability to functionalization with diverse molecules/compounds further enhance their specificity and efficacy against a broad spectrum of pathogens, including the notorious drug-resistant strains (Ghirardello et al., 2021; Guo et al., 2022).

4.3.5 Sensors

The interaction of CDs with diverse chemical substances can result in fluorescence quenching or amplification, mediated via photosensitization or FRET. This distinctive ability establishes CDs as effective fluorescence sensors for the detection of various chemical compounds, metal ions, pH levels, and additional parameters (da Silva & Gonçalves, 2011; Xiao & Sun, 2018).

As a result, considerable efforts have been focused on enhancing the detection sensitivity and selectivity of CDs. Moreover, their extraordinary biocompatibility, capacity to emit various luminous hues, excellent photostability, notable water solubility, and advanced surface modification capabilities have rendered them increasingly appealing for sensor applications (Choudhury, 2021; Ji et al., 2020).

The advantages of CDs include facile manufacture, cost-effective synthesis methods, straightforward functionalization processes, exceptional brightness, high quantum yield, chemical stability, and low toxicity (Zhou et al., 2021; Zhang et al., 2020).

4.3.5.1 Biosensors

CDs serve as biosensor carriers due to their exceptional water solubility, adaptability in surface modification, nontoxic nature, excitation-dependent multicolor emission, superior biocompatibility, effective cell permeability, and remarkable photostability. CD-based biosensors provide the visual monitoring of glucose (Shi et al., 2011), cellular copper (Zhu et al., 2011), phosphate (Zhao et al., 2011), iron (Zhu et al.,

2013), potassium (Wei et al., 2012), pH (Shi et al., 2012), and nucleic acids (Li et al., 2011).

A groundbreaking work conducted in 2011 revealed the capacity of CDs to identify nucleic acids (Li et al., 2011). Consequently, CDs have been acknowledged as superior fluorescence sensors for the detection of biomolecules. This detection is generally accomplished by FRET and fluorescence quenching, which convert intangible analyte information into discernible optical signals, like variations in fluorescence intensity, wavelength, or hue. Single-stranded DNA exhibits a pronounced affinity for adsorption onto the surface of carbon quantum dots, leading to fluorescence quenching as a result of π–π interactions. When single-stranded DNA hybridizes with its complementary strand to produce double-stranded DNA, the π–π interactions diminish, resulting in the detachment of double-stranded DNA from the CD surface and a progressive recovery of fluorescence. This characteristic facilitates the identification of target DNA (Li et al., 2011; Loo et al., 2016).

A versatile and resilient CD-based fluorescence resonance energy transfer (FRET) sensor for the detection and imaging of mitochondrial H_2O_2 was demonstrated. The CQDs function as the energy transfer donor and the carrier for the sensing system. A boronate-based H_2O_2 recognition element, boronate-protected fluorescein, was covalently attached to the CDs (Du et al., 2014). This method can be employed to monitor exogenous H_2O_2 levels in L929 cells and to visualize endogenously generated H_2O_2 in RAW 264.7 macrophage cells.

Dopamine is an essential neurotransmitter that plays a critical function in the central nervous system and is associated with neurological diseases, including Parkinson's disease. Tammina et al. (2019) presented N, P codoped CDs for the precise detection of dopamine in blood and urine, achieving a low limit of detection (LOD) of 0.021 mM without interference from possible impurities. Dopamine is oxidized to dopamine–quinone during the detection process in the presence of O_2 or a basic environment. This oxidation influences the recombination kinetics of the separated electrons and holes in CDs, resulting in the suppression of CQD fluorescence.

Cheng et al. (2018) recently developed a photo-electrochemical apta-sensor for thrombin detection with a CDs/TiO_2/ITO electrode functionalized with carboxyl-functionalized thrombin. In this configuration, CDs function as a photoactive matrix to augment the visible light consumption of the apta-sensor. Thrombin can be identified with a low limit of detection of 0.83 pM by observing variations in photocurrent.

Hu et al. (2022) developed a pH-responsive fluorescence sensor utilizing N-CDs for the assessment of milk freshness. The mechanism of fluorescence quenching transitions gradually from dynamic to static with a fall in pH. This alteration is ascribed to the inhibition of H+ transport from nitrogen atoms to the conjugated framework. As a result, N-CDs were affixed to filter papers or starch-gel membranes to develop portable fluorescence sensors. These sensors produce human-readable signals that transition from vibrant yellow-green to deep yellow and ultimately to blue when milk freshness diminishes under UV light exposure.

4.3.5.2 Chemical Sensors

CDs have demonstrated versatility as chemical sensors for the detection of pH levels and metal ions. In these applications, the energy transfer mechanisms between the analytes and the surface groups of CDs result in the generation of new electron–hole pairs and subsequent modifications in the fluorescence of CDs (Ghosh et al., 2022; Wang et al., 2021; Ma et al., 2023).

Nayak et al. (2021) demonstrated a visually discernible pH-dependent modulation of PL associated with the aggregation-induced emission of CDs. CDs, passivated with amine and carboxylic groups, aggregated into micron-sized three-dimensional clusters upon heating, generating a vivid yellow light at neutral pH. Under acidic conditions, emission quenching transpired due to hydrogen bonding linked to the amide group, promoting the nonradiative recombination of excited species. In contrast, in an alkaline environment, the interconnected CDs were deconstructed due to the hydrolysis of the amine-conjugated network. This disintegration yielded free CDs with amino, carboxylic, and hydroxyl groups on their surfaces, leading to blue emission.

CDs have been engineered as highly efficient chemical sensors for various metal ions, utilizing the static quenching effect that arises from the production of nonfluorescent complexes between CDs and metal ions (Wang et al., 2017).

Typically, surface-oxidized CDs are utilized for the detection of Cu^{2+} or Fe^{3+}, although highly toxic heavy metal ions, including Hg^{2+}, Ag^{+}, Pb^{2+}, and Cr (VI), can be detected with great sensitivity by heteroatom-doped CDs (Luo et al., 2018). N, S codoped CDs have been demonstrated to facilitate electron transfer and augment the coordination between CDs and Hg^{2+}, resulting in enhanced fluorescence quenching and attaining an exceptional limit of detection for Hg^{2+} as low as 2.0 μM (Li et al., 2015).

4.3.6 Energy Storage

4.3.6.1 Solar Cells

Solar cells (SCs) have experienced three generations of development to date. The initial generation of solar cells consists of crystalline SiO_2, featuring a core of monocrystalline SiO_2. Nonetheless, crystalline SiO_2 is prohibitively expensive, and the manufacture of cells incurs significant costs. SiO_2, an indirect band-gap material, exhibits a significantly lower absorption coefficient in the visible spectrum compared to other optoelectronic materials (Green, 2002). Due to the deficiencies in crystalline SiO_2 solar cells, researchers commenced investigations into thin-film solar cells, marking the onset of the second phase of development. Materials possessing a high absorption coefficient and a direct band-gap configuration are utilized in thin-film solar cells. The advantages of thin-film solar cells encompass their reduced thickness, cost-effectiveness, and diverse varieties. Currently, there exist GaAs thin-film solar cells, CdTe thin-film solar cells, and others (Imamzai et al., 2012). Thin-film solar cells, conversely, possess disadvantages like restricted reserves of rare elements, challenging extraction processes, and certain materials that are environmentally hazardous. The evolution of solar cells has reached its third generation, encompassing dye-sensitized solar cells (DSSCs), polymer solar cells, perovskite solar cells, and quantum dot solar cells. Typically, manufacturing cost, performance, and durability are utilized to evaluate solar devices. The primary challenge is to equilibrate these parameters in solar devices to ensure their contribution to the global energy sector. Notwithstanding the numerous advantages of PSCs, more enhancements will be highly beneficial for their potential commercial use. Nontoxic materials, diminished charge carrier recombination, large-scale manufacturing, and enhanced stability under environmental conditions must also be addressed (Shi & Jayatissa, 2018).

4.3.6.2 Dye-Sensitized Solar Cells

DSCs have garnered significant attention owing to their diversity, affordability, and ease of processing (Zhang et al., 2013). Despite the advantages of dye-sensitized solar cells (DSCs) stemming from the variety of organic dyes and their commendable efficiency, issues such as the photobleaching of organic dyes, the expense and toxicity of ruthenium-based dyes, and the volatility of electrolytes may impede their widespread implementation. Stable light-absorbing CDs derived from abundant and inexpensive sources demonstrate their potential in dye-sensitized solar cells (DSCs).

Consistent optical absorption of CDs derived from several economical materials demonstrates potential in DSCs (Riaz et al., 2019). A CD-interconnected dye and semiconductor complex system was established in 2013 for the advancement of highly efficient photoelectric conversion technologies. Due to the integrated hyperchromic effect between Rhodamine B (RhB) and CDs, CDs enhance the UV-visible absorption of RhB solutions. The CDs function as an electron transfer mediator, facilitating the movement of photogenerated electrons and mitigating their recombination, hence significantly enhancing photoelectric conversion efficiency. The incorporation of CDs into the dye and semiconductor complex enhanced the complex's photoelectric conversion efficiency sevenfold (Ma et al., 2013).

Geleta and Imae (2021) documented the utilization of n-type ZnO and p-type NiO semiconductor nanocomposites as photo-anodes in dye-sensitized solar cells to enhance power conversion efficiency. The influence of CDs on the n-ZnO and p-NiO nanocomposite was examined. The use of 8 wt% NiO resulted in an efficiency that was 1.58 times greater than that of ZnO. Doping CDs in ZnO/NiO nanocomposites (8 wt %) enhanced the efficiency by a factor of 3.8 compared to the undoped ZnO/NiO nanocomposite.

4.3.6.3 Organic Solar Cells

Organic solar cells (OSCs) are also referred to as polymer solar cells. They can be extensively manufactured through the roll-to-roll method and are characterized by the device's flexibility and reduced cost (Vercelli, 2021). CDs have recently been employed in organic solar cells as an electron acceptor (Cui et al., 2018), active-layer dopant (Liu et al., 2014), hole transport layer (HTL) (Li et al., 2013), electron transport layer (ETL) (Pali et al., 2018), interfacial modification layer (Lin et al.,

2016), and dopant for the buffer layer (Yan et al., 2016), demonstrating their remarkable utility in organic solar cells (OSCs).

Nonetheless, it is crucial to acknowledge that most CD synthetic approaches for OSCs are hindered by rigorous operational parameters, potentially limiting their broader practical application. Nonetheless, although numerous researchers have concentrated on undoped and N-doped carbon-based nanomaterials for application in organic solar cells (OSCs), the findings document the hydrothermal synthesis of N,SCQDs (Riaz & Park, 2022).

Wang et al. (2019a, b) documented that N,S-CD-modified ZnO ETL cells had a superior power conversion efficiency (PCE) of 9.31%, surpassing that of the unmodified ZnO cells (Wang et al., 2019a). Zhang et al. (2018) documented the superior performance of inverted poly-SCs utilizing a bilayer ZnO/CQDs electron extraction layer (EEL). The inverted polymer composite blend solar cells exhibit a notable increase in power conversion efficiency (PCE) of 9.64%, representing a 27% improvement over the control cell's efficiency of 7.59% (Zhang et al., 2018). Conversely, the spectral ranges of both photoluminescence and excitation of the CD-filled polysiloxane composite align effectively with the harvest spectrum (480–650 nm) and the unused optical spectrum (380–480 nm) of the bulk heterojunction (BHJ) solar cell (Zhang et al., 2017), as the CD-filled polysiloxane composite is capable of emitting light with wavelengths between 400 and 650 nm.

4.3.6.4 Supercapacitors

Supercapacitors differ from batteries since they are energy-storage devices that maintain high output power over prolonged durations (Raza et al., 2018). The creation of a nanomaterial-based electrode with enhanced specific capacitance and improved long-term stability represents a superior option for supercapacitors.

CDs are frequently utilized in supercapacitors due to their tunable size and structure, as well as their superior electrical conductivity. Consequently, employing metal oxide/sulfide and CDs to fabricate nanohybrid structures has proven to be an effective method for producing efficient and reliable supercapacitors. A sustainable method for synthesizing CD-MnO_2 nanostructures from an eco-friendly waste source to achieve elevated power density and specific capacitance has been published. CD-MnO_2 exhibited a superior surface area and improved electrical conductivity relative to pristine MnO_2, as demonstrated by the highly conductive CDs in the structural analysis (Xu et al. 2017a, b). CDs can augment the electrical conductivity of metal oxide/sulfide hybrids and simultaneously enhance the electrochemical characteristics of energy storage systems.

The use of CDs-based ionic liquid (CQDIL) as a conductive agent and binder in an activated carbon electrode enhances supercapacitor performance by improving electrolyte wettability on the CQDIL-functionalized activated carbon electrode, as well as increasing the number of active components and conductivity in the CQDIL/AC electrode. A power density of 1.64 kW kg^{-1} and an energy density of 83.2 Wh kg^{1} were achieved (Wei et al., 2013).

4.4 Conclusion and Future Perspectives

In recent years, carbon dots (CDs) have gained popularity owing to their unique characteristics. This chapter provides a deep insight into the production techniques of CDs using conventional and AI and ML-assisted protocols, along with their application in varied fields such as bioimaging, drug delivery, gene therapy, combating COVID-19, sensors, and energy storage devices. Among the most sought-after properties, QY is regarded as a significant characteristic that is contingent upon several factors, including structure, surface functionality, size, morphology, and structural defects. Nonetheless, nanoparticle synthesis often entails several chemicals and interrelated experimental parameters, including reagent concentrations, reaction duration, pH conditions, precursor involved, temperature, and mixing efficacy, making accurate synthesis difficult. By use of a dataset including experimental parameters and reaction conditions, ML models are trained to identify ideal conditions for the synthesis of CDs with high quantum yields. The ML approaches were proven to be more efficient and reliable than conventional trial-and-error techniques. In the future, machine learning is expected to play a more crucial role in the synthesis of CDs. As technology advances with enhanced computing capabilities and more intricate algorithms, it is expected that the CDs with desired precursor, high QY, and desired emission wavelength (color) can be prepared in an economical and eco-friendly manner.

References

Bao, L., Zhang, Z. L., Tian, Z. Q., Zhang, L., Liu, C., Lin, Y., Qi, B., & Pang, D. W. (2011). Electrochemical tuning of luminescent carbon nanodots: From preparation to luminescence mechanism. *Advanced Materials (Deerfield Beach, Fla.), 23*(48), 5801–5806. https://doi.org/10.1002/adma.201102866

Bao, L., Liu, C., Zhang, Z. L., & Pang, D. W. (2015). Photoluminescence-tunable carbon nanodots: Surface-state energy-gap tuning. *Advanced Materials (Deerfield Beach, Fla.), 27*(10), 1663–1667. https://doi.org/10.1002/adma.201405070

Bian, Z., Bao, T., Sun, X., Wang, N., Mu, Q., Jiang, T., Yu, Z., Ding, J., Wang, T., & Zhou, Q. (2024). Machine learning tools to assist the synthesis of antibacterial carbon dots. *International Journal of Nanomedicine, 19*, 5213. https://doi.org/10.2147/IJN.S451680

Bozrova, S. V., Baryshnikova, M. A., Sokolova, Z. A., Nabiev, I. R., & Sukhanova, A. V. (2018). In vitro cytotoxicity of CdSe/ZnS quantum dots and their interaction with biological systems. *KnE Energy, 3*(2), 58–63. https://doi.org/10.18502/ken.v3i2.1792

Breiman, L. (2001). Random forests. *Machine Learning, 45*, 5–32. https://doi.org/10.1023/A:1010933404324

Cao, L., Wang, X., Meziani, M. J., Lu, F., Wang, H., Luo, P. G., Lin, Y., Harruff, B. A., Veca, L. M., Murray, D., Xie, S. Y., & Sun, Y. P. (2007). Carbon dots for multiphoton bioimaging. *Journal of the American Chemical Society, 129*(37), 11318–11319. https://doi.org/10.1021/ja073527l

Cao, L., Meziani, M. J., Sahu, S., & Sun, Y. P. (2013). Photoluminescence properties of graphene versus other carbon nanomaterials. *Accounts of Chemical Research, 46*(1), 171–180. https://doi.org/10.1021/ar300128j

Chae, A., Choi, Y., Jo, S., Nur'aeni, N., Paoprasert, P., Park, S. Y., & In, I. (2017). Microwave-assisted synthesis of fluorescent carbon quantum dots from an A2/B3 monomer set. *RSC Advances, 7*, 12663–12669. https://doi.org/10.1039/C6RA28176A

Chai, S., Zhou, L., Pei, S., Zhu, Z., & Chen, B. (2021). P-Doped carbon quantum dots with antibacterial activity. *Micromachines, 12*, 1116. https://doi.org/10.3390/mi12091116

Chandrasekaran, P., Sivaraman, G., Rasala, S., Sethuraman, M. G., Kotla, N. G., & Rochev, Y. (2022). Quercetin conjugated fluorescent nitrogen-doped carbon dots for targeted cancer therapy application. *Soft Matter, 18*(30), 5645–5653. https://doi.org/10.1039/d2sm00747a

Chen, Y., Lian, H., Wei, Y., He, X., Chen, Y., Wang, B., Zeng, Q., & Lin, J. (2018). Concentration-induced multi-colored emissions in carbon dots: Origination from triple fluorescent centers. *Nanoscale, 10*(14), 6734–6743. https://doi.org/10.1039/c8nr00204e

Chen, Y., Cao, Y., Ma, C., & Zhu, J. J. (2020). Carbon-based dots for electrochemiluminescence sensing. *Materials Chemistry Frontiers, 4*(2), 369–385. https://doi.org/10.1039/C9QM00572B

Chen, H.-H., Lin, C.-J., Anand, A., Lin, H.-J., Lin, H.-Y., Mao, J.-Y., Wang, P.-H., Tseng, Y. J., Tzou, W.-S., Huang, C.-C., & Wang, R. Y. L. (2022). Development of antiviral carbon quantum dots that target the Japanese encephalitis virus envelope protein. *Journal of Biological Chemistry, 298*, 101957. https://doi.org/10.1016/j.jbc.2022.101957

Chen, J., Luo, J. B., Hu, M. Y., Zhou, J., Huang, C. Z., & Liu, H. (2023). Controlled synthesis of multicolor carbon dots assisted by machine learning. *Advanced Functional Materials, 33*(2), 2210095. https://doi.org/10.1002/adfm.202210095

Chen, H., Geng, X., Ning, Q., Shi, L., Zhang, N., He, S., Zhao, M., Zhang, J., Li, Z., Shi, J., & Li, J. (2024). Biophilic positive carbon dot exerts antifungal activity and augments corneal permeation for fungal keratitis. *Nano Letters, 24*, 4044–4053. https://doi.org/10.1021/acs.nanolett.4c01042

Cheng, L., Li, Y., Zhai, X., Xu, B., Cao, Z., & Liu, W. (2014). Polycation-b-polyzwitterion copolymer grafted luminescent carbon dots as a multifunctional platform for serum-resistant gene delivery and bioimaging. *ACS Applied Materials & Interfaces, 6*(22), 20487–20497. https://doi.org/10.1021/am506076r

Cheng, W., Pan, J., Yang, J., Zheng, Z., Lu, F., Chen, Y., & Gao, W. (2018). A photoelectrochemical aptasensor for thrombin based on the use of carbon quantum dot-sensitized TiO_2 and visible-light photoelectrochemical activity. *Mikrochimica Acta, 185*(5), 263. https://doi.org/10.1007/s00604-018-2800-z

Choi, Y., Kim, S., Choi, M. H., Ryoo, S. R., Park, J., Min, D. H., & Kim, B. S. (2014). Highly biocompatible carbon nanodots for simultaneous bioimaging and targeted photodynamic therapy in vitro and in vivo. *Advanced Functional Materials, 24*(37), 5781–5789.

Choudhury, S. D. (2021). Synthesis, properties and applications of luminescent carbon dots. In A. K. Tyagi & R. S. Ningthoujam (Eds.), *Handbook on synthesis strategies for advanced materials* (Indian Institute of Metals Series) (pp. 421–460). Springer. https://doi.org/10.1007/978-981-16-1892-5_11

Criminisi, A., Shotton, J., & Konukoglu, E. (2012). Decision forests: A unified framework for classification, regression, density estimation, manifold learning and semi-supervised learning. *Now: Foundations and Trends in Computer Graphics, 7*, 81–227. http://ieeexplore.ieee.org/document/8187032

Cui, B., Yan, L., Gu, H., Yang, Y., Liu, X., Ma, C.-Q., Chen, Y., & Jia, H. (2018). Fluorescent carbon quantum dots synthesized by chemical vapor deposition: An alternative candidate for electron acceptor in polymer solar cells. *Optical Materials, 75*, 166–173. https://doi.org/10.1016/j.optmat.2017.10.010

da Silva, J. C. E., & Gonçalves, H. M. (2011). Analytical and bioanalytical applications of carbon dots. *TrAC Trends in Analytical Chemistry, 30*(8), 1327–1336. https://doi.org/10.1016/j.trac.2011.04.009

Dager, A., Uchida, T., Maekawa, T., & Tachibana, M. (2019). Synthesis and characterization of mono-disperse carbon quantum dots from fennel seeds: Photoluminescence analysis using machine learning. *Scientific Reports, 9*(1), 14004. https://doi.org/10.1038/s41598-019-50397-5

Deng, T., Wang, J., Bao, F., Song, X., Deng, D., Gu, Y., & Zhang, R. (2018). Carbon dots-cluster-DOX nanocomposites fabricated by a co-self-assembly strategy for tumor-targeted bioimaging and therapy. *Particle & Particle Systems Characterization, 35*(9), 1800190. https://doi.org/10.1002/ppsc.201800190

Dhenadhayalan, N., Lin, K. C., Suresh, R., & Ramamurthy, P. (2016). Unravelling the multiple emissive states in citric acid-derived carbon-dots. *Journal of Physical Chemistry C, 120*(2), 1252–1261. https://doi.org/10.1021/acs.jpcc.5b08516

Ding, H., Yu, S. B., Wei, J. S., & Xiong, H. M. (2016). Full-color light-emitting carbon dots with a surface-state-controlled luminescence mechanism. *ACS Nano, 10*(1), 484–491. https://doi.org/10.1021/acsnano.5b05406

Du, F., Min, Y., Zeng, F., Yu, C., & Wu, S. (2014). A targeted and FRET-based ratiometric fluorescent nanoprobe for imaging mitochondrial hydrogen peroxide in living cells. *Small (Weinheim an der Bergstrasse, Germany), 10*(5), 964–972. https://doi.org/10.1002/smll.201302036

Du, X., Zhang, M., Ma, Y., Wang, X., Liu, Y., Huang, H., & Kang, Z. (2023). Size-dependent antibacterial of carbon dots by selective absorption and differential oxidative stress of bacteria. *Journal of Colloid and Interface Science, 634*, 44–53. https://doi.org/10.1016/j.jcis.2022.12.025

Duché, G., & Sanderson, J. M. (2024). The chemical reactivity of membrane lipids. *Chemical Reviews, 124*, 3284–3330. https://doi.org/10.1021/acs.chemrev.3c00608

Dunbar, C. E., High, K. A., Joung, J. K., Kohn, D. B., Ozawa, K., & Sadelain, M. (2018). Gene therapy comes of age. *Science (New York, N.Y.), 359*(6372), eaan4672. https://doi.org/10.1126/science.aan4672

Esmaeili Koutamehr, M., Moradi, M., Tajik, H., Molaei, R., Khakbaz Heshmati, M., & Alizadeh, A. (2023). Sour whey-derived carbon dots; synthesis, characterization, antioxidant activity and antimicrobial performance on foodborne pathogens. *LWT, 184*, 114978. https://doi.org/10.1016/j.lwt.2023.114978

Fan, Z., Li, Y., Li, X., Fan, L., Zhou, S., Fang, D., & Yang, S. (2014). Surrounding media sensitive photoluminescence of boron-doped graphene quantum dots for highly fluorescent dyed crystals, chemical sensing and bioimaging. *Carbon, 70*, 149–156. https://doi.org/10.1016/j.carbon.2013.12.085

Fibriani, A., Taharuddin, A. A. P., Yamahoki, N., Stephanie, R., Laurelia, J., Agustiyanti, D. F., Wisnuwardhani, P. H., Angelina, M., Rubiyana, Y., Ningrum, R. A., Wardiana, A., Desriani, D., Iskandar, F., Permatasari, F. A., & Giri-Rachman, E. A. (2023). Porphyrin-derived carbon dots for an enhanced antiviral activity targeting the CTD of SARS-CoV-2 nucleocapsid. *Journal of Genetic Engineering and Biotechnology, 21*, 93. https://doi.org/10.1186/s43141-023-00548-z

Friedman, J. H. (2001). Greedy function approximation A gradient boosting machine. *The Annals of Statistics, 29*(5), 1189–1232. http://www.jstor.org/stable/2699986

Friedman, J. H. (2002). Stochastic gradient boosting. *Computational Statistics & Data Analysis, 38*(4), 367–378. https://doi.org/10.1016/S0167-9473(01)00065-2

Gajic, I., Kekic, D., Jankovic, M., Tomic, N., Skoric, M., Petrovic, M., Mitic Culafic, D., Opavski, N., Ristivojevic, P., Krstic Ristivojevic, M., & Lukovic, B. (2025). Nature's arsenal: Uncovering antibacterial agents against antimicrobial resistance. *Antibiotics, 14*, 253. https://doi.org/10.3390/antibiotics14030253

Gao, W., & Zhang, L. (2021). Nanomaterials arising amid antibiotic resistance. *Nature Reviews Microbiology, 19*, 5–6. https://doi.org/10.1038/s41579-020-00469-5

Gao, L., Zhao, X., Wang, J., Wang, Y., Yu, L., Peng, H., & Zhu, J. (2018). Multiple functionalized carbon quantum dots for targeting glioma and tissue imaging. *Optical Materials, 75*, 764–769. https://doi.org/10.1016/j.optmat.2017.11.044

Ge, G., Li, L., Wang, D., Chen, M., Zeng, Z., Xiong, W., Wu, X., & Guo, C. (2021). Carbon dots: Synthesis, properties and biomedical applications. *Journal of Materials Chemistry. B, 9*(33), 6553–6575. https://doi.org/10.1039/d1tb01077h

Geleta, T. A., & Imae, T. (2021). Nanocomposite photoanodes consisting of p-NiO/n-ZnO heterojunction and carbon quantum dot additive for dye-sensitized solar cells. *ACS: Applied Nano Materials, 4*, 236–249. https://doi.org/10.1021/acsanm.0c02547

Gharpure, S., & Ankamwar, B. (2021). Use of nanotechnology in combating coronavirus. *3 Biotech, 11*, 358. https://doi.org/10.1007/s13205-021-02905-6

Ghirardello, M., Ramos-Soriano, J., & Galan, M. C. (2021). Carbon dots as an emergent class of antimicrobial agents. *Nanomaterials, 11*, 1877. https://doi.org/10.3390/nano11081877

Ghosh, S., Ghosal, K., Mohammad, S. A., & Sarkar, K. (2019). Dendrimer functionalized carbon quantum dot for selective detection of breast cancer and gene therapy. *Chemical Engineering Journal, 373*, 468–484. https://doi.org/10.1016/j.cej.2019.05.023

Ghosh, T., Das, T. K., Das, P., Banerji, P., & Das, N. C. (2022). Current scenario and recent advancement of doped carbon dots: A short review scientocracy update (2013–2022). *Carbon Letters, 32*(4), 953–977. https://doi.org/10.1007/s42823-022-00339-5

Green, M. A. (2002). Third generation photovoltaics: Solar cells for 2020 and beyond. *Physica E: Low-dimensional Systems and Nanostructures, 14*, 65–70. https://doi.org/10.1016/S1386-9477(02)00361-2

Guo, B., Liu, G., Hu, C., Lei, B., & Liu, Y. (2022). The structural characteristics and mechanisms of antimicrobial carbon dots: A mini review. *Materials Advances, 3*, 7726–7741. https://doi.org/10.1039/D2MA00625A

Guo, H., Lu, Y., Lei, Z., Bao, H., Zhang, M., Wang, Z., Guan, C., Tang, B., Liu, Z., & Wang, L. (2024). Machine learning-guided realization of full-color high-quantum-yield carbon quantum dots. *Nature Communications, 15*(1), 4843. https://doi.org/10.1038/s41467-024-49172-6

Han, J., & Na, K. (2019). Transfection of the TRAIL gene into human mesenchymal stem cells using biocompatible polyethyleneimine carbon dots for cancer gene therapy. *Journal of Industrial and Engineering Chemistry, 80*, 722–728. https://doi.org/10.1016/j.jiec.2019.02.015

Han, Y., Tang, B., Wang, L., Bao, H., Lu, Y., Guan, C., & Wu, M. (2020). Machine-learning-driven synthesis of carbon dots with enhanced quantum yields. *ACS Nano, 14*(11), 14761–14768. https://doi.org/10.1021/acsnano.0c01899

Hansen, N. (2006). Towards a new evolutionary computation. In J. A. Lozano, P. Larrañaga, I. Inza, & E. Bengoetxea (Eds.), *Studies in fuzziness and soft computing* (Vol. 192, pp. 75–102). Springer.

Hasanzadeh, A., Jahromi, M. A. M., Abdoli, A., Mohammad-Beigi, H., Fatahi, Y., Nourizadeh, H., & Karimi, M. (2021). Photoluminescent carbon quantum dot/poly-l-Lysine core-shell nanoparticles: A novel candidate for gene delivery. *Journal of Drug Delivery Science and Technology, 61*, 102118. https://doi.org/10.1016/j.jddst.2020.102118

Hou, Y., Lu, Q., Deng, J., Li, H., & Zhang, Y. (2015). One-pot electrochemical synthesis of functionalized fluorescent carbon dots and their selective sensing for mercury ion. *Analytica Chimica Acta, 866*, 69–74. https://doi.org/10.1016/j.aca.2015.01.039

Hu, X., Zhang, X., Li, Y., Shi, J., Huang, X., Li, Z., Zhang, J., Li, W., Xu, Y., & Zou, X. (2022). Easy-to-use visual sensing system for milk freshness, sensitized with acidity-responsive N-doped carbon quantum dots. *Foods (Basel, Switzerland), 11*(13), 1855. https://doi.org/10.3390/foods11131855

Huang, X., Zhang, F., Zhu, L., Choi, K. Y., Guo, N., Guo, J., Tackett, K., Anilkumar, P., Liu, G., Quan, Q., Choi, H. S., Niu, G., Sun, Y. P., Lee, S., & Chen, X. (2013). Effect of injection routes on the biodistribution, clearance, and tumor uptake of carbon dots. *ACS Nano, 7*(7), 5684–5693. https://doi.org/10.1021/nn401911k

Hussen, N. H., Hasan, A. H., FaqiKhedr, Y. M., Bogoyavlenskiy, A., Bhat, A. R., & Jamalis, J. (2024). Carbon dot based carbon nanoparticles as potent antimicrobial, antiviral, and anticancer agents. *ACS Omega, 9*, 9849–9864. https://doi.org/10.1021/acsomega.3c05537

IALP. (2025). Conference report on Antimicrobial resistance (AMR) Research Priorities & Action Plan. *Journal of Laboratory Physicians, 17*, 4. https://doi.org/10.25259/JLP_214_2024

Imamzai, M., Aghaei, M., Thayoob, Y. H. M., & Forouzanfar, M. (2012). A review on comparison between traditional silicon solar cells and thin-film CdTe solar cells. In *Proceedings of national graduate conference (Nat-Grad)* (pp. 1–5).

Ji, C., Zhou, Y., Leblanc, R. M., & Peng, Z. (2020). Recent developments of carbon dots in biosensing: a review. *ACS Sensors, 5*(9), 2724–2741. https://doi.org/10.1021/acssensors.0c01556

Jia, X., Pei, M., Zhao, X., Tian, K., Zhou, T., & Liu, P. (2016). PEGylated oxidized alginate-DOX prodrug conjugate nanoparticles cross-linked with fluorescent carbon dots for tumor theranostics. *ACS Biomaterials Science & Engineering, 2*(9), 1641–1648. https://doi.org/10.1021/acsbiomaterials.6b00443

Jiang, F., Chen, D., Li, R., Wang, Y., Zhang, G., Li, S., Zheng, J., Huang, N., Gu, Y., Wang, C., & Shu, C. (2013). Eco-friendly synthesis of size-controllable amine-functionalized graphene quantum dots with antimycoplasma properties. *Nanoscale, 5*(3), 1137–1142. https://doi.org/10.1039/c2nr33191h

Jiang, K., Sun, S., Zhang, L., Lu, Y., Wu, A., Cai, C., & Lin, H. (2015). Red, green, and blue luminescence by carbon dots: Full-color emission tuning and multicolor cellular imaging. *Angewandte Chemie, 54*(18), 5360–5363. https://doi.org/10.1002/anie.201501193

Jiang, K., Wang, Y., Gao, X., Cai, C., & Lin, H. (2018). Facile, quick, and gram-scale synthesis of ultralong-lifetime room-temperature-phosphorescent carbon dots by microwave irradiation. *Angewandte Chemie (International ed. in English), 57*(21), 6216–6220. https://doi.org/10.1002/anie.201802441

Jie, G., Zhou, Q., & Jie, G. (2019). Graphene quantum dots-based electrochemiluminescence detection of DNA using multiple cycling amplification strategy. *Talanta, 194*, 658–663. https://doi.org/10.1016/j.talanta.2018.10.098

Jing, Y., Zhu, Y., Yang, X., Shen, J., & Li, C. (2011). Ultrasound-triggered smart drug release from multifunctional core-shell capsules one-step fabricated by coaxial electrospray method. *Langmuir: The ACS Journal of Surfaces and Colloids, 27*(3), 1175–1180. https://doi.org/10.1021/la1042734

Ju, B., Nie, H., Liu, Z., Xu, H., Li, M., Wu, C., Wang, H., & Zhang, S. X. (2017). Full-colour carbon dots: Integration of multiple emission centres into single particles. *Nanoscale, 9*(35), 13326–13333. https://doi.org/10.1039/c7nr04576j

Juan, C. A., Pérez de la Lastra, J. M., Plou, F. J., & Pérez-Lebeña, E. (2021). The Chemistry of reactive oxygen species (ROS) revisited: Outlining their role in biological macromolecules (DNA, lipids and proteins) and induced pathologies. *International Journal of Molecular Sciences, 22*, 4642. https://doi.org/10.3390/ijms22094642

Kalkal, A., Allawadhi, P., Pradhan, R., Khurana, A., Bharani, K. K., & Packirisamy, G. (2021). *Allium sativum* derived carbon dots as a potential theranostic agent to combat the COVID-19 crisis. *Sensors International, 2*, 100102. https://doi.org/10.1016/j.sintl.2021.100102

Kim, S., Hwang, S. W., Kim, M. K., Shin, D. Y., Shin, D. H., Kim, C. O., Yang, S. B., Park, J. H., Hwang, E., Choi, S. H., Ko, G., Sim, S., Sone, C., Choi, H. J., Bae, S., & Hong, B. H. (2012). Anomalous behaviors of visible luminescence from graphene quantum dots: Interplay between size and shape. *ACS Nano, 6*(9), 8203–8208. https://doi.org/10.1021/nn302878r

Kim, S., Song, Y., & Heller, M. J. (2017). Seamless aqueous arc discharge process for producing graphitic carbon nanostructures. *Carbon, 120*, 83–88. https://doi.org/10.1016/j.carbon.2017.04.006

Konwar, A. N., Hazarika, S. N., Bharadwaj, P., & Thakur, D. (2022). Emerging non-traditional approaches to combat antibiotic resistance. *Current Microbiology, 79*, 330. https://doi.org/10.1007/s00284-022-03029-7

Kortel, M., Mansuriya, B. D., Vargas Santana, N., & Altintas, Z. (2020). Graphene quantum dots as flourishing nanomaterials for bio-imaging, therapy development, and micro-supercapacitors. *Micromachines, 11*(9), 866. https://doi.org/10.3390/mi11090866

LeCun, Y., Bengio, Y., & Hinton, G. (2015). Deep learning. *Nature, 521*(7553), 436–444. https://doi.org/10.1038/nature14539

León-Buitimea, A., Garza-Cárdenas, C. R., Garza-Cervantes, J. A., Lerma-Escalera, J. A., & Morones-Ramírez, J. R. (2020). The demand for new antibiotics: Antimicrobial peptides, nanoparticles, and combinatorial therapies as future strategies in antibacterial agent design. *Frontiers in Microbiology, 11*. https://doi.org/10.3389/fmicb.2020.01669

Li, H., He, X., Kang, Z., Huang, H., Liu, Y., Liu, J., Lian, S., Tsang, C. H., Yang, X., & Lee, S. T. (2010). Water-soluble fluorescent carbon quantum dots and photocatalyst design. *Angewandte Chemie, 49*(26), 4430–4434. https://doi.org/10.1002/anie.200906154

Li, H., Zhang, Y., Wang, L., Tian, J., & Sun, X. (2011). Nucleic acid detection using carbon nanoparticles as a fluorescent sensing platform. *Chemical Communications (Cambridge, England), 47*(3), 961–963. https://doi.org/10.1039/c0cc04326e

Li, M., Ni, W., Kan, B., Wan, X., Zhang, L., Zhang, Q., Long, G., Zuo, Y., & Chen, Y. (2013). Graphene quantum dots as the hole transport layer material for high-performance organic solar cells. *Physical Chemistry Chemical Physics: PCCP, 15*(43), 18973–18978. https://doi.org/10.1039/c3cp53283f

Li, L., Yu, B., & You, T. (2015). Nitrogen and sulfur co-doped carbon dots for highly selective and sensitive detection of Hg (II) ions. *Biosensors & Bioelectronics, 74*, 263–269. https://doi.org/10.1016/j.bios.2015.06.050

Li, H., Huang, J., Song, Y., Zhang, M., Wang, H., Lu, F., Huang, H., Liu, Y., Dai, X., Gu, Z., Yang, Z., Zhou, R., & Kang, Z. (2018). Degradable carbon dots with broad-spectrum antibacterial activity. *ACS Applied Materials & Interfaces, 10*, 26936–26946. https://doi.org/10.1021/acsami.8b08832

Li, X., Huang, R., Tang, F.-K., Li, W.-C., Wong, S. S. W., Leung, K. C.-F., & Jin, L. (2019). Red-emissive guanylated polyene-functionalized carbon dots arm oral epithelia against invasive fungal infections. *ACS Applied Materials & Interfaces, 11*, 46591–46603. https://doi.org/10.1021/acsami.9b18003

Li, H., Zhang, Y., Ding, J., Wu, T., Cai, S., Zhang, W., Cai, R., Chen, C., & Yang, R. (2022). Synthesis of carbon quantum dots for application of alleviating amyloid-β mediated neurotoxicity. *Colloids and Surfaces. B, Biointerfaces, 212*, 112373. https://doi.org/10.1016/j.colsurfb.2022.112373

Li, Z., Ye, B., Fang, J., Li, M., Xiong, Y., Xiong, P., Zhou, Y., Guo, Z., Zhong, H., & Liu, Z. (2024). Engineering the functional surface of carbon dots for antibacterial, bacterial bioimaging and sensing applications. *New Journal of Chemistry, 48*, 6020–6038. https://doi.org/10.1039/D3NJ05964B

Li, J., Wu, S., Shi, X., Cao, Y., Hao, H., Wang, J., & Han, Q. (2025). Machine learning-assisted biomass-derived carbon dots as fluorescent sensor array for discrimination of warfarin and its metabolites. *Langmuir: The ACS Journal of Surfaces and Colloids, 41*(3), 1694–1702. https://doi.org/10.1021/acs.langmuir.4c03945

Lim, S. Y., Shen, W., & Gao, Z. (2015). Carbon quantum dots and their applications. *Chemical Society Reviews, 44*(1), 362–381. https://doi.org/10.1039/C4CS00269E

Lin, X., Yang, Y., Nian, L., Su, H., Ou, J., Yuan, Z., Xie, F., Hong, W., Yu, D., Zhang, M., Ma, Y., & Chen, X. (2016). Interfacial modification layers based on carbon dots for efficient inverted polymer solar cells exceeding 10% power conversion efficiency. *Nano Energy, 26*, 216–223. https://doi.org/10.1016/j.nanoen.2016.05.011

Lin, F., Bao, Y.-W., & Wu, F.-G. (2019a). Carbon dots for sensing and killing microorganisms. *C Journal of Carbon Research, 5*(2), 33. https://doi.org/10.3390/c5020033

Lin, C. J., Chang, L., Chu, H. W., Lin, H. J., Chang, P. C., Wang, R. Y. L., Unnikrishnan, B., Mao, J. Y., Chen, S. Y., & Huang, C. C. (2019b). High amplification of the antiviral activity of curcumin through transformation into carbon quantum dots. *Small (Weinheim an der Bergstrasse, Germany), 15*(41), e1902641. https://doi.org/10.1002/smll.201902641

Lin, C., Chang, L., Chu, H., Lin, H., Chang, P., Wang, R. Y. L., Unnikrishnan, B., Mao, J., Chen, S., & Huang, C. (2019c). Antiviral carbon dots: High amplification of the antiviral activity of curcumin through transformation into carbon quantum dots. *Small, 15*. https://doi.org/10.1002/smll.201970219

Liu, C., Zhang, P., Zhai, X., Tian, F., Li, W., Yang, J., Liu, Y., Wang, H., Wang, W., & Liu, W. (2012). Nano-carrier for gene delivery and bioimaging based on carbon dots with PEI-passivation enhanced fluorescence. *Biomaterials, 33*(13), 3604–3613. https://doi.org/10.1016/j.biomaterials.2012.01.052

Liu, F., Jang, M. H., Ha, H. D., Kim, J. H., Cho, Y. H., & Seo, T. S. (2013). Facile synthetic method for pristine graphene quantum dots and graphene oxide quantum dots: Origin of blue and green luminescence. *Advanced Materials (Deerfield Beach, Fla.), 25*(27), 3657–3662. https://doi.org/10.1002/adma.201300233

Liu, C., Chang, K., Guo, W., Li, H., Shen, L., Chen, W., & Yan, D. (2014). Improving charge transport property and energy transfer with carbon quantum dots in inverted polymer solar cells. *Applied Physics Letters, 105*(7). https://doi.org/10.1063/1.4893994

Liu, H., He, Z., Jiang, L. P., & Zhu, J. J. (2015a). Microwave-assisted synthesis of wavelength-tunable photoluminescent carbon nanodots and their potential applications. *ACS Applied Materials & Interfaces, 7*(8), 4913–4920. https://doi.org/10.1021/am508994w

Liu, Y., Xu, L. P., Dai, W., Dong, H., Wen, Y., & Zhang, X. (2015b). Graphene quantum dots for the inhibition of β amyloid aggregation. *Nanoscale, 7*(45), 19060–19065. https://doi.org/10.1039/c5nr06282a

Liu, M. L., Yang, L., Li, R. S., Chen, B. B., Liu, H., & Huang, C. Z. (2017). Large-scale simultaneous synthesis of highly photoluminescent green amorphous carbon nanodots and yellow crystalline graphene quantum dots at room temperature. *Green Chemistry, 19*(15), 3611–3617. https://doi.org/10.1039/C7GC01236E

Liu, J. H., Li, R. S., Yuan, B., Wang, J., Li, Y. F., & Huang, C. Z. (2018). Mitochondria-targeting single-layered graphene quantum dots with dual recognition sites for ATP imaging in living cells. *Nanoscale, 10*(36), 17402–17408. https://doi.org/10.1039/c8nr06061d

Liu, M. L., Chen, B. B., Li, C. M., & Huang, C. Z. (2019). Carbon dots: Synthesis, formation mechanism, fluorescence origin and sensing applications. *Green Chemistry, 21*(3), 449–471. https://doi.org/10.1039/C8GC02736F

Liu, J., Geng, Y., Li, D., Yao, H., Huo, Z., Li, Y., Zhang, K., Zhu, S., Wei, H., Xu, W., Jiang, J., & Yang, B. (2020a). Deep red emissive carbonized polymer dots with unprecedented narrow full width at half maximum. *Advanced Materials (Deerfield Beach, Fla.), 32*(17), e1906641. https://doi.org/10.1002/adma.201906641

Liu, Y., Jain, A., Eng, C., Way, D. H., Lee, K., Bui, P., Kanada, K., de Oliveira Marinho, G., Gallegos, J., Gabriele, S., Gupta, V., Singh, N., Natarajan, V., Hofmann-Wellenhof, R., Corrado, G. S., Peng, L. H., Webster, D. R., Ai, D., Huang, S. J., Liu, Y., Dun, R. C., & Coz, D. (2020b). A deep learning system for differential diagnosis of skin diseases. *Nature Medicine, 26*(6), 900–908. https://doi.org/10.1038/s41591-020-0842-3

Liu, J., Li, R., & Yang, B. (2020c). Carbon Dots: A new type of carbon-based nanomaterial with wide applications. *ACS Central Science, 6*, 2179–2195. https://doi.org/10.1021/acscentsci.0c01306

Łoczechin, A., Séron, K., Barras, A., Giovanelli, E., Belouzard, S., Chen, Y. T., Metzler-Nolte, N., Boukherroub, R., Dubuisson, J., & Szunerits, S. (2019). Functional carbon quantum dots as medical countermeasures to human coronavirus. *ACS Applied Materials & Interfaces, 11*(46), 42964–42974. https://doi.org/10.1021/acsami.9b15032

Loo, A. H., Sofer, Z., Bouša, D., Ulbrich, P., Bonanni, A., & Pumera, M. (2016). Carboxylic carbon quantum dots as a fluorescent sensing platform for DNA detection. *ACS Applied Materials & Interfaces, 8*(3), 1951–1957. https://doi.org/10.1021/acsami.5b10160

Lu, Y., Zhang, L., & Lin, H. (2014). The use of a microreactor for rapid screening of the reaction conditions and investigation of the photoluminescence mechanism of carbon dots. *Chemistry (Weinheim an der Bergstrasse, Germany), 20*(15), 4246–4250. https://doi.org/10.1002/chem.201304358

Luo, L., Wang, P., Wang, Y., & Wang, F. (2018). pH-assisted selective detection of Hg (II) and Ag (I) based on nitrogen-rich carbon dots. *Sensors and Actuators B: Chemical, 273*, 1640–1647. https://doi.org/10.1016/j.snb.2018.07.090

Luo, J. B., Chen, J., Liu, H., Huang, C. Z., & Zhou, J. (2022). High-efficiency synthesis of red carbon dots using machine learning. *Chemical Communications, 58*(64), 9014–9017. https://doi.org/10.1039/d2cc03473e

Ma, Z., Zhang, Y. L., Wang, L., Ming, H., Li, H., Zhang, X., Wang, F., Liu, Y., Kang, Z., & Lee, S. T. (2013). Bioinspired photoelectric conversion system based on carbon-quantum-dot-doped dye-semiconductor complex. *ACS Applied Materials & Interfaces, 5*(11), 5080–5084. https://doi.org/10.1021/am400930h

Ma, H., Guan, L., Chen, M., Zhang, Y., Wu, Y., Liu, Z., & Li, X. (2023). Synthesis and enhancement of carbon quantum dots from Mopan persimmons for Fe^{3+} sensing and anti-counterfeiting applications. *Chemical Engineering Journal, 453*, 139906. https://doi.org/10.1016/j.cej.2022.139906

Mahani, M., Pourrahmani-Sarbanani, M., Yoosefian, M., Divsar, F., Mousavi, S. M., & Nomani, A. (2021). Doxorubicin delivery to breast cancer cells with transferrin targeted carbon quantum dots: An in vitro and in silico study. *Journal of Drug Delivery Science and Technology, 62*, 102342. https://doi.org/10.1016/j.jddst.2021.102342

Marković, Z. M., Labudová, M., Danko, M., Matijašević, D., Mičušík, M., Nádaždy, V., Kováčová, M., Kleinová, A., Špitalský, Z., Pavlović, V., Milivojević, D. D., Medić, M., & Todorović Marković, B. M. (2020). Highly efficient antioxidant F- and Cl-doped carbon quantum dots for bioimaging. *ACS Sustainable Chemistry & Engineering, 8*(43), 16327–16338. https://doi.org/10.1021/acssuschemeng.0c06260

Miao, S., Liang, K., Zhu, J., Yang, B., Zhao, D., & Kong, B. (2020). Hetero-atom-doped carbon dots: Doping strategies, properties and applications. *Nano Today, 33*, 100879. https://doi.org/10.1016/j.nantod.2020.100879

Molaei, M. J. (2019). A review on nanostructured carbon quantum dots and their applications in biotechnology, sensors, and chemiluminescence. *Talanta, 196*, 456–478. https://doi.org/10.1016/j.talanta.2018.12.042

Moradlou, O., Rabiei, Z., & Delavari, N. (2019). Antibacterial effects of carbon quantum dots@hematite nanostructures deposited on titanium against Gram-positive and Gram-negative bacteria. *Journal of Photochemistry and Photobiology A: Chemistry, 379*, 144–149. https://doi.org/10.1016/j.jphotochem.2019.04.047

Murray, C. J. L., Ikuta, K. S., Sharara, F., Swetschinski, L., Robles Aguilar, G., Gray, A., Han, C., Bisignano, C., Rao, P., Wool, E., Johnson, S. C., Browne, A. J., Chipeta, M. G., Fell, F., Hackett, S., Haines-Woodhouse, G., Kashef Hamadani, B. H., Kumaran, E. A. P., McManigal, B., Achalapong, S., Agarwal, R., Akech, S., Albertson, S., Amuasi, J., Andrews, J., Aravkin, A., Ashley, E., Babin, F.-X., Bailey, F., Baker, S., Basnyat, B., Bekker, A., Bender, R., Berkley, J. A., Bethou, A., Bielicki, J., Boonkasidecha, S., Bukosia, J., Carvalheiro, C., Castañeda-Orjuela, C., Chansamouth, V., Chaurasia, S., Chiurchiù, S., Chowdhury, F., Clotaire Donatien, R., Cook, A. J., Cooper, B., Cressey, T. R., Criollo-Mora, E., Cunningham, M., Darboe, S., Day, N. P. J., De Luca, M., Dokova, K., Dramowski, A., Dunachie, S. J., Duong Bich, T., Eckmanns, T., Eibach, D., Emami, A., Feasey, N., Fisher-Pearson, N., Forrest, K., Garcia, C., Garrett, D., Gastmeier, P., Giref, A. Z., Greer, R. C., Gupta, V., Haller, S., Haselbeck, A., Hay, S. I., Holm, M., Hopkins, S., Hsia, Y., Iregbu, K. C., Jacobs, J., Jarovsky, D., Javanmardi, F., Jenney, A. W. J., Khorana, M., Khusuwan, S., Kissoon, N., Kobeissi, E., Kostyanev, T., Krapp, F., Krumkamp, R., Kumar, A., Kyu, H. H., Lim, C., Lim, K., Limmathurotsakul, D., Loftus, M. J., Lunn, M., Ma, J., Manoharan, A., Marks, F., May, J., Mayxay, M., Mturi, N., Munera-Huertas, T., Musicha, P., Musila, L. A., Mussi-Pinhata, M. M., Naidu, R. N., Nakamura, T., Nanavati, R., Nangia, S., Newton, P., Ngoun, C., Novotney, A., Nwakanma, D., Obiero, C. W., Ochoa, T. J., Olivas-Martinez, A., Olliaro, P., Ooko, E., Ortiz-Brizuela, E., Ounchanum, P., Pak, G. D., Paredes, J. L., Peleg, A. Y., Perrone, C., Phe, T., Phommasone, K., Plakkal, N., Ponce-de-Leon, A., Raad, M., Ramdin, T., Rattanavong, S., Riddell, A., Roberts, T., Robotham, J. V., Roca, A., Rosenthal, V. D., Rudd, K. E., Russell, N., Sader, H. S., Saengchan, W., Schnall, J., Scott, J. A. G., Seekaew, S., Sharland, M., Shivamallappa, M., Sifuentes-Osornio, J., Simpson, A. J., Steenkeste, N., Stewardson, A. J., Stoeva, T., Tasak, N., Thaiprakong, A., Thwaites, G., Tigoi,

C., Turner, C., Turner, P., van Doorn, H. R., Velaphi, S., Vongpradith, A., Vongsouvath, M., Vu, H., Walsh, T., Walson, J. L., Waner, S., Wangrangsimakul, T., Wannapinij, P., Wozniak, T., Young Sharma, T. E. M. W., Yu, K. C., Zheng, P., Sartorius, B., Lopez, A. D., Stergachis, A., Moore, C., Dolecek, C., & Naghavi, M. (2022). Global burden of bacterial antimicrobial resistance in 2019: A systematic analysis. *Lancet, 399*, 629–655. https://doi.org/10.1016/S0140-6736(21)02724-0

Naik, K., Chaudhary, S., Ye, L., & Parmar, A. S. (2022). A strategic review on carbon quantum dots for cancer-diagnostics and treatment. *Frontiers in Bioengineering and Biotechnology, 10*, 882100. https://doi.org/10.3389/fbioe.2022.882100

Nayak, S., Das, P., & Singh, M. K. (2021). Carbon dot with aggregation induced emission and pH triggered disintegration. *Colloid and Interface Science Communications, 45*, 100537. https://doi.org/10.1016/j.colcom.2021.100537

Nguyen, D. N., Kim, M., Baeck, U., Lim, J., Shin, N., Kim, J., Choi, H., Park, H. S., Sim, U., & Kim, J. K. (2022). Machine learning assisted design and optimization of transition metal-incorporated carbon quantum dot catalysts for hydrogen evolution reaction. *ArXiv.*. https://arxiv.org/abs/2210.14701

Pali, L. S., Gupta, S. K., & Garg, A. (2018). Organic solar cells on Al electroded opaque substrates: Assessing the need of ZnO as electron transport layer. *Solar Energy, 160*, 396–403. https://doi.org/10.1016/j.solener.2017.12.028

Pant, M., Kumar, S., Kiran, K., Bisht, N. S., Pande, V., & Dandapat, A. (2023). A universal green approach for the synthesis of NPS-codoped carbon quantum dots with enhanced broad-spectrum antibacterial and antioxidant activities. *RSC Advances, 13*, 9186–9194. https://doi.org/10.1039/D2RA08103B

Peng, J., Gao, W., Gupta, B. K., Liu, Z., Romero-Aburto, R., Ge, L., Song, L., Alemany, L. B., Zhan, X., Gao, G., Vithayathil, S. A., Kaipparettu, B. A., Marti, A. A., Hayashi, T., Zhu, J. J., & Ajayan, P. M. (2012). Graphene quantum dots derived from carbon fibers. *Nano Letters, 12*(2), 844–849. https://doi.org/10.1021/nl2038979

Pham, Q. D. M., Thomson, S. M., Schaible, B. N., Mills, K. D., Atala, A., Porada, C. D., & Almeida-Porada, G. (2023). Acceptability of prenatal diagnosis and prenatal treatment of haemophilia using cell and gene therapies within US haemophilia community. *Haemophilia: The Official Journal of the World Federation of Hemophilia, 29*(4), 1024–1031. https://doi.org/10.1111/hae.14805

Priyadarshini, E., Rawat, K., Prasad, T., & Bohidar, H. B. (2018). Antifungal efficacy of Au@ carbon dots nanoconjugates against opportunistic fungal pathogen, *Candida albicans. Colloids Surfaces B Biointerfaces, 163*, 355–361. https://doi.org/10.1016/j.colsurfb.2018.01.006

Qu, S., Zhou, D., Li, D., Ji, W., Jing, P., Han, D., Liu, L., Zeng, H., & Shen, D. (2016). Toward efficient orange emissive carbon nanodots through conjugated sp(2)-domain controlling and surface charges engineering. *Advanced Materials (Deerfield Beach, Fla.), 28*(18), 3516–3521. https://doi.org/10.1002/adma.201504891

Raza, W., Ali, F., Raza, N., Luo, Y., Kim, K. H., Yang, J., Kumar, S., Mehmood, A., & Kwon, E. E. (2018). Recent advancements in supercapacitor technology. *Nano Energy, 52*, 441–473. https://doi.org/10.1016/j.nanoen.2018.08.013

Rezaei, A., & Hashemi, E. (2021). A pseudohomogeneous nanocarrier based on carbon quantum dots decorated with arginine as an efficient gene delivery vehicle. *Scientific Reports, 11*(1), 13790. https://doi.org/10.1038/s41598-021-93153-4

Riaz, S., & Park, S. J. (2022). Thioacetamide-derived nitrogen and sulfur co-doped carbon quantum dots for "green" quantum dot solar cells. *Journal of Industrial and Engineering Chemistry, 105*, 111–120. https://doi.org/10.1016/j.jiec.2021.09.009

Riaz, R., Ali, M., Maiyalagan, T., Anjum, A. S., Lee, S., Ko, M. J., & Jeong, S. H. (2019). Dye-sensitized solar cell (DSSC) coated with energy down shift layer of nitrogen-doped carbon quantum dots (N-CQDs) for enhanced current density and stability. *Applied Surface Science., 483*, 425–431. https://doi.org/10.1016/j.apsusc.2019.03.236

Ruiz-Pérez, R., Newman-Portela, A. M., & Ruiz-Fresneda, M. A. (2024). Emerging global trends in next-generation alternatives to classic antibiotics for combatting multidrug-resistant bacteria. *Journal of Cleaner Production, 478*, 143895. https://doi.org/10.1016/j.jclepro.2024.143895

Russell, S. J., & Norvig, P. (2020). *Artificial intelligence: A modern approach*. Prentice Hall.

Salam, M. A., Al-Amin, M. Y., Salam, M. T., Pawar, J. S., Akhter, N., Rabaan, A. A., & Alqumber, M. A. A. (2023). Antimicrobial resistance: A growing serious threat for global public health. *Healthcare, 11*, 1946. https://doi.org/10.3390/healthcare11131946

Salvi, A., Kharbanda, S., Thakur, P., Shandilya, M., & Thakur, A. (2024). Biomedical application of carbon quantum dots: A review. *Carbon Trends, 17*, 100407. https://doi.org/10.1016/j.cartre.2024.100407

Samphire, J., Takebayashi, Y., Hill, S. A., Hill, N., Heesom, K. J., Lewis, P. A., Alibhai, D., Bragginton, E. C., Dorh, J., Dorh, N., Spencer, J., & Galan, M. C. (2022). Green fluorescent carbon dots as targeting probes for LED-dependent bacterial killing. *Nano Select, 3*, 662–672. https://doi.org/10.1002/nano.202100183

Santiago, S. R., Lin, T. N., Yuan, C. T., Shen, J. L., Huang, H. Y., & Lin, C. A. (2016). Origin of tunable photoluminescence from graphene quantum dots synthesized via pulsed laser ablation. *Physical Chemistry Chemical Physics: PCCP, 18*(32), 22599–22605. https://doi.org/10.1039/c6cp03159e

Sawpari, R., Samanta, S., Banerjee, J., Das, S., Dash, S. S., Ahmed, R., Giri, B., & Dash, S. K. (2023). Recent advances and futuristic potentials of nano-tailored doxorubicin for prostate cancer therapy. *Journal of Drug Delivery Science and Technology, 81*, 104212. https://doi.org/10.1016/j.jddst.2023.104212

Senanayake, R. D., Yao, X., Froehlich, C. E., Cahill, M. S., Sheldon, T. R., McIntire, M., Haynes, C. L., & Hernandez, R. (2022). Machine learning-assisted carbon dot synthesis: Prediction of emission color and wavelength. *Journal of Chemical Information and Modeling, 62*(23), 5918–5928. https://doi.org/10.1021/acs.jcim.2c01007

Senior, A. W., Evans, R., Jumper, J., Kirkpatrick, J., Sifre, L., Green, T., Qin, C., Žídek, A., Nelson, A. W. R., Bridgland, A., Penedones, H., Petersen, S., Simonyan, K., Crossan, S., Kohli, P., Jones, D. T., Silver, D., Kavukcuoglu, K., & Hassabis, D. (2020). Improved protein structure prediction using potentials from deep learning. *Nature, 577*(7792), 706–710. https://doi.org/10.1038/s41586-019-1923-7

Serag, E., Helal, M., & El Nemr, A. (2024). Curcumin loaded onto folic acid carbon dots as a potent drug delivery system for antibacterial and anticancer applications. *Journal of Cluster Science, 35*, 519–532. https://doi.org/10.1007/s10876-023-02491-y

Shahriari, B., Swersky, K., Wang, Z., Adams, R. P., & De Freitas, N. (2016). Taking the human out of the loop: A review of Bayesian optimization. *Proceedings of the IEEE, 104*(1), 148–175. Article 7352306. https://doi.org/10.1109/JPROC.2015.2494218

Sharma, S., Chauhan, A., Ranjan, A., Mathkor, D. M., Haque, S., Ramniwas, S., Tuli, H. S., Jindal, T., & Yadav, V. (2024). Emerging challenges in antimicrobial resistance: Implications for pathogenic microorganisms, novel antibiotics, and their impact on sustainability. *Frontiers in Microbiology, 15*. https://doi.org/10.3389/fmicb.2024.1403168

Shen, J., Zhu, Y., Yang, X., & Li, C. (2012). Graphene quantum dots: Emergent nanolights for bioimaging, sensors, catalysis and photovoltaic devices. *Chemical Communications (Cambridge, England), 48*(31), 3686–3699. https://doi.org/10.1039/c2cc00110a

Shen, C. L., Lou, Q., Zang, J. H., Liu, K. K., Qu, S. N., Dong, L., & Shan, C. X. (2020). Near-infrared chemiluminescent carbon nanodots and their application in reactive oxygen species bioimaging. *Advanced Science (Weinheim, Baden-Wurttemberg, Germany), 7*(8), 1903525. https://doi.org/10.1002/advs.201903525

Shi, Z., & Jayatissa, A. H. (2018). Perovskites-based solar cells: A review of recent progress, materials and processing methods. *Materials (Basel, Switzerland), 11*(5), 729. https://doi.org/10.3390/ma11050729

Shi, W., Wang, Q., Long, Y., Cheng, Z., Chen, S., Zheng, H., & Huang, Y. (2011). Carbon nanodots as peroxidase mimetics and their applications to glucose detection. *Chemical Communications, 47*(23), 6695–6697. https://doi.org/10.1039/c1cc11943e

Shi, W., Li, X., & Ma, H. (2012). A tunable ratiometric pH sensor based on carbon nanodots for the quantitative measurement of the intracellular pH of whole cells. *Angewandte Chemie, 51*(26), 6432–6435. https://doi.org/10.1002/anie.201202533

Silver, D., Schrittwieser, J., Simonyan, K., Antonoglou, I., Huang, A., Guez, A., Hubert, T., Baker, L., Lai, M., Bolton, A., Chen, Y., Lillicrap, T., Hui, F., Sifre, L., van den Driessche, G., Graepel, T., & Hassabis, D. (2017). Mastering the game of Go without human knowledge. *Nature, 550*(7676), 354–359. https://doi.org/10.1038/nature24270

Singh, P., Arpita, K. S., Kumar, P., Kataria, N., Bhankar, V., Kumar, K., Kumar, R., Hsieh, C. T., & Khoo, K. S. (2023a). Assessment of biomass-derived carbon dots as highly sensitive and selective templates for the sensing of hazardous ions. *Nanoscale, 15*(40), 16241–16267. https://doi.org/10.1039/d3nr01966g

Singh, P., Rani, N., Kumar, S., Kumar, P., Mohan, B., Pallavi, B. V., Kataria, N., Kumar, R., & Kumar, K. (2023b). Assessing the biomass-based carbon dots and their composites for photocatalytic treatment of wastewater. *Journal of Cleaner Production, 413*, 137474. https://doi.org/10.1016/j.jclepro.2023.137474

Song, S. H., Jang, M. H., Chung, J., Jin, S. H., Kim, B. H., Hur, S. H., Yoo, S., Cho, Y. H., & Jeon, S. (2014). Highly efficient light-emitting diode of graphene quantum dots fabricated from graphite intercalation compounds. *Advanced Optical Materials, 2*(11), 1016–1023. https://doi.org/10.1002/adom.201400184

Song, Y. B., Zhu, S. J., Zhang, S. T., Fu, Y., Wang, L., Zhao, X., & Yang, B. (2015). Investigation from chemical structure to photoluminescent mechanism: A type of carbon dots from the pyrolysis of citric acid and an amine. *Journal of Materials Chemistry C, 3*(23), 2050–7526. https://doi.org/10.1039/C5TC00813A

Song, Y., Yan, X., Li, Z., Qu, L., Zhu, C., Ye, R., Li, S., Du, D., & Lin, Y. (2018). Highly photoluminescent carbon dots derived from linseed and their applications in cellular imaging and sensing. *Journal of Materials Chemistry, B, 6*(19), 3181–3187. https://doi.org/10.1039/c8tb00116b

Song, W., Wang, X., Nong, S., Wang, M., Kang, S., Wang, F., & Xu, L. (2024). D-cysteine-derived carbon dots for selective discrimination, imaging, and synergistic elimination of gram-positive bacteria and fungi. *Advanced Functional Materials, 34*. https://doi.org/10.1002/adfm.202402761

Srivastava, S., Awasthi, R., Tripathi, D., Rai, M. K., Agarwal, V., Agrawal, V., Gajbhiye, N. S., & Gupta, R. K. (2012). Magnetic-nanoparticle-doped carbogenic nanocomposite: an effective magnetic resonance/fluorescence multimodal imaging probe. *Small (Weinheim an der Bergstrasse, Germany), 8*(7), 1099–1109. https://doi.org/10.1002/smll.201101863

Sturabotti, E., Camilli, A., Georgian Moldoveanu, V., Bonincontro, G., Simonetti, G., Valletta, A., Serangeli, I., Miranda, E., Amato, F., Giacomo Marrani, A., Migneco, L. M., Sennato, S., Simonis, B., Vetica, F., & Leonelli, F. (2024a). Targeting the antifungal activity of carbon dots against *Candida albicans* biofilm formation by tailoring their surface functional groups. *Chemistry: A European Journal, 30*. https://doi.org/10.1002/chem.202303631

Sturabotti, E., Camilli, A., Leonelli, F., & Vetica, F. (2024b). Carbon dots as bioactive antifungal nanomaterials. *ChemMedChem, 19*. https://doi.org/10.1002/cmdc.202400463

Sun, Y. P., Zhou, B., Lin, Y., Wang, W., Fernando, K. A., Pathak, P., Meziani, M. J., Harruff, B. A., Wang, X., Wang, H., Luo, P. G., Yang, H., Kose, M. E., Chen, B., Veca, L. M., & Xie, S. Y. (2006). Quantum-sized carbon dots for bright and colorful photoluminescence. *Journal of the American Chemical Society, 128*(24), 7756–7757. https://doi.org/10.1021/ja062677d

Sun, B., Wu, F., Zhang, Q., Chu, X., Wang, Z., Huang, X., Li, J., Yao, C., Zhou, N., & Shen, J. (2021). Insight into the effect of particle size distribution differences on the antibacterial activity of carbon dots. *Journal of Colloid and Interface Science, 584*, 505–519. https://doi.org/10.1016/j.jcis.2020.10.015

Tammina, S. K., Yang, D., Koppala, S., Cheng, C., & Yang, Y. (2019). Highly photoluminescent N, P doped carbon quantum dots as a fluorescent sensor for the detection of dopamine and temperature. *Journal of Photochemistry and Photobiology B, Biology, 194*, 61–70. https://doi.org/10.1016/j.jphotobiol.2019.01.004

Tao, S., Lu, S., Geng, Y., Zhu, S., Redfern, S. A. T., Song, Y., Feng, T., Xu, W., & Yang, B. (2018). Design of metal-free polymer carbon dots: A new class of room-temperature phosphorescent materials. *Angewandte Chemie, 57*(9), 2393–2398. https://doi.org/10.1002/anie.201712662

Tao, H., Wu, T., Aldeghi, M., Wu, T. C., Aspuru-Guzik, A., & Kumacheva, E. (2021). Nanoparticle synthesis assisted by machine learning. *Nature Reviews Materials, 6*(8), 701–716. https://doi.org/10.1038/s41578-021-00337-5

Tariq, M., Khan, M. A., Hasan, H., Yadav, S., Sahoo, A. K., & Sk, M. P. (2024). Biologically active dual functional zinc-doped biomass-derived carbon dots. *RSC Sustainability, 2*, 3114–3122. https://doi.org/10.1039/D4SU00439F

Tong, T., Hu, H., Zhou, J., Deng, S., Zhang, X., Tang, W., Fang, L., Xiao, S., & Liang, J. (2020). Antiviral carbon dots: Glycyrrhizic-acid-based carbon dots with high antiviral activity by multisite inhibition mechanisms. *Small, 16*. https://doi.org/10.1002/smll.202070068

Tumilaar, S. G., Hardianto, A., Dohi, H., & Kurnia, D. (2024). A comprehensive review of free radicals, oxidative stress, and antioxidants: Overview, clinical applications, global perspectives, future directions, and mechanisms of antioxidant activity of flavonoid compounds. *Journal of Chemistry, 2024*, 1–21. https://doi.org/10.1155/2024/5594386

Vadakkan, K., Ghodake, G. S., Gunasekaran, S., Lai, C. W., & Rumjit, N. P. (2025). Carbon dots as nano-pioneers: A critical review on advances in fabrication and antibacterial mechanism. *Inorganic Chemistry Communications, 179*, 114883. https://doi.org/10.1016/j.inoche.2025.114883

Vercelli, B. (2021). The role of carbon quantum dots in organic photovoltaics: A short overview. *Coatings, 11*(2), 232. https://doi.org/10.3390/coatings11020232

Waltraud, H., & Arnold, N. (2008). SNOBFIT- Stable noisy optimization by branch and fit. *ACM Transactions on Mathematical Softwares (TMOS), 35*(2), 1–25. Article 9. https://doi.org/10.1145/1377612.1377613

Wang, L., Zhu, S. J., Wang, H. Y., Qu, S. N., Zhang, Y. L., Zhang, J. H., Chen, Q. D., Xu, H. L., Han, W., Yang, B., & Sun, H. B. (2014). Common origin of green luminescence in carbon nanodots and graphene quantum dots. *ACS Nano, 8*(3), 2541–2547. https://doi.org/10.1021/nn500368m

Wang, Y., Chang, Q., & Hu, S. (2017). Carbon dots with concentration-tunable multicolored photoluminescence for simultaneous detection of Fe^{3+} and Cu^{2+} ions. *Sensors and Actuators B: Chemical, 253*, 928–933. https://doi.org/10.1016/j.snb.2017.07.031

Wang, Y., Yan, L., Ji, G., Wang, C., Gu, H., Luo, Q., Chen, Q., Chen, L., Yang, Y., Ma, C. Q., & Liu, X. (2019a). Synthesis of N, S-doped carbon quantum dots for use in organic solar cells as the zno modifier to eliminate the light-soaking effect. *ACS Applied Materials & Interfaces, 11*(2), 2243–2253. https://doi.org/10.1021/acsami.8b17128

Wang, X., Yang, P., Feng, Q., Meng, T., Wei, J., Xu, C., & Han, J. (2019b). Green preparation of fluorescent carbon quantum dots from cyanobacteria for biological imaging. *Polymers, 11*(4), 616. https://doi.org/10.3390/polym11040616

Wang, H., Zhang, M., Ma, Y., Wang, B., Shao, M., Huang, H., Liu, Y., & Kang, Z. (2020). Selective inactivation of Gram-negative bacteria by carbon dots derived from natural biomass: *Artemisia argyi* leaves. *Journal of Materials Chemistry B, 8*, 2666–2672. https://doi.org/10.1039/C9TB02735A

Wang, P., Meziani, M. J., Fu, Y., Bunker, C. E., Hou, X., Yang, L., Msellek, H., Zaharias, M., Darby, J. P., & Sun, Y. P. (2021). Carbon dots *versus* nano-carbon/organic hybrids- Dramatically different behaviors in fluorescence sensing of metal cations with structural and mechanistic implications. *Nanoscale Advances, 3*(8), 2316–2324. https://doi.org/10.1039/d1na00002k

Wang, S., Wang, D., Wang, G., Zhang, M., Sun, Y., & Ding, J. (2025). Antibacterial carbon dots. *Materials Today Bio, 30*, 101383. https://doi.org/10.1016/j.mtbio.2024.101383

Wei, W., Xu, C., Ren, J., Xu, B., & Qu, X. (2012). Sensing metal ions with ion selectivity of a crown ether and fluorescence resonance energy transfer between carbon dots and graphene. *Chemical Communications, 48*(9), 1284–1286. https://doi.org/10.1039/c2cc16481g

Wei, Y., Liu, H., Jin, Y., Cai, K., Li, H., Liu, Y., Kang, Z., & Zhang, Q. (2013). Carbon nanoparticle ionic liquid functionalized activated carbon hybrid electrode for efficiency enhancement in supercapacitors. *New Journal of Chemistry, 37*(4), 886–889. https://doi.org/10.1039/C3NJ41151F

Xia, J., Chen, S., Zou, G. Y., Yu, Y. L., & Wang, J. H. (2018). Synthesis of highly stable red-emissive carbon polymer dots by modulated polymerization: From the mechanism to application in intracellular pH imaging. *Nanoscale, 10*(47), 22484–22492. https://doi.org/10.1039/c8nr08208a

Xia, C., Zhu, S., Feng, T., Yang, M., & Yang, B. (2019). Evolution and synthesis of carbon dots: From carbon dots to carbonized polymer dots. *Advanced Science (Weinheim, Baden-Wurttemberg, Germany), 6*(23), 1901316. https://doi.org/10.1002/advs.201901316

Xiao, L., & Sun, H. (2018). Novel properties and applications of carbon nanodots. *Nanoscale Horizons, 3*(6), 565–597. https://doi.org/10.1039/c8nh00106e

Xing, C., Chen, G., Zhu, X., An, J., Bao, J., Wang, X., & Xu, X. (2024). Synthesis of carbon dots with predictable photoluminescence by the aid of machine learning. *Nano Research, 17*(3), 1984–1989. https://doi.org/10.1007/s12274-023-5893-6

Xu, X., Ray, R., Gu, Y., Ploehn, H. J., Gearheart, L., Raker, K., & Scrivens, W. A. (2004). Electrophoretic analysis and purification of fluorescent single-walled carbon nanotube fragments. *Journal of the American Chemical Society, 126*(40), 12736–12737. https://doi.org/10.1021/ja040082h

Xu, J., Hou, K., Ju, Z., Ma, C., Wang, W., Wang, C., & Chen, Z. (2017a). Synthesis and electrochemical properties of carbon dots/manganese dioxide (CQDs/MnO2) nanoflowers for supercapacitor applications. *Journal of the Electrochemical Society, 164*(2), A430. https://doi.org/10.1149/2.1241702jes

Xu, X., Tang, W., Zhou, Y., Bao, Z., Su, Y., Hu, J., & Zeng, H. (2017b). Steering photoelectrons excited in carbon dots into platinum cluster catalyst for solar-driven hydrogen production. *Advanced Science (Weinheim, Baden-Wurttemberg, Germany), 4*(12), 1700273. https://doi.org/10.1002/advs.201700273

Xu, K., Wang, Z., Cui, M., Jiang, Y., Li, C., Wang, Z.-X., Li, L., Jia, C., Zhang, L., & Wu, F. (2025). Turning waste into treasure: functionalized biomass-derived carbon dots for superselective visualization and eradication of gram-positive bacteria. *Advanced Science, 12*. https://doi.org/10.1002/advs.202411084

Xue, Y., Liu, C., Andrews, G., Wang, J., & Ge, Y. (2022). Recent advances in carbon quantum dots for virus detection, as well as inhibition and treatment of viral infection. *Nano Convergence, 9*(1), 15. https://doi.org/10.1186/s40580-022-00307-9

Yadav, N., Mudgal, D., Mishra, A., Shukla, S., Malik, T., & Mishra, V. (2024). Harnessing fluorescent carbon quantum dots from natural resource for advancing sweat latent fingerprint recognition with machine learning algorithms for enhanced human identification. *Plos One, 19*(1), e0296270. https://doi.org/10.1371/journal.pone.0296270

Yahaya Pudza, M., Zainal Abidin, Z., Abdul Rashid, S., Md Yasin, F., Noor, A. S. M., & Issa, M. A. (2019). Sustainable synthesis processes for carbon dots through response surface methodology and artificial neural network. *Processes, 7*(10), 704. https://doi.org/10.3390/pr7100704

Yan, X., Li, B., & Li, L. S. (2013). Colloidal graphene quantum dots with well-defined structures. *Accounts of Chemical Research, 46*(10), 2254–2262. https://doi.org/10.1021/ar300137p

Yan, L., Yang, Y., Ma, C., Liu, X., Wang, H., & Xu, B. (2016). Synthesis of carbon quantum dots by chemical vapor deposition approach for use in polymer solar cell as the electrode buffer layer. *Carbon, 109*, 598–607. https://doi.org/10.1016/j.carbon.2016.08.058

Yang, S. T., Wang, X., Wang, H., Lu, F., Luo, P. G., Cao, L., Meziani, M. J., Liu, J. H., Liu, Y., Chen, M., Huang, Y., & Sun, Y. P. (2009). Carbon dots as nontoxic and high-performance fluorescence imaging agents. *The Journal of Physical Chemistry. C, Nanomaterials and Interfaces, 113*(42), 18110–18114. https://doi.org/10.1021/jp9085969

Yang, Y., Cui, J., Zheng, M., Hu, C., Tan, S., Xiao, Y., Yang, Q., & Liu, Y. (2012). One-step synthesis of amino-functionalized fluorescent carbon nanoparticles by hydrothermal carbonization of chitosan. *Chemical Communications (Cambridge, England), 48*(3), 380–382. https://doi.org/10.1039/c1cc15678k

Yang, Y., Yan, Z., Wei, D., Zhong, J., Liu, L., Zhang, L., Wang, F., Wei, X., Xie, C., Lu, W., & He, D. (2013). Tumor-penetrating peptide functionalization enhances the anti-glioblastoma effect of doxorubicin liposomes. *Nanotechnology, 24*(40), 405101. https://doi.org/10.1088/0957-4484/24/40/405101

Yang, L., Wang, Z., Wang, J., Jiang, W., Jiang, X., Bai, Z., He, Y., Jiang, J., Wang, D., & Yang, L. (2016). Doxorubicin conjugated functionalizable carbon dots for nucleus targeted delivery and enhanced therapeutic efficacy. *Nanoscale, 8*(12), 6801–6809. https://doi.org/10.1039/c6nr00247a

Yang, M., Tang, Q., Meng, Y., Liu, J., Feng, T., Zhao, X., Zhu, S., Yu, W., & Yang, B. (2018). Reversible "off-on" fluorescence of Zn^{2+}-passivated carbon dots: Mechanism and potential for the detection of EDTA and Zn^2. *Langmuir: The ACS Journal of Surfaces and Colloids, 34*(26), 7767–7775. https://doi.org/10.1021/acs.langmuir.8b00947

Yu, M., Li, P., Huang, R., Xu, C., Zhang, S., Wang, Y., Gong, X., & Xing, X. (2023). Antibacterial and antibiofilm mechanisms of carbon dots: A review. *Journal of Materials Chemistry B, 11*, 734–754. https://doi.org/10.1039/D2TB01977A

Yuan, F., Wang, Z., Li, X., Li, Y., Tan, Z., Fan, L., & Yang, S. (2017). Bright multicolor bandgap fluorescent carbon quantum dots for electroluminescent light-emitting diodes. *Advanced Materials (Deerfield Beach, Fla.), 29*(3). https://doi.org/10.1002/adma.201604436

Zhang, Y., Ma, D., Zhang, Y., Chen, W., & Huang, S. (2013). N-doped carbon quantum dots for TiO_2-based photocatalysts and dye-sensitized solar cells. *Nano Energy, 2*, 545–552. https://doi.org/10.1140/EPJS/S11734-022-00503-Y

Zhang, X., Liu, C., Li, Z., Guo, J., Shen, L., Guo, W., Zhang, L., Ruan, S., & Long, Y. (2017). An easily prepared carbon quantum dots and employment for inverted organic photovoltaic devices. *Chemical Engineering Journal, 315*, 621–629. https://doi.org/10.1016/j.cej.2017.01.067

Zhang, R., Zhao, M., Wang, Z., Wang, Z., Zhao, B., Miao, Y., Zhou, Y., Wang, H., Hao, Y., Chen, G., & Zhu, F. (2018). Solution-processable ZnO/carbon quantum dots electron extraction layer for highly efficient polymer solar cells. *ACS Applied Materials & Interfaces, 10*(5), 4895–4903. https://doi.org/10.1021/acsami.7b17969

Zhang, W., Wu, B., Li, Z., Wang, Y., Zhou, J., & Li, Y. (2020). Carbon quantum dots as fluorescence sensors for label-free detection of folic acid in biological samples. *Spectrochimica Acta. Part A, Molecular and Biomolecular Spectroscopy, 229*, 117931. https://doi.org/10.1016/j.saa.2019.117931

Zhang, X., Wu, P., Hao, X., Liu, J., Huang, Z., Weng, S., Chen, W., Huang, L., & Huang, J. (2023). Quaternized carbon dots with enhanced antimicrobial ability towards Gram-negative bacteria for the treatment of acute peritonitis caused by *E. coli*. *Journal of Materials Chemistry B, 11*, 7696–7706. https://doi.org/10.1039/D3TB00889D

Zhang, Q., Li, X., Yu, L., Wang, L., Wen, Z., Su, P., Sun, Z., & Wang, S. (2025). Machine learning-assisted fluorescence visualization for sequential quantitative detection of aluminum and fluoride ions. *Journal of Environmental Sciences, 149*, 68–78. https://doi.org/10.1016/j.jes.2024.01.023

Zhao, H. X., Liu, L. Q., Liu, Z. D., Wang, Y., Zhao, X. J., & Huang, C. Z. (2011). Highly selective detection of phosphate in very complicated matrixes with an off-on fluorescent probe of europium-adjusted carbon dots. *Chemical Communications (Cambridge, England), 47*(9), 2604–2606. https://doi.org/10.1039/c0cc04399k

Zhao, C., Wang, X., Yu, L., Wu, L., Hao, X., Liu, Q., Lin, L., Huang, Z., Ruan, Z., Weng, S., Liu, A., & Lin, X. (2022). Quaternized carbon quantum dots with broad-spectrum antibacterial activity for the treatment of wounds infected with mixed bacteria. *Acta Biomaterials, 138*, 528–544. https://doi.org/10.1016/j.actbio.2021.11.010

Zhao, X., Zhang, S., Zhang, M., Zhang, Z., Zhou, M., & Cao, J. (2025). Antifungal performance and mechanisms of carbon quantum dots in cellulosic materials. *ACS Nano, 19*, 14121–14136. https://doi.org/10.1021/acsnano.5c00052

Zheng, M., Ruan, S., Liu, S., Sun, T., Qu, D., Zhao, H., Xie, Z., Gao, H., Jing, X., & Sun, Z. (2015). Self-targeting fluorescent carbon dots for diagnosis of brain cancer cells. *ACS Nano, 9*(11), 11455–11461. https://doi.org/10.1021/acsnano.5b05575

Zhou, X., & Ye, Q. (2021). Cellular immune response to COVID-19 and potential immune modulators. *Frontiers in Immunology, 12*, 646333. https://doi.org/10.3389/fimmu.2021.646333

Zhou, C., Wu, S., Qi, S., Song, W., & Sun, C. (2021). Facile and high-yield synthesis of n-doped carbon quantum dots from biomass quinoa saponin for the detection of Co^2. *Journal of Analytical Methods in Chemistry, 2021*, 9732364. https://doi.org/10.1155/2021/9732364

Zhu, S., Zhang, J., Qiao, C., Tang, S., Li, Y., Yuan, W., Li, B., Tian, L., Liu, F., Hu, R., Gao, H., Wei, H., Zhang, H., Sun, H., & Yang, B. (2011). Strongly green-photoluminescent graphene quantum dots for bioimaging applications. *Chemical Communications (Cambridge, England), 47*(24), 6858–6860. https://doi.org/10.1039/c1cc11122a

Zhu, A., Qu, Q., Shao, X., Kong, B., & Tian, Y. (2012). Carbon-dot-based dual-emission nanohybrid produces a ratiometric fluorescent sensor for in vivo imaging of cellular copper ions. *Angewandte Chemie, 51*(29), 7185–7189. https://doi.org/10.1002/anie.201109089

Zhu, S., Meng, Q., Wang, L., Zhang, J., Song, Y., Jin, H., Zhang, K., Sun, H., Wang, H., & Yang, B. (2013). Highly photoluminescent carbon dots for multicolor patterning, sensors, and bioimaging. *Angewandte Chemie, 52*(14), 3953–3957. https://doi.org/10.1002/anie.201300519

Zhuo, S., Shao, M., & Lee, S. T. (2012). Upconversion and downconversion fluorescent graphene quantum dots: Ultrasonic preparation and photocatalysis. *ACS Nano, 6*(2), 1059–1064. https://doi.org/10.1021/nn2040395

Analytical Techniques for Characterization of Carbon Dots

5

Duyen H. H. Nguyen, Arjun Muthu, Áron Béni, and József Prokisch

Abstract

Carbon dots (CDs) are a class of carbon nanomaterials smaller than 10 nm. Thanks to their functionalization, high surface area, biocompatibility, stability, and green, cost-effective production methods, they have been effectively utilized in various applications, such as drug delivery, bioimaging, sensors, and effluent treatments. Although these properties make CDs useful for multiple purposes, there are complexities in their characterization. Various characterization techniques exist for characterizing CDs, ranging from basic methods to confirm the presence of CDs to advanced methods that require an understanding of their

D. H. H. Nguyen (✉)
Institute of Animal Science, Faculty of Agricultural and Food Sciences and Environmental Management, Biotechnology and Nature Conservation, University of Debrecen, Debrecen, Hungary

Institute of Life Sciences, Vietnam Academy of Science and Technology, Ho Chi Minh City, Vietnam

Doctoral School of Nutrition and Food Science, University of Debrecen, Debrecen, Hungary
e-mail: nguyen.huu.huong.duyen@agr.unideb.hu

A. Muthu
Doctoral School of Nutrition and Food Science, University of Debrecen, Debrecen, Hungary

Institute of Agricultural Chemistry and Soil Science, Faculty of Agricultural and Food Sciences and Environmental Management, University of Debrecen, Debrecen, Hungary

Á. Béni
Institute of Agricultural Chemistry and Soil Science, Faculty of Agricultural and Food Sciences and Environmental Management, University of Debrecen, Debrecen, Hungary

J. Prokisch
Institute of Animal Science, Faculty of Agricultural and Food Sciences and Environmental Management, Biotechnology and Nature Conservation, University of Debrecen, Debrecen, Hungary

M. Rai et al. (eds.), *Carbon Nanodots*, https://doi.org/10.1007/978-981-95-4705-0_5

structure. This chapter will comprehensively discuss all techniques used to understand the structure of CDs, including UV-Vis spectroscopy, photoluminescence (PL), scanning electron microscopy (SEM), transmission electron microscopy (TEM), atomic force microscopy (AFM), dynamic light scattering (DLS), Raman spectroscopy, Fourier transform infrared spectroscopy (FTIR), X-ray photoelectron spectroscopy (XPS), nuclear magnetic resonance spectroscopy (NMR), zeta potential, X-ray diffraction (XRD), electron spin resonance spectroscopy (ESR), and thermogravimetric analysis (TGA). This collective data will assist researchers in selecting appropriate characterization methods and making informed choices for applications.

Keywords

Nanomaterials · Carbon nanodots · Characterization · Analytical method

5.1 Introduction

CDs are well-known for their fluorescent properties, nano-size, biocompatibility, and active surface area, which can be modified depending on the precursors. Various analytical techniques were applied to the characterization of CDs, which are crucial for providing critical insights into their structural and functional characteristics, facilitating their optimization for diverse applications. Additionally, thousands of CDs have been synthesized from various precursors, resulting in diverse CDs that require further analytical techniques for a deeper understanding of CDs (Nguyen et al., 2024a). Mintz et al. (2021) recommended characterizing CDs by analyzing the combined data from different techniques to obtain the overall structure of CDs, particularly during the initial phase of the synthesis of CDs.

Standardizing the structure of CDs is crucial for consistent evaluation of their properties and performance. Currently, the recommended characterization stages for thoroughly characterized CDs, based on the purpose of the analytical technique, include (1) evaluating the optical properties to confirm fluorescence characteristics, (2) assessing size and morphology, (3) analyzing surface functionality, and (4) examining core structure. Table 5.1 lists and categorizes analytical techniques for characterizing CDs, followed by the recommended characterization stage. Certain methods are classified as "fundamental techniques," which provide essential characterization information, and "supporting techniques," which offer deeper insights into the properties of CDs due to their capacity to yield essential and highly relevant information (Table 5.1). These techniques should be prioritized based on their reliability, sensitivity, and relevance to the specific research objectives, while supplementary techniques can complement the analysis as necessary.

The fundamental techniques to obtain the primary information of CDs include photoluminescence (PL) spectroscopy and UV-visible spectroscopy for optical properties; transmission electron microscopy (TEM) for size and morphology; Fourier-transform infrared (FTIR) and Raman spectroscopy for surface

Table 5.1 Analytical techniques for characterization of carbon dots (CDs)

Purpose of analytical techniques	Analytical techniques	Analytical approaches	Key information obtained
Optical properties	UV-visible (UV-Vis) spectroscopy	Fundamental technique	Bandgap energy, optical transitions
	Photoluminescence (PL)	Fundamental technique	Quantum yield, fluorescence lifetime, peak shifts
Size and morphology	Transmission electron microscopy (TEM)	Fundamental technique	Particle size, shape at the atomic level
	Scanning electron microscopy (SEM)	Supporting technique	Particle size, shape
	Atomic force microscopy (AFM)	Supporting technique	Nanostructure topography, roughness
	Dynamic light scattering (DLS)	Fundamental technique	Size distribution, aggregation, or stability over time
Surface functionality	Raman spectroscopy	Fundamental technique	Carbon bonding structure, defect analysis
	Fourier-transform infrared spectroscopy (FTIR)	Fundamental technique	Presence of surface functional groups
	X-ray photoelectron spectroscopy (XPS)	Supporting technique	Surface elemental composition, bonding states
	Solution-state nuclear magnetic resonance (NMR)	Supporting technique	Surface functionalization, organic ligands
	Zeta potential analysis	Supporting technique	Surface charge, colloidal stability
Core structure	X-ray diffraction (XRD)	Fundamental technique	Crystalline characteristics of materials
	Solid-state nuclear magnetic resonance (NMR)	Supporting technique	Carbon core structure, sp^2/sp^3 hybridization, graphitic/amorphous nature
	Electron spin resonance (ESR)	Supporting technique	Monitoring free radical activity or reaction mechanisms
	Thermogravimetric analysis (TGA)	Supporting technique	Thermal stability, degradation behavior, or compositional analysis

functionality; and X-ray Diffraction (XRD) for core structure (González-González et al., 2022; Mintz et al., 2021). Other analytical techniques can provide various characterization insights. The choice of analytical techniques depends on available facilities and study goals.

This chapter examines analytical techniques for characterizing carbon dots (CDs), detailing their fundamental principles, advantages, and potential challenges within the analytical process.

5.2 Optical Property Analysis

5.2.1 UV-Visible (UV-Vis) Spectroscopy

UV-Visible (UV-Vis) spectroscopy is a powerful optical technique used to characterize CDs, a class of nanomaterials with unique optical properties. Although the structure of CDs varies, they typically consist of two primary parts: (1) the carbon core and (2) the carbon shell, which contains diverse functional groups that bind to the surface. The UV-Vis spectra provide insights into the overall structure of CDs, as shown in Fig. 5.1a. CDs typically exhibit strong absorption in the UV region (10–400 nm), with some extending into the visible range (400–700 nm). This absorption is often linked to the presence of surface defects and band bending, which contribute to their photoluminescence properties (Jabbar Abbas & Atia Muhson, 2019; Zhao et al., 2015). To understand UV-Vis spectra of CDs, the spectra can be analyzed into four main parts (Mintz et al., 2021), including:

- The 200–300 nm region of the UV-Vis spectrum corresponds to the π–π^* transition of the C=C bond and sp^2 networks found in carbon structures, which are common across all types of CDs.

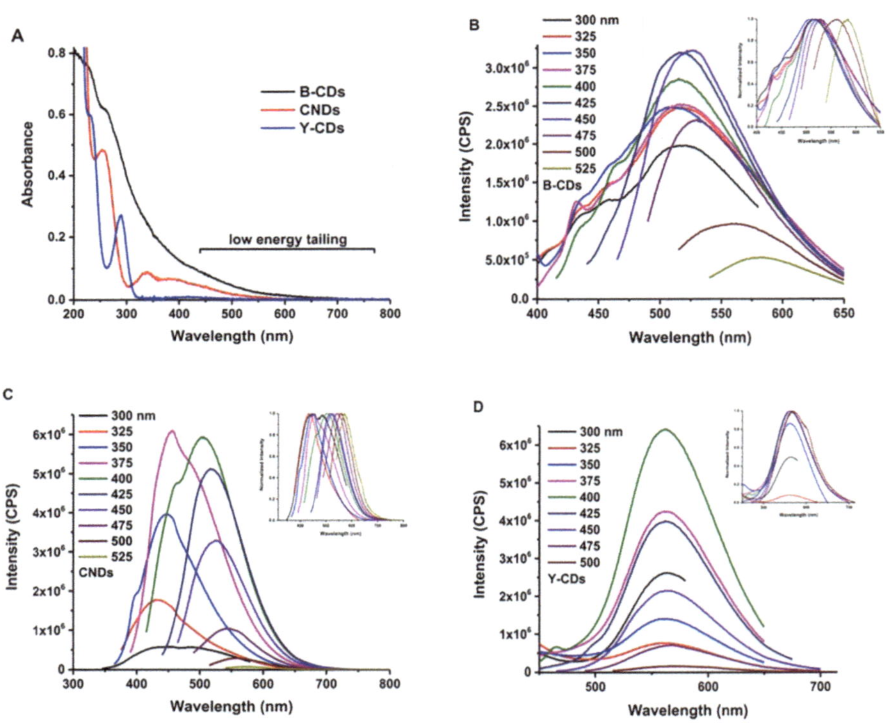

Fig. 5.1 (**a**) UV- Vis spectra and (**b**–**d**) PL spectra of three different CDs. (Adapted from Mintz et al., 2021)

- The 300–400 nm region of the UV-Vis spectrum displays the n–π* transition of the C=O or C=N bond, depending on the type of CDs.
- Above 400 nm (visible region) indicates the presence of a functional group on the carbon shell, the surface of CDs.
- A low-energy tailing can exhibit the π electronic bands in a nanostructure and the degree of order in a system.

5.2.2 Photoluminescence (PL)

Photoluminescence (PL) is a key characteristic of CDs that makes them stand out among other carbon-based nanomaterials, offering promising applications in bioimaging, optoelectronics, and sensing. Unlike traditional carbon-based materials, CDs can emit light when excited, making them a compelling alternative to semiconductor quantum dots and organic dyes.

Various optical analyses can be conducted to understand the PL properties of CDs (Fig. 5.1b–d), listed below deeply:

- ***Excitation (λ_{ex}) and emission (λ_{em}) spectra***: Determines the wavelengths at which CDs absorb and emit light.
- **Excitation-dependent emission**: Identifies if the emission shifts with different excitation wavelengths (common in CDs due to multiple emissive centers).
- **Emission intensity**: Evaluates fluorescence brightness and stability under different conditions (pH, solvent, functionalization).
- ***Quantum yield (QY)***: Compares the fluorescence efficiency of CDs to a reference fluorophore. Quinine sulfate is the most commonly used reference due to its similar excitation and emission characteristics to CDs. The QY is calculated using the following simplified equation (Crosby & Demas, 1971):

$$QY_x = QY_{std}\left(\frac{K_x}{K_{std}}\right)\left(\frac{RI_x}{RI_{std}}\right)^2$$

where the subscripts "*x*" and "std" refer to the measured sample and the standard fluorophore; QY, the quantum yield; *K*, gradients from linear regression; RI, the refractive index of the solution; and the value RI_x/RI_{std} is 1 for extremely low concentration solutions. For quinine sulfate as a standard fluorophore, the QY is 0.54.

The photoluminescent (PL) properties of CDs vary based on their structural diversity. Different precursors can create CDs with distinct structures. Consequently, the exact mechanisms behind their fluorescence are still being investigated (Ai et al., 2021). The PL properties of CDs are affected by factors such as the presence of sp^2 domains, oxidized surface defects, solvent polarity, hydrogen bonding interactions, and the choice of precursors. The emission mechanism is influenced by both the graphitic core and surface defects, where blue fluorescence typically results from π–π^* transitions and n–π^* charge transfer (Kailasa & Hussai, 2022). Multiple mechanisms influence the PL of CDs, including:

- ***Quantum Confinement and Conjugated π-Domains***: The structure of the carbon core, particularly its conjugated π-domains, significantly impacts the PL mechanism and facilitates the quantum confinement effect (Ai et al., 2021; Molaei, 2019; Zhu et al., 2015).
- ***Surface State***: The surface state, shaped by the hybridization of the carbon backbone and attached chemical groups, plays a key role in determining the PL properties. Changes in surface oxidation can result in shifts in emission peaks (Ai et al., 2021; Ding et al., 2016; Yan et al., 2019).
- ***Molecular State and Crosslink-Enhanced Emission (CEE)***: The presence of fluorescent molecules either on the surface or within the CDs, along with the CEE effect, further contributes to the PL characteristics (Ai et al., 2021; Wan et al., 2022; Zhu et al., 2015).

5.3 Size and Morphological Characterization

5.3.1 Scanning and Transmission Electron Microscopy

Scanning electron microscopy (SEM) and transmission electron microscopy (TEM) are both valuable methods for examining the size and shape of nanosized materials, specifically carbon dots (CDs). These techniques yield in-depth information regarding the structural and morphological features and the distribution of CDs, which is vital for understanding and enhancing their distinctive optical and chemical properties. SEM excels in surface analysis and ease of use, while TEM is unmatched in revealing atomic-scale details and internal structures. Both techniques complement each other and are often utilized in tandem to offer a comprehensive understanding of CDs.

At its core, TEM operates by transmitting a beam of electrons through an ultrathin specimen (IUPAC, 2019). The interaction of these electrons with the material produces various signals that can be used to construct high-resolution images, allowing for the observation of features at the atomic scale (Howe et al., 2012). This capability is particularly valuable for studying CDs, as it precisely measures their size, shape, and distribution. TEM analysis has revealed that CDs typically exhibit a spherical morphology with varying size distributions depending on their precursors and synthesis method (Nguyen et al., 2024a). For instance, cationic polymer-derived carbon dots have been shown to possess sizes of 1.8 ± 0.4 nm and 5.4 ± 2 nm, depending on the specific polymer used (He et al., 2019). Similarly, CDs synthesized via a Maillard reaction model were found to have an average particle size of 2.5 ± 0.5 nm (Nguyen et al., 2024b). Furthermore, CDs derived from pyrolytic carbon black exhibit differences in photoluminescence properties due

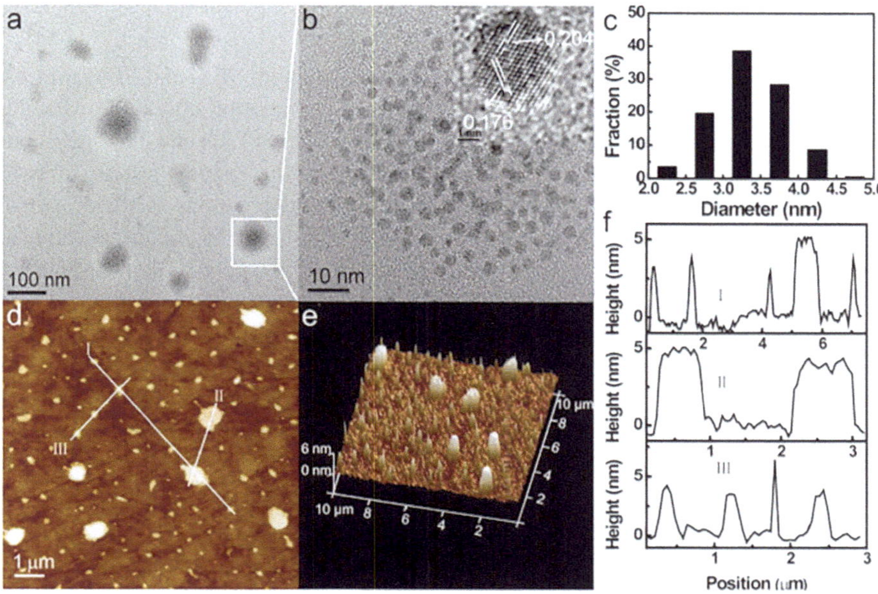

Fig. 5.2 (**a**) TEM, (**b**) TEM and HR-TEM (top left corner), (**c**) size distribution, (**d**) 2D AFM image, (**e**) 3D AFM image, and (**f**) height profile. (Adapted from Huang et al., 2014)

to variations in oxidation levels, as reflected in their TEM images (González-González et al., 2022).

Beyond basic imaging, high-resolution TEM (HR-TEM) can provide further insights into the detailed structural characteristics of CDs (Mehta et al., 2014a). This includes the visualization of graphitic layers and surface functional groups, which are crucial for their fluorescence properties and applications in sensing (Selva Sharma et al., 2023). Figure 5.2a, b reveals the typical TEM and HR-TEM of CDs, showing the high resolution of HR-TEM images compared with TEM images. TEM, in combination with other techniques like X-ray diffraction (XRD) and Raman spectroscopy, helps confirm the monodispersity and uniformity of CDs, which is essential for achieving consistent optical properties (Ding et al., 2013; Mintz et al., 2021).

One of the main challenges in the characterization of CDs is achieving uniformity in their size and shape, which is vital for their consistent performance in various applications. TEM provides valuable insights into these aspects, thereby guiding the optimization of synthesis processes (González-González et al., 2022). Resolution and contrast in TEM imaging present the greatest challenges for future research aimed at achieving a deeper understanding of the complex structures of CDs, particularly those with doped elements or intricate surface modifications (Kitchawengkul et al., 2021). Improved imaging techniques will not only aid in the fundamental understanding of these materials but also enhance their application potential in fields ranging from technology to biomedicine.

5.3.2 Atomic Force Microscopy (AFM)

While the size and morphology of CDs can be observed with SEM and TEM images, atomic force microscopy (AFM) can provide more information about CDs' height, roughness, and mechanical properties. Figure 5.2d, e displays 2D and 3D topographical AFM maps of CDs, revealing the surface properties by scanning a sharp tip over the specimen's surface. The height profile at different positions is illustrated in Fig. 5.2f (Xu et al., 2015). It has been mentioned that the AFM technique can also provide insights into the surface characteristics of CDs, such as the presence of hydroxyl groups, which contribute to their high fluorescent quantum yield and photostability. Due to advancements in this technique, various ongoing research projects are being developed to improve the resolution and sensitivity of AFM probes. Carbon nanotube tips, for example, provide higher resolution and are being explored to enhance AFM's capabilities in characterizing complex nanostructures (Wilson & Macpherson, 2009).

5.3.3 Dynamic Light Scattering (DLS)

Dynamic light scattering (DLS) is a widely used technique for characterizing the size distribution of particles, such as CDs, in a solution (Mehta et al., 2014a; Rojas-Valencia et al., 2021). This method is particularly useful for analyzing nanoparticles due to its noninvasive nature and ability to provide rapid results. DLS offers several advantages, such as being a fast and convenient tool for particle size analysis. Figure 5.3c shows the DLS of CDs extracted from *Saccharum officinarum* juice with a measured average diameter of 2.71 nm. However, it is sensitive to factors like particle interaction, polydispersity, and light absorption, which can affect the accuracy of the measurements (Hassan et al., 2015). Recent advancements in DLS, such as adaptive and statistically driven data analysis, have improved the precision of measurements by addressing issues related to transient size fractions and sample aggregation (Malm & Corbett, 2019).

DLS offers several advantages, such as being a fast and convenient tool for particle size analysis. However, it is sensitive to influences such as particle interaction, polydispersity, and light absorption, which may compromise measurement accuracy (Hassan et al., 2015). Recent developments in DLS, namely adaptive and statistically driven data analysis, have enhanced measurement precision by tackling issues related to transient size fractions and sample aggregation (Malm & Corbett, 2019). While DLS is a powerful method for quickly assessing the particle distribution of nanoparticles, it encounters considerable challenges in measuring CDs due to their strong fluorescent characteristics.

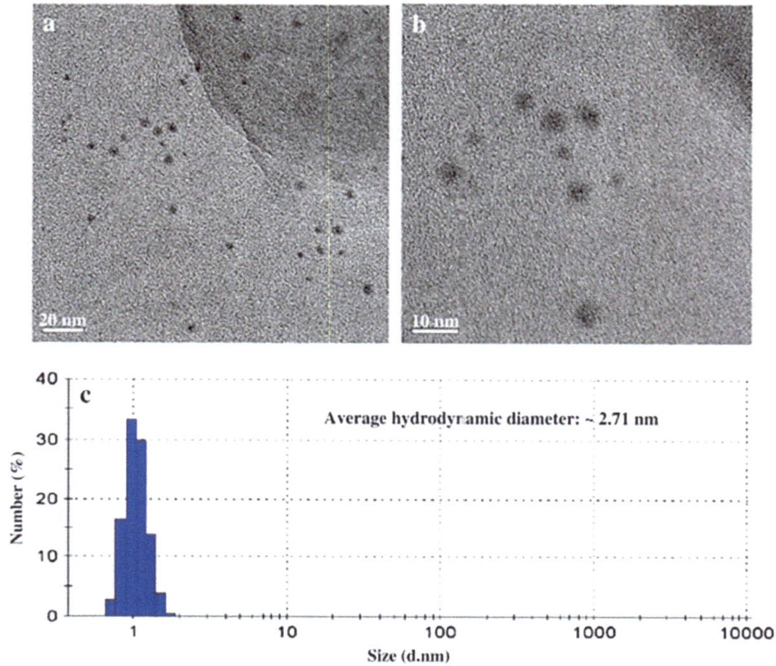

Fig. 5.3 (**a** and **b**) HR TEM of carbon dots (CDs) and (**c**) dynamic light scattering (DLS) of CDs. Adapted from Mehta et al. (2014a). One-pot green synthesis of carbon dots by using *Saccharum officinarum* juice for fluorescent imaging of bacteria (*Escherichia coli*) and yeast (*Saccharomyces cerevisiae*) cells. Materials Science and Engineering: C, 38, 20-27. (Mehta et al., 2014a)

5.4 Surface Functionality and Chemical Composition

5.4.1 Raman Spectroscopy

Raman spectroscopy is a powerful, nondestructive, and rapid technique that provides you with any material's chemical and structural details. It is based on the inelastic scattering of light, also known as Raman scattering. When a molecule interacts with monochromatic light, most photons scatter with no energy change, i.e., Rayleigh's scattering. Some fractions of photons can undergo Raman scattering, where the molecules shift their energy to molecular vibrations (Reich & Thomsen, 2004; Wu et al., 2018). Each molecule has its molecular shift, leading to individual characterization. Equipment consists of three different lasers as their source, such as a green (532 nm), near-infrared (785 nm), and an infrared (1064 nm) laser. Each laser has its advantages and applications depending on the properties of a sample. The Raman spectrophotometer, along with the Raman spectra of CDs illustrated in Fig. 5.4.

Raman spectroscopy is good at revealing how the samples are, whether graphitized or carbonized. It provides vital information for understanding the

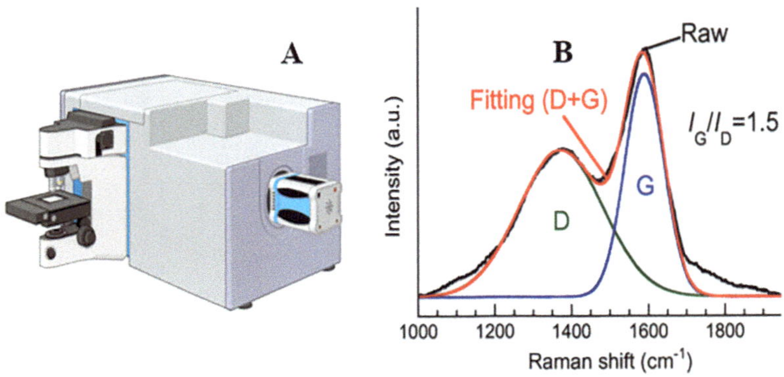

Fig. 5.4 (**a**) Raman spectrometer; (**b**) Raman spectra of glycothermally synthesized CDs (Yoshinaga et al., 2021)

different synthesis methods and is helpful in the optimization of CDs for various applications or processes. The most common laser used for carbon-based materials and CDs is 532 nm, where strong signals can be detected with this type of laser. However, fluorescent interference becomes the main disadvantage when applying this method to CD samples. This overlap led to a weaker Raman signal. Several techniques could be applied for overcoming this drawback, like micro-scale spatially offset Raman spectroscopy (Micro-SORS) (Conti et al., 2016) or time-resolved Raman spectroscopy (Wang & Fang, 2024).

Raman spectroscopy can characterize materials with crystal size, chemical functionality, and doped with multiple elements (Dervishi et al., 2019; Mintz et al., 2021). Zhao et al. reported that the peaks at 1580 cm^{-1} (G band) show the graphitic domain with sp^2 hybridization in a 2D hexagonal lattice linked with the graphite's E_{2g} mode. Peak at 1350 cm^{-1} (D band) shows the defects or disordered regions, corresponding to the sp^3 carbon hybridizations in the termination plane of glassy carbon (Liu et al., 2003; Zhao et al., 2025). Similarly, reports confirm the graphite structure in CDs with the sp^2 hybridization (G band) and sp^3 hybridization (D band). Boron doping gives rise to higher surface defects, increases the chemiluminescence of the CDs, and reduces the G band peak's intensity (Li et al., 2022; Xiao et al., 2025). Singh et al. reported similar results of a narrow G band at 1570 cm^{-1} and a broader D-band at 1343 cm^{-1}, respectively (Singh et al., 2025). The relative intensity ratio (I_D/I_G) values were found to be 0.92, which shows that the samples have moderate defects in the lattices. When biomass is carbonized, higher-density defects occur at the amorphous carbon structure. Other bands, such as the 2D band at 2700 cm^{-1}, represent the multi-layered graphene structures, and the D+G band at 2900 cm^{-1} represents the combination modes, indicating the disorders in the CDs. The higher D band intensity usually comes from the CDs with low H_2 content; in some cases, it may arise from the particles in sub-nanometric sizes or due to polymers (Bian et al., 2025; Dervishi et al., 2019).

5.4.2 Fourier-Transform Infrared Spectroscopy (FTIR)

FTIR spectrometry provides real-time analytical measurement and a nondestructive technique that facilitates the detection and quantification of the functional groups present in unknown samples, including the organic and inorganic components in different states of matter. Understanding the surface functionality is an essential element in CD characterization. Each molecule has its own molecular vibration. When an infrared light interacts with the sample, each molecule will start to vibrate uniquely. These molecular vibrations are reflected in the absorption spectra across different wavelength ranges and use the Fourier transformation to convert this spectrum into a readable spectrum. The FTIR spectra of CDs are influenced by the raw materials, synthesizing techniques, and reaction parameters, resulting in varying functional groups (–C=O, –COOH, –OH, $-NH_2$) on their surfaces (Roy et al., 2015; Zheng et al., 2017). Figure 5.5 illustrates the FTIR spectrophotometer and an example of FTIR spectra of CDs. The functional group on the surface can act as a binder for the other molecules. FTIR, along with other techniques, helps in analyzing these effects, such as differences in photoluminescence properties, which are influenced by the degree of oxidation and the presence of specific functional groups (González-González et al., 2022).

Generally, –C=C absorption peaks will appear from 1450 to 1600 cm^{-1} (Ţucureanu et al., 2016). Baruah et al. reported the bending of –C–H peak at 877.14 cm^{-1} and –C=C peak at 672 cm^{-1}. Peaks at 1056.8 cm^{-1} and 1618.89 cm^{-1} reveal that the *Mesua Ferrea* CD samples contain the –C–O–C and –C=O stretching, giving insights into the ether and carbonyl groups in the surface of the carbon structure (Baruah et al., 2025). This understanding helps in the development of the optical sensing film used for the detection of heavy metal ions. Wu et al. found the peaks of 1640 cm^{-1} (–C=O), 1023 cm^{-1} (–C–O–C), and 2940 cm^{-1} (–C–H) stretching in the CD sample from the polysaccharides of the

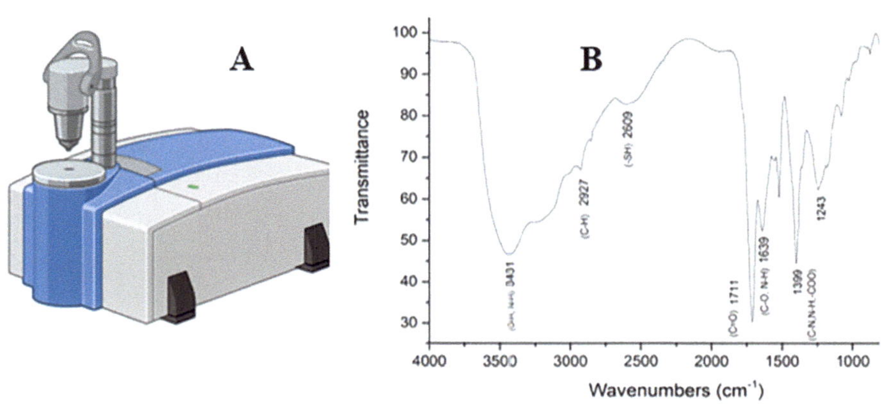

Fig. 5.5 (**a**) Fourier-transform infrared spectrometer (FTIR); (**b**) FTIR spectra of carbon dots (CDs) (Lin et al., 2019)

Stropharia rugosoannulata waste (Wu et al., 2025). Nitrogen (N)-doped CDs produced from the pyrolysis have the peaks at 1400 cm^{-1} (–C–N), 1537 cm^{-1} (–C=C), 1213 cm^{-1} (–C–O), and 1718 cm^{-1} (–C=O) explaining the CD surface vibrations, which help them act as an effective inhibitor against the corrosion for the carbon steel (Ma et al., 2025). Similarly, the N-doped CDs showed peaks at 1296 cm^{-1} (–C=C) and 1375 cm^{-1} (–C-OH), which can be related to the vibrational stretching of carbon at aromatic structures, which reveals the presence of hydroxyl, amine, and carbonyl functional groups, which help to irrigate lettuce and pepper plantations compounds (Salha et al., 2025). Castelo et al. synthesized the CD from *Caryocar brasiliensi* by microwave pyrolysis method, where the FTIR analysis reveals the presence of peaks at 1391 cm^{-1} (–C=C), 1654 cm^{-1} (–C=O), and 2930 cm^{-1} (–C–H), which corresponded to the immunosensing of *Salmonella typhimurium* with the limit of detection (LOD) up to 1 CFU/mL (Castelo et al., 2025). Bian et al. mentioned that the FTIR can give critical insights into the CDs' functional groups, carbonization, and surface chemistry, making it a powerful tool in characterizing CDs (Bian et al., 2025).

5.4.3 X-Ray Photoelectron Spectroscopy (XPS)

X-ray photoelectron spectroscopy (XPS) is used to analyze the surface composition and chemical states of CDs. The XPS spectrophotometer with the typical XPS spectra of CDs is shown in Fig. 5.6. It identifies the presence of functional groups and elements such as carbon, oxygen, and nitrogen on the surface of CDs. XPS is a sensitive tool for surface irradiation, penetrating the sample up to a few micrometers. It allows depth profiling to understand the chemical distribution from the surface to

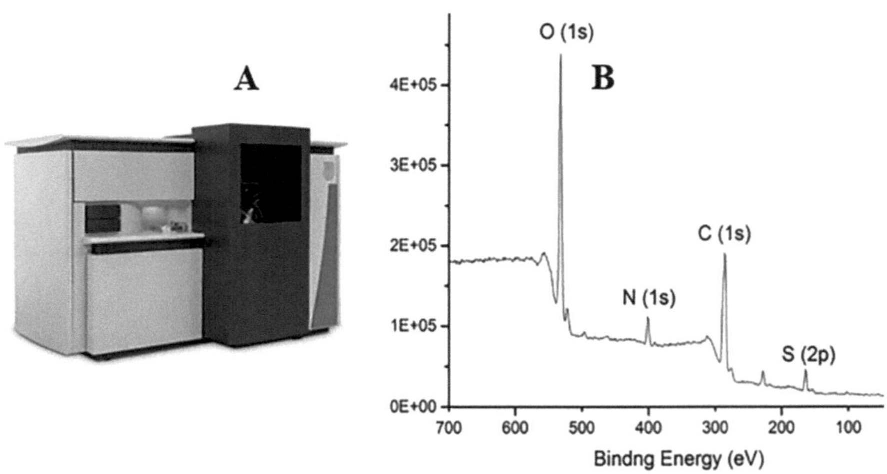

Fig. 5.6 (**a**) X-ray photoelectron spectrometer (XPS); (**b**) XPS spectra of carbon dots (CDs) from orange peel (Zaca-Moran et al., 2024)

the core of CDs. The detector detects the resultant electron with lesser energy, which provides the binding energy of the element in different intensity peaks to identify the elements present on the surface (Greczynski et al., 2023; Stevie & Donley, 2020). For example, measuring XPSs of birch leaves-derived CDs, which comprise ~79% carbon, ~21% oxygen, and 0.7% nitrogen, provides chemical bonding information including the 285.0 eV (C–C), 286.4 eV (C–N/C–O), 289.4 eV (COOH), and 288.1 eV (C=O) (Tang et al., 2023, 2025).

For instance, XPS analysis has revealed the presence of oxidized surface defects and functional groups like phenolic –OH and –COOH, which are crucial for the photoluminescence properties of CDs (Nguyen et al., 2020; Papagiannouli et al., 2018). Additionally, XPS can detect changes in the O/C atomic ratio due to chemical oxidation processes, affecting photostability and fluorescence of CDs (Tan et al., 2013). XPS is also used to study the effects of doping and surface modifications on CDs. For example, N-doped CDs synthesized by microwave show a significant difference in oxygen and carbon content by analyzing XPS data (Ren et al., 2025). As the microwave processing time increased from 120 to 180 min, the carbon percentage rose from 73% to 87%, the oxygen percentage dropped from 22% to 9%, and the nitrogen percentage slightly decreased from 5% to 4%. In another case, XPS spectra of CDs co-doped with sulfur (S) and nitrogen (N) reveal ~63% carbon, ~24% oxygen, ~11% nitrogen, and ~3% sulfur (Zhu et al., 2025). High-resolution XPS of C1s shows peaks at 288.2 eV (C=O/C=N), 287.2 eV (C–S), 284.6 eV (C=C/C–C), and 285.8 eV (C–N/C–O). N1s high-resolution XPS shows 398.65 eV (Pyridinic N), 400.5 eV (Graphitic N), and 399.72 eV (Pyrrolic N), indicating nitrogen mainly resides in the carbon core. These CDs are useful for imaging, optical sensing, display technologies, and quantitative heme estimation in biological samples.

5.4.4 Nuclear Magnetic Resonance (NMR) Spectroscopy

Nuclear magnetic resonance (NMR) spectroscopy can provide detailed information about the functional group, chemical bonding, and chemical structure on the sample's surface. It can give precise data about how structural changes can happen at the synthesizing stage, indicating whether carbonization or polymerization is involved in the chemical alteration. NMR is based on the nuclear spin resonance principle, in which a sample is placed in a high magnetic field, and nuclei with a spin (such as ^{1}H or ^{13}C) interact with the field and absorb certain radiofrequency radiation. It leads to a transition in the energy levels. As the nuclei return to their original state, they emit the signal, which can be detected with a spectrometer, which provides insights into their molecular structure and chemical composition of the sample. CDs can be characterized mainly using ^{13}C NMR and ^{1}H NMR (Arroyave et al., 2021; Tang et al., 2022). The NMR spectrometer with a typical ^{1}H NMR spectrum of carbon dots (CDs) is shown in Fig. 5.7.

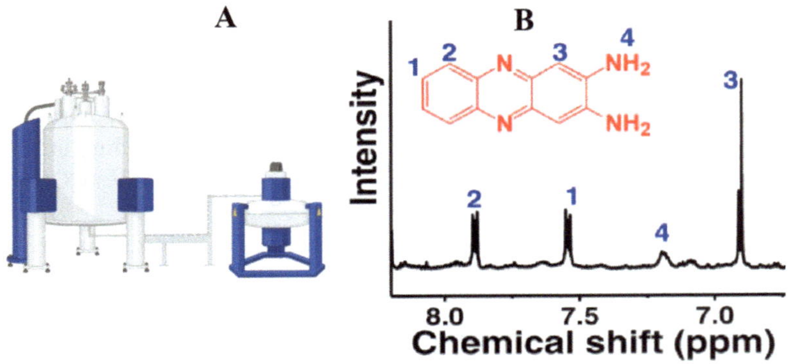

Fig. 5.7 (**a**) Nuclear magnetic resonance (NMR) spectrometer; (**b**) focused ^{1}H NMR spectra of carbon dots (CDs) from hydrothermal synthesis (Zhang et al., 2021)

NMR can identify and quantify fluorophores present in the CD samples, influencing the fluorescence property (Duan et al., 2020). ^{13}C NMR helps in identifying the carbon hybridizations such as carbonyl, carboxyl, and amorphous carbon domains (sp^2 aliphatic and sp^2 aromatic carbons) (Philippidis et al., 2013). Evans et al. reported that NMR can be utilized as a tool to avoid false-positive results arising from confusion between CDs and oligomers or other small molecular residues, particularly when CDs are present at concentrations below the micromolar level (Evans et al., 2018). Sousa et al. reported that CDs are detected with broader peaks, representing larger molecules or aggregates due to slower molecular tumbling and stronger dipole–dipole interactions. In contrast, sharp, narrow peaks represent the residues, or small molecular impurities present in the sample, where the purification step must be done (Sousa et al., 2024). The purification step must be effective until the peaks appear broader. Broader peaks show the absence of minor molecular impurities in the LOD of NMR. For better sensitivity of ^{1}H NMR, the sample size of a 500 Da organic molecule should have a minimum of 1 mg to have a standard signal-to-noise ratio. It also corresponds to a millimolar concentration in a 500-μL NMR tube. Bartolomei et al. reported the results of ^{1}H NMR analysis, which revealed the chiroptical properties of nonpurified CDs, purified CDs, and their chiral precursors. The outcome shows that the chirality is from the unreacted precursor, not from the CDs (Bartolomei et al., 2022). Another essential technique is 2D NMR, namely DOSY (diffusion-ordered spectroscopy) NMR, which can separate the signal based on the molecular diffusion rates rather than chemical shifts. It can differentiate CDs from other small molecular byproducts. Usually, the CD diameter size ranges between 2 and 10 nm. It comprises various functional groups on its surface and the carbon core, which are more significant than the organic molecules and typically less than 1 nm. This size difference can lead to slower diffusion in the solution of CDs, making them appear in 2D DOSY spectra as differential signal clusters. CDs in water at ambient temperature with the size of 2–10 nm must have the diffusion coefficient of 2.0–5.0 × 10^{10} m^2 s^{-1} (Wang et al., 2011).

5.4.5 Zeta Potential Measurement

Zeta potential (ζ) provides the electrochemical stability of solid particles in the stagnant or mobile liquid phase. Considering the particles' charge density, two layers, the Stern layer and the diffusion layer, play an essential role in estimating zeta potential, which is usually represented in millivolts (mV). It is also called the electrokinetic potential in the interfacial plane generated between the diffuse and Stern layers. CDs' zeta potential depends on CDs' concentration, size, solvent, temperature, electroconductivity, and pH. It gives the colloidal stability of the system (Lunardi et al., 2021; Serrano-Lotina et al., 2023). Ye et al. synthesized CDs utilizing lanthanum nitrate and folic acid as precursors, where they increased the lanthanum nitrate concentration to use CDs as an effective inhibitor for glioblastoma cells (Ye et al., 2025). Zeta potential measurement of the samples reveals that they could achieve adequate stability in the suspension at 30% lanthanum nitrate. Stachowaka et al. reported that the zeta potential is affected by the negative group present on the CD surface (Stachowska et al., 2021). Citric acid and urea in the ratio of 1:6 have the ζ value of ~−35 mV, which shows better stability in their aqueous suspension. Researchers utilize this technique to check the stability of their nanosuspension by incorporating certain other nanoparticles for improving/analyzing the CDs' affinity toward the specific target molecules (Dinç et al., 2025). Reports show that the 5-fluorouracil and hyaluronic acid were loaded with CDs as active tumor Wnt/β-catenin inhibitors (Yang et al., 2025). CDs with 5-fluorouracil have a ζ value of ~ −5 mV. ζ value increased to ~−23 mV when loaded with hyaluronic acid, which makes it more affinity toward the tumor cell CD44 receptor. Jardan et al. showed the mechanism of the electrostatic repulsion happening in their sample matrix using the zeta potential analysis (Jardan et al., 2025). They mentioned that at higher pH, there is an increase in negative charge in the system, which tends to cause the CDs to aggregate, causing the fluorescence quenching effect. Results also indicate that the CDs carried high positive charges at low pH, which caused the adequate disaggregation of CDs. In turn, they increase the fluorescence due to their higher electrostatic repulsion. These data are similar to the results of Setianto et al. (2024). This study gives us a complete understanding of how zeta potential can help us to characterize the CDs by providing sufficient evidence for the electrostatic repulsion involved in the anti-fluorescence quenching and disaggregation of CDs. The Zetasizer analysis and the corresponding zeta potential graph of the CDs are illustrated in Fig. 5.8.

5.5 Core Structure Analysis

5.5.1 X-Ray Diffraction (XRD)

X-ray diffraction (XRD) is commonly used to determine the crystalline structure of carbon dots. It helps in identifying the presence of graphitic structures within the CDs, which is essential for understanding their optical properties and potential

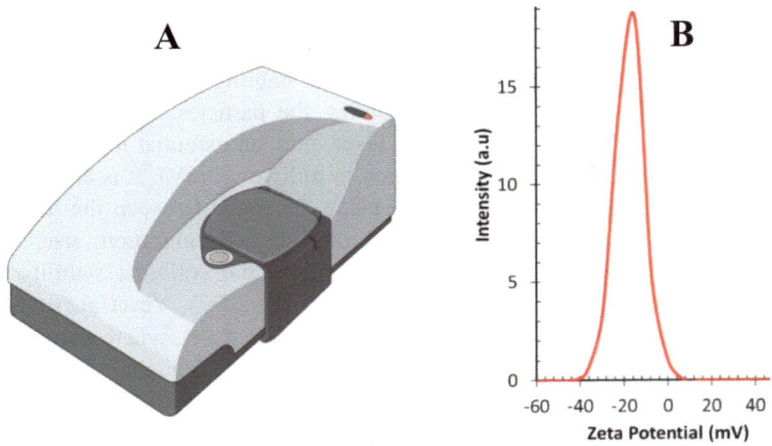

Fig. 5.8 (**a**) Zetasizer; (**b**) zeta potential of carbon dots (CDs) from hydrothermal synthesis (Setianto et al., 2024)

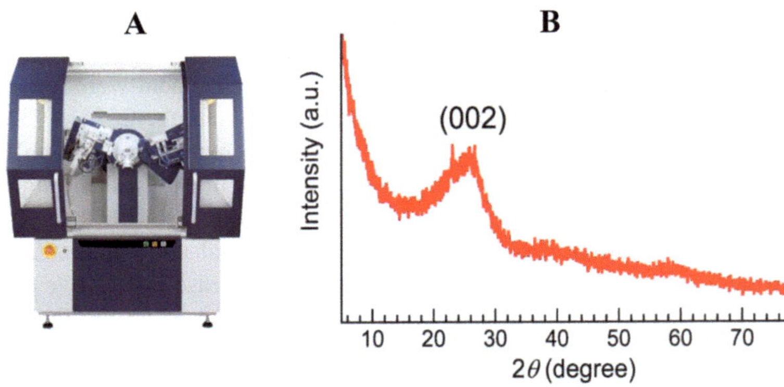

Fig. 5.9 (**a**) X-ray diffractometer; (**b**) X-ray diffraction (XRD) spectra of glycothermally synthesized carbon dots (CDs) (Yoshinaga et al., 2021)

applications. For instance, XRD analysis has been used to confirm the structural properties of CDs synthesized from various precursors, revealing the presence of graphitic carbon cores and other structural features (Mehta et al., 2014b; Shamsipur et al., 2018).

The X-ray diffractometer and typical XRD spectra are illustrated in Fig. 5.9. Peaks are generally produced due to the X-ray diffracted by the sample's lattice planes from the constructive interference of a monochromatic X-ray beam (Bunaciu et al., 2015; Epp, 2016). The peak's intensity mainly depends on the atom's distribution within the lattice. Bragg's law relates the wavelength of electromagnetic radiation to the diffraction angle and lattice spacing in a crystalline sample. The typical radiation source is Cu-Kα ($\lambda = 1.5406$ Å) (Mintz et al., 2021). By scanning the sample across a range of 2θ angles, the detector can capture all the potential diffractions of the lattice. The scanning range (2θ) typically ranges from 5° to 80°, with a focus mainly between 10° and 40° for carbon-based materials. A slower scan speed of 1° per minute ensures high-quality data, and a step size of 0.02–0.05 is maintained to achieve better resolutions (Lee et al., 2014). Converting these diffracted peaks into d-spacings enables the identification of the compound, as each compound possesses a unique set of d-spacings (Holder & Schaak, 2019; Khan et al., 2020; Mintz et al., 2021).

CDs usually peak between 20° and 30° (2θ), representing the (002) plane of graphitic carbon, representing a disordered turbostratic carbon structure. CDs, which contain crystalline phases as they are doped with metal ions or inorganic materials, produce sharp peaks showing the presence of crystalline domains. Due to its surface oxidation or functionalization, CDs exhibit a d-spacing layer higher than that of graphite (~0.34 nm) (Hui, 2023; Pal et al., 2020). The XRD pattern of nitrogen and sulfur co-doped CDs from persimmon fruit shows broader peaks, representing amorphous materials (Krishnaiah et al., 2025). Peaks with $2\theta = 24°$ and 44° related to (002) and (100) planes are carbon-based nanomaterials, which are more typical of graphitic carbon structures (Jung et al., 2024). Yang et al. found that the XRD pattern can also be characterized by an interlayer spacing of 0.36 nm connected to the (002) plane and 0.21 nm linked with the (100) plane (Yang et al., 2014). Madadi et al. reported the XRD patterns of copper cobaltite bimetallic oxide with graphene quantum dot composite (Madadi et al., 2025). The doping of graphene quantum dots into the copper cobaltite does not change the crystal structure. A subtle shift toward lower angles was observed due to increased interplanar spacing (d-spacing) within the crystal lattice, which also suggests the creation of larger proton-conducting channels within the nanocomposite (Xu et al., 2021; Zahedi & Shabani-Nooshabadi, 2023).

5.5.2 Electron Spin Resonance (ESR)

Electron spin resonance (ESR) spectroscopy, often called electron paramagnetic resonance (EPR) spectroscopy, is a robust analytical method designed to examine materials that contain unpaired electrons. A continuous microwave beam of fixed frequency irradiates the samples, while the magnetic field is fixed. This technique is more valuable for analyzing paramagnetic entities, including free radicals, transition metal complexes, and specific defects present in solid materials (He et al., 2014). Different types of microwave frequencies are employed, which are as follows: 95 GHz (W-band), 35 GHz (Q-band), 20 GHz (K-band), 9.25 GHz (X-band), and 3.5 GHz (S-band) (Weckhuysen et al., 2004). The most used band frequency is X-band (9.25 GHz) with a corresponding resonance field of 0.330 T for $g = 2$. This technique can characterize the transition metal ions present on the surface and lattice and is also coordinated with the extra-lattice ligands or oxygen atoms (Eaton & Eaton, 2002). It can examine the paramagnetic clusters and adsorbed paramagnetic molecules to investigate Lewis acidity and defects. The ESR spectra exhibit significant interactions compared to the electron Zeeman interaction. It makes the first-order perturbation theory for spectral evaluations (Naveed et al., 2018). CNDs exhibit different g-values depending on surface functional groups and doping effects. The typical *g*-value for graphitic/defective carbon ranges between ~2.002 and 2.005 (Pham et al., 2014). Cao et al. utilized the EPS method to evaluate the photocatalytic production of lactic acid from biomass sugars using N-doped CDs and an ultra-thin carbon nitride layer (Cao et al., 2025). EPR spectra show the presence of several radicals, including charge carriers (e^- and h^+), superoxide (O_2^-), singlet oxygen ($^1O_2$), and hydroxyl (OH). The free radical quenching experiments were performed, and the results showed the O_2^- radical in the catalytic conversion process. EPR spectra of hypoxic molybdenum oxide-doped safflower CDs exhibit enhanced phototherapy against cancer cells (Wang et al., 2025). When the CDs encounter H_2O_2, sextet and triplet signals are observed, with the production of O_2^- and $^1O_2$. It can also be inferred that the peroxidase enzyme-like activity was observed when the hypoxic molybdenum oxide-doped safflower CDs produced O_2^- and OH to induce the enhanced reactive oxygen species production. Similarly, researchers indicated that the hydrophilic CDs doped with Fe^{3+} ion and carboxyl groups show the capability of recycling the natural surface pollution. Zhang et al. reported that EPR spectra show the OH signals, which in turn show the ability of the synthesized material in the efficient photo-induced electron transfer (Zhang et al., 2025).

5.5.3 Thermogravimetric Analysis (TGA)

Thermogravimetric analysis (TGA) is a quantitative analytical method where the sample is put in a high-temperature furnace with a steady temperature rise over 1600 °C and constant or variable gas flow rates where the mass change is recorded. The weight loss of the samples is caused by decomposition of the sample, evaporation of volatile constituents, or reduction in the oxidation states due to the high temperature. The weight of the transition metals increases in high oxygen atmospheres. The technique can identify the polymorphic reactions, transformations, and phase transitions without weight change. Differential scanning calorimetry (DSC) curves can be coupled with TGA for a simple understanding. This can differentiate between the physical and chemical changes in the sample (Loganathan et al., 2017; Rami et al., 2021; Saadatkhah et al., 2020). Zhang et al. explored various applications of CDs by examining the doping compositions. They also discussed the functional groups of the CDs and the purity of the precursors used in the fabrication process (Shishodia et al., 2021; Zhang et al., 2019). Xiao et al. synthesized CDs in waterborne polyurethane, and N-doped CDs exhibit the formation of a thermally stable network structure with an increase in decomposition temperature (Xiao et al., 2020). R. Zhu et al. compared functionalized CDs with a carbonyl and an amino group. Results indicate that the amino-functionalized CDs are more stable than the carbonyl CDs (Zainal Abidin et al., 2020; Zhu et al., 2019a). Rahmani and Gheamy found that the CDs with gum tragacanth (carbohydrate polymer) show enhanced

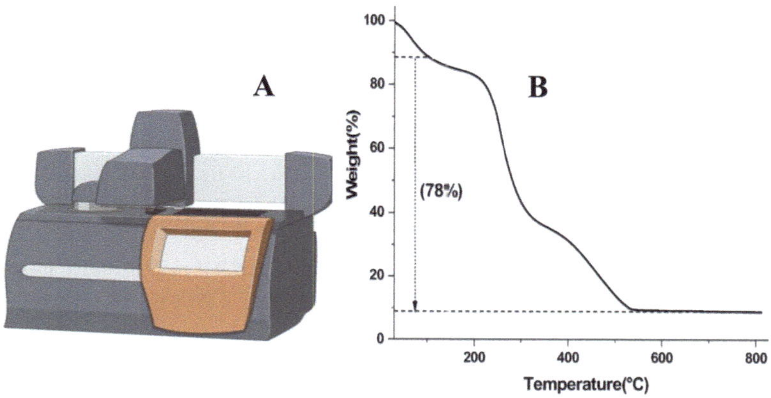

Fig. 5.10 (**a**) Thermal analyzer; (**b**) thermogravimetric analysis (TGA) of carbon dots coated with cobalt zinc ferrite (Malitha et al., 2024)

thermal stability. TGA results indicate the notable influence of the CD's size on its stability, which is linked to its pseudo-graphene structure (Rahmani & Ghaemy, 2019). Similarly, CDs from pine fruit and oleic acid show higher thermal stability (Rimal et al., 2020; Shahba & Sabet, 2020). Zhu et al. showed that the amphiphilic CDs integrated with the polyvinyl alcohol composite film exhibit a higher T_g value than the pure film (Zhu et al., 2019b). This is due to the hydrogen bonding from the interfacial interactions between the PVA and amphiphilic CDs, which creates makeshift stability. Hydrogen bonding plays an important role in the thermal stability of CDs. Ferrer-Ruiz et al. derived CDs from glutamic acid that show constant thermal stability up to 140 °C. Above that temperature, the graphitization begins slowly, so the thermal stability remains until 220 °C. Full decomposition occurs at 400 °C, 150 °C higher than the precursor temperature (Ferrer-Ruiz et al., 2025). The thermal analyzer and typical TGA graph are illustrated in Fig. 5.10.

5.6 Conclusion

The characterization of carbon dots (CDs) involves a multifaceted approach using various analytical techniques. Each method, from optical properties analysis and size and morphology characterization to surface functionality, chemical composition, and core structure analysis, provides unique insights into the nature of CDs. Techniques like UV-Visible spectroscopy, photoluminescence (PL), transmission electron microscopy (TEM), and Raman spectroscopy are fundamental in understanding the basic properties and functionality of CDs. Supporting techniques such as atomic force microscopy (AFM), dynamic light scattering (DLS), X-ray photoelectron spectroscopy (XPS), nuclear magnetic resonance (NMR) spectroscopy, and zeta potential measurement contribute to a more comprehensive understanding. Advanced methods like electron spin resonance (ESR) and thermogravimetric analysis (TGA) further enhance our knowledge of the structural and thermal properties of CDs. By integrating these techniques, researchers can achieve a holistic characterization, optimizing CDs for various applications in technology, biomedicine, and environmental science.

References

Ai, L., Yang, Y., Wang, B., Chang, J., Tang, Z., Yang, B., & Lu, S. (2021). Insights into photoluminescence mechanisms of carbon dots: Advances and perspectives. *Science Bulletin, 66*(8), 839–856. https://doi.org/10.1016/j.scib.2020.12.015

Arroyave, J. M., Ambrusi, R. E., Robein, Y., Pronsato, M. E., Brizuela, G., Di Nezio, M. S., & Centurión, M. E. (2021). Carbon dots structural characterization by solution-state NMR and UV–visible spectroscopy and DFT modeling. *Applied Surface Science, 564*, 150195. https://doi.org/10.1016/j.apsusc.2021.150195

Bartolomei, B., Bogo, A., Amato, F., Ragazzon, G., & Prato, M. (2022). Nuclear magnetic resonance reveals molecular species in carbon nanodot samples disclosing flaws. *Angewandte Chemie International Edition, 61*(20), e202200038. https://doi.org/10.1002/anie.202200038

Baruah, S., Hazarika, P., Konwar, A., Hazarika, K. K., & Hazarika, S. (2025). Synergistic optical sensing of heavy metal ions using *Mesua Ferrea* derived carbon dot embedded Alginate based biopolymeric film as a sensor platform. *International Journal of Biological Macromolecules, 286*, 138227. https://doi.org/10.1016/j.ijbiomac.2024.138227

Bian, Z., Gomez, E., Gruebele, M., Levine, G. B., Link, S., Mehmood, A., & Nie, S. (2025). Bottom-up carbon dots: Purification, single-particle dynamics, and electronic structure. *Chemical Science.* https://doi.org/10.1039/D4SC05843G

Bunaciu, A. A., Udriştioiu, E. G., & Aboul-Enein, H. Y. (2015). X-ray diffraction: Instrumentation and applications. *Critical Reviews in Analytical Chemistry, 45*(4), 289–299. https://doi.org/10.1080/10408347.2014.949616

Cao, X., Wang, X., He, Y., Li, J., Yang, H., Zhang, P., Yang, S., Yu, H., Xu, S., Yuan, H., & Gu, L. (2025). Boosting lactic acid photocatalytic synthesis from biomass sugars: The synergistic effect of N-doped carbon dots on ultrathin carbon nitride. *Food Chemistry, 464*, 141645. https://doi.org/10.1016/j.foodchem.2024.141645

Castelo, R. M., de Albuquerque Oliveira, M., Furtado, R. F., de Oliveira, B. P., Martoni, L. V. L., Machado, T. F., Muniz, C. R., da Silva Abreu, F. O. M., Machado, S. A. S., Melo, A. M. A., Cheng, H. N., Biswas, A., & Alves, C. R. (2025). Carbon-dot pequi-nut in the development of immunosensor to detect pathogenic bacteria. *Brazilian Journal of Microbiology.* https://doi.org/10.1007/s42770-025-01612-1

Conti, C., Botteon, A., Colombo, C., Realini, M., & Matousek, P. (2016). Fluorescence suppression using micro-scale spatially offset Raman spectroscopy. *The Analyst, 141*(18), 5374–5381. https://doi.org/10.1039/C6AN00852F

Crosby, G. A., & Demas, J. N. (1971). Measurement of photoluminescence quantum yields. *Review. The Journal of Physical Chemistry, 75*(8), 991–1024. https://doi.org/10.1021/j100678a001

Dervishi, E., Ji, Z., Htoon, H., Sykora, M., & Doorn, S. K. (2019). Raman spectroscopy of bottom-up synthesized graphene quantum dots: Size and structure dependence. *Nanoscale, 11*(35), 16571–16581. https://doi.org/10.1039/C9NR05345J

Dinç, S., Tjardts, T., Maschkowitz, G., Lukaszczuk, V., Taghi Gharibzahedi, S. M., & Altintas, Z. (2025). Highly sensitive detection of Campylobacter jejuni using a carbon dot-embedded nanoMIPs fluorescent sensor. *Microchemical Journal, 208*, 112301. https://doi.org/10.1016/j.microc.2024.112301

Ding, H., Cheng, L.-W., Ma, Y.-Y., Kong, J.-L., & Xiong, H.-M. (2013). Luminescent carbon quantum dots and their application in cell imaging. *New Journal of Chemistry, 37*(8), 2515. https://doi.org/10.1039/c3nj00366c

Ding, H., Yu, S.-B., Wei, J.-S., & Xiong, H.-M. (2016). Full-color light-emitting carbon dots with a surface-state-controlled luminescence mechanism. *ACS Nano, 10*(1), 484–491.

Duan, P., Zhi, B., Coburn, L., Haynes, C. L., & Schmidt-Rohr, K. (2020). A molecular fluorophore in citric acid/ethylenediamine carbon dots identified and quantified by multinuclear solid-state nuclear magnetic resonance. *Magnetic Resonance in Chemistry, 58*(11), 1130–1138. https://doi.org/10.1002/mrc.4985

Eaton, S. S., & Eaton, G. R. (2002). Electron paramagnetic resonance spectroscopy. In *Characterization of materials*. Wiley. https://doi.org/10.1002/0471266965.com067

Epp, J. (2016). 4—X-ray diffraction (XRD) techniques for materials characterization. In G. Hübschen, I. Altpeter, R. Tschuncky, & H.-G. Herrmann (Eds.), *Materials characterization using nondestructive evaluation (NDE) methods* (pp. 81–124). Woodhead Publishing. https://doi.org/10.1016/B978-0-08-100040-3.00004-3

Evans, R., Dal Poggetto, G., Nilsson, M., & Morris, G. A. (2018). Improving the interpretation of small molecule diffusion coefficients. *Analytical Chemistry, 90*(6), 3987–3994. https://doi.org/10.1021/acs.analchem.7b05032

Ferrer-Ruiz, A., Rodríguez-Pérez, L., Martín, N., & Herranz, M. Á. (2025). Post-synthetic graphitization and photoluminescence tuning of carbon dots from L-glutamic acid. *Small Structures*, 2400532. https://doi.org/10.1002/sstr.202400532

González-González, R. B., González, L. T., Madou, M., Leyva-Porras, C., Martinez-Chapa, S. O., & Mendoza, A. (2022). Synthesis, purification, and characterization of carbon dots from non-activated and activated pyrolytic carbon black. *Nanomaterials, 12*(3), 298. https://doi.org/10.3390/nano12030298

Greczynski, G., Haasch, R. T., Hellgren, N., Lewin, E., & Hultman, L. (2023). X-ray photoelectron spectroscopy of thin films. *Nature Reviews Methods Primers, 3*(1), 40. https://doi.org/10.1038/s43586-023-00225-y

Hassan, P. A., Rana, S., & Verma, G. (2015). Making sense of Brownian motion: colloid characterization by dynamic light scattering. *Langmuir, 31*(1), 3–12. https://doi.org/10.1021/la501789z

He, W., Liu, Y., Wamer, W. G., & Yin, J.-J. (2014). Electron spin resonance spectroscopy for the study of nanomaterial-mediated generation of reactive oxygen species. *Journal of Food and Drug Analysis, 22*(1), 49–63. https://doi.org/10.1016/j.jfda.2014.01.004

He, X., Chen, P., Zhang, J., Luo, T.-Y., Wang, H.-J., Liu, Y.-H., & Yu, X.-Q. (2019). Cationic polymer-derived carbon dots for enhanced gene delivery and cell imaging. *Biomaterials Science, 7*(5), 1940–1948. https://doi.org/10.1039/C8BM01578C

Holder, C. F., & Schaak, R. E. (2019). Tutorial on powder X-ray diffraction for characterizing nanoscale materials. *ACS Nano, 13*(7), 7359–7365. https://doi.org/10.1021/acsnano.9b05157

Howe, J. M., Fultz, B., & Miao, S. (2012). Transmission electron microscopy. In E. N. Kaufmann (Ed.), *Characterization of materials* (1st ed., pp. 1–46). Wiley. https://doi.org/10.1002/0471266965.com082.pub2

Huang, J. J., Zhong, Z. F., Rong, M. Z., Zhou, X., Chen, X. D., & Zhang, M. Q. (2014). An easy approach of preparing strongly luminescent carbon dots and their polymer-based composites for enhancing solar cell efficiency. *Carbon, 70*, 190–198.

Hui, S. (2023). Carbon dots (CDs): Basics, recent potential biomedical applications, challenges, and future perspectives. *Journal of Nanoparticle Research, 25*(4), 68. https://doi.org/10.1007/s11051-023-05701-w

IUPAC. (2019). *Transmission electron microscopy*. International Union of Pure and Applied Chemistry (IUPAC), Edition: 3.0.1. https://doi.org/10.1351/goldbook.T06481

Jabbar Abbas, S., & Atia Muhson, A. (2019). Preparation and characterization of the carbon quantum dots. *Journal of Kufa-Physics, 11*(02), 45–51. https://doi.org/10.31257/2018/JKP/2019/110207

Jardan, Y. A. B., El-Wekil, M. M., Elmasry, M. R., Mahmoud, A. M., & Ali, A.-M. B. H. (2025). Development of a charge-mediated disaggregation-induced fluorescence enhancement platform using NS-carbon dots for metformin quantification and pharmacokinetic studies. *Journal of Fluorescence.* https://doi.org/10.1007/s10895-025-04170-9

Jung, J. Y., Park, J. Y., Hong, W. T., Kim, Y. K., & Yang, H. K. (2024). Luminescent composite material fabricated from carbon dots and red phosphors for application to UV-LED chips and metal ion detection. *Ceramics International, 50*(5), 7908–7918. https://doi.org/10.1016/j.ceramint.2023.12.119

Kailasa, S. K. K., & Hussai, C. M. (2022). *Carbon dots in analytical chemistry: Detection and imaging*. Elsevier.

Khan, H., Yerramilli, A. S., D'Oliveira, A., Alford, T. L., Boffito, D. C., & Patience, G. S. (2020). Experimental methods in chemical engineering: X-ray diffraction spectroscopy—XRD. *The Canadian Journal of Chemical Engineering, 98*(6), 1255–1266. https://doi.org/10.1002/cjce.23747

Kitchawengkul, N., Prakobkij, A., Anutrasakda, W., Yodsin, N., Jungsuttiwong, S., Chunta, S., Amatatongchai, M., & Jarujamrus, P. (2021). Mimicking peroxidase-like activity of nitrogen-doped carbon dots (N-CDs) coupled with a laminated three-dimensional microfluidic paper-based analytical device (laminated 3D-μPAD) for smart sensing of total cholesterol from whole blood. *Analytical Chemistry, 93*(18), 6989–6999. https://doi.org/10.1021/acs.analchem.0c05459

Krishnaiah, P., Atchudan, R., Gangadaran, P., Perumal, S., Rajendran, R. L., Sundramoorthy, A. K., Suresh Kumar, R., Ramalingam, S., Ahn, B.-C., Lee, S. W., & Lee, Y. R. (2025). A sustainable synthesis and applications of biomass waste-derived tunable fluorescent carbon dots: *In vitro* and *in vivo* fluorescent imaging. *Journal of Photochemistry and Photobiology A: Chemistry, 458*, 115944. https://doi.org/10.1016/j.jphotochem.2024.115944

Lee, S.-Y., Yoo, H.-M., Rhee, K. Y., & Park, S.-J. (2014). Synthesis, characterization, and KOH activation of nanoporous carbon for increasing CO2 adsorption capacity. *Research on Chemical Intermediates, 40*(7), 2535–2542. https://doi.org/10.1007/s11164-014-1665-y

Li, T., Wu, C., Yang, M., Li, B., Yan, X., Zhu, X., Yu, H., Hu, M., & Yang, J. (2022). Long-lived color-tunable room-temperature phosphorescence of boron-doped carbon dots. *Langmuir, 38*(7), 2287–2293. https://doi.org/10.1021/acs.langmuir.1c02973

Lin, H., Huang, J., & Ding, L. (2019). Preparation of Carbon Dots with High-Fluorescence Quantum Yield and Their Application in Dopamine Fluorescence Probe and Cellular Imaging. Journal of Nanomaterials, 2019(1), 5037243. https://doi.org/10.1155/2019/5037243

Liu, J., Shao, M., Chen, X., Yu, W., Liu, X., & Qian, Y. (2003). Large-scale synthesis of carbon nanotubes by an ethanol thermal reduction process. *Journal of the American Chemical Society, 125*(27), 8088–8089. https://doi.org/10.1021/ja035763b

Loganathan, S., Valapa, R. B., Mishra, R. K., Pugazhenthi, G., & Thomas, S. (2017). Chapter 4 – Thermogravimetric analysis for characterization of nanomaterials. In S. Thomas, R. Thomas, A. K. Zachariah, & R. K. Mishra (Eds.), *Thermal and rheological measurement techniques for nanomaterials characterization* (pp. 67–108). Elsevier. https://doi.org/10.1016/B978-0-323-46139-9.00004-9

Lunardi, C. N., Gomes, A. J., Rocha, F. S., De Tommaso, J., & Patience, G. S. (2021). Experimental methods in chemical engineering: Zeta potential. *The Canadian Journal of Chemical Engineering, 99*(3), 627–639. https://doi.org/10.1002/cjce.23914

Ma, X., Ma, Y., Dong, L., Chang, X., Yang, T., Cao, X., Feng, L., Hu, Z., & Hu, T. (2025). Novel hydroxyl and carboxyl group-enriched nitrogen-doped carbon dots N-CDs as highly efficient corrosion inhibitors for carbon steel Q235 in 1 M HCl. *Inorganic Chemistry Communications, 175*, 114167. https://doi.org/10.1016/j.inoche.2025.114167

Madadi, M., Rohani Moghadam, M., Salarizadeh, P., Bazmandegan-Shamili, A., & Shahbakhsh, M. (2025). Sensitive electrochemical detection of dopamine using $CuCo_2O_4$/graphene quantum dots-modified carbon paste electrode. *Inorganic Chemistry Communications, 171*, 113566. https://doi.org/10.1016/j.inoche.2024.113566

Malitha, Md. D., Molla, Md. T. H., Bashar, Md. A., Chandra, D., & Ahsan, Md. S. (2024). Fabrication of a reusable carbon quantum dots (CQDs) modified nanocomposite with enhanced visible light photocatalytic activity. Scientific Reports, 14(1), 17976. https://doi.org/10.1038/s41598-024-66046-5

Malm, A. V., & Corbett, J. C. W. (2019). Improved Dynamic Light Scattering using an adaptive andstatistically driven time resolved treatment of correlation data. *Scientific Reports, 9*(1), 13519. https://doi.org/10.1038/s41598-019-50077-4

Mehta, V. N., Jha, S., & Kailasa, S. K. (2014a). One-pot green synthesis of carbon dots by using Saccharum officinarum juice for fluorescent imaging of bacteria (Escherichia coli) and yeast (Saccharomyces cerevisiae) cells. *Materials Science and Engineering: C, 38*, 20–27. https://doi.org/10.1016/j.msec.2014.01.038

Mehta, V. N., Jha, S., Singhal, R. K., & Kailasa, S. K. (2014b). Preparation of multicolor emitting carbon dots for HeLa cell imaging. *New Journal of Chemistry, 38*(12), 6152–6160. https://doi.org/10.1039/C4NJ00840E

Mintz, K. J., Bartoli, M., Rovere, M., Zhou, Y., Hettiarachchi, S. D., Paudyal, S., Chen, J., Domena, J. B., Liyanage, P. Y., Sampson, R., Khadka, D., Pandey, R. R., Huang, S., Chusuei, C. C., Tagliaferro, A., & Leblanc, R. M. (2021). A deep investigation into the structure of carbon dots. *Carbon, 173*, 433–447. https://doi.org/10.1016/j.carbon.2020.11.017

Molaei, M. J. (2019). A review on nanostructured carbon quantum dots and their applications in biotechnology, sensors, and chemiluminescence. *Talanta, 196*, 456–478. https://doi.org/10.1016/j.talanta.2018.12.042

Naveed, K.-R., Wang, L., Yu, H., Ullah, R. S., Haroon, M., Fahad, S., Li, J., Elshaarani, T., Khan, R. U., & Nazir, A. (2018). Recent progress in the electron paramagnetic resonance study of polymers. *Polymer Chemistry, 9*(24), 3306–3335. https://doi.org/10.1039/C8PY00689J

Nguyen, H. A., Srivastava, I., Pan, D., & Gruebele, M. (2020). Unraveling the fluorescence mechanism of carbon dots with *sub* -single-particle resolution. *ACS Nano, 14*(5), 6127–6137. https://doi.org/10.1021/acsnano.0c01924

Nguyen, D., El-Ramady, H., & Prokisch, J. (2024a). Food safety aspects of carbon dots: A review. *Environmental Chemistry Letters.* https://doi.org/10.1007/s10311-024-01779-3

Nguyen, D., Muthu, A., El-Ramady, H., Daróczi, L., Nagy, L., Kéki, S., Béni, Á., Csarnovics, I., & Prokisch, J. (2024b). Optimization of extraction conditions to synthesize green carbon nanodots using the Maillard reaction. *Materials Advances.* https://doi.org/10.1039/D4MA00037D

Pal, A., Palashuddin Sk, M., & Chattopadhyay, A. (2020). Recent advances in crystalline carbon dots for superior application potential. *Materials Advances, 1*(4), 525–553. https://doi.org/10.1039/D0MA00108B

Papagiannouli, I., Patanen, M., Blanchet, V., Bozek, J. D., De Anda Villa, M., Huttula, M., Kokkonen, E., Lamour, E., Mevel, E., Pelimanni, E., Scalabre, A., Trassinelli, M., Bassani, D. M., Lévy, A., & Gaudin, J. (2018). Depth profiling of the chemical composition of free-standing carbon dots using X-ray photoelectron spectroscopy. *The Journal of Physical Chemistry C, 122*(26), 14889–14897. https://doi.org/10.1021/acs.jpcc.8b03800

Pham, C. V., Krueger, M., Eck, M., Weber, S., & Erdem, E. (2014). Comparative electron paramagnetic resonance investigation of reduced graphene oxide and carbon nanotubes with different chemical functionalities for quantum dot attachment. *Applied Physics Letters, 104*(13), 132102. https://doi.org/10.1063/1.4870297

Philippidis, A., Spyros, A., Anglos, D., Bourlinos, A. B., Zbořil, R., & Giannelis, E. P. (2013). Carbon-dot organic surface modifier analysis by solution-state NMR spectroscopy. *Journal of Nanoparticle Research, 15*(7), 1777. https://doi.org/10.1007/s11051-013-1777-0

Rahmani, Z., & Ghaemy, M. (2019). One-step hydrothermal-assisted synthesis of highly fluorescent N-doped carbon dots from gum tragacanth: Luminescent stability and sensitive probe for Au3+ ions. *Optical Materials, 97*, 109356. https://doi.org/10.1016/j.optmat.2019.109356

Rami, J. M., Patel, C. D., Patel, C. M., & Patel, M. V. (2021). Thermogravimetric analysis (TGA) of some synthesized metal oxide nanoparticles. *Materials Today: Proceedings, 43*, 655–659. https://doi.org/10.1016/j.matpr.2020.12.554

Reich, S., & Thomsen, C. (2004). Raman spectroscopy of graphite. *Philosophical Transactions of the Royal Society of London. Series A: Mathematical, Physical and Engineering Sciences, 362*(1824), 2271–2288. https://doi.org/10.1098/rsta.2004.1454

Ren, C., Tian, C., Zhang, M., Li, F., Li, Y., Zhang, F., Zhang, J., Chen, G., & Tang, J. (2025). The fluorescence properties of nitrogen-doped carbon dots by microwave green approaches. *Journal of Molecular Structure, 1319*, 139364. https://doi.org/10.1016/j.molstruc.2024.139364

Rimal, V., Shishodia, S., & Srivastava, P. K. (2020). Novel synthesis of high-thermal stability carbon dots and nanocomposites from oleic acid as an organic substrate. *Applied Nanoscience, 10*(2), 455–464. https://doi.org/10.1007/s13204-019-01178-z

Rojas-Valencia, O. G., Regules-Carrasco, M., Hernández-Fuentes, J., Germán, C. M. R.-S., Estrada-Flores, M., & Villagarcía-Chávez, E. (2021). Synthesis of blue emissive carbon quantum dots from Hibiscus Sabdariffa flower: Surface functionalization analysis by FT-IR spectroscopy. *Materialia, 19*, 101182. https://doi.org/10.1016/j.mtla.2021.101182

Roy, P., Chen, P.-C., Periasamy, A. P., Chen, Y.-N., & Chang, H.-T. (2015). Photoluminescent carbon nanodots: Synthesis, physicochemical properties and analytical applications. *Materials Today, 18*(8), 447–458. https://doi.org/10.1016/j.mattod.2015.04.005

Saadatkhah, N., Carillo Garcia, A., Ackermann, S., Leclerc, P., Latifi, M., Samih, S., Patience, G. S., & Chaouki, J. (2020). Experimental methods in chemical engineering: Thermogravimetric analysis – TGA. *The Canadian Journal of Chemical Engineering, 98*(1), 34–43. https://doi.org/10.1002/cjce.23673

Salha, B. A., Maruthapandi, M., Perelshtein, I., Luong, J. H. T., & Gedanken, A. (2025). Synergistic effects of nitrogen-doped carbon dots on pepper and lettuce growth via irrigation enhancement. *Diamond and Related Materials, 152*, 111998. https://doi.org/10.1016/j.diamond.2025.111998

Selva Sharma, A., Suresh Nair, S., Varghese, A. W., Usha, A., Varghese, R. E., Joseph, R., & Thekkuveettil, A. (2023). Dual-emissive carbon dots: exploring their fluorescence properties for sensitive turn-off–on recognition of ferric and pyrophosphate ions and its application in fluorometric detection of the loop-mediated isothermal amplification reaction. *Langmuir, 39*(16), 5779–5792. https://doi.org/10.1021/acs.langmuir.3c00041

Serrano-Lotina, A., Portela, R., Baeza, P., Alcolea-Rodriguez, V., Villarroel, M., & Ávila, P. (2023). Zeta potential as a tool for functional materials development. *Catalysis Today, 423*, 113862. https://doi.org/10.1016/j.cattod.2022.08.004

Setianto, S., Men, L. K., Bahtiar, A., Panatarani, C., & Joni, I. M. (2024). Carbon quantum dots with honeycomb structure: A novel synthesis approach utilizing cigarette smoke precursors. *Scientific Reports, 14*(1), 1996. https://doi.org/10.1038/s41598-024-52106-3

Shahba, H., & Sabet, M. (2020). Two-step and green synthesis of highly fluorescent carbon quantum dots and carbon nanofibers from pine fruit. *Journal of Fluorescence, 30*(4), 927–938. https://doi.org/10.1007/s10895-020-02562-7

Shamsipur, M., Molaei, K., Molaabasi, F., Alipour, M., Alizadeh, N., Hosseinkhani, S., & Hosseini, M. (2018). Facile preparation and characterization of new green emitting carbon dots for sensitive and selective off/on detection of Fe3+ ion and ascorbic acid in water and urine samples and intracellular imaging in living cells. *Talanta, 183*, 122–130. https://doi.org/10.1016/j.talanta.2018.02.042

Shishodia, S., Rimal, V., & Srivastava, P. K. (2021). Synthesis of tunable high-thermal stability carbon dots via functionalization for applications in high-temperature environment. *Applied Nanoscience, 11*(5), 1691–1706. https://doi.org/10.1007/s13204-021-01814-7

Singh, P., Kumar, S., & Kumar, K. (2025). Biogenic synthesis of Allium cepa derived magnetic carbon dots for enhanced photocatalytic degradation of methylene blue and rhodamine B dyes. *Biomass Conversion and Biorefinery, 15*(1), 869–887. https://doi.org/10.1007/s13399-023-05047-2

Sousa, D. A., Berberan-Santos, M. N., & Prata, J. V. (2024). Are "Carbon dots" always carbon dots? Evidence for their supramolecular nature from structural and dynamic studies in solution and in the pure solid. *Chemistry – A European Journal, 30*(3), e202302955. https://doi.org/10.1002/chem.202302955

Stachowska, J. D., Murphy, A., Mellor, C., Fernandes, D., Gibbons, E. N., Krysmann, M. J., Kelarakis, A., Burgaz, E., Moore, J., & Yeates, S. G. (2021). A rich gallery of carbon dots based photoluminescent suspensions and powders derived by citric acid/urea. *Scientific Reports, 11*(1), 10554. https://doi.org/10.1038/s41598-021-89984-w

Stevie, F. A., & Donley, C. L. (2020). Introduction to x-ray photoelectron spectroscopy. *Journal of Vacuum Science & Technology A: Vacuum, Surfaces, and Films, 38*(6), 063204. https://doi.org/10.1116/6.0000412

Tan, M., Zhang, L., Tang, R., Song, X., Li, Y., Wu, H., Wang, Y., Lv, G., Liu, W., & Ma, X. (2013). Enhanced photoluminescence and characterization of multicolor carbon dots using plant soot as a carbon source. *Talanta, 115*, 950–956. https://doi.org/10.1016/j.talanta.2013.06.061

Tang, H., Tang, Y., Xiao, M., Zhu, H., & Guo, M. (2022). Study on microwave synthesis mechanism of carbon dots based on NMR characterization. *Colloids and Surfaces A: Physicochemical and Engineering Aspects, 651*, 129564. https://doi.org/10.1016/j.colsurfa.2022.129564

Tang, S., Liu, Y., Opoku, H., Gregorsson, M., Zhang, P., Auroux, E., Dang, D., Mudring, A.-V., Wågberg, T., Edman, L., & Wang, J. (2023). Fluorescent carbon dots from birch leaves for sustainable electroluminescent devices. *Green Chemistry, 25*(23), 9884–9895. https://doi.org/10.1039/D3GC03827K

Tang, S., Liu, Y., Opoku, H., Gregorsson, M., Zhang, P., Auroux, E., Dang, D., Mudring, A.-V., Wågberg, T., Edman, L., & Wang, J. (2025). Correction: Fluorescent carbon dots from birch leaves for sustainable electroluminescent devices. *Green Chemistry*. https://doi.org/10.1039/D4GC90134G

Ţucureanu, V., Matei, A., & Avram, A. M. (2016). FTIR spectroscopy for carbon family study. *Critical Reviews in Analytical Chemistry, 46*(6), 502–520. https://doi.org/10.1080/10408347.2016.1157013

Wan, J., Zhang, X., Jiang, Y., Xu, S., Li, J., Yu, M., Zhang, K., & Su, Z. (2022). Regulation of multi-color fluorescence of carbonized polymer dots by multiple contributions of effective conjugate size, surface state, and molecular fluorescence. *Journal of Materials Chemistry B, 10*(36), 6991–7002. https://doi.org/10.1039/D2TB01330D

Wang, Z., & Fang, G. (2024). Sensitivity enhancement of time-gated Raman spectroscopy by reducing the systematic noise of residual fluorescence background. *Optics Letters, 49*(24), 7086. https://doi.org/10.1364/OL.540236

Wang, F., Xie, Z., Zhang, H., Liu, C., & Zhang, Y. (2011). Highly luminescent organosilane-functionalized carbon dots. *Advanced Functional Materials, 21*(6), 1027–1031. https://doi.org/10.1002/adfm.201002279

Wang, W., Sheng, X., Wang, Y., Yu, M., Shen, Y., Xia, Y., Li, T., Cao, S., Zhang, M., Wang, W., & Yang, Y. (2025). A Mo-doped carbon dot nanozyme for enhanced phototherapy in vitro. *Nanoscale Advances*. https://doi.org/10.1039/D5NA00028A

Weckhuysen, B. M., Heidler, R., & Schoonheydt, R. A. (2004). Electron spin resonance spectroscopy. In H. G. Karge & J. Weitkamp (Eds.), *Characterization I* (pp. 295–335). Springer. https://doi.org/10.1007/b94238

Wilson, N. R., & Macpherson, J. V. (2009). Carbon nanotube tips for atomic force microscopy. *Nature Nanotechnology, 4*(8), 483–491. https://doi.org/10.1038/nnano.2009.154

Wu, J.-B., Lin, M.-L., Cong, X., Liu, H.-N., & Tan, P.-H. (2018). Raman spectroscopy of graphene-based materials and its applications in related devices. *Chemical Society Reviews, 47*(5), 1822–1873. https://doi.org/10.1039/C6CS00915H

Wu, J., Yan, Z., Shan, P., Li, W., Wang, K., & Li, H. (2025). Characterization of gelatin-based composite films loaded with polysaccharide and carbon dots from Stropharia rugosoannulata and their application in pork fresh-keeping. *International Journal of Biological Macromolecules, 297*, 139702. https://doi.org/10.1016/j.ijbiomac.2025.139702

Xiao, L., Shi, J., Nan, B., Chen, W., Zhang, Q., Zhang, E., & Lu, M. (2020). Highly sensitive detection of Fe3+ ions using waterborne polyurethane-carbon dots self-healable fluorescence film. *Macromolecular Materials and Engineering, 305*(3), 1900810. https://doi.org/10.1002/mame.201900810

Xiao, Y., Zhao, Y., Chen, Q., Long, X., Li, Y., Ma, M., Ming, Q., Mai, Y., Shi, Q., Yang, X., & Zhang, X. (2025). Chemiluminescence of potassium permanganate and carbon dots and its analytical applications. *Journal of Food Composition and Analysis, 139*, 107156. https://doi.org/10.1016/j.jfca.2024.107156

Xu, H., Yang, X., Li, G., Zhao, C., & Liao, X. (2015). Green synthesis of fluorescent carbon dots for selective detection of tartrazine in food samples. *Journal of Agricultural and Food Chemistry, 63*(30), 6707–6714.

Xu, G., Wu, Z., Xie, Z., Wei, Z., Li, J., Qu, K., Li, Y., & Cai, W. (2021). Graphene quantum dot reinforced hyperbranched polyamide proton exchange membrane for direct methanol fuel cell. *International Journal of Hydrogen Energy, 46*(15), 9782–9789. https://doi.org/10.1016/j.ijhydene.2020.06.303

Yan, Y., Xia, L., & Ma, L. (2019). Solvent-controlled synthesis of multicolor photoluminescent carbon dots for bioimaging. *RSC Advances, 9*(42), 24057–24065. https://doi.org/10.1039/C9RA04241E

Yang, X., Zhuo, Y., Zhu, S., Luo, Y., Feng, Y., & Dou, Y. (2014). Novel and green synthesis of high-fluorescent carbon dots originated from honey for sensing and imaging. *Biosensors and Bioelectronics, 60*, 292–298. https://doi.org/10.1016/j.bios.2014.04.046

Yang, Z., Zhou, M., Yin, T., Wang, C.-Y., Zhu, G.-Y., Bai, L.-P., Jiang, Z.-H., & Zhang, W. (2025). Functionalized carbon dots with intrinsic Wnt/β-catenin inhibition to synergistically Promote 5-fluorouracil chemotherapy. *International Journal of Nanomedicine, 20*(null), 1951–1964. https://doi.org/10.2147/IJN.S503540

Ye, X., Zhong, H., Liu, L., Huang, J., Xia, Z., Tang, Z., Wei, W., Huang, W., Ye, Y., & Jiang, Q. (2025). A novel and high-performance tumor inhibitor of La, N co-doped carbon dots for U251 and LN229 cells. *Colloids and Surfaces B: Biointerfaces, 249*, 114520. https://doi.org/10.1016/j.colsurfb.2025.114520

Yoshinaga, T., Shinoda, M., Iso, Y., Isobe, T., Ogura, A., & Takao, K. (2021). Glycothermally Synthesized Carbon Dots with Narrow-Bandwidth and Color-Tunable Solvatochromic Fluorescence for Wide-Color-Gamut Displays. ACS Omega, 6(2), 1741–1750. https://doi.org/10.1021/acsomega.0c05993

Zaca-Moran, O., Díaz-Monge, F., Rodríguez-Juárez, A., Gómez-Muñoz, C. L., Zaca-Moran, P., Secundino-Sánchez, O., & Díaz-Reyes, J. (2024). Synthesis and characterization of fluorescent carbon dots obtained from citrus x sinensis by an eco-friendly method. Results in Chemistry, 11, 101788. https://doi.org/10.1016/j.rechem.2024.101788

Zahedi, F., & Shabani-Nooshabadi, M. (2023). Porous structure Ni/$CuCo_2O_4$ core–shell as a novel type of three-dimensional electrode with facile fabrication and binder-free toward enhanced methanol oxidation and supercapacitor performances. *Fuel, 335*, 127083. https://doi.org/10.1016/j.fuel.2022.127083

Zainal Abidin, N. H., Wongso, V., Hui, K. C., Cho, K., Sambudi, N. S., Ang, W. L., & Saad, B. (2020). The effect of functionalization on rice-husks derived carbon quantum dots properties and cadmium removal. *Journal of Water Process Engineering, 38*, 101634. https://doi.org/10.1016/j.jwpe.2020.101634

Zhang, J., Liu, J., Wang, X., Mai, J., Zhao, W., Ding, Z., & Fang, Y. (2019). Construction of Z-scheme tungsten trioxide nanosheets-nitrogen-doped carbon dots composites for the enhanced photothermal synergistic catalytic oxidation of cyclohexane. *Applied Catalysis B: Environmental, 259*, 118063. https://doi.org/10.1016/j.apcatb.2019.118063

Zhang, Q., Wang, R., Feng, B., Zhong, X., & Ostrikov, K. (Ken). (2021). Photoluminescence mechanism of carbon dots: Triggering high-color-purity red fluorescence emission through edge amino protonation. *Nature Communications, 12*(1), 6856. https://doi.org/10.1038/s41467-021-27071-4

Zhang, T., Huang, B., Huang, H., Yan, A., Lu, S., & Qian, X. (2025). Visible light boosted Fenton-like reaction of carbon dot-Fe(III) complex: Kinetics and mechanism insights. *Chinese Chemical Letters, 110885*. https://doi.org/10.1016/j.cclet.2025.110885

Zhao, Y., Liu, X., Yang, Y., Kang, L., Yang, Z., Liu, W., & Chen, L. (2015). Carbon dots: from intense absorption in visible range to excitation-independent and excitation-dependent photoluminescence. *Fullerenes, Nanotubes and Carbon Nanostructures, 23*(11), 922–929. https://doi.org/10.1080/1536383X.2015.1018413

Zhao, W., Lu, Z., Song, F., Han, J., Zhang, Q., Cong, Y., Lu, A., & Gao, T. (2025). One-step electrolytic synthesis of nitrogen-doped carbon dots using graphite electrode in ionic liquid and their LED applications. *Colloids and Surfaces A: Physicochemical and Engineering Aspects, 712*, 136459. https://doi.org/10.1016/j.colsurfa.2025.136459

Zheng, J., Wang, Y., Zhang, F., Yang, Y., Liu, X., Guo, K., Wang, H., & Xu, B. (2017). Microwave-assisted hydrothermal synthesis of solid-state carbon dots with intensive emission for white light-emitting devices. *Journal of Materials Chemistry C, 5*(32), 8105–8111. https://doi.org/10.1039/C7TC01701D

Zhu, S., Song, Y., Zhao, X., Shao, J., Zhang, J., & Yang, B. (2015). The photoluminescence mechanism in carbon dots (graphene quantum dots, carbon nanodots, and polymer dots): Current state and future perspective. *Nano Research, 8*(2), 355–381. https://doi.org/10.1007/s12274-014-0644-3

Zhu, R., Huang, W., Ma, X., Zhang, Y., Yue, C., Fang, W., Hu, Y., Wang, J., Dang, J., Zhao, H., & Li, Z. (2019a). Nitrogen-doped carbon dots-V2O5 nanobelts sensing platform for sensitive detection of ascorbic acid and alkaline phosphatase activity. *Analytica Chimica Acta, 1089*, 131–143. https://doi.org/10.1016/j.aca.2019.08.061

Zhu, Z., Yang, P., Chen, M., Zhang, T., Cao, Y., Zhang, W., Zhou, X., & Chen, W. (2019b). Microwave synthesis of amphiphilic carbon dots from xylose and construction of luminescent composites with shape recovery performance. *Journal of Luminescence, 213*, 474–481. https://doi.org/10.1016/j.jlumin.2019.05.006

Zhu, J., Zhao, S., Guan, R., & Zhang, H. (2025). Full-spectrum luminescence of nitrogen-sulfur co-doped carbon dots and their applications in LED lighting and organelle-lysosome labeling. *Dyes and Pigments, 237*, 112686. https://doi.org/10.1016/j.dyepig.2025.112686

Part II

Carbon Nanodots in Human Health

Applications of Carbon Dots in Diagnostics and Bioimaging

6

Mohammadhassan Gholami-Shabani, Katayoon Nematpour, Sareh Sohrabi, Masoomeh Shams-Ghahfarokhi, Shiva Mahmoudi-Gahroui, Reihaneh Seiad-Ahmadnezhad, Ali Eslamifar, Fatemehsadat Jamzivar, and Mehdi Razzaghi-Abyaneh

Abstract

Carbon dots (CDs), which are characterized by the nanoscale size, high fluorescence, and spectacular biocompatibility, occupy a very unique position in diagnostics and bioimaging. Because of their unique optical characteristics, functional surface groups, and ease in manufacturing with a variety of natural sources, CDs are promising for use in healthcare, food safety, and agriculture. This chapter highlights recent advances in eco-production of CDs and their practical application in bioimaging, medical diagnostics, therapy, food safety monitoring, and sustainable agriculture technologies. It also discusses the complexity and possible growth in the future of CD technology for practical use.

M. Gholami-Shabani
Department of Mycology, Pasteur Institute of Iran, Tehran, Iran

Department of Labor Environment, Health, Safety, Environment and Energy, Ministry of Industry, Mine and Trade, Tehran, Iran

K. Nematpour
Department of Labor Environment, Health, Safety, Environment and Energy, Ministry of Industry, Mine and Trade, Tehran, Iran

S. Sohrabi
Department of Medical Biotechnology, School of Advanced Medical Sciences and Technologies, Hamadan University of Medical Sciences, Hamadan, Iran

Samen Pharmaceutical Company, Mashhad, Iran

M. Shams-Ghahfarokhi
Department of Mycology, Faculty of Medical Sciences, Tarbiat Modares University, Tehran, Iran

S. Mahmoudi-Gahroui · R. Seiad-Ahmadnezhad · F. Jamzivar · M. Razzaghi-Abyaneh (✉)
Department of Mycology, Pasteur Institute of Iran, Tehran, Iran
e-mail: mrab442@pasteur.ac.ir

A. Eslamifar
Department of Clinical Research, Pasteur Institute of Iran, Tehran, Iran

M. Rai et al. (eds.), *Carbon Nanodots*, https://doi.org/10.1007/978-981-95-4705-0_6

Keywords

Carbon dots · Diagnostics · Bioimaging · Green synthesis · Photodynamic therapy

6.1 Introduction

The advent of nanotechnology has resulted in the identification of many innovative materials, each with unique properties that can be utilized for a wide range of applications (Malik et al., 2023; Gholami-Shabani et al., 2023). Currently, carbon dots (CDs) have generated notable interest, finding themselves at the center of materials science and nanobiotechnology (Parambil & Rajamani, 2024). CDs started as an unwanted byproduct of carbon nanotube purification in 2004 while searching for something else; CDs were initially considered as impurities (Hu et al., 2024a). But further investigation revealed their unique optical properties, particularly their high photoluminescence, rapidly moving them into the forefront of nanomaterials study. CDs are sphere-like, nanometer-scale size C-based nanoparticles, generally have a diameter of less than 10 nm (Liu et al., 2020). The core of CDs consists of sp^2/sp^3 hybridized carbon atoms, which can be amorphous or crystalline, and have various surface functional groups such as carboxylic (Shabbir et al., 2023). This structural adaptability means that such dots have a rich set of tunable properties, which enable their utilization for a wide array of fields. Photoluminescence features of CDs are the most attractive properties of them (Elugoke et al., 2024). Fluorescence of CDs is present in a variety of colors, and the color changes depending on excitation wavelength—a feature known as excitation-dependent emission. The tunable fluorescence of CDs makes it an exceptional bioimaging agent, fluorescence detector, and optoelectronic device (Yan et al., 2019). Other than their optical qualities, CDs exhibit a variety of extraordinary traits. Since they are non-reactive and nontoxic, they are trusted in vivo for biomedicinal use (Unnikrishnan et al., 2025). Their high solubility in water further justifies their capacity to be integrated into biological situations, and they therefore can diffuse effectively in physiologic solutions. Because of these qualities, CDs have become useful materials for the use in drug delivery, photothermal therapy and biosensing in bioresearch (Wang et al., 2022a). Moreover, the wide range of surface functionalization allows researchers to attach many biomolecules, polymers, or targeting ligands to the protein of interest covalently. Such versatility in surface modification means that CDs can be easily manipulated for targeted purposes such as cancer cell targeting, metal detection or molecular recognition with specific proteins or nucleic acids (Debnath et al., 2025). As the field advances, the area now holds a critical role in focusing on CD synthesis methods with the perspective of the environmental dimension (Long et al., 2021). The standard entrapment synthesis pathways conventionally require the use of strong chemicals and a capacity to expend much energy, which are impractical for eco-friendly manufacturing (Jing et al., 2023). As a result, there has been an increasing number of researchers who attempt to establish eco-friendly production methods based upon the use of recyclable and cost-effective starting materials (Jing et al., 2023). These eco-friendly sources include orange peels, peels from bananas, lemons, herbal preparations, food wastes, fungi materials, and those obtained from agro-processing industries (Singh et al., 2025a). The use of these "green" precursors reduces the environmental burden associated with nanomaterial synthesis and helps to make the use of waste materials economically viable (Gholami-Shabani et al., 2016; Khan et al., 2024). The implementation of such a strategy favors the circular economy, as the wastes are integrated into the production cycle, reducing landfilled wastes, and conserving the scarce resources (Gholami-Shabani et al., 2017; Singh et al., 2025a). It is because of compatibility with living organisms as well as being cost-efficient and least polluting, that CDs find ever-increasing application in practical life well beyond the lab. In medicine, CDs are being investigated in order to have real-time imaging of cellular activity and early detection of diseases and are potentially applied as therapeutic tools (Qureshi et al., 2024). For food safety purposes, CDs play a fundamental role in the construction of sensors that measure pesticide presence, heavy metals, or microbial pathogens (Singh et al., 2025b). In addition, researchers are checking CDs for use in precision agriculture in the evaluation of soil health, monitoring plant stress and enhancing nutrient delivery (Chaudhary et al., 2024). The property in which CDs are able to fluoresce provides a setting for dynamic observation, important for decision-making in both medicine and agriculture. Further studies will add a lot to the possible options for carbon dot applications (Manzoor et al., 2023). The scientists are working actively in an attempt to refine quantum efficiency, to provide better size uniformity, and to optimize CDs for therapeutic effects or diagnosis. Special mention should be made of theranostic CDs, CDs that combine diagnostic functions and therapeutics, which are particularly important in oncology because of their dual potential (Pandey et al., 2024). The integration of CDs with other nanomaterials, including metal nanoparticles or derivatives from graphene, could develop hybrid systems that have cooperative functionalities (Kouloumpis et al., 2017). On a global scale, however, CDs are very versatile and sustainable as nanomaterials that have the potential to contribute a lot in many areas of disciplines. The story of CD transformation from serendipity to crucial components of modern nanotechnology emphasizes the continually shaping and interdisciplinary character of the progress of mankind. As scientists explore more viable manufacturing approaches and find innovative applications, the importance of CDs in promoting sustainability and technology is likely to increase.

6.2 Synthesis of Carbon Dots

6.2.1 Top-Down and Bottom-Up Approaches

Through two main approaches (top-down and bottom-up), carbon dots (CDs) may be synthesized. Variations in the approach also have their unique advantages/limitations depending on the demands of the character of the product, the size of production, the financial expense of the product, and the environmental effect. It is critical to understand both approaches in order to customize CDs successfully for bioimaging, sensing, catalysis and other advanced technical areas (Bruno et al., 2021). There are a number of techniques created for synthesizing CDs for a wide range of applications and some are shown in Fig. 6.1.

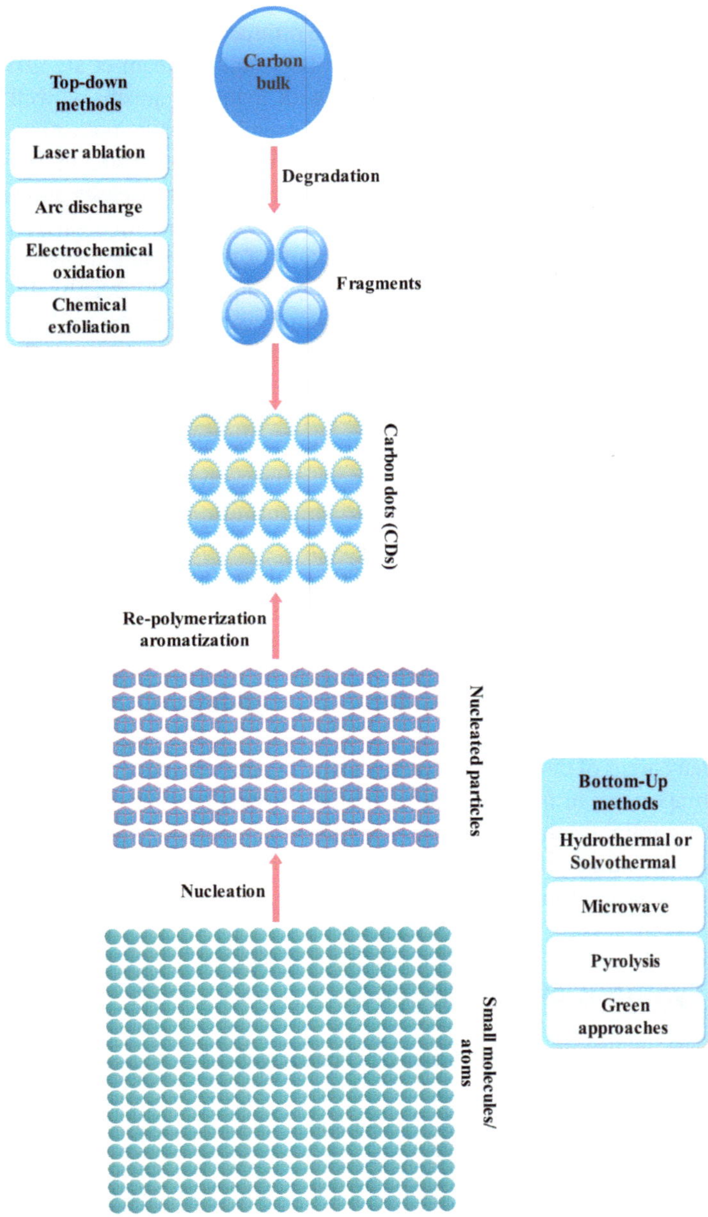

Fig. 6.1 Techniques created for synthesis CDs

6.2.1.1 Top-Down Approaches

In top-down synthesis, bulk carbonaceous materials are fractured into nanoscale particles, using physical, chemical or electrochemical fragmentation methods. In essence, from this approach, extensive carbon materials such as graphite, carbon nanotubes, graphene, or activated carbon are degraded to yield ultrasmall CDs (Liu et al., 2024). Common top-down techniques include:

(a) Laser ablation: A controlled environment, such as one that contains water, inert gas, or passivating agents, is used to impinge on a carbon target with high-energy laser pulses. The laser-created energy breaks down the carbon material through nanoparticles, which can be removed and processed for purification (Shorgar et al., 2023).
(b) Arc discharge: In arc discharge, there is a high current in a controlled environment filled with inert gas between graphite electrodes. When vaporized by the plasma arc, the product carbon is condensed, from which nanoparticles may be isolated, including CDs (Khayal et al., 2021).
(c) Electrochemical oxidation: An electrochemical cell is utilized to catalyze the oxidation of graphite electrodes immersed in an aqueous medium. A slow exfoliation of carbon takes place, with the production of nano-sized particles which are then isolated and stabilized (Rocco et al., 2023).
(d) Chemical exfoliation: Due to its commonness as a bulk carbon source, graphite can undergo chemical oxidation and exfoliation using sheer power acids or chemicals such as nitric acid, sulfuric acid, or potassium permanganate. Graphite's layered structure breaks into isolated nanosheets or CDs (Cui et al., 2021).

High-purity CDs are capable of being obtained with desired graphitic structures by top-down techniques, but these are energy-intensive operations and sometimes require resorting to harsh chemicals or complicated purifications. There are also doubts about scaling these approaches, as well as environmental issues arising from reagent consumption and waste that increase the search for greener approaches (Liu et al., 2024).

6.2.1.2 Bottom-Up Approaches

Complementary to the top-down approach, bottom-up techniques concentrate on CD accumulation from molecular or atomic-based starting materials. CDs are prepared from small organic precursors such as citric acid, sugar, amino acids or natural polymers. Such materials are subjected to thermal or chemical transformations, which stimulate carbonization and nucleation and form CDs (Liu et al., 2024). Key bottom-up techniques include:

(a) Hydrothermal and solvothermal synthesis: For these techniques, the precursors have to undergo high temperatures and pressures inside closed autoclaves. Hydrothermal processes operate with water as an aqueous solvent, which contrasts with solvothermal procedures that use organic solvents. High temperatures accelerate the conversion of organic precursors into CDs with those characteristics that can be adjusted through controlling temperature, time and choice of the starting materials (Newman Monday et al., 2021).

(b) Microwave-assisted synthesis: When using microwave irradiation, this method offers a quick and energy-efficient means of homogeneously heating the reaction mixture; otherwise, the carbonization process is accelerated. Sustainable production of CDs on a larger scale is possible in a relatively short period of time using microwave-assisted synthesis (Nazar et al., 2024).

(c) Pyrolytic synthesis: The process entails the pyrolysis of precursors under no exposure to air or an inert gas condition. The use of the process of heat polymerization converts the organic molecules into carbon nanoparticles; in most cases, producing CDs with higher fluorescence (Wang et al., 2022c).

(d) Green synthesis approaches with utilization of carbon feedstocks derived through biomass or waste: In recent years, materials obtained from biowaste such as fruit peel, leaves, coffee grounds, and dairy products, have come to the fore as a significant resource for synthesizing CDs using a bottom—up approach. By doing this, the approach reduces the negative impact on the environment and at the same time makes the waste materials valuable nanomaterials in the course of a circular economy (Cai et al., 2025).

Bottom-up processes become popular due to their simplicity, flexibility and minimal damage to the environment. They enable more precise control over particle size and surface functional groups, as well as the optical properties. Further, these approaches allow creation of CD structures customized for separate tasks such as adaptation to bio-medical imaging and enhancement of sensitivity in sensing devices (Etefa et al., 2024).

6.2.2 Green and Sustainable Synthesis

In response to the market demand for eco-friendly nanomaterials, there is a significant increase in research projects on the sustainable manufacture of carbon dots (CDs). Traditional methods of synthesis are great at producing good CDs but tend to rely on dangerous reagents, massive amounts of energy, and complex purification procedures. These issues are especially noticeable when the production is to be used in an industrial, large-scale application. In contrast, green and sustainable synthesis methodologies focus on the protection of the environment and CD preservation function within the remit of green chemistry and circular economy principles. Table 6.1 shows some examples of Green Synthesis Routes for CDs (Jing et al., 2023).

6.2.2.1 Principles of Green Synthesis

Green synthesis encourages the making of chemical compounds and approaches that do not employ or seriously limit the utilization and emission of hazardous substances (Gholami-Shabani et al., 2016). Green synthesis of the preparation of CDs is

Table 6.1 Green synthesis routes for carbon dots

Source material	Synthesis method	Key properties	References
Orange peel	Hydrothermal	Blue fluorescence, antioxidant activity	Sadeghi-Chahnasir et al. (2024)
Peanut shells	Microwave-assisted	High quantum yield, stable under UV light	Ma et al. (2017)
Rice husk	Thermal-assisted carbonization	Detection of Sn (II) ions	Ngu et al. (2016)
Ginger *(Zingiber officinale)*	Hydrothermal	Antifungal, antibacterial, antioxidant properties	Bhavikatti et al. (2024)
Fungal biomass (*Ganoderma lucidum*)	Hydrothermal	Imaging and detecting 2,4-dinitrophenol and 4-nitrophenol	Tu et al. (2020)
Lemon juice	Hydrothermal	High sensitivity, high precision, and good selectivity for Hg^{2+}, fluorescent probe for multicolor live cell imaging	Tadesse et al. (2020)
Banana peel	Pyrolysis, hydrothermal	Eco-friendly, stable in water, showed blue emission under UV light	Wijayanti et al. (2024)
Medicinal herb	Herbal carbonization	Hemostatic effect, anti-gouty effect, anti-inflammatory effect, hypoglycemic effect, hepatoprotective effect, anti-psoriasis effect, anti-liver cancer activity, cell imaging	Li et al. (2022)
Green tea extract	Hydrothermal	Antioxidant, antimicrobial properties	Wen et al. (2023)
Chestnut shell	Hydrothermal	Sensitive detection of phosphate and tetracycline hydrochloride	Yin et al. (2025)

concerned with the use of nontoxic, renewable materials such as biomasses and food waste, along with water-based, environmentally friendly solvents (Jing et al., 2023). In addition, energy conservation is obtained through practices such as microwave heating, and the intent is to minimize the necessity to purify after synthesis (Hu et al., 2019). Such a way contributes to the further development of eco-friendly nanotechnologies through a reduction in the usage of fossil fuels and minimization of environmental contamination (Kumar et al., 2024).

6.2.2.2 Renewable Precursors and Waste Valorization

A variety of natural and waste-derived materials can be used as precursors in the green synthesis of CDs, including agricultural and food waste such as rice husks, sugarcane bagasse, and coffee grounds; herbal and medicinal plant extracts; biopolymers like chitosan, cellulose, and starch; and fungal or microbial biomass (Aksu & Güzdemir, 2025). These materials are inherently rich in carbon, oxygen, nitrogen, and other heteroatoms, which are beneficial for doping CDs and enhancing their photoluminescence and functional properties. By utilizing such waste materials, green synthesis not only lowers production costs but also contributes to environmental sustainability by reducing waste and enabling value-added recycling (Aksu & Güzdemir, 2025).

6.2.2.3 Green Synthesis Techniques

New energy-efficient and sustainable approaches have been adopted to produce carbon dots (CDs) in an environmentally friendly approach (Khan et al., 2024). In these approaches, biomass-derived precursor is heated under airtight conditions in water or mild solvents to create hydrothermal and solvothermal synthesis, respectively (Usman & Cheng, 2024). This method is very similar to the natural carbonization of biomass and yields CDs synthesized in mild conditions. Hydrothermal synthesis stands out for its easy process, the ability to scale it up easily and its use of water as a sustainable solvent (Newman Monday et al., 2021). The quick and uniform heating by using microwave-assisted synthesis markedly shortens reaction times, bringing several hours-long processes down to mere minutes (Nazar et al., 2024). This approach will favor green chemistry, allowing little energy and CD's exhibiting superior optical performance. Ultrasound-assisted and mechanochemical techniques, such as ultrasonic irradiation and mechanical grinding, can improve reaction rates and permit synthesis at low temperatures, supplying green options to conventional high-temperature processes (Kumar et al., 2020). Pyrolysis and combustion-free carbonization are the terms for thermal decomposition of organic waste in the absence of oxygen under stable atmospheric conditions (Wang et al., 2022c). These approaches ensure CDs are manufactured efficiently when optimized to oversee the production of dangerous byproducts. Also, the use of low-temperature carbonization is under investigation with a view to achieving a cleaner, more efficient way of producing CDs (Wu et al., 2025a). Further, there is increasing interest in using microorganisms to accomplish CDs synthesis as a sustainable method (Moradi et al., 2023). Certain bacteria, fungi and algae can transform organic matter into carbon-based nanomaterials in milder aqueous conditions. Catalytic CD synthesis using biological systems represents a more environmentally friendly route with lower energy requirements and minimal toxic residues, thus enhancing the environmental footprint of this process.

6.2.2.4 Advantages of Green Synthesized CDs

Synthetic CDs produced using green processes will, most of the time, exhibit performance that is comparable with or surpasses that of the conventional types of synthetics. Key benefits include:

(a) Biocompatibility and low toxicity: Starting from being derived from nature, CDs are generally inert to heavy metals and toxic residuals, which makes them an excellent candidate for biomedical applications such as drug delivery, imaging, and biosensing (Fathi-karkan et al., 2025).
(b) Surface functionalization: Researchers observe that water solubility and their ability to be further modified improve in CDs from biomass, as they are typically

composed of several different functional groups (–OH, –COOH, and –NH_2) (Fathi-karkan et al., 2025).

(c) Photoluminescent properties: Green CDs often exhibit strong, tunable photoluminescence with high quantum efficiencies, making them ideal materials for sensing and imaging applications (Hashemi & Mousazadeh, 2021).

(d) Cost-effectiveness and scalability: Using cheap and easy-to-find precursors, the manufacturing costs are lowered; in turn, techniques like microwave or hydrothermal synthesis will allow the large-scale manufacturing (Hashemi & Mousazadeh, 2021).

6.3 Applications in Health

6.3.1 Bioimaging and Diagnostics

Carbon dots (CDs) have quickly become critical tools for imaging and diagnostics in biomedical work (Fig. 6.2). Their bright, tunable fluorescence, as well as their nanoscale-dimension size and good biocompatibility, nontoxicity, and ease of functionalization, allows for their multiple use in imaging and diagnostic fields (Boakye-Yiadom et al., 2019). The different avenues through which CDs improve bioimaging and diagnostics are explored in the subsequent sections below. Table 6.2 shows some examples of CDs for Biomedical Imaging and Diagnostics.

6.3.1.1 Cellular Imaging

CDs play an important role in live-cell imaging due to their characteristics. Since CDs are in a size range (typically <10 nm) that easily intrudes into cells and settles in various compartments such as cytoplasm, nucleus, or mitochondria, without perturbing the cell, their entry is seamless (Niu et al., 2017). Consequently, these properties make them ideal fluorescent probes for:

Track during the real-time evolution of intracellular processes such as endocytosis, exocytosis, apoptosis, or organelle activity. Scientists can perform multicolor imaging with the help of CDs, as these nanoparticles demonstrate fluorescence, which depends on the excitation, or can be modified by altering surface modification; this allows tuning from blue to red. Sustained fluorescence microscopy

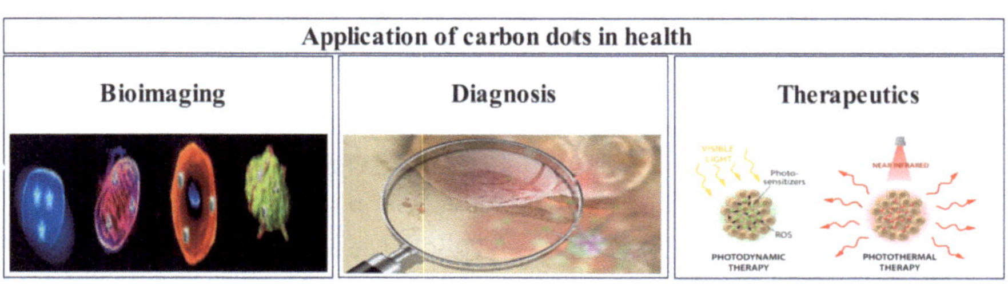

Fig. 6.2 Applications of CDs in health

Table 6.2 Carbon dots for biomedical imaging and diagnostics

Application	CD source	Feature	References
In vitro cell imaging	Citric acid	Blue fluorescence in HeLa cells	Jana et al. (2019)
Brain tumor detection	Glucose	NIR imaging via BBB penetration	Yang et al. (2023)
Zebrafish whole-body imaging	Herbal CDs	Deep-tissue penetration	Gollavelli et al. (2022)
Pathogen detection	Bacteria-modified CDs	Specific targeting of *Escherichia coli*	Sun et al. (2024)
Cancer marker sensing	Aptamer-CD conjugates	HER2 detection in breast cancer cells	Motaghi et al. (2017) and Wu et al. (2003)
Multicolor bioimaging	Fungal CDs	Tunable emission colors	Qin et al. (2023)
Kidney cell imaging	Medicinal herb CDs	Target-specific fluorescence	Paveethra et al. (2024)
Real-time diagnostics	pH-sensitive CDs	Live-cell pH monitoring	Yang et al. (2020)
Blood serum analysis	Dopamine CDs	Laccase detection	Wu et al. (2022)
Viral load monitoring	CD-antibody hybrids	Quantification of viral antigens	Innocenzi et al. (2025)

CD carbon dot

observations are possible using CDs, due to their great photostability, which is superior to that of the organic dyes and quantum dots. Additionally, the possibility of attaching CDs to peptides or nuclear localization signals makes it possible to control attachments of these probes to specific organelles so that fluorescence imaging becomes more detailed and specific (Mancini et al., 2023).

6.3.1.2 Animal Models

In vivo imaging of small animal types such as zebrafish, mice, and rats may have a lot to gain from the application of CDs (Sharma et al., 2024). When entering the bloodstream, CDs travel around the circulatory network but tend to accumulate in specific tissues, organs or tumor-sites depending on their surface decoration, size, and altered structure, respectively (Truskewycz et al., 2022). Advantages of using CDs for in vivo animal imaging include:

(a) Non-invasive visualization: Using NIR-emitting CDs, the researchers are able to have clearer views of deeper tissues, due to minimal scattering and light absorption in living organisms (Bao et al., 2018; Hu et al., 2024b).
(b) Reduced photobleaching and autofluorescence: The photostability exhibited by CDs is excellent in comparison to traditional fluorescent dyes or proteins to give longer life to their fluorescence, although exposed to continuous illumination (Jorns & Pappas, 2021).

(c) Extended circulation time: Increased residence time of CDs compared to small-molecule dyes in the bloodstream further facilitates the prospect of real-time imaging applications (Calabrese et al., 2021; Bartkowski et al., 2024)
(d) Renal clearance and low toxicity: As the body can easily remove many CDs quickly via the kidneys, there is little chance for long-term buildup or toxicity (Hussen et al., 2024).

These properties make CDs a highly viable substitution for traditional imaging agents in applications that target tumor localization, organ mapping, and drug distribution analysis (Bartkowski et al., 2024).

6.3.1.3 Targeted Imaging and Diagnostics

Most futuristic usage of the CDs has been in the case of targeted diagnostics, in which their surfaces are functionalized by ligands that selectively connect to given cells or molecular targets. It makes CDs particularly suitable for early detection of such diseases as cancer or infections (Bhattacharya et al., 2024). Key strategies include:

(a) Antibody or aptamer conjugation: A connection of CDs to antibodies or DNA/RNA aptamers, which specifically recognize tumor-associated antigens, surface proteins of bacteria or viral particles can be constructed. The presence and location of the target cells are marked by the fluorescence signal of these functionalized CDs when they bind to their targets (Motaghi et al., 2017).
(b) Biomarker detection: CDs can also be designed to sense pH-change, reactive oxygen species, enzymes or target metabolites, becoming "smart probes" for detecting biomarkers related to disease. For example, CDs that are susceptible to glutathione levels can be employed to track oxidative stress of cancer or neurodegenerative diseases (Chen et al., 2025).
(c) Pathogen tracking: Functionalized CDs are capable of fast binding of fungi, bacteria or viruses, providing a fast and sensitive detection in the clinical samples. This is faster diagnostics for infections than other traditional culture-based methods (Kumari et al., 2024).

Thanks to their excellent signal-to-noise ratio, CDs are often more specific and can better suppress background than traditional probes, an admirable characteristic for qualitative imaging as well as for quantitative diagnostic assays such as point-of-care (Mohammadi et al., 2022).

6.3.2 Therapeutic Applications

Apart from many applications in diagnostics and bioimaging, carbon dots (CDs) are coming to the fore in their therapeutic properties. Owing to their outstanding biocompatibility, tunable surface chemistry, and interaction with biomolecules and light, different medical treatments are being proposed for CDs. Key therapeutic applications under current investigation are discussed below (Parashar et al., 2023).

6.3.2.1 Photothermal Therapy (PTT)

Photothermal therapy (PTT) represents a novel, viable, and minimally invasive cancer treatment; it uses agents that can convert light energy, especially in the near infrared (NIR) region, to generate localized heat. CDs are turning out to be promising candidates for this application because of a variety of distinctive benefits (Balou et al., 2022).

In addition, CDs hold considerable promise for use in photodynamic and photothermal antibacterial therapies due to their unique optical properties, biocompatibility, and functionalizability (Wang et al., 2024).

(a) NIR absorption: Some CDs, especially heteroatom-doped ones (nitrogen, sulfur, metals), can absorb NIR light (650–900 nm), a part of the spectrum that is deeply penetrating for biological tissue and harmful only to the tissues that emit it (Chien et al., 2019).
(b) Localized heating: CDs aggregates in tumor tissues upon being exposed to the NIR light can convert absorbed energy to heat rapidly. This heat ablates cancerous cells selectively, by increasing the local temperature (usually over 42 °C), causing protein denaturation and subsequent cell kill (Jing et al., 2025).
(c) Minimal side effects: As compared to chemotherapy or radiotherapy, PTT with CDs is very localized, and thus the perspective of systemic toxicity is excluded. CDs are also characterized by low photobleaching and good biocompatibility with tissues; the risks of the appearance of long-term consequences are minimized (Bhattacharya et al., 2023).
(d) Synergistic use: CDs could be made as co-loaded with chemotherapeutic drugs or photosensitizers to enable combined therapy methods (for example, chemo-PTT), for increased treatment efficiency and decreased drug resistance (Pal et al., 2024; Bodele et al., 2025).

This impairs PTT with the use of CDs as an attractive approach for targeted, image-guided cancer therapy.

6.3.2.2 Photodynamic Therapy (PDT)

Photodynamic therapy (PDT) is another light-induced cancer therapy technique whereby a photosensitizer produces reactive oxygen species (ROS) under light activation. ROS, like singlet oxygen, can trigger oxidative stress to harm cell membranes, organelles, and DNA, pulling at their death's strings (apoptosis, programmed cell death) (Wang et al., 2024). CDs are particularly appealing for the reasons below:

(a) Efficient ROS generation: Under the light that can be seen/or NIR light, CDs, particularly doped with nitrogen, sulfur or transition metals, may act as effective photosensitizers. They take in the light energy and provide it to oxygen molecules for the production of ROS in situ (Sun et al., 2024).

(b) Targeted cell killing: CDs can be functionalized with ligands or peptides that drive them to cancer cells, so ROS production remains localized to malignant tissues, thus sparing healthy cells (Kashyap et al., 2023; Sun et al., 2024).
(c) Hypoxia adaptability: Certain cutting-edge CDs are designed to respond also in low-oxygen conditions (hypoxia), which are characteristic of solid tumors, thus increasing PDT effectiveness (Sun et al., 2024).
(d) Fluorescence-guided therapy: The innate fluorescence of CDs also facilitates simultaneous diagnosis and treatment such that doctors can use images of the tumor sites to both track the treatment progress and assess it in real-time (Li et al., 2023a).

As PDT is repeatable and has zero invasiveness, CDs represent a new generation of theranostic (therapeutic + diagnostic) agents for this emerging discipline.

6.3.2.3 Hemostatic Activity

One promising and new area of application of CDs is hemostatic application—the use of materials that promote clotting of blood in order to stop bleeding. This is especially helpful in trauma care, surgery, and battlefield medicine (Han et al., 2023). Recent reports, particularly when herbal or plant-derived CDs are used, have indicated that:

(a) Intrinsic coagulation activation: Some CDs can trigger the intrinsic arm of the coagulation cascade (which is sometimes called the contact activation pathway), causing rapid generation of fibrin —the protein crucial for clot formation (Wu et al., 2025b).
(b) Increased fibrinogen levels: CDs may cause the synthesis or release of fibrinogen, a precursor of fibrin, which increases the rate of thrombosis and shortens bleeding time (Lee et al., 2020; Wu et al., 2025b).
(c) Biocompatibility and safety: Herb-extracted CDs (many of which are synthesized from ginger, turmeric or Chinese herbs) have low toxicity and natural biodegradability as well as improved compatibility with the human body, compared to synthetic hemostatic agents (Zhao et al., 2025a).
(d) Topical and injectable Use: CDs can be used as powders, gels or even be made part of bandages in order to check external bleeding. They may also be engineered for injectable formulations in order to treat internal bleeding or to assist in surgical recovery (Zhang et al., 2020; Yang et al., 2021).

Multifunctional by nature, CDs with hemostatic action would be a next-generation bioactive nanomaterial for working in emergency medicine, orthopedics and other fields (Ghosal & Ghosh, 2019).

6.4 Applications in the Food Industry

Food safety is one of the major issues in the world today, especially as demands for freshness, sustainability, and low food waste increase. As outstanding optical materials, with high chemical reactivity and biocompatibility, carbon dots (CDs) are being promoted as a potentially successful tool for quality monitoring of food and the preservation of shelf life. Figure 6.3 shows applications of CDs in the food industry. The following are two of the most ingenious applications of CDs in this area (Naseer et al., 2025). Table 6.3 shows CDs in food safety and spoilage detection.

6.4.1 Smart Packaging

Smart packaging is the term used to define a type of packaging materials that contain sensors or indicators that give real-time data on the condition of the food products. CDs are especially convenient for such a purpose because of their fluorescence, sensitivity to environmental changes, and ability for polymer or film integration (Kumar & Gaikwad, 2023).

6.4.1.1 How It Works

(a) Sensing pH or spoilage compounds: When food degrades, particularly protein-rich foods such as meat or seafood, they emit volatile substances including ammonia, trimethylamine and hydrogen sulfide. Such compounds alter the local pH of the food (Koshy et al., 2021).

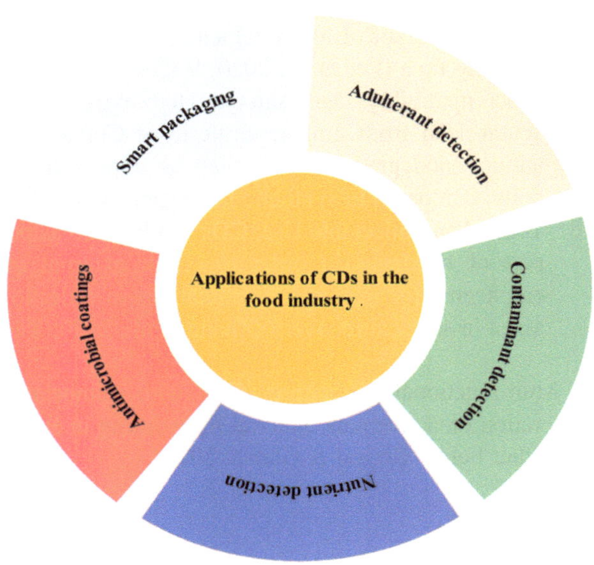

Fig. 6.3 Applications of CDs in the food industry

Table 6.3 Carbon dots in food safety and spoilage detection

Application	CD source	Key feature	References
Meat and seafood spoilage indicator	CDs + gelatin film	Color change in response to pH	Zhao et al. (2025b) and Marappan et al. (2025)
Edible fruit coating	Herbal CDs	Antimicrobial under UV light	Alas et al. (2022)
Dairy contamination detection	Milk-based CDs	Fluorescence response to adulterants	Ghohestani et al. (2024)
Pathogen detection on surfaces	CDs from citric acid	Visual detection of *Salmonella*	Wang et al. (2015)
Freshness sensor in packaging	CDs with zeolites	Ammonia vapor sensitivity	Travlou et al. (2017)
Rapid pH sensor	Herbal-based CDs	Optical change under spoilage conditions	Choudhary et al. (2021)
Fungal toxin detection	CDs from green tea	Response to aflatoxins	Mohammed and Omer (2024)
Pesticide residue analysis	CDs functionalized	Quenching of fluorescence by pesticides	Singh et al. (2025b)
Meat quality imaging	Curcumin CDs	Imaging oxidation in processed meat	Nguyen et al. (2025)
Seafood freshness analysis	Shrimp shell CDs	Colorimetric changes with spoilage	Zhang et al. (2025)

CDs carbon dots

(b) CDs as fluorescence sensors: Some of these CDs are pH-sensitive, that is, they are a fluorescence color or intensity changer in response to the pH change. For example, under UV rays, the CDs in the packaging film may change from bright blue to green or dimmer as a visual spoilage indication (Liu et al., 2019).

(c) Colorimetric or optical cues: The motion of the emission can be observed visually or using a smartphone camera from a second screen, allowing for convenient and non-intrusive monitoring (Jia et al., 2023).

(d) Case study example: Fruits that had been wrapped in a polymer film containing CDs displayed slower fungal growth and fluorescence under UV light, which became non-existent as the fruit declined. Not only was this a preservation of shelf life but also an implemented spoilage alarm (Chaudhary et al., 2025).

6.4.1.2 Advantages

Smart packing that uses carbon dots (CDs) confers multiple superior benefits in relation to food safety and quality monitoring. It makes use of nontoxic edible products safe to incorporate materials, which are safe for food use (Malchenko et al., 2025). This technology allows for real-time monitoring of food freshness, on-site and without the use of elaborate instrumentation, a process that is easily accessible and friendly in its use. Such packaging also helps create consumer engagement and trust since it visibly describes product quality (Khan et al., 2025a). In addition, it contributes to decreasing food waste by separating between literal spoilage and experimental date of expiration. As a whole, smart packaging incorporating CDs is a significant step towards the improvement of the field of smart food management: safer consumption practice and reduced environmental impact (Kumar & Gaikwad, 2023).

6.4.2 Antimicrobial Coatings

Not only can CDs detect food, but CDs can also maintain food by acting as antimicrobial agents themselves (Priyadarshini et al., 2024). Multiple CDs—especially those obtained from some sources of plants or doped with heteroatoms (N, S or metal ions)—exhibit inherent antibacterial activity which can still be amplified upon illumination (Saravanan et al., 2024).

6.4.2.1 Mechanism of Action

(a) Direct bacterial inhibition: CDs bind to bacterial cell walls, causing loss of internal membrane integrity and leakage of internal contents and death (Bing et al., 2016).
(b) ROS generation under light: Under conditions of visible or UV light, CDs can produce reactive oxygen species (ROS)—highly reactive molecules that can oxidize proteins, lipids and DNA in bacterial cells, thus speeding up bacterial death. This photo-activated antimicrobial characteristic is most potent against the common foodborne pathogens, including: *Escherichia coli* (*E. coli*), *Listeria monocytogenes*, *Salmonella* spp (Dong et al., 2020).
(c) Biofilm inhibition: CDs can also inhibit the development of biofilms that are protective coverings formed by bacteria, making them harder to be cleaned with cleansing substances or antibiotics. This property is very important in maintaining food surfaces hygienically (Yu et al., 2023).

6.4.2.2 Applications

(a) Food wrapping materials: CDs are embedded in thin films or coatings on meat, produce, and dairy processing containers (Khan et al., 2025b).
(b) Surface sprays or edible coatings: CD-based formulations can be sprayed on top of the surfaces of fruits and vegetables as methods to retard microbial contamination, without affecting taste or security (Priyadarshi et al., 2024).
(c) Active packaging inserts: One can put disks or pads holding CDs in food trays to function as antimicrobial storage tanks (Ezati et al., 2024).

6.4.2.3 Benefits

Carbon dots (CDs) present promising antimicrobial properties contributing to better food protection and security (Ezati et al., 2024). They help microbial load reduction and food products' shelf life extension, and serve as an all-natural alternative to synthetic preservatives (Zhao et al., 2023). CDs can also demonstrate light-adjusted activity that allows "on-demand" sterilization following the specific lighting conditions (Nagay et al., 2025). When produced from food-grade precursors, they are edible for humans, presenting them with an available and sustainable way of improving the quality and safety of food (Ezati et al., 2024).

6.4.3 Nutrient and Adulterant Detection

It is of public health and the food industry's interest to guarantee the nutritional quality and safety of food products. Conventional methods associated with nutrient analysis and detection of adulteration necessitate the use of expensive instruments, skilled workers, and time-consuming processes. Instead, the Carbon Dots (CDs) provide an effective, affordable and sensitive alternative, involving fluorescent nanoscale sensors for the detection of particular types of biomolecules and hazardous impurities (Jose et al., 2024).

6.4.3.1 Nutrient Sensing

The term nutrient sensing is the determination and quantification of important biomolecules, for example, vitamins, amino acids, antioxidants, and small peptides, in food or biological samples. CDs can be functionalized with certain recognition elements, such as enzymes, aptamers, or functional groups, with targeted nutrients interacting selectively (Redondo-Fernandez et al., 2023).

Mechanisms and Examples

Fluorescence quenching/enhancement: If any nutrient adsorbs or associates with the surface of the CDs, it can quench or enhance the brightness of the CDs. Such a change is quantifiable with high sensitivity and may be used for the quantification of nutrient levels (Akhila et al., 2023).

Glutathione (GSH) detection: A major antioxidant in food and biological systems is GSH. Sulfur- or nitrogen-doped CDs discriminate GSH by fluorescence quenching through the mechanism of a redox reaction or inner filter that does not invade GSH supply and is deemed to report on oxidative stress or antioxidant content (Wang et al., 2022c).

Vitamin C (ascorbic acid): As a powerful reducing agent, vitamin C may affect the oxidation state of CDs or complex with surface-bound ions (such as Fe^{3+}), and changes in fluorescence can be quantified. This makes it possible to detect it in juices and supplements, as well as in fortification studies (Sahu et al., 2024).

Amino acids and peptides: CDs can be designed to attach to specific amino acids, such as tryptophan, cysteine, or histidine, by electrostatic and hydrogen bonds; this influences their emission profile. Some sensors can discern between a variety of amino acids at once (Kolanowska et al., 2022).

Advantages

Carbon dot sensors have high sensitivity, capable of extending detection limits to the nanomolar levels, making them suitable for trace-level analysis. They give straightforward and easy-to-read outputs through fluorescent measurements, making for rapid and precise detection. Such sensors are portable and amenable to adaptation for field testing and inclusion into point-of-care diagnostic devices. Their flexibility of usage makes them useful tools in different applications such as food science, nutritional studies, and health monitoring (Lodha et al., 2024).

6.4.4 Contaminant Detection

Contaminant detection is a key application of CD, in particular, in detecting adulterants, toxins and heavy metals in food matrices. CDs provide rapid, selective, and cost-effective techniques relative to both conventional chromatographic and spectrometric processes (Singh et al., 2025b).

6.4.4.1 Key Targets and Detection Mechanisms

Melamine: A noxious chemical occasionally illegally added to milk or protein products with the intent to misrepresent the amounts of nitrogen content. Although the possibility of CDs detecting melamine as well as that of carbon, hydrogen, oxygen and nitrogen particles had been previously discussed, CDs and melamine could interact through the formation of hydrogen bonds or ionic interactions between the functional groups on the CD surface and melamine, producing significant fluorescence quenching (Wang & Lu, 2022).

Malachite green: A synthetic dye that is prohibited by many nations because of its cancer-causing properties, but used illegally in treating fungi on fish. CDs are able to interact with the aromatic rings of such dyes through π–π stacking or electrostatic interactions that change the intensity of emission (Latiza et al., 2025).

Heavy metals (Pb^{2+}, Hg^{2+}, Cd^{2+}, etc.): CDs, particularly those having carboxyl, hydroxyl, or amine groups, exhibit a high affinity for simultaneous association with metal ions. The binding usually induces a detectable quenching of the fluorescence. Even some CDs could separate metal ions on the basis of response patterns (Echeverry-Gonzalez & Kouznetsov, 2021).

Pesticides and antibiotics: Such functionalized CDs can also be created to bind particular molecules, such as organophosphates or sulfonamides. When two signals interact, signal change can be related to contaminant concentration (Chu et al., 2020).

6.4.4.2 Practical Applications

Carbon dot-based technologies allow fast screening of food products during production or packaging so that quality control can be timed. These systems can be assembled onto test strips, portable fluorimeters, which make them portable and usable in the field. They are effective in sophisticated food matrices of milk, meat, vegetables, beverages, etc., which is a property for versatile applications. Also, they act as advanced warning systems in testing contaminated or adulterated foodstuffs, adding food safety and consumer protection (Szczepankowska et al., 2023).

6.4.4.3 Benefits

Carbon dots (CDs) possess outstanding performance in analytical purposes because their detection limits are very low and range from parts per million (ppm) to parts per billion (ppb). They have a surface that can be tailored to provide high selectivity to targets, thus improving the detection accuracy (Sendão et al., 2023). With green synthesis methods used for their production, CDs provide an eco-friendly alternative to normal sensors. Also, their cost-effectiveness eliminates dependence on sophisticated laboratory setups and makes them easily suitable for multitudinous settings (Dhariwal et al., 2024).

6.5 Applications in Agriculture

Nanotechnology is gaining wider application in agriculture in the effort to increase productivity and sustainability and mitigate the environmental impact. Among several nanomaterials, carbon dots (CDs) have become promising tools by virtue of their high degree of miniaturization coupled with their ability to have fluorescence turned on or off demand, which allows them to carry out their tasks with great consistency (Mansuriya & Altintas, 2021). They can also be rendered biocompatible, be functionalized, and work in tandem with other compatible. The versatility of their properties allows both biostimulant and sensorial uses, which are applicable in both traditional and precision agriculture (Li et al., 2023b; Maholiya et al., 2023). Figure 6.4 and Table 6.4 show CDs in agricultural applications.

6.5.1 Precision Agriculture

Precision agriculture entails the real-time control over the inputs of crops to enhance outputs and eliminate much of the waste. CDs provide state-of-the-art sensing and diagnostic equipment for smarter agriculture (Murali et al., 2022).

6.5.1.1 Real-Time Monitoring

Carbon dots (CDs) can be integrated into nanosensors and smart delivery systems to monitor essential soil and plant health parameters, enabling precision agriculture (Wahab et al., 2024). For instance, CDs embedded in sensor films can respond to changes in humidity by altering their fluorescence, providing real-time data on soil moisture to guide efficient irrigation scheduling (Bharti et al., 2024). pH-sensitive CDs can detect soil acidity or alkalinity through changes in their emission properties, while those functionalized with ion-specific receptors are capable of identifying key nutrients such as nitrogen (NH_4^+/NO_3^-), phosphorus, and potassium (Liu et al., 2025). Additionally, CDs applied as foliar sprays or absorbed through roots can be used for plant health monitoring, as they enable bioimaging and spectral analysis of metabolic changes, signaling stress or nutrient deficiencies (Pan et al., 2024). These advanced tools support data-driven decision-making, leading to optimized use of agricultural inputs, cost savings, and enhanced environmental sustainability.

Fig. 6.4 Applications of CDs in agriculture

6.5.1.2 Pesticide and Pathogen Detection

Early detection of diseases and chemical residues is important for effective crop protection as well as food safety and carbon dots (CDs) have become valuable rapid sensors in this area. CDs are able to interact with molecules such as organophosphates or neonicotinoids, which lead to changes in fluorescence that can be followed: changes in emission or quenching. Such changes can readily be detected in the field and can even be used on smartphones for diagnostic platforms. For pathogen detection, Conjugated CDS with antibodies, aptamers or enzymes can be chosen to selectively identify microbial infections or viral markers; timely intervention can thus be made, additionally corroborating IPM implication. These CD-based sensors are non-destructive, cost-effective, and can provide results in minutes, enabling farmers to take informed decisions and avoid crop losses (Chu et al., 2020).

Table 6.4 Carbon dots in agricultural applications

Crop/ application	CD type	Effect	References
Soybean	Herbal CDs	Increased protein content under drought	Ji et al. (2023)
Tomato	Fluorescent CDs	Growth under saline stress	Chen et al. (2022)
Maize	CDs with NPK	Improved nutrient uptake	Zhao et al. (2021)
Wheat	CDs from manure	Soil pH buffering and better germination	Tripathi and Sarkar (2015)
Lettuce	Chlorophyll-like CDs	Better chloroplast formation	Hu et al. (2022)
Arabidopsis	Light-responsive CDs	Gene expression related to photosynthesis	Reyes and Chua (2004)
Mung	ND	Growth of mung bean sprouts, including root elongation, stem elongation and biomass.	Wang et al. (2015)

ND not determined, *CDs* carbon dots

6.6 Future Prospects

While the world is searching for smarter, safer, and sustainable technologies, carbon dots (CDs) pop up as one of the most versatile nanomaterials of the century (Manzoor et al., 2023). Their specific physicochemical properties—tunable fluorescence, biocompatibility, eco-friendliness, and functionality-friendliness place them at the edge of discoveries for various industries. In the future, CDs will be on the verge of transforming a variety of applications, specifically in healthcare, agriculture, food safety, and environmental control (Sistani & Shekarchizadeh, 2022). One of the key opportunities here is in developing a formidable biosafety and regulation framework. Despite the relatively low level of cytotoxicity and biodegradability of CDs from the primary tests, the way to their clinical and agricultural acceptance requires whole-scale, long-term studies (Mohammadinejad et al., 2019). Knowing their environmental fate, accumulation and their interactions within intricate biotic and ecological systems is crucial towards public acceptability and policy conjunction (Li et al., 2023b).

In order to achieve their commercial value, emphasis should shift to production systems that are scalable and standardized. Existing green synthesis methods utilizing wastes of biomass or natural precursors appear promising; however, reproducibility, purity control, and improvements in yield requirements for industrial applications need to be enhanced. Future energy should be invested in the creation of energy-efficient, cheap and ecologically sound manufacturing pipelines. One of the very promising fields in the integration of CDs with other advanced materials and digital platforms. Hybrid systems like CDs conjugated with metal oxides, graphene derivatives, or quantum dots can open synergistic features of highly efficient biosensors, diagnostics, and the treatment of PD. The combination of the emerging digital technologies such as IoT (Internet of Things), AI (artificial intelligence), and wearable electronics, with CDs can lead to real-time, remote and unattended sensing applications. These intelligent platforms can identify pathogens, environmental toxins, or nutrient misbalances so precisely and fast that it has never been possible before (Fu et al., 2020; Zanbili & Poursattar Marjani, 2025).

The collusion between nanotechnology and data science opens up the doors for personalized healthcare and precision agriculture. In medicine, they may be incorporated in ingestible or implantable sensors to conduct continuous monitoring of metabolic markers for the early detection of diseases or in responsive therapy. In agriculture, they could run climate-resilient smart farming systems that would be able to change the level of irrigation, fertilization or pest suppression in function of the real-time field conditions. To sum up, the future of CDs lies in multidisciplinary cooperation, responsible innovation, and democratization of the technology. With biosafety concerns satisfied, scaling green synthesis, and integration with smart systems, CDs can trigger the breakthroughs of the industries—we are moving forward to a more sustainable, wise and healthy world (Khan et al., 2024).

6.7 Conclusion

The advent of Carbon Dots (CDs) as a revolutionary class of nanomaterials exhibiting brilliance in the realm of the optical, as well as an eco-friendly nature, has qualified them to be leading contenders in a broad range of applications in healthcare, food safety, and agriculture. Such unique structural and photophysical traits of having tunable fluorescence, nanoscale dimensions, water solubility, and low cytotoxicity, among others, place them as a next generation of tools in diagnostics, therapeutics, sensing, and precision monitoring. While their functional versatility is one of the major distinguishing points for CDs, the ability of their synthesis from renewable, low-cost biomass in accordance with the principles of green chemistry and circular economy is an essential point in their favor as well (Mintz et al., 2019). In health, CDs are re-inventing the bioimaging and diagnosis terrain with their real-time, multiple color, high-resolution imaging applications from cellular to whole organisms. They have also exhibited outstanding potential in therapeutic settings, including photothermal and photodynamic therapy for cancer treatment, and even in such uses as hemostasis, where those derived from plants exhibit clot-building properties. At the same time, in the food industry, CDs provide creative approaches to identifying the spoilage, sensing the nutritive content, and combating the microbial pollution with smart packaging and antimicrobial covering (Phan & Cho, 2022). In the fields of agriculture, the CDs are not only increasing the photosynthetic efficiency and crop yields but are also increasing the plant-resilience against abiotic stresses like drought and salinity. Their contribution to seed germination, nutrient sensing, and pathogen detection aptly points towards their potential for developing sustainable and precision agriculture practices. In addition, their incorporation with smart technologies, including nanosensors and Internet of Things (IoT) platforms, is making real-time decision-making possible, informing resource optimization and stewardship of the environment (Zaman et al., 2025). Despite these advances, challenges remain. A better understanding of the long-term biological and environmental effects of CDs is very important for biosafety. In addition, scalable and standardized production methods need to be established to meet the requirements of the industrial and clinical uses. Regulatory frameworks will also have to change in tandem to enable safe and responsible deployment (Tavan et al., 2025). As for the future, the future of CDs is at the intersection of material sciences, biotechnology, and digital innovation. Their hooking up with artificial intelligence, wearable devices, and hybrid nanomaterials may result in revolutionary tools of personalized medicine, smart diagnostics, and climate-proof agriculture. If interdisciplinary teamwork and research and development expenditures remain, CDs can be primary building blocks involved in building a more environmentally friendly, health-conscious society and one that is more tech-savvy.

Author Contributions MG-S and MRA conceived and designed the study. MG-S, KN, SS, MS-G, SM-G, RS-A, AE, FJ, and MRA collected the data and contributed to the data analysis. All authors approved the final version of the chapter. MRA supervised the study.

Conflict of Interest Authors declare no conflict of interest.

References

Akhila, M. J., Subodh, G., Sibi, K. S., & Biju, V. (2023). Folic acid-derived luminescent carbon dots: Effect of Ag inclusion and switch ON sensing of sulfide ions. *Journal of Molecular Liquids, 385*, 122396. https://doi.org/10.1016/j.molliq.2023.122396

Aksu, M., & Güzdemir, Ö. (2025). Food waste-derived carbon quantum dots and their applications in food technology: A critical review. *Food and Bioprocess Technology*, 1–26. https://doi.org/10.1007/s11947-025-03854-1

Alas, M. O., Dogan, G., Yalcin, M. S., Ozdemir, S., & Genc, R. (2022). Multicolor emitting carbon dot-reinforced PVA composites as edible food packaging films and coatings with antimicrobial and UV-blocking properties. *ACS Omega, 7*(34), 29967–29983. https://doi.org/10.1021/acsomega.2c02984

Balou, S., Shandilya, P., & Priye, A. (2022). Carbon dots for photothermal applications. *Frontiers in Chemistry, 10*, 1023602. https://doi.org/10.3389/fchem.2022.1023602

Bao, X., Yuan, Y., Chen, J., Zhang, B., Li, D., Zhou, D., Jing, P., Xu, G., Wang, Y., Holá, K., Shen, D., Wu, C., Song, L., Liu, C., Zbořil, R., & Qu, S. (2018). *In vivo* theranostics with near-infrared-emitting carbon dots—Highly efficient photothermal therapy based on passive targeting after intravenous administration. *Light Science & Applications, 7*(1). https://doi.org/10.1038/s41377-018-0090-1

Bartkowski, M., Zhou, Y., Mustafa, M. N. A., Eustace, A. J., & Giordani, S. (2024). CARBON DOTS: Bioimaging and anticancer drug delivery. *Chemistry – A European Journal, 30*(19), e202303982. https://doi.org/10.1002/chem.202303982

Bharti, A., Jain, U., & Chauhan, N. (2024). From lab to field: Nano-biosensors for real-time plant nutrient tracking. *Plant Nano Biology*, 100079. https://doi.org/10.1016/j.plana.2024.100079

Bhattacharya, T., Shin, G. H., & Kim, J. T. (2023). Carbon dots: Opportunities and challenges in cancer therapy. *Pharmaceutics, 15*(3), 1019. https://doi.org/10.3390/pharmaceutics15031019

Bhattacharya, T., Preetam, S., Mukherjee, S., Kar, S., Roy, D. S., Singh, H., et al. (2024). Anticancer activity of quantum size carbon dots: Opportunities and challenges. *Discover Nano, 19*(1), 122. https://doi.org/10.1186/s11671-024-04069-7

Bhavikatti, S. K., Zainuddin, S. L., Ramli, R. B., Nadaf, S. J., Dandge, P. B., Khalate, M., & Karobari, M. I. (2024). Green-synthesized carbon dots from ginger: Multifunctional agents against oral pathogens with biocompatibility in human gingival fibroblast cells. *Current Plant Biology, 40*, 100392. https://doi.org/10.1016/j.cpb.2024.100392

Bing, W., Sun, H., Yan, Z., Ren, J., & Qu, X. (2016). Programmed bacteria death induced by carbon dots with different surface charge. *Small, 12*(34), 4713–4718. https://doi.org/10.1002/smll.201600294

Boakye-Yiadom, K. O., Kesse, S., Opoku-Damoah, Y., Filli, M. S., Aquib, M., Joelle, M. M., Farooq, M. A., Mavlyanova, R., Raza, F., Bavi, R., & Wang, B. (2019). Carbon dots: Applications in bioimaging and theranostics. *International Journal of Pharmaceutics, 564*, 308–317. https://doi.org/10.1016/j.ijpharm.2019.04.055

Bodele, V. G., Lade, S. N., Undirwade, D. S., Umekar, M. J., Burle, S. S., Hanmante, P. S., & Lohiya, R. T. (2025). Transforming oncology with carbon quantum dots: Synthesis, properties, and therapeutic potential. *Next Nanotechnology, 7*, 100181. https://doi.org/10.1016/j.nxnano.2025.100181

Bruno, F., Sciortino, A., Buscarino, G., Soriano, M. L., Ríos, Á., Cannas, M., Gelardi, F., Messina, F., & Agnello, S. (2021). A comparative study of top-down and bottom-up carbon nanodots and their interaction with mercury ions. *Nanomaterials, 11*(5), 1265. https://doi.org/10.3390/nano11051265

Cai, D., Zhong, X., Xu, L., Xiong, Y., Deng, W., Zou, G., Hou, H., & Ji, X. (2025). Biomass-derived carbon dots: Synthesis, modification and application in batteries. *Chemical Science, 16*(12), 4937–4970. https://doi.org/10.1039/D4SC08659G

Calabrese, G., De Luca, G., Nocito, G., Rizzo, M. G., Lombardo, S. P., Chisari, G., Forte, S., Sciuto, E. L., & Conoci, S. (2021). Carbon dots: An innovative tool for drug delivery in brain tumors. *International Journal of Molecular Sciences, 22*(21), 11783. https://doi.org/10.3390/ijms222111783

Chaudhary, M., Singh, P., Singh, G. P., & Rathi, B. (2024). Structural features of carbon dots and their agricultural potential. *ACS Omega, 9*(4), 4166–4185. https://doi.org/10.1021/acsomega.3c04638

Chaudhary, A., Siddiqui, M. W., Gaikwad, K. K., Ahmad, M. S., Homa, F., Singh, D. R., Lata, D., Alam, T., Saurabh, A., Deshmukh, R., & Khaneghah, A. M. (2025). Novel applications of carbon dots (CDs) for enhanced preservation of postharvest fruit and vegetables. *Food Control*, 111182. https://doi.org/10.1016/j.foodcont.2025.111182

Chen, Q., Cao, X., Li, Y., Sun, Q., Dai, L., Li, J., Guo, Z., Zhang, L., & Ci, L. (2022). Functional carbon nanodots improve soil quality and tomato tolerance in saline-alkali soils. *Science of the Total Environment, 830*, 154817. https://doi.org/10.1016/j.scitotenv.2022.154817

Chen, J., Chen, M., & Yu, X. (2025). Fluorescent probes in autoimmune disease research: Current status and future prospects. *Journal of Translational Medicine, 23*(1), 411. https://doi.org/10.1186/s12967-025-06430-5

Chien, Y. H., Chan, K. K., Anderson, T., Kong, K. V., Ng, B. K., & Yong, K. T. (2019). Advanced near infrared light responsive nanomaterials as therapeutic platforms for cancer therapy. *Advanced Therapeutics, 2*(3), 1800090. https://doi.org/10.1002/adtp.201800090

Choudhary, S., Joshi, B., & Joshi, A. (2021). Translation of carbon dot biosensors into an embedded optical setup for spoilage and adulteration detection. *ACS Food Science & Technology, 1*(6), 1068–1076. https://doi.org/10.1021/acsfoodscitech.1c00089

Chu, H. W., Unnikrishnan, B., Anand, A., Lin, Y. W., & Huang, C. C. (2020). Carbon quantum dots for the detection of antibiotics and pesticides. *Journal of Food and Drug Analysis, 28*(4), 539. https://doi.org/10.38212/2224-6614.1269

Cui, L., Ren, X., Sun, M., Liu, H., & Xia, L. (2021). Carbon dots: Synthesis, properties and applications. *Nanomaterials, 11*(12), 3419. https://doi.org/10.3390/nano11123419

Debnath, R., Ikbal, A. M., Ravi, N. K., Kargarzadeh, H., Palit, P., & Thomas, S. (2025). Carbon nanodots-based polymer nanocomposite: A potential drug delivery armament of phytopharmaceuticals. *Polymers, 17*(3), 365. https://doi.org/10.3390/polym17030365

Dhariwal, J., Rao, G. K., & Vaya, D. (2024). Recent advancements towards the green synthesis of carbon quantum dots as an innovative and eco-friendly solution for metal ion sensing and monitoring. *RSC Sustainability, 2*(1), 11–36. https://doi.org/10.1039/D3SU00375B

Dong, X., Liang, W., Meziani, M. J., Sun, Y. P., & Yang, L. (2020). Carbon dots as potent antimicrobial agents. *Theranostics, 10*(2), 671. https://doi.org/10.7150/thno.39863

Echeverry-Gonzalez, C. A., & Kouznetsov, V. V. (2021). Carbon Dots: An insight into their application in heavy metal sensing. *Recent Progress in Materials, 3*(2), 1–7. https://doi.org/10.21926/rpm.2102015

Elugoke, S. E., Uwaya, G. E., Quadri, T. W., & Ebenso, E. E. (2024). Carbon quantum dots: Basics, properties, and fundamentals. In *Carbon dots: Recent developments and future perspectives* (pp. 3–42). American Chemical Society. https://doi.org/10.1021/bk-2024-1465.ch001

Etefa, H. F., Tessema, A. A., & Dejene, F. B. (2024). Carbon dots for future prospects: Synthesis, characterizations and recent applications: A review (2019–2023), *10*(3), 60. https://doi.org/10.3390/c10030060

Ezati, P., Khan, A., Bhattacharya, T., Zaitoon, A., Zhang, W., Roy, S., Rhim, J. W., & Lim, L. T. (2024). New opportunities and recent advances in carbon dots for sustainable and intelligent food packaging. *Food and Bioprocess Technology*, 1–27. https://doi.org/10.1007/s11947-024-03731-3

Fathi-karkan, S., Nasiri, S., Hasannia, M., Rahdar, A., Nouri, M., Najafi, M., Basiri, A., Tehrani, F. S., Hashemi, S. M., & Romanholo Ferreira, L. F. (2025). Green carbon dots and theranostic applications. *Journal of Drug Delivery Science and Technology, 108*. https://doi.org/10.1016/j.jddst.2025.106861

Fu, Y., Zeng, G., Lai, C., Huang, D., Qin, L., Yi, H., Liu, X., Zhang, M., Li, B., Liu, S., & Li, L. (2020). Hybrid architectures based on noble metals and carbon-based dots nanomaterials: A review of recent progress in synthesis and applications. *Chemical Engineering Journal, 399*, 125743. https://doi.org/10.1016/j.cej.2020.125743

Ghohestani, E., Tashkhourian, J., & Hemmateenejad, B. (2024). Rapid detection and quantification of milk adulterants using a nanoclusters-based fluorescent optical tongue. *Food Chemistry, 456*, 139973. https://doi.org/10.1016/j.foodchem.2024.139973

Gholami-Shabani, M., Shams-Ghahfarokhi, M., Gholami-Shabani, Z., & Razzaghi-Abyaneh, M. (2016). Microbial enzymes: Current features and potential applications in nanobiotechnology. In R. Prasad (Ed.), *Advances and applications through fungal nanobiotechnology* (pp. 75–98). Springer. https://doi.org/10.1007/978-3-319-42990-8_5

Gholami-Shabani, M., Gholami-Shabani, Z., Shams-Ghahfarokhi, M., Jamzivar, F., & Razzaghi-Abyaneh, M. (2017). Green nanotechnology: Biomimetic synthesis of metal nanoparticles using plants and their application in agriculture and forestry. In *Nanotechnology: An agricultural paradigm* (pp. 133–175). https://doi.org/10.1007/978-981-10-4573-8_8

Gholami-Shabani, M., Shams-Ghahfarokhi, M., Jamzivarl, F., & Razzaghi-Abyaneh, M. (2023). Fungal biofilms in nanotechnology era diagnosis, drug delivery, and treatment strategies. In *Myconanotechnology* (pp. 129–168). CRC Press.

Ghosal, K., & Ghosh, A. (2019). Carbon dots: The next generation platform for biomedical applications. *Materials Science and Engineering: C, 96*, 887–903. https://doi.org/10.1016/j.msec.2018.11.060

Gollavelli, G., Ghule, A. V., & Ling, Y. C. (2022). Multimodal imaging and phototherapy of cancer and bacterial infection by graphene and related nanocomposites. *Molecules, 27*(17), 5588. https://doi.org/10.3390/molecules27175588

Han, B., Shen, L., Xie, H., Huang, Q., Zhao, D., Huang, X., Chen, X., & Li, J. (2023). Synthesis of carbon dots with hemostatic effects using traditional Chinese medicine as a biomass carbon source. *ACS Omega, 8*(3), 3176–3183. https://doi.org/10.1021/acsomega.2c06600

Hashemi, N., & Mousazadeh, M. H. (2021). Green synthesis of photoluminescent carbon dots derived from red beetroot as a selective probe for Pd^{2+} detection. *Journal of Photochemistry and Photobiology A: Chemistry, 421*, 113534. https://doi.org/10.1016/j.jphotochem.2021.113534

Hu, C., Li, M., Qiu, J., & Sun, Y. P. (2019). Design and fabrication of carbon dots for energy conversion and storage. *Chemical Society Reviews, 48*(8), 2315–2337. https://doi.org/10.1039/C8CS00750K

Hu, J., Jia, W., Wu, X., Zhang, H., Wang, Y., Liu, J., Yang, Y., Tao, S., & Wang, X. (2022). Carbon dots can strongly promote photosynthesis in lettuce (*Lactuca sativa* L.). *Environmental Science: Nano, 9*(4), 1530–1540. https://doi.org/10.1039/D1EN00948F

Hu, Y., Seivert, O., Tang, Y., Karahan, H. E., & Bianco, A. (2024a). Carbon dot synthesis and purification: Trends, challenges and recommendations. *Angewandte Chemie International Edition, 63*(48), e202412341. https://doi.org/10.1002/anie.202412341

Hu, J., Zheng, Z., Yang, Y., Chen, L., & Kang, W. (2024b). Advance of near infrared emissive carbon dots in diagnosis and therapy: Synthesis, luminescence, and application. *Advanced Healthcare Materials, 13*(30), 2401513. https://doi.org/10.1002/adhm.202401513

Hussen, N. H., Hasan, A. H., FaqiKhedr, Y. M., Bogoyavlenskiy, A., Bhat, A. R., & Jamalis, J. (2024). Carbon dot based carbon nanoparticles as potent antimicrobial, antiviral, and anticancer agents. *ACS Omega, 9*(9), 9849–9864. https://doi.org/10.1021/acsomega.3c05537

Innocenzi, P., De Forni, D., & Lori, F. (2025). Antiviral activity of carbon dots: Strategies and mechanisms of action. *Small Structures, 6*(3), 2400401. https://doi.org/10.1002/sstr.202400401

Jana, J., Lee, H. J., Chung, J. S., Kim, M. H., & Hur, S. H. (2019). Blue emitting nitrogen-doped carbon dots as a fluorescent probe for nitrite ion sensing and cell-imaging. *Analytica Chimica Acta, 1079*, 212–219. https://doi.org/10.1016/j.aca.2019.06.064

Ji, Y., Yue, L., Cao, X., Chen, F., Li, J., Zhang, J., Wang, C., Wang, Z., & Xing, B. (2023). Carbon dots promoted soybean photosynthesis and amino acid biosynthesis under drought stress: Reactive oxygen species scavenging and nitrogen metabolism. *Science of the Total Environment, 856*, 159125. https://doi.org/10.1016/j.scitotenv.2022.159125

Jia, Z., Liu, Y., Cheng, L., Deng, Z., Zhang, M., & Tuo, H. (2023). Carbon dots with light-responsive oxidase-like activity for colorimetric detection of dopamine and the catalytic mechanism. *Frontiers in Chemistry, 11*, 1288418. https://doi.org/10.3389/fchem.2023.1288418

Jing, H. H., Bardakci, F., Akgöl, S., Kusat, K., Adnan, M., Alam, M. J., Gupta, R., Sahreen, S., Chen, Y., Gopinath, S. C., & Sasidharan, S. (2023). Green carbon dots: Synthesis, characterization, properties and biomedical applications. *Journal of Functional Biomaterials, 14*(1), 27. https://doi.org/10.3390/jfb14010027

Jing, H. H., Shati, A. A., Alfaifi, M. Y., Elbehairi, S. E., & Sasidharan, S. (2025). The future of plant based green carbon dots as cancer nanomedicine: From current progress to future perspectives and beyond. *Journal of Advanced Research, 67*, 133–159. https://doi.org/10.1016/j.jare.2024.01.034

Jorns, M., & Pappas, D. (2021). A review of fluorescent carbon dots, their synthesis, physical and chemical characteristics, and applications. *Nanomaterials, 11*(6), 1448. https://doi.org/10.3390/nano11061448

Jose, J., Rangaswamy, M., Shamnamol, G. K., & Greeshma, K. P. (2024). A short review on natural precursors-Plant-based fluorescent carbon dots for the targeted detection of metal ions. *Sustainable Chemistry for the Environment*, 100114. https://doi.org/10.1016/j.scenv.2024.100114

Kashyap, B. K., Singh, V. V., Solanki, M. K., Kumar, A., Ruokolainen, J., & Kesari, K. K. (2023). Smart nanomaterials in cancer theranostics: Challenges and opportunities. *ACS Omega, 8*(16), 14290–14320. https://doi.org/10.1021/acsomega.2c07840

Khan, A. U., Liu, Y., Wang, S., Ullah, M. W., Chen, Q., Zhang, D., Kang, Z., & Mao, B. (2024). Advancements in the green synthesis of carbon dots for sustainable development. *Sustainable Materials and Technologies*, e01004. https://doi.org/10.1016/j.susmat.2024.e01004

Khan, R. N., Zubair, M., Azeem, F., Nadeem, H., Muzammil, S., Tabassum, A., & Rasul, I. (2025a). Biosensors in food packaging. In *Applications of biosensors in healthcare* (pp. 711–730). Academic Press. https://doi.org/10.1016/B978-0-443-21592-6.00030-6

Khan, A., Ezati, P., Tammina, S. K., Riahi, Z., & Priyadarshi, R. (2025b). Carbon-based nanomaterials in food packaging and preservation. In *Functional nanomaterials and nanocomposites for biodegradable food packaging* (pp. 109–128). Springer. https://doi.org/10.1007/978-981-97-9084-5_5

Khayal, A., Dawane, V., Amin, M. A., Tirth, V., Yadav, V. K., Algahtani, A., Khan, S. H., Islam, S., Yadav, K. K., & Jeon, B. H. (2021). Advances in the methods for the synthesis of carbon dots and their emerging applications. *Polymers, 13*(18), 3190. https://doi.org/10.3390/polym13183190

Kolanowska, A., Dzido, G., Krzywiecki, M., Tomczyk, M. M., Łukowiec, D., Ruczka, S., & Boncel, S. (2022). Carbon quantum dots from amino acids revisited: Survey of renewable precursors toward high quantum-yield blue and green fluorescence. *ACS Omega, 7*(45), 41165–41176. https://doi.org/10.1021/acsomega.2c04751

Koshy, R. R., Koshy, J. T., Mary, S. K., Sadanandan, S., Jisha, S., & Pothan, L. A. (2021). Preparation of pH sensitive film based on starch/carbon nano dots incorporating anthocyanin for monitoring spoilage of pork. *Food Control, 126*, 108039. https://doi.org/10.1016/j.foodcont.2021.108039

Kouloumpis, A., Thomou, E., Chalmpes, N., Dimos, K., Spyrou, K., Bourlinos, A. B., Koutselas, I., Gournis, D., & Rudolf, P. (2017). Graphene/carbon dot hybrid thin films prepared by a modified Langmuir–Schaefer method. *ACS Omega, 2*(5), 2090–2099. https://doi.org/10.1021/acsomega.7b00107

Kumar, L., & Gaikwad, K. K. (2023). Carbon dots for food packaging applications. *Sustainable Food Technology, 1*(2), 185–199. https://doi.org/10.1039/D2FB00020B

Kumar, R., Kumar, V. B., & Gedanken, A. (2020). Sonochemical synthesis of carbon dots, mechanism, effect of parameters, and catalytic, energy, biomedical and tissue engineering applications. *Ultrasonics Sonochemistry, 64*, 105009. https://doi.org/10.1016/j.ultsonch.2020.105009

Kumar, A., Tyagi, P. K., Tyagi, S., & Ghorbanpour, M. (2024). Integrating green nanotechnology with sustainable development goals: A pathway to sustainable innovation. *Discover Sustainability, 5*(1), 364. https://doi.org/10.1007/s43621-024-00610-x

Kumari, S., Nehra, M., Jain, S., Kumar, A., Dilbaghi, N., Marrazza, G., Chaudhary, G. R., & Kumar, S. (2024). Carbon dots for pathogen detection and imaging: Recent breakthroughs and future trends. *Microchimica Acta, 191*(11), 684. https://doi.org/10.1007/s00604-024-06762-x

Latiza, R. J., Olay, J., Eguico, C., Yan, R. J., & Rubi, R. V. (2025). Environmental applications of carbon dots: Addressing microplastics, air and water pollution. *Journal of Hazardous Materials Advances*, 100591. https://doi.org/10.1016/j.hazadv.2025.100591

Lee, T. Y., Jayakumar, T., Thanasekaran, P., Lin, K. C., Chen, H. M., Veerakumar, P., & Sheu, J. R. (2020). Carbon dot nanoparticles exert inhibitory effects on human platelets and reduce mortality in mice with acute pulmonary thromboembolism. *Nanomaterials, 10*(7), 1254. https://doi.org/10.3390/nano10071254

Li, D., Xu, K. Y., Zhao, W. P., Liu, M. F., Feng, R., Li, D. Q., Bai, J., & Du, W. L. (2022). Chinese medicinal herb-derived carbon dots for common diseases: Efficacies and potential mechanisms. *Frontiers in Pharmacology, 13*, 815479. https://doi.org/10.3389/fphar.2022.815479

Li, C., Huang, J., Yuan, L., Xie, W., Ying, Y., Li, C., Yu, Y., Pan, Y., Qu, W., Hao, H., & Algharib, S. A. (2023a). Recent progress of emitting long-wavelength carbon dots and their merits for visualization tracking, target delivery and theranostics. *Theranostics, 13*(9), 3064. https://doi.org/10.7150/thno.80579

Li, G., Xu, J., & Xu, K. (2023b). Physiological functions of carbon dots and their applications in agriculture: A review. *Nanomaterials, 13*(19), 2684. https://doi.org/10.3390/nano13192684

Liu, M. L., Chen, B. B., Li, C. M., & Huang, C. Z. (2019). Carbon dots: Synthesis, formation mechanism, fluorescence origin and sensing applications. *Green Chemistry, 21*(3), 449–471. https://doi.org/10.1039/C8GC02736F

Liu, J., Li, R., & Yang, B. (2020). Carbon dots: A new type of carbon-based nanomaterial with wide applications. *ACS Central Science, 6*(12), 2179–2195. https://doi.org/10.1021/acscentsci.0c01306

Liu, H., Zhong, X., Pan, Q., Zhang, Y., Deng, W., Zou, G., Hou, H., & Ji, X. (2024). A review of carbon dots in synthesis strategy. *Coordination Chemistry Reviews, 498*, 215468. https://doi.org/10.1016/j.ccr.2023.215468

Liu, Y., Fang, M., Wu, Z., Wang, H., Li, Y., Dong, C., Shuang, S., & Gong, X. (2025). Multifunctional carbon dots for pH sensing, intracellular imaging, amino acid classification, and test strip development. *Spectrochimica Acta, Part A: Molecular and Biomolecular Spectroscopy*, 125868. https://doi.org/10.1016/j.saa.2025.125868

Lodha, S. R., Merchant, J. G., Pillai, A. J., Gore, A. H., Patil, P. O., Nangare, S. N., Kalyankar, G. G., Shah, S. A., Shah, D. R., & Patole, S. P. (2024). Carbon dot-based fluorescent sensors for pharmaceutical detection: Current innovations, challenges, and future prospects. *Heliyon, 10*(24), e41020. https://doi.org/10.1016/j.heliyon.2024.e41020

Long, C., Jiang, Z., Shangguan, J., Qing, T., Zhang, P., & Feng, B. (2021). Applications of carbon dots in environmental pollution control: A review. *Chemical Engineering Journal, 40*, 126848. https://doi.org/10.1016/j.cej.2020.126848

Ma, X., Dong, Y., Sun, H., & Chen, N. (2017). Highly fluorescent carbon dots from peanut shells as potential probes for copper ion: The optimization and analysis of the synthetic process. *Materials Today Chemistry, 5*, 1–10. https://doi.org/10.1016/j.mtchem.2017.04.004

Maholiya, A., Ranjan, P., Khan, R., Murali, S., Nainwal, R. C., Chauhan, P. S., Sathish, N., Chaurasia, J. P., & Srivastava, A. K. (2023). An insight into the role of carbon dots in the agriculture system: A review. *Environmental Science: Nano, 10*(4), 959–995. https://doi.org/10.1039/D2EN00954D

Malchenko, V. A., Tracey, C. T., Krivoshapkina, E. F., & Nazarova, E. A. (2025). Are food-derived carbon dots really edible? A potential harm versus potential benefits analysis. *Trends in Food Science & Technology*, 105031. https://doi.org/10.1016/j.tifs.2025.105031

Malik, S., Muhammad, K., & Waheed, Y. (2023). Nanotechnology: A revolution in modern industry. *Molecules, 28*(2), 661. https://doi.org/10.3390/molecules28020661

Mancini, F., Menichetti, A., Degli Esposti, L., Montesi, M., Panseri, S., Bassi, G., Montalti, M., Lazzarini, L., Adamiano, A., & Iafisco, M. (2023). Fluorescent carbon dots from food industry by-products for cell imaging. *Journal of Functional Biomaterials, 14*(2), 90. https://doi.org/10.3390/jfb14020090

Mansuriya, B. D., & Altintas, Z. (2021). Carbon Dots: Classification, properties, synthesis, characterization, and applications in health care—An updated review (2018–2021). *Nanomaterials, 11*(10), 2525. https://doi.org/10.3390/nano11102525

Manzoor, S., Dar, A. H., Dash, K. K., Pandey, V. K., Srivastava, S., Bashir, I., & Khan, S. A. (2023). Carbon dots applications for development of sustainable technologies for food safety: A comprehensive review. *Applied Food Research, 3*(1), 100263. https://doi.org/10.1016/j.afres.2023.100263

Marappan, G., Tahir, H. E., Karim, N., Lakshmanan, A., Shishir, M. R., Hashim, S., Khogly, A., Khan, S., Huang, X., Sivalingam, Y., & Zhai, X. (2025). Natural pigment-based pH/gas-sensitive intelligent packaging film for freshness monitoring of meat and seafood: Influencing factors, technological advances, and future perspectives. *Food Reviews International*, 1–38. https://doi.org/10.1080/87559129.2025.2473026

Mintz, K. J., Zhou, Y., & Leblanc, R. M. (2019). Recent development of carbon quantum dots regarding their optical properties, photoluminescence mechanism, and core structure. *Nanoscale, 11*(11), 4634–4652. https://doi.org/10.1039/C8NR10059D

Mohammadi, R., Naderi-Manesh, H., Farzin, L., Vaezi, Z., Ayarri, N., Samandari, L., & Shamsipur, M. (2022). Fluorescence sensing and imaging with carbon-based quantum dots for early diagnosis of cancer: A review. *Journal of Pharmaceutical and Biomedical Analysis, 212*, 114628. https://doi.org/10.1016/j.jpba.2022.114628

Mohammadinejad, R., Dadashzadeh, A., Moghassemi, S., Ashrafizadeh, M., Dehshahri, A., Pardakhty, A., Sassan, H., Sohrevardi, S. M., & Mandegary, A. (2019). Shedding light on gene therapy: Carbon dots for the minimally invasive image-guided delivery of plasmids and noncoding RNAs-A review. *Journal of Advanced Research, 18*, 81–93. https://doi.org/10.1016/j.jare.2019.01.004

Mohammed, L. J., & Omer, K. M. (2024). Recent advances in detection of aflatoxins using carbon dots: A review. *Microchemical Journal*, 111708. https://doi.org/10.1016/j.microc.2024.111708

Moradi, M., Molaei, R., Kousheh, S. A., Guimarães, J., & McClements, D. J. (2023). Carbon dots synthesized from microorganisms and food by-products: Active and smart food packaging applications. *Critical Reviews in Food Science and Nutrition, 63*(14), 1943–1959. https://doi.org/10.1080/10408398.2021.2015283

Motaghi, H., Mehrgardi, M. A., & Bouvet, P. (2017). Carbon dots-AS1411 aptamer nanoconjugate for ultrasensitive spectrofluorometric detection of cancer cells. *Scientific Reports, 7*(1), 10513. https://doi.org/10.1038/s41598-017-11087-2

Murali, S., Khan, R., Gogoi, S., Morchhale, R. K., & Singhal, A. (2022). Future prospects of carbon dots application in agriculture. In *Carbon dots in agricultural systems* (pp. 263–285). Academic Press. https://doi.org/10.1016/B978-0-323-90260-1.00011-5

Nagay, B. E., Malheiros, S. S., Borges, M. H., Aparicio, C., van den Beucken, J. J., & Barão, V. A. (2025). Progress in visible-light-activated photocatalytic coatings to combat implant-related infections: From mechanistic to translational roadmap. *Bioactive Materials, 51*, 83–137. https://doi.org/10.1016/j.bioactmat.2025.04.037

Naseer, M. S., Imran, A., Jameel, Q. Y., Bishoyi, A. K., Ahmed, F., Elawady, A., Rajput, P., Islam, F., Zahoor, T., Kinki, A. B., & Naz, S. (2025). Application of carbon quantum dots in food business: A comprehensive review. *eFood, 6*(3), e70054. https://doi.org/10.1002/efd2.70054

Nazar, M., Hasan, M., Wirjosentono, B., Gani, B. A., & Nada, C. E. (2024). Microwave synthesis of carbon quantum dots from arabica coffee ground for fluorescence detection of Fe^{3+}, Pb^{2+}, and Cr^{3+}. *ACS Omega, 9*(18), 20571–20581. https://doi.org/10.1021/acsomega.4c02254

Newman Monday, Y., Abdullah, J., Yusof, N. A., Abdul Rashid, S., & Shueb, R. H. (2021). Facile hydrothermal and solvothermal synthesis and characterization of nitrogen-doped carbon dots from palm kernel shell precursor. *Applied Sciences, 11*(4), 1630. https://doi.org/10.3390/app11041630

Ngu, P. Z., Chia, S. P., Fong, J. F., & Ng, S. M. (2016). Synthesis of carbon nanoparticles from waste rice husk used for the optical sensing of metal ions. *New Carbon Materials, 31*(2), 135–143. https://doi.org/10.1016/S1872-5805(16)60008-2

Nguyen, D. H., El-Ramady, H., & Prokisch, J. (2025). Food safety aspects of carbon dots: A review. *Environmental Chemistry Letters, 23*(1), 337–360. https://doi.org/10.1007/s10311-024-01779-3

Niu, N., Ma, Z., He, F., Li, S., Li, J., Liu, S., & Yang, P. (2017). Preparation of carbon dots for cellular imaging by the molecular aggregation of cellulolytic enzyme lignin. *Langmuir, 33*(23), 5786–5795. https://doi.org/10.1021/acs.langmuir.7b00617

Pal, S., Mondal, R., & Das, P. K. (2024). Nanogel–carbon dot conjugates: A synergistic approach for enhanced cancer treatment through combination therapy. *ACS Medicinal Chemistry Letters, 15*(10), 1668–1676. https://doi.org/10.1021/acsmedchemlett.4c00194

Pan, Z., Zang, H., Li, Y., Wang, X., Xia, N., Liu, C., Li, Z., Han, Y., Tang, Z., & Sun, J. (2024). Foliar application of carbon dots enhances nitrogen uptake and assimilation through CEPD1-dependent signaling in plants. *Plant Physiology and Biochemistry, 217*, 109229. https://doi.org/10.1016/j.plaphy.2024.109229

Pandey, V. K., Tripathi, A., Taufeeq, A., Dar, A. H., Samrot, A. V., Rustagi, S., Malik, S., Bhattacharya, T., Kovacs, B., & Shaikh, A. M. (2024). Significance and applications of carbon dots in anti cancerous nanodrug conjugate development: A review. *Applied Surface Science Advances, 19*, 100550. https://doi.org/10.1016/j.apsadv.2023.100550

Parambil, A. M., & Rajamani, P. (2024). Carbon dots: A promising path towards environmental sustainability. *Environmental Science: Advances, 3*(11), 1513–1523. https://doi.org/10.1039/D4VA00273C

Parashar, A. K., Verma, K. K., Kumar, R., & Arora, V. (2023). A concise review of carbon dots and their pharmaceutical and biomedical applications. *Recent Advances in Drug Delivery and Formulation: Formerly Recent Patents on Drug Delivery & Formulation, 17*(3), 183–192. https://doi.org/10.2174/0126673878237423230919070049

Paveethra, S., Manisekaran, H., & Sasidharan, S. (2024). Medicinal plants derived green carbon dots: Synthesis, characterization and their potential applications in cancer therapy. *Asian Pacific Journal of Cancer Prevention: APJCP, 25*(10), 3393. https://doi.org/10.31557/APJCP.2024.25.10.3393

Phan, L. M., & Cho, S. (2022). [Retracted] Fluorescent carbon dot-supported imaging-based biomedicine: A comprehensive review. *Bioinorganic Chemistry and Applications, 2022*(1), 9303703. https://doi.org/10.1155/2022/9303703

Priyadarshi, R., Uzun, S., & Rhim, J. W. (2024). Edible coating using carbon quantum dots for fresh produce preservation: A review of safety perspectives. *Advances in Colloid and Interface Science*, 103211. https://doi.org/10.1016/j.cis.2024.103211

Priyadarshini, E., Kumar, R., Balakrishnan, K., Pandit, S., Kumar, R., Jha, N. K., & Gupta, P. K. (2024). Biofilm inhibition on medical devices and implants using carbon dots: An updated review. *ACS Applied Bio Materials, 7*(5), 2604–2619. https://doi.org/10.1021/acsabm.4c00024

Qin, F., Bai, J., He, P., Wang, X., Wu, S., Yu, X., Yao, Y., & Ren, L. (2023). Light on multicolor fluorescent carbon dots from a single biomass precursor through simply tuning the solvothermal temperature and time strategy. *Materials Today Chemistry, 34*, 101816. https://doi.org/10.1016/j.mtchem.2023.101816

Qureshi, Z. A., Dabash, H., Ponnamma, D., & Abbas, M. K. (2024). Carbon dots as versatile nanomaterials in sensing and imaging: Efficiency and beyond. *Heliyon, 10*(11), e31634. https://doi.org/10.1016/j.heliyon.2024.e31634

Redondo-Fernandez, G., Canga, J. C., Soldado, A., Encinar, J. R., & Costa-Fernandez, J. M. (2023). Functionalized heteroatom-doped carbon dots for biomedical applications: A review. *Analytica Chimica Acta, 1284*, 341874. https://doi.org/10.1016/j.aca.2023.341874

Reyes, J. L., & Chua, N. H. (2004). Interactions between light and carbon signaling pathways in *Arabidopsis*. *Genome Biology, 5*, 1–3. https://doi.org/10.1186/gb-2004-5-3-213

Rocco, D., Moldoveanu, V. G., Feroci, M., Bortolami, M., & Vetica, F. (2023). Electrochemical synthesis of carbon quantum dots. *ChemElectroChem, 10*(3), e202201104. https://doi.org/10.1002/celc.202201104

Sadeghi-Chahnasir, F., Amiripour, F., & Ghasemi, S. (2024). Orange peel-derived carbon dots/Cu-MOF nanohybrid for fluorescence determination of l-ascorbic acid and Fe^{3+}. *Analytica Chimica Acta, 1287*, 342066. https://doi.org/10.1016/j.aca.2023.342066

Sahu, G., Chawre, Y., Kujur, A. B., Miri, P., Sinha, A., Nagwanshi, R., Karbhal, I., Ghosh, K. K., Jena, V. K., & Satnami, M. L. (2024). Nitrogen doped carbon quantum dots as fluorescence "turn-off-on" sensor for detection of fe^{3+} ions and ascorbic acid in *Moringa oleifera* and citrus lemon. *Journal of Fluorescence, 8*, 1–3. https://doi.org/10.1007/s10895-024-04012-0

Saravanan, A., Das, P., Maruthapandi, M., Aryal, S., Michaeli, S., Mastai, Y., Luong, J. H., & Gedanken, A. (2024). Heteroatom co-doping (N, NS, NB) on carbon dots and their antibacterial and antioxidant properties. *Surfaces and Interfaces, 46*, 103857. https://doi.org/10.1016/j.surfin.2024.103857

Sendão, R. M., Esteves da Silva, J. C., & Pinto da Silva, L. (2023). Polyfluoroalkyl substances (PFASs) detection via carbon dots: A review. *Sustainable Chemistry, 4*(4), 339–362. https://doi.org/10.3390/suschem4040024

Shabbir, H., Csapó, E., & Wojnicki, M. (2023). Carbon quantum dots: The role of surface functional groups and proposed mechanisms for metal ion sensing. *Inorganics, 11*(6), 262. https://doi.org/10.3390/inorganics11060262

Sharma, M., Kumar, C., Arya, S. K., Puri, S., & Khatri, M. (2024). Neurological effects of carbon quantum dots on zebrafish: A review. *Neuroscience, 560*, 334–346. https://doi.org/10.1016/j.neuroscience.2024.10.016

Shorgar, N., Bhati, I., & Jhalora, P. (2023). Laser ablation synthesis of quantum dots. In *Elsevier eBooks* (pp. 53–75). https://doi.org/10.1016/b978-0-12-824153-0.00004-5

Singh, G. B., Sethi, A., Kumari, M., Panigrahi, P., Chaudhary, P., & Sharma, A. (2025a). Biorefining agro-industrial waste into green nanomaterials for sustainable agriculture. In *Nano-microbiology for sustainable development* (pp. 105–137). Springer Nature Switzerland. https://doi.org/10.1007/978-3-031-78845-1_5

Singh, H., Aggarwal, V., Kachore, A., Bala, E., Kumar, R., Sharma, R. K., Selvaraj, M., Albormani, O., & Verma, P. K. (2025b). Carbon dots: An emerging food analysis nanoprobes for detection of contaminants. *Food Chemistry., 4*, 143180. https://doi.org/10.1016/j.foodchem.2025.143180

Sistani, S., & Shekarchizadeh, H. (2022). Applications of carbon quantum dots in detection and packaging of foods. *Journal of Food Science and Technology (Iran), 19*(127), 193–209. https://doi.org/10.22034/FSCT.19.127.193

Sun, L., Zhao, Y., Peng, H., Zhou, J., Zhang, Q., Yan, J., Liu, Y., Guo, S., Wu, X., & Li, B. (2024). Carbon dots as a novel photosensitizer for photodynamic therapy of cancer and bacterial infectious diseases: Recent advances. *Journal of Nanobiotechnology, 22*(1), 210. https://doi.org/10.1186/s12951-024-02479-4

Szczepankowska, J., Khachatryan, G., Khachatryan, K., & Krystyjan, M. (2023). Carbon dots—Types, obtaining and application in biotechnology and food technology. *International Journal of Molecular Sciences, 24*(19), 14984. https://doi.org/10.3390/ijms241914984

Tadesse, A., Hagos, M., RamaDevi, D., Basavaiah, K., & Belachew, N. (2020). Fluorescent-nitrogen-doped carbon quantum dots derived from citrus lemon juice: Green synthesis, mercury (II) ion sensing, and live cell imaging. *ACS Omega, 5*(8), 3889–3898. https://doi.org/10.1021/acsomega.9b03175

Tavan, M., Yousefian, Z., Bakhtiar, Z., Rahmandoust, M., & Mirjalili, M. H. (2025). Carbon quantum dots: Multifunctional fluorescent nanomaterials for sustainable advances in biomedicine and agriculture. *Industrial Crops and Products, 231*, 121207. https://doi.org/10.1016/j.indcrop.2025.121207

Travlou, N. A., Secor, J., & Bandosz, T. J. (2017). Highly luminescent S-doped carbon dots for the selective detection of ammonia. *Carbon, 114*, 544–556. https://doi.org/10.1016/j.carbon.2016.12.035

Tripathi, S., & Sarkar, S. (2015). Influence of water soluble carbon dots on the growth of wheat plant. *Applied Nanoscience, 5*, 609–616. https://doi.org/10.1007/s13204-014-0355-9

Truskewycz, A., Yin, H., Halberg, N., Lai, D. T., Ball, A. S., Truong, V. K., Rybicka, A. M., & Cole, I. (2022). Carbon dot therapeutic platforms: Administration, distribution, metabolism, excretion, toxicity, and therapeutic potential. *Small, 18*(16), 2106342. https://doi.org/10.1002/smll.202106342

Tu, Y., Wang, S., Yuan, X., Wei, Y., Qin, K., Zhang, Q., Chen, X., & Ji, X. (2020). A novel fluorescent nitrogen, phosphorus-doped carbon dots derived from *Ganoderma Lucidum* for bioimaging and high selective two nitrophenols detection. *Dyes and Pigments, 178*, 108316. https://doi.org/10.1016/j.dyepig.2020.108316

Unnikrishnan, B., Anand, A., Lin, C. J., Lee, C. Y., Nain, A., Srivastava, P., Wu, R. S., Chu, H. W., Wang, C. Y., Shi, R. H., & Lee, K. H. (2025). Exploring cellular dynamics: Engineered fluorescent carbon dots for organelle staining and cellular response analysis. *Coordination Chemistry Reviews, 534*, 216552. https://doi.org/10.1016/j.ccr.2025.216552

Usman, M., & Cheng, S. (2024). Recent trends and advancements in green synthesis of biomass-derived carbon dots. *Eng, 5*(3), 2223–2263. https://doi.org/10.3390/eng5030116

Wahab, A., Muhammad, M., Ullah, S., Abdi, G., Shah, G. M., Zaman, W., & Ayaz, A. (2024). Agriculture and environmental management through nanotechnology: Eco-friendly nanomaterial synthesis for soil-plant systems, food safety, and sustainability. *Science of the Total Environment, 23*, 171862. https://doi.org/10.1016/j.scitotenv.2024.171862

Wang, B., & Lu, S. (2022). The light of carbon dots: From mechanism to applications. *Matter, 5*(1), 110–149. https://doi.org/10.1016/j.matt.2021.10.016
Wang, R., Xu, Y., Zhang, T., & Jiang, Y. (2015). Rapid and sensitive detection of *Salmonella typhimurium* using aptamer-conjugated carbon dots as fluorescence probe. *Analytical Methods, 7*(5), 1701–1706. https://doi.org/10.1039/C4AY02880E
Wang, D., Meng, Y. T., Zhang, Y., Wang, Q., Lu, W. J., Shuang, S. M., & Dong, C. (2022a). A specific discriminating GSH from Cys/Hcy fluorescence nanosensor: The carbon dots-MnO_2 nanocomposites. *Sensors and Actuators B: Chemical, 367*, 132135. https://doi.org/10.1016/j.snb.2022.132135
Wang, B., Cai, H., Waterhouse, G. I., Qu, X., Yang, B., & Lu, S. (2022b). Carbon dots in bioimaging, biosensing and therapeutics: A comprehensive review. *Small Science, 2*(6), 2200012. https://doi.org/10.1002/smsc.202200012
Wang, C., Yang, M., Shi, H., Yao, Z., Liu, E., Hu, X., Guo, P., Xue, W., & Fan, J. (2022c). Carbon quantum dots prepared by pyrolysis: Investigation of the luminescence mechanism and application as fluorescent probes. *Dyes and Pigments, 204*, 110431. https://doi.org/10.1016/j.dyepig.2022.110431
Wang, S., McCoy, C. P., Li, P., Li, Y., Zhao, Y., Andrews, G. P., Wylie, M. P., & Ge, Y. (2024). Carbon dots in photodynamic/photothermal antimicrobial therapy. *Nanomaterials, 14*(15), 1250. https://doi.org/10.3390/nano14151250
Wen, F., Li, P., Zhang, Y., Zhong, H., Yan, H., & Su, W. (2023). Preparation, characterization of green tea carbon quantum dots/curcumin antioxidant and antibacterial nanocomposites. *Journal of Molecular Structure, 1273*, 134247. https://doi.org/10.1016/j.molstruc.2022.134247
Wijayanti, N. P., Permatasari, F. A., Damayanti, S., Anggadiredja, K., Iskandar, F., Wibowo, I., & Rachmawati, H. (2024). Toxicity assessment and bioimaging potential of carbon dots synthesized from banana peel in zebrafish model. *Narra Journal, 4*(3), e1228. https://doi.org/10.52225/narra.v4i3.1228
Wu, X., Liu, H., Liu, J., Haley, K. N., Treadway, J. A., Larson, J. P., Ge, N., Peale, F., & Bruchez, M. P. (2003). Immunofluorescent labeling of cancer marker Her2 and other cellular targets with semiconductor quantum dots. *Nature Biotechnology, 21*(1), 41–46. https://doi.org/10.1038/nbt764
Wu, R., Yu, S., Chen, S., Dang, Y., Wen, S. H., Tang, J., Zhou, Y., & Zhu, J. J. (2022). A carbon dots-enhanced laccase-based electrochemical sensor for highly sensitive detection of dopamine in human serum. *Analytica Chimica Acta, 1229*, 340365. https://doi.org/10.1016/j.aca.2022.340365
Wu, W., Mu, L., Luo, X., Zhang, Y., Huang, J., Peng, H., & Tao, J. (2025a). Effect of temperature on synthesis of carbon quantum dots and biochar through one-step hydrothermal treatment of distillers' grains. *Industrial Crops and Products, 227*, 120832. https://doi.org/10.1016/j.indcrop.2025.120832
Wu, X., Yao, M., Qiao, X., Li, L., Meng, Z., Liu, S., Sun, Y., Gan, H., Zhu, X., Wu, Z., & Gu, R. (2025b). Silkworm Cocoon—Derived carbon dots for post-trauma hemostasis and tissue repair. *Pharmaceuticals, 18*(5), 603. https://doi.org/10.3390/ph18050603
Yan, F., Sun, Z., Zhang, H., Sun, X., Jiang, Y., & Bai, Z. (2019). The fluorescence mechanism of carbon dots, and methods for tuning their emission color: A review. *Microchimica Acta, 186*, 1–37. https://doi.org/10.1007/s00604-019-3688-y
Yang, Y. Z., Xiao, N., Liu, S. G., Han, L., Li, N. B., & Luo, H. Q. (2020). pH-induced aggregation of hydrophilic carbon dots for fluorescence detection of acidic amino acid and intracellular pH imaging. *Materials Science and Engineering: C, 108*, 110401. https://doi.org/10.1016/j.msec.2019.110401
Yang, X., Li, P., Tang, W., Du, S., Yu, M., Lu, H., Tan, H., & Xing, X. (2021). A facile injectable carbon dot/oxidative polysaccharide hydrogel with potent self-healing and high antibacterial activity. *Carbohydrate Polymers, 251*, 117040. https://doi.org/10.1016/j.carbpol.2020.117040

Yang, S., Chen, Z., Zhou, P., Xia, J., Deng, T., & Yu, C. (2023). Carbon dots based on endogenous nutrients with visible and NIR fluorescence to penetrate blood-brain barrier. *Carbon, 202*, 130–140. https://doi.org/10.1016/j.carbon.2022.10.067

Yin, Z., Wang, H., Tang, X., Mou, M., Yu, H., & Wang, H. (2025). Exploration of biomass-derived carbon dots based on chestnut shell for the sensitive detection of phosphate and tetracycline hydrochloride. *Spectrochimica Acta Part A: Molecular and Biomolecular Spectroscopy, 330*, 125746.

Yu, M., Li, P., Huang, R., Xu, C., Zhang, S., Wang, Y., Gong, X., & Xing, X. (2023). Antibacterial and antibiofilm mechanisms of carbon dots: A review. *Journal of Materials Chemistry B, 11*(4), 734–754. https://doi.org/10.1039/D2TB01977A

Zaman, W., Ayaz, A., & Park, S. (2025). Nanomaterials in agriculture: A pathway to enhanced plant growth and abiotic stress resistance. *Plants, 14*(5), 716. https://doi.org/10.3390/plants14050716

Zanbili, F., & Poursattar Marjani, A. (2025). Innovative green and bio-based approaches for photosensitive nanoparticle synthesis: A review on methodologies, characterization, and applications. *Micro and Nano Systems Letters, 13*(1), 3. https://doi.org/10.1186/s40486-025-00223-7

Zhang, Q., Li, Z., Zhang, M., Wang, W., Shen, J., Ye, Z., & Zhou, N. (2020). Injectable in situ self-cross-linking hydrogels based on hemoglobin, carbon quantum dots, and sodium alginate for real-time detection of wound bacterial infection and efficient postoperative prevention of tumor recurrence. *Langmuir, 36*(44), 13263–13273. https://doi.org/10.1021/acs.langmuir.0c02219

Zhang, W., Yang, Z. Q., Xiao, L., Xu, X., Guan, T., Zhou, W., Gong, X., & Hu, Q. (2025). Colorimetric and fluorescent paper indicator for seafood freshness monitoring using gas-sensitive carbon dots-based paper indicator. *Journal of Food Composition and Analysis, 6*, 107321. https://doi.org/10.1016/j.jfca.2025.107321

Zhao, F., Xin, X., Cao, Y., Su, D., Ji, P., Zhu, Z., & He, Z. (2021). Use of carbon nanoparticles to improve soil fertility, crop growth and nutrient uptake by corn (*Zea mays* L.). *Nanomaterials, 11*(10), 2717. https://doi.org/10.3390/nano11102717

Zhao, L., Zhang, M., Mujumdar, A. S., & Wang, H. (2023). Application of carbon dots in food preservation: A critical review for packaging enhancers and food preservatives. *Critical Reviews in Food Science and Nutrition, 63*(24), 6738–6756. https://doi.org/10.1080/10408398.2022.2039896

Zhao, Y., Li, Y., Li, D., Yuan, H., & Shen, C. (2025a). Eco-friendly synthesized carbon dots from Chinese herbal medicine: A review. *International Journal of Nanomedicine, 31*, 3045–3065. https://doi.org/10.2147/IJN.S497892

Zhao, H., Liu, W., Min, C., Qi, Y., Chen, X., & Zhang, H. (2025b). pH-responsive color indicator film based on gelatin/chitosan cross-linking with anthocyanin-Fe^{2+} chelate for pork freshness monitoring. *Food Hydrocolloids, 162*, 110895. https://doi.org/10.1016/j.foodhyd.2024.110895

Carbon Dots and Nanobiosensors for Disease Detection

7

K. M. Gijith Mohan and Keerti Jain

Abstract

Carbon Dots (CDs) are nanomaterials of size range 1–10 nm with unique optical, chemical and biological properties, which make them suitable for numerous applications in the biomedical field. Biomedical engineering, which integrates nanotechnology and material science, has significantly increased the use of CDs in medical research, especially in the development of nanobiosensors for the detection of diseases. A nanobiosensor is an analytical device that consists of a nanocarrier integrated with a detection bioelement, such as DNA, antibody, cells or other biomolecules that can bind to the biological counterpart, which is present in the samples. This enables the detection and quantification of biomarkers, which are highly specific for each disease. CDs possess good thermal conductivity, electron transfer property, small size, and high surface/volume ratio, essentially making them appropriate candidates for nanobiosensor development. By these properties, CDs exhibit excellent detection limits, higher sensitivity and selectivity, and good biocompatibility and are used as the sensing component of a nanobiosensor. Utilisation of nanobiosensors in the early diagnosis of diseases provides patients with a greater opportunity for successful recovery, especially in cases of life-threatening diseases. This chapter summarizes CD-based nanobiosensors, their applications, associated challenges and future perspectives for use in the biomedical field.

Keywords

Carbon dots · Biomedical engineering · Nanobiosensor · Transducer · Biomarker

---

K. M. Gijith Mohan · K. Jain (✉)
Drug Delivery and Nanomedicine Research Laboratory, Department of Pharmaceutics, National Institute of Pharmaceutical Education and Research (NIPER)-Raebareli, Lucknow, India
e-mail: keertijain.02@niperraebareli.edu.in

M. Rai et al. (eds.), *Carbon Nanodots*, https://doi.org/10.1007/978-981-95-4705-0_7

Abbreviations

AFP: alpha-fetoprotein
CDs: carbon dots
CEA: carcinoembryonic antigen
CVDs: cardiovascular diseases
FRET Förster resonance energy transfer
PCB: printed circuit board

7.1 Introduction

Nanobiosensors are nanoscale analytical devices used to detect or measure biochemical molecules that serve as biomarkers of diseases. Upon interaction with such molecules, nanobiosensors produce a physical or chemical change that is converted into quantifiable digital signals. Nanobiosensors have high sensitivity and accuracy in detecting biomarkers, which makes them more advantageous in disease detection as compared to conventional methods. This enables the early detection of diseases to attain higher treatment success rates while reducing the complexity and cost of the treatment. Nanobiosensors have proved to be fast and accurate in the detection of critical diseases like cancers, Alzheimer's disease, microbial infections, cardiovascular diseases, psychiatric disorders and blood glucose monitoring in diabetes, which provided valuable outcomes in therapy. Moreover, this approach is applied to point-of-care testing, where the clinical evaluation can be done in a hospital setting or at home without the need for centralized laboratory facilities. This recent trend turned out to be the most valuable utility of nanobiosensors in enhancing the quality of healthcare globally (Suthar and Jain, 2023; Suthar et al., 2023; Can Demirdöğen, 2021).

CDs are quasispherical zero-dimensional nanoparticles of carbon having a size range of 1–10 nm in diameter and possess unique inherent electronic, thermal, chemical and fluorescent properties (Patel et al., 2024; Sharma et al., 2016). CDs can be administered into biological systems due to their better stability, solubility, biocompatibility and less toxicity. The exceptional electron donor and acceptor behaviour of CDs makes them luminescent and provides high quantum yield, making them well-suited in fabricating nanobiosensors (Sonam et al., 2024; Pourmadadi et al., 2023). Luminescence of CDs is associated with either chemical or electrochemical changes that can be detected and evaluated. This forms the basis of the evolution of CDs as a biosensing module in the design of nanobiosensors for disease detection (Sonam et al., 2024; Kulkarni et al., 2022). CDs can be functionalised by attaching molecules to the surface functional groups to provide sensitivity and selectivity in binding with disease biomarkers (Patel et al., 2024; Ahmad et al., 2022). Biomedical engineering explored the competency of CDs to sense biomarkers and integrated them into nanobiosensors to apply for the detection of diseases (Wang & Hu, 2014).

7.2 Carbon Dot-Integrated Nanobiosensors

CDs integrated with nanobiosensors possess distinct optical, electric, and catalytic properties that can be translated to quantitative and qualitative data for biomedical applications. When the biomolecule recognition component binds with an analyte molecule, the resultant variations in these properties can be observed and recorded. Fluorescence and photoluminescence of CDs are largely used for the analysis of biomarkers that are helpful in disease detection. Being in the nanoscale size range, the quantum confinement of CDs provides a specific emission range as per the needs of the application. Electric and catalytic properties are imparted by the tunable surface properties of CDs that provide a myriad of applications in biomedical research (Chellachamy Anbalagan et al., 2024; Ge et al., 2015; Mehra et al., 2015). Figure 7.1 explains the assembly of a CD-based nanobiosensor. A nanobiosensor works with three components,

(a) Biomolecule recognition component.
(b) Transducer component.
(c) Signal processing component.

The biomolecule recognition component comprises detection elements that can specifically bind to the target analytes present in the sample. This component can provide either the presence or the amount of the analyte in the sample. The common biomolecules used as detection elements include receptors, DNA, antibodies, proteins, cells or microorganisms, etc., that can detect and bind to the counterpart bioanalyte through various binding mechanisms. Upon binding with the analyte, it creates certain measurable changes in the transducer part, which is composed of an

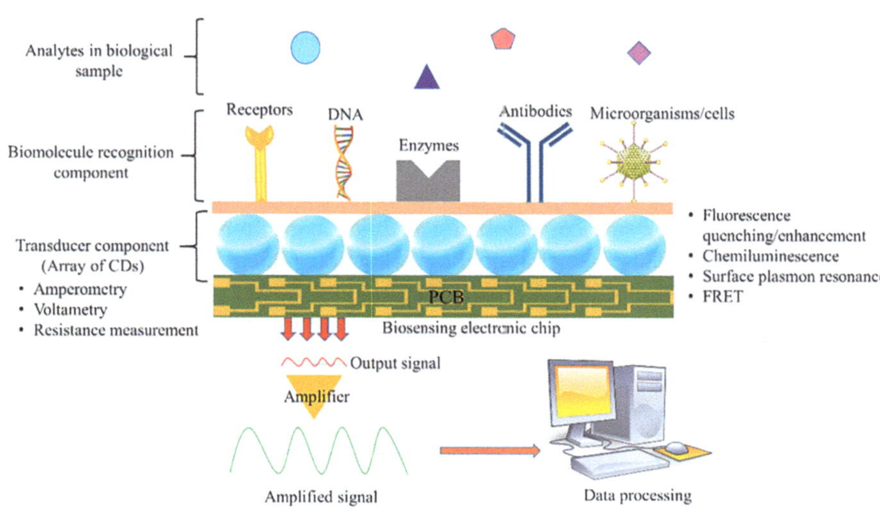

Fig. 7.1 Schematic diagram of the workflow of CDs-based nanobiosensor

array of CDs (Zhang & Du, 2020). There are several mechanisms like quenching or fluorescence enhancement, Förster resonance energy transfer (FRET), surface plasmon resonance and chemiluminescence, which cause certain changes in the transducer to generate signals (Miguel-Rojas & Pérez-De-Luque, 2023). The signals obtained from the transducer are transferred to a printed circuit board (PCB) and then translated in terms of current, voltage or resistance. The PCB forms the interface with other electronic modules, like an amplifier and a microcontroller-based data acquisition device. An amplifier turns the feeble signal into a strong one and then passes it to a processing unit, where the data is evaluated and recorded. The processing unit consists of hardware made up of a microcontroller chip, which converts the analogue signals to digital signals that can be processed and stored in a computer for further analysis (Hercog & Gergič, 2014).

7.3 Carbon Dot-Based Nanobiosensors in Disease Detection

CDs integrated nanobiosensors are highly sensitive in detecting specific biomarkers, making them highly suitable for early disease detection. Due to the ability to enhance signal transduction, an improved detection accuracy at the molecular level can be achieved using CDs. In biosensing, CDs significantly enhance the performance by improving the detection limits, sensitivity, and selectivity. CDs can detect pathogens and biomarkers at a very low concentrations, making them valuable tools for early diagnosis. There are several characteristic physicochemical properties that provide numerous functional advantages to CDs in biosensing applications.

(a) *Improved Signal Processing*

Being inherently fluorescent, CDs can generate signals upon binding with bioanalytes, thereby enhancing the sensitivity of optical nanobiosensors. CDs possess lower detection limits because of their high quantum yield, which can generate intense fluorescence even for analytes at low concentrations (Qureshi et al., 2024). Moreover, the fluorescence of CDs can be tuned by the selection of an appropriate carbon precursor, which offers a greater flexibility in nanobiosensor design (Ge et al., 2015). CDs have outstanding electrical conductivity due to the abundance of free electrons on the surface, facilitating an efficient electron transfer between the sensor and analytes. This enables a superior interaction, which makes CDs to be used as an appropriate biosensing component (Mishra et al., 2021). All these properties enable CDs to provide a good quality output, which can be utilized for the analysis of biomarkers of various diseases.

(b) *Higher Sensitivity*

Extremely small size and surface functionalization capabilities provide CDs with a higher sensitivity in detecting bioanalytes (Echabaane et al., 2021). The high surface-to-volume ratio of CDs imparts a quantum confinement effect, which makes a more sensitive interaction with analyte molecules to produce a high fluorescence outcome. Surface properties allow CDs to get modified with a range of functional groups, making them more sensitive towards specific analyte molecules (Kaur et al., 2017). The functional groups present over the surface of the CDs can sense changes in the microenvironment, and this is reflected in the fluorescence intensity, which can be quantified for the evaluation of the analyte (Yahav et al., 2023). Sasikumar K et al. performed bilirubin detection with a limit of detection of 34 nM, highlighting the sensitivity of CDs-based nanobiosensor to detect low concentrations of analytes in the samples (Sasikumar et al., 2023). The sensitivity and specificity of biosensors can be further improved by keeping the signal-to-noise ratio higher by applying techniques like sandwich assays, which use more than one sensing component (Prabowo et al., 2021).

(c) *Choice of Transduction Mechanisms*

CDs offer a range of mechanisms for the transduction of signals obtained from the interaction with the bioanalyte. Based on this, the detection mechanisms can be selected according to the properties of the bioanalyte to enhance the quality of results. CDs can work based on fluorescence, electrochemical, electro-chemiluminescence, photoelectrochemical, thermal and colorimetric transduction principles, suiting the properties of analyte molecules (Molaei, 2020). Further, there are several mechanisms utilized, such as FRET, which is a dipole-dipole-based interaction between donor and acceptor resulting in non-radioactive energy transfer that is translated and analysed (Kayani et al., 2024). Surface Plasmon resonance applies the resonance of the free electrons over the CDs, which causes a change in the refractive index of the sensor due to the binding process and allows continuous evaluation of the analytes (Shrivastav et al., 2021). In a quenching-based process, variations in the fluorescence caused by the sensor-analyte interaction are evaluated to obtain a transduction signal, which is further processed to obtain sensed data (Kamal et al., 2025). Briefly, CDs provide significant flexibility and efficiency in fabricating nanobiosensors as compared to other sensing modules.

(d) *Portability and Miniaturisation*

As nanobiosensors have a nanoscale size range, they enable us to incorporate them into small portable devices that can be used for point-of-care testing and remote testing.

7.3.1 Cancer Detection

Early detection is the key to successful cancer treatment as it supports timely and personalized therapeutic interventions. As cancer biomarkers are present at very low concentrations in blood at initial stages, it necessitates more sensitive and accurate sensors for the detection (Patel et al., 2023). In oncology research CDs have been successfully used to detect cancer biomarkers like alpha-fetoprotein (AFP) and carcinoembryonic antigen (CEA). AFP is a biomarker for hepatocellular carcinoma and can be detected using bioengineered antibodies or aptamers. These biomolecules exhibit a high degree of selectivity in binding towards AFP, inducing a measurable change in the fluorescence of CDs that is explored for the generation of quantitative data. Recently, researchers constructed a ratiometric fluorescence enzyme-linked immunosorbent assay (RF-ELISA) for the early detection of AFP in hepatocellular carcinoma using carbon dots@SiO2@CdTe quantum dots. The results showed that the sensitivity and accuracy of this method were comparable with those of the commercial ELIZA with a low detection limit of 8.7 pg $mL^{-1,}$ highlighting the potential of CDs for biosensing applications (Liang et al., 2022). In a similar study, Chellachamy et al. developed a nanobiosensor used for the detection of CEA using CDs modified with HRP-conjugated CEA antibody. The research reported good specificity, stability, reproducibility and low detection limit (10 pg/ml), which signifies CDs' utility in cancer detection (Chellachamy Anbalagan et al., 2024). To sum up, CD-based nanobiosensors can meet the rising demand for novel approaches in cancer therapy.

7.3.2 Infectious Diseases

Bioengineered CDs can be used to sense many pathogens like bacteria, viruses, fungi and their associated antigens or nucleic acids. This feature has been utilized for the detection of many infectious diseases like COVID-19, SARS-CoV, MERS-CoV, human immunodeficiency virus disease, hepatitis, tuberculosis, candidiasis, etc. Infectious diseases are hugely associated with high rates of morbidity and mortality if therapeutic interventions are delayed. CD-based nanobiosensors have been established for high sensitivity and selectivity for pathogen detection, facilitating early diagnosis to enhance treatment outcomes and also have a great impact in controlling disease spread. Other advantages, like cost effectiveness, portability and ease of use, make them more suitable for point-of-care testing approaches, helping to cut down delays in traditional lab assessments (Kim & Min, 2024). Scientists designed a simple, FRET-based, rapid and quantitative method using CDs to detect SARS-CoV-2 virus. CDs modified with SARS-CoV-2 antibodies sensed the spike protein variants with a limit of detection of 0.05 ng/mL in 12 min. This depicts the relevance of nanobiosensors in timely detection and prevention of transmission (Li et al., 2023).

7.3.3 Neurological Disorders

Major neurological disorders like Alzheimer's disease (AD), Parkinson's disease (PD), and multiple sclerosis (MS) can be managed more efficiently if detected at early stages, which can slow down disease progression, reduce symptoms and improve the quality of life of the patient. CD-based nanobiosensors offer an efficient platform to identify the early biomarkers. AD is characterized by the development of beta amyloid peptides, which cause neurotoxicity and oxidative stress in the brain, leading to neurodegeneration and dementia. These peptides can also serve as an early biomarker, which can be detected using nanobiosensors to facilitate prompt therapeutic intervention (Suthar et al., 2023; Kaushik et al., 2016). Apart from beta amyloid peptides, there are other biomarkers like tau protein and apolipoprotein E4, which are targeted for the detection of AD using nanobiosensors (Sonwani et al., 2025; Suthar & Jain, 2023).

PD is an idiopathic disorder characterized by ambiguous diagnostic procedures and treatment options and has no cure with current strategies. It is detected only when patients are presented with symptoms like difficulty in movement and tremors resulting from the irreversible brain damage that decreases the efficiency of clinical therapies. Nanobiosensors can effectively overcome these hurdles as they enable early diagnosis and development of precision medicine for PD patients (Dhahi et al., 2024). α-synuclein is one of the focused biomarkers utilized for the early detection of PD using nanobiosensors (Khatami et al., 2025). Electrochemical, optical, and seed amplification assays are the major strategies used by nanobiosensors for the sensitive and selective detection of α-synuclein (Maddocks et al., 2024). MS is a chronic neurodegenerative autoimmune disease affecting the central nervous system. Unlike other neurodegenerative diseases, MS diagnosis is greatly affected in its accuracy and the current strategies are expensive and time-consuming, demanding new techniques for disease detection and monitoring. Nanobiosensors that can detect specific biomarkers associated with MS, such as myelin basic protein, neurofilament light chain, glial fibrillary acidic protein and specific microRNAs like miR-155 in the samples, give an early indication of the commencement of disease. It has been reported that the development of carbon-based non-label electrochemical nanobiosensors that are of low cost and sensitive enough to detect miR-155. The sensor exhibited an exceptionally low detection limit of 10 aM for miR-155, thanks to the superior electrical conductivity of CDs. The remarkable efficiency of CDs in detecting biomarkers is promising and can bring a breakthrough in MS therapy (Tarannum and Jain, 2023; Dobson & Giovannoni, 2019).

7.3.4 Diabetes Management

Effective diabetes management requires continuous monitoring of variations in blood glucose levels and adjustment of the medication regimen to achieve the therapeutic goal. CD-based nanobiosensors are developed to have a rapid, real-time evaluation of blood glucose. This strategy offers not only early detection of diabetes but also prevents further complications, which are developed due to uncontrolled blood glucose levels. Diabetes management often imparts a high degree of distress in patients, mainly due to the necessity of frequent blood sampling. This can have an impact on adherence to treatment plans, potentially leading to poor control over blood glucose levels (Khosravi Ardakani et al., 2022). CD-based nanbiosensors provide a platform for non-invasive blood sampling, thereby improving the quality of life of diabetes patients. Apart from blood, body fluids like saliva, tears, and urine can also act as substrates for the nanobiosensors to have an accurate quantification of blood glucose (Makaram et al., 2014). Cost-effectiveness is another advantage offered by nanbiosensors, which promotes the integration into routine clinical practice to enhance diabetic care. Alshareef M et al. developed fluorescent nitrogen-doped CDs to detect acetone in samples, which can provide a correlation with blood glucose levels. This strategy is applied for glucose monitoring through breath analysis that can become an alternative to conventional blood sampling, thereby improving the quality of diabetes patients (Alshareef et al., 2022). Thus, by incorporating nanobiosensing, conventional diabetes treatments will be redefined to meet emerging therapeutic targets and develop personalized treatment approaches.

7.3.5 Cardiovascular Diseases

Cardiovascular diseases (CVDs) top the list of major causes of death worldwide. Amidst advancements in technologies, the death toll is increasing every year due to the silent nature of the disease, and most of the interventions are not successful at a later phase. Nanobiosensors prove to be an effective agent for the early detection of CVDs as there are many biomarkers such as C-reactive protein, cardiac troponin I or T (cTnI/T), myoglobin, lipoprotein-associated phospholipase A(2), interlukin-6 (IL-6), interlukin-1 (IL-1), low-density lipoprotein, myeloperoxidase and tumour necrosis factor alpha, which can forecast cardiovascular events well in advance. Nanobiosensors bioengineered with aptamers and antibodies offer a proficient, inexpensive and sensitive platform for the detection of multiple cardiac biomarkers in clinical samples, even at low concentrations (A. Qureshi et al., 2012). Chen J et al. equipped a CD-based nanobiosensor to detect myoglobin, which serves as a biomarker for the detection of myocardial infarction. The study reported that CDs functionalized with myoglobin aptamer, when treated with myoglobin, cause a recovery of quenched fluorescence, which can be quantified with a detection limit as low as 20 pgmL^{-1}. This significantly low detection limit projects the utility of CDs-based nanobiosensor in the effective management of CVDs (Chen et al., 2019). Table 7.1 outlines important nanobiosensors and biomarkers used for disease detection.

7.4 Challenges

Despite several advantages, there are certain challenges to be resolved for the enhancement of the efficiency of CDs in the fabrication of nanobiosensors. Nanoscale size range offers multiple benefits to CDs but also presents difficulties in characterisation and optimisation due to non-uniform size range, structural complexities, aggregation behaviour, etc. Some of the current techniques lack that sensitivity to obtain consistent results in the evaluation of CDs, demanding novel

Table 7.1 Nanobiosensors used for the detection of different diseases

Disease	Device	Nanobiosensor type	Target biomarker/ analyte	Reference
Cancer	MoSERS	Optical	Extracellular vesicles, miRNA	Jalali et al. (2023)
Diabetes	Eversense, Noviosense	Electrochemical, enzymatic	Glucose, insulin	Ahmed et al. (2022) and Geelhoed-Duijvestijn et al. (2021)
Infectious diseases	SALIVERA, NFluidEX	Electrochemical, optical	SARS-CoV-2 N-protein, miRNA	Mahshid (2023)
Cardiovascular diseases	Atellica® VTLi	Optical	Troponin I, CRP, BNP	Atellica® VTLi Patient-Side Immunoassay Analyzer – Siemens Healthineers India (n.d.)

approaches for characterisation. The size and shape of CDs can be varied easily due to environmental conditions, which affect the properties significantly. This in turn affects the functionalization, scalability and stability of the biosensor. The absence of a standardised synthesis protocol is one of the major challenges that affect the reproducibility of fluorescence and the physical properties of CDs. This has a direct impact on the scalability and mass production of CDs. Photo-bleaching is a common drawback associated with CD-based nanobiosensors that affects the fluorescence intensity and stability, preventing long-term applications. In some cases, the fluorescent signals from CD-based biosensors turn out to be complex, making the analysis and interpretation of data difficult. These factors limit the establishment of standard protocols for the evaluation to ensure the accuracy and reliability of CDs-based nanobiosensors (Sen & Bose, 2025). To develop portable, economic and simple nanobiosensors, there is a need for the integration with other technologies, such as electronics and microfluidics, which poses great difficulty for the researchers. Moreover, compliance with the regulatory frameworks to ensure the safety of patients is of prime importance, especially in biomedical applications. These processes for nanobiosensors are time-consuming and laborious. The data regarding toxicity and long-term effects of CDs in the human body is limited and presents difficulty in the establishment of standard clinical protocols for biosensing applications (Malik et al., 2023). To meet these challenges, researchers are working on novel methods of synthesis, functionalization and design of CDs-based biosensors to enhance the sensitivity, stability and performance in biosensing applications.

7.5 Conclusion and Future Perspectives

CDs hold enormous potential in the field of biomedical research due to innovative technology, practical clinical use, and strong market prospects. Being highly efficient sensing agents, CDs integrated nanobiosensors can transform future healthcare. Biosensors integrated with CDs as transducers are powerful tools for the early detection of diseases via targeting and detecting biomarkers with a high degree of sensitivity and specificity. This opens profound possibilities for early therapeutic interventions, which can bring high treatment success rates. Especially patients with cancers, Alzheimer's disease and cardiovascular diseases are mostly benefited with this approach, as CDs contribute to advanced biosensor development, which can improve detection accuracy, thus leading to better disease management. Constant efforts are made to improve the sensitivity, selectivity and reliability of CDs-based nanobiosensors to provide more efficient sensing applications. Integration of nanobiosensors with wearable accessories is another approach holding potential for continuous disease monitoring. Engineering of high-throughput, multiplexed detection systems that allow simultaneous detection of multiple biomarkers could significantly extend the use of CDs in biomedical studies. Incorporation of artificial intelligence and the Internet of Things with nanobiosensors is a rapidly developing segment in healthcare to achieve a multifold enhancement of diagnostic capabilities. Thus, with the advancements in bioengineering, the efficiency of biosensors is enhanced to meet the rising demand for new therapeutic strategies. With continuous improvements, CD-based nanobiosensors will revolutionize disease diagnosis, thereby supporting point-of-care testing and the development of personalized medicine, ultimately contributing to improving healthcare outcomes.

Acknowledgements We acknowledge the Department of Pharmaceuticals (DOP), Ministry of Chemicals and Fertilizers, Government of India, for providing facilities for writing this book chapter. The (NIPER)-Raebareli communication number for this chapter is (NIPER)-Raebareli / Communication/810.

References

Ahmad, J., Rizwanullah, M., Suthar, T., Albarqi, H. A., Ahmad, M. Z., Vuddanda, P. R., Khan, M. A., & Jain, K. (2022). Receptor-targeted surface-engineered nanomaterials for breast cancer imaging and Theranostic applications. *Critical Reviews™ in Therapeutic Drug Carrier Systems, 39*(6), 1–44. https://doi.org/10.1615/CritRevTherDrugCarrierSyst.2022040686

Ahmed, I., Jiang, N., Shao, X., Elsherif, M., Alam, F., Salih, A., Butta, H., & Yetisen, A. K. (2022). Recent advances in optical sensors for continuous glucose monitoring. *Sensors and Diagnostics, 1*(6), 1098–1125. https://doi.org/10.1039/D1SD00030F

Alshareef, M., Snari, R. M., Alaysuy, O., Aldawsari, A. M., Abumelha, H. M., Katouah, H., & El-Metwaly, N. M. (2022). Optical detection of acetone using "turn-off" fluorescent Rice straw based cellulose carbon dots imprinted onto paper dipstick for diabetes monitoring. *ACS Omega, 7*(19), 16766–16777. https://doi.org/10.1021/ACSOMEGA.2C01492/ASSET/IMAGES/LARGE/AO2C01492_0022.JPEG

Atellica® VTLi Patient-side Immunoassay Analyzer – Siemens Healthineers India. (n.d.). Retrieved April 21, 2025, from https://www.siemens-healthineers.com/en-in/cardiac/cardiac-systems/atellica-vtli

Can Demirdöğen, B. (2021). A literature review of biosensors for multiple sclerosis: Towards personalized medicine and point-of-care testing. *Multiple Sclerosis and Related Disorders, 48*, 102675. https://doi.org/10.1016/J.MSARD.2020.102675

Chellachamy Anbalagan, A., Korram, J., Doble, M., & Sawant, S. N. (2024). Bio-functionalized carbon dots for signaling immuno-reaction of carcinoembryonic antigen in an electrochemical biosensor for cancer biomarker detection. *Discover Nano, 19*(1), 1–15. https://doi.org/10.1186/S11671-024-03980-3/FIGURES/9

Chen, J., Ran, F., Chen, Q., Luo, D., Ma, W., Han, T., Wang, C., & Wang, C. (2019). A fluorescent biosensor for cardiac biomarker myoglobin detection based on carbon dots and deoxyribonuclease I-aided target recycling signal amplification. *RSC Advances, 9*(8), 4463–4468. https://doi.org/10.1039/C8RA09459D

Dhahi, T. S., Yousif Dafhalla, A. K., Al-Mufti, A. W., Elobaid, M. E., Adam, T., & Gopinath, S. C. B. (2024). Application of Nanobiosensor engineering in the diagnosis of neurodegenerative disorders. *Results in Engineering, 24*, 102790. https://doi.org/10.1016/J.RINENG.2024.102790

Dobson, R., & Giovannoni, G. (2019). Multiple sclerosis—A review. *European Journal of Neurology, 26*(1), 27–40. https://doi.org/10.1111/ENE.13819

Echabaane, M., Hfaiedh, S., Smiri, B., Saidi, F., & Dridi, C. (2021). Development of an impedimetric sensor based on carbon dots and chitosan nanocomposite modified electrode for cu(II) detection in water. *Journal of Solid State Electrochemistry, 25*(6), 1797–1806. https://doi.org/10.1007/S10008-021-04949-3/METRICS

Ge, J., Jia, Q., Liu, W., Guo, L., Liu, Q., Lan, M., Zhang, H., Meng, X., Wang, P., Ge, J., Jia, Q., Liu, W., Guo, L., Lan, M., Zhang, H., Meng, X., Wang, P., & Liu, Q. (2015). Red-emissive carbon dots for fluorescent, photoacoustic, and thermal Theranostics in living mice. *Advanced Materials, 27*(28), 4169–4177. https://doi.org/10.1002/ADMA.201500323

Geelhoed-Duijvestijn, P., Vegelyte, D., Kownacka, A., Anton, N., Joosse, M., & Wilson, C. (2021). Performance of the prototype NovioSense noninvasive biosensor for tear glucose in type 1 diabetes. *Journal of Diabetes Science and Technology, 15*(6), 1320–1325. https://doi.org/10.1177/1932296820964844

Hercog, D., & Gergič, B. (2014). A flexible microcontroller-based data acquisition device. *Sensors (Basel), 14*(6), 9755–9775. https://doi.org/10.3390/S140609755

Jalali, M., del Real Mata, C., Montermini, L., Jeanne, O., Hosseini, I., Gu, Z., Spinelli, C., Lu, Y., Tawil, N., Guiot, M. C., He, Z., Wachsmann-Hogiu, S., Zhou, R., Petrecca, K., Reisner, W. W., Rak, J., & Mahshid, S. (2023). MoS2-Plasmonic Nanocavities for Raman spectra of single extracellular vesicles reveal molecular progression in glioblastoma. *ACS Nano, 17*(13), 12052–12071. https://doi.org/10.1021/ACSNANO.2C09222/ASSET/IMAGES/LARGE/NN2C09222_0007.JPEG

Kamal, A., Hong, S., & Ju, H. (2025). Carbon quantum dots: Synthesis, characteristics, and quenching as biocompatible fluorescent probes. *Biosensors, 15*(2), 99. https://doi.org/10.3390/BIOS15020099

Kaur, S., Mehra, N. K., Jain, K., & Jain, N. K. (2017). Development and evaluation of targeting ligand-anchored CNTs as prospective targeted drug delivery system. *Artificial Cells, Nanomedicine, and Biotechnology, 45*(2), 242–250. https://doi.org/10.3109/21691401.2016.1146728

Kaushik, A., Jayant, R. D., Tiwari, S., Vashist, A., & Nair, M. (2016). Nano-biosensors to detect beta-amyloid for Alzheimer's disease management. *Biosensors and Bioelectronics, 80*, 273–287. https://doi.org/10.1016/J.BIOS.2016.01.065

Kayani, K. F., Ghafoor, D., Mohammed, S. J., & Shatery, O. B. A. (2024). Carbon dots: Synthesis, sensing mechanisms, and potential applications as promising materials for glucose sensors. *Nanoscale Advances, 7*(1), 42–59. https://doi.org/10.1039/D4NA00763H

Khatami, S. H., Khanifar, H., Movahedpour, A., Taheri-Anganeh, M., Ehtiati, S., Khanifar, H., & Asadi, A. (2025). Electrochemical biosensors in early detection of Parkinson disease. *Clinica Chimica Acta, 565*, 120001. https://doi.org/10.1016/J.CCA.2024.120001

Khosravi Ardakani, H., Gerami, M., Chashmpoosh, M., Omidifar, N., & Gholami, A. (2022). Recent Progress in Nanobiosensors for precise detection of blood glucose level. *Biochemistry Research International, 2022*, 2964705. https://doi.org/10.1155/2022/2964705

Kim, Y. J., & Min, J. (2024). Advances in nanobiosensors during the COVID-19 pandemic and future perspectives for the post-COVID era. *Nano Convergence, 11*(1), 1–57. https://doi.org/10.1186/S40580-023-00410-5

Kulkarni, M. B., Ayachit, N. H., & Aminabhavi, T. M. (2022). Recent advancements in Nanobiosensors: Current trends, challenges, applications, and future scope. *Biosensors, 12*(10), 892. https://doi.org/10.3390/BIOS12100892

Li, Y., Ren, Y., Yi, Z., Han, S., Liu, S., Long, F., & Zhu, A. (2023). Detection of SARS-CoV-2 S protein based on FRET between carbon quantum dots and gold nanoparticles. *Heliyon, 9*(11), e22674. https://doi.org/10.1016/j.heliyon.2023.e22674

Liang, X., Lin, Z., Li, L., Tang, D., & Kong, J. (2022). Ratiometric fluorescence enzyme-linked immunosorbent assay based on carbon dots@SiO 2 @CdTe quantum dots with dual functionalities for alpha-fetoprotein. *Analyst, 147*(12), 2851–2858. https://doi.org/10.1039/D2AN00691J

Maddocks, G. M., Eisenstein, M., & Soh, H. T. (2024). Biosensors for Parkinson's disease: Where are we now, and where do we need to go? *ACS Sensors, 9*(9), 4307–4327. https://doi.org/10.1021/acssensors.4c00790

Mahshid, S. (2023). (invited)). translational applications of nanostructured biosensors: Diagnostics at the point of care. *Meeting Abstracts, MA2023-01*(53), 2638–2638. https://doi.org/10.1149/MA2023-01532638MTGABS

Makaram, P., Owens, D., & Aceros, J. (2014). Trends in nanomaterial-based non-invasive diabetes sensing technologies. *Diagnostics, 4*(2), 27–46. https://doi.org/10.3390/diagnostics4020027

Malik, S., Singh, J., Goyat, R., Saharan, Y., Chaudhry, V., Umar, A., Ibrahim, A. A., Akbar, S., Ameen, S., & Baskoutas, S. (2023). Nanomaterials-based biosensor and their applications: A review. *Heliyon, 9*(9), e19929. https://doi.org/10.1016/J.HELIYON.2023.E19929

Mehra, N. K., Jain, K., & Jain, N. K. (2015). Pharmaceutical and biomedical applications of surface engineered carbon nanotubes. *Drug Discovery Today, 20*(6), 750. https://doi.org/10.1016/J.DRUDIS.2015.01.006

Miguel-Rojas, C., & Pérez-De-Luque, A. (2023). Nanobiosensors and nanoformulations in agriculture: New advances and challenges for sustainable agriculture. *Emerging Topics in Life Sciences, 7*(2), 229–238. https://doi.org/10.1042/ETLS20230070

Mishra, L., Behera, R. K., Mondal, S., Panigrahi, A., & Sarangi, M. K. (2021). Surface charge alteration in carbon dots governs the interfacial electron transfer and transport. *Journal of Physical Chemistry C, 125*(42), 23398–23408. https://doi.org/10.1021/ACS.JPCC.1C06459/ASSET/IMAGES/LARGE/JP1C06459_0008.JPEG

Molaei, M. J. (2020). Principles, mechanisms, and application of carbon quantum dots in sensors: A review. *Analytical Methods, 12*(10), 1266–1287. https://doi.org/10.1039/C9AY02696G

Patel, P., Kumar, K., Jain, V. K., Popli, H., Yadav, A. K., & Jain, K. (2023). Nanotheranostics for diagnosis and treatment of breast cancer. *Current Pharmaceutical Design, 29*(10), 732–747. https://doi.org/10.2174/1381612829666230329122911

Sonwani ,A., Pathak, A., Jain, K. (2025). Development and characterization of multifunctional dendrimeric nanoconjugates for delivery of rutin: in vitro characterization for potential neuroprotective application. *Nanomedicine (Lond),* 1–14. https://doi.org/10.1080/17435889.2025.2576463

Patel, P., Pathak, A., Jain, K. (2024). Novel Ligand Decorated Theranostic Zein Nanoparticles Coloaded with Paclitaxel and Carbon Quantum Dots: Formulation and Optimization. *Nanomedicine, 19(5),* 367–382. https://doi.org/10.2217/nnm-2023-0197

Pourmadadi, M., Rahmani, E., Rajabzadeh-Khosroshahi, M., Samadi, A., Behzadmehr, R., Rahdar, A., & Ferreira, L. F. R. (2023). Properties and application of carbon quantum dots (CQDs) in biosensors for disease detection: A comprehensive review. *Journal of Drug Delivery Science and Technology, 80*, 104156. https://doi.org/10.1016/j.jddst.2023.104156

Prabowo, B. A., Cabral, P. D., Freitas, P., & Fernandes, E. (2021). The challenges of developing biosensors for clinical assessment: A review. *Chemosensors, 9*(11), 299. https://doi.org/10.3390/CHEMOSENSORS9110299

Qureshi, A., Gurbuz, Y., & Niazi, J. H. (2012). Biosensors for cardiac biomarkers detection: A review. *Sensors and Actuators B: Chemical, 171–172*, 62–76. https://doi.org/10.1016/J.SNB.2012.05.077

Qureshi, Z. A., Dabash, H., Ponnamma, D., & Abbas, M. K. G. (2024). Carbon dots as versatile nanomaterials in sensing and imaging: Efficiency and beyond. *Heliyon, 10*(11), e31634. https://doi.org/10.1016/J.HELIYON.2024.E31634

Sasikumar, K., Rajamanikandan, R., & Ju, H. (2023). Fluorescent carbon dots for highly sensitive bilirubin sensing with excellent selectivity. *Journal of Science: Advanced Materials and Devices, 8*(3), 100599. https://doi.org/10.1016/J.JSAMD.2023.100599

Sen, S., & Bose, A. (2025). Carbon dots: A review of innovations, applications, challenges, and future prospects. *Inorganic Chemistry Communications, 173*, 113852. https://doi.org/10.1016/J.INOCHE.2024.113852

Sharma, P., Kumar Mehra, N., Jain, K., & Jain, N. K. (2016). Biomedical applications of carbon nanotubes: A critical review. *Current Drug Delivery, 13*(6), 796–817. https://doi.org/10.2174/1567201813666160623091814

Shrivastav, A. M., Cvelbar, U., & Abdulhalim, I. (2021). A comprehensive review on plasmonic-based biosensors used in viral diagnostics. *Communications Biology, 4*(1), 1–12. https://doi.org/10.1038/s42003-020-01615-8

Sonam, S., Patel, P., Jain, K. (2024). PAMAM dendrimer and carbon quantum dots complexes as theranostic nanocarrier: Synthesis, optimization and photophysical characterization. *Optics & Laser Technology, 175(110761)*, 1–12. https://doi.org/10.1016/j.optlastec.2024.110761

Suthar, T., & Jain, K. (2023). Nanostructure-based molecules as diagnostic and Theranostic tools in Alzheimer's disease. In *Natural product-based synthetic drug molecules in Alzheimer's disease* (pp. 133–153). Springer. https://doi.org/10.1007/978-981-99-6038-5_6

Suthar, T., Patel, P., Singh, P., Datusalia, A.K., Yadav, A.K., Jain, K. (2023) Hesperidin Microemulsion: Formulation Optimization, Characterization, and *in vitro* Evaluation. *Journal of Drug Delivery Science and Technology, 80(104166)*, 1–12. https://doi.org/10.1016/j.jddst.2023.104166

Tarannum, S., Jain, K. (2023). Drug Delivery Strategies in Multiple Sclerosis, Huntington's Disease and Other Neurodegenerative Diseases. In: Mishra, A., Kulhari, H. (eds) Drug Delivery Strategies in Neurological Disorders: Challenges and Opportunities. Springer, Singapore. https://doi.org/10.1007/978-981-99-6807-7_16

Wang, Y., & Hu, A. (2014). Carbon quantum dots: Synthesis, properties and applications. *Journal of Materials Chemistry C, 2*(34), 6921–6939. https://doi.org/10.1039/C4TC00988F

Yahav, G., Pawar, S., Lipovsky, A., Gupta, A., Gedanken, A., Duadi, H., & Fixler, D. (2023). Probing polarity and pH sensitivity of carbon dots in Escherichia coli through time-resolved fluorescence analyses. *Nanomaterials (Basel), 13*(14), 2068. https://doi.org/10.3390/NANO13142068

Zhang, C., & Du, X. (2020). Electrochemical sensors based on carbon nanomaterial used in diagnosing metabolic disease. *Frontiers in Chemistry, 8*, 547911. https://doi.org/10.3389/FCHEM.2020.00651/BIBTEX

Harnessing the Prospects of Carbon Nanodots for Next-Generation Drug Delivery Applications

8

Namish Manchanda, Mohini Diwan, Honey Goel, and Sushama Talegaonkar

Abstract

With the bloom of nanotechnology, carbon-derived nanomaterials have revolutionized the landscape of diverse sectors, especially with drug delivery therapeutics, biomedical and theranostic applications, owing to their inimitable characteristics such as high surface area, tunable electronic properties, and mechanical strength. Among the diverse carbon nanomaterials, carbon nanodots (CNDs; being a subset of the carbon nano-allotrope family) have carved a niche in the field of drug delivery as quintessential nanocarriers primarily defined by their remarkable fundamental physicochemical features, superior biological compatibility, easy synthetic pathways, and tailored surface attributes. Chemically functionalized CNDs have been successfully adapted into more polar derivatives, thereby overcoming major impediments when it comes to their application in biological media.

It is a matter of fact that the inevitable consequence of extensive production and widespread application of carbon-derived nanomaterials is their consequential presence in the environment. Therefore, it becomes indispensable to critically assess the aspects related to their toxicity, biological distribution, and production feasibility for unleashing novel fixes for imminent issues. Although the scientific

N. Manchanda
Centre for Pharmaceutical Nanotechnology, Department of Pharmaceutics, National Institute of Pharmaceutical Education & Research (NIPER), Mohali, Punjab, India

M. Diwan · S. Talegaonkar (✉)
Department of Pharmaceutics, School of Pharmaceutical Sciences, Delhi Pharmaceutical Sciences and Research University (DPSRU), Government of NCT of Delhi, Delhi, India
e-mail: stalegaonkar@dpsru.edu.in

H. Goel (✉)
Department of Pharmaceutics, University Institute of Pharmaceutical Sciences and Research, Baba Farid University of Health Sciences (BFUHS), Faridkot, Punjab, India
e-mail: honeygoel@gmail.com

M. Rai et al. (eds.), *Carbon Nanodots*, https://doi.org/10.1007/978-981-95-4705-0_8

community has produced a vital amount of results in the realm of drug delivery systems along with their potential as vehicles for therapeutic drug loading, stabilization, and biodistribution employing CNDs so far, however, it is no wonder that this venture is still in its infancy until reaching its zenith. Thus, this current outlay of literature provides an in-depth discussion of the synthesis and functional aspects of CNDs, for improving therapeutic outcomes via CND-driven targeted delivery approaches. Further, this chapter also lays huge emphasis on the potential applications, challenges, and future avenues of CNDs for their effective clinical translation into the pharmaceutical settings.

Keywords

Carbon nanodots · Nanomaterials · Graphene · Cancer · Drug delivery · Bottom-up approaches · Top-up approaches

8.1 Introduction

Carbon nanodots (CNDs) are the quasi-spherical, biocompatible, eco-friendly, and low-toxicity carbon-derived nanoparticles (typically ranging from 2–8 nm) which exhibit excellent fluorescence and physicochemical properties, making them an attractive drug delivery platform for various biomedical applications (Sameer et al., 2023; Zaidi et al., 2022). CNDs are particularly promising for drug delivery due to their polar nature, chemical inertness, and ability to be functionalized for targeted delivery as nanocarriers for chemotherapeutics, antibiotics, and genes, potentially improving drug efficacy and reducing side effects (Cohen et al., 2020; Zaidi et al., 2022). Additionally, CNDs exhibit applications in biosensing, bioimaging, and theranostics, with their intrinsic photoluminescence facilitating these functions (Cohen et al., 2020; Zaidi et al., 2022; Debnath et al., 2024). The versatility of CNDs extends to their use in treating various ailments such as cancer, ocular diseases, infectious diseases, and brain disorders (Kaurav et al., 2023).

CNDs were first identified serendipitously in 2004 by Xu et al. during the electrophoretic purification of single-walled carbon nanotubes, where they observed fluorescent carbonaceous nanoparticles as by-products (Xu et al., 2004). This discovery was followed by the work of Sun et al. in 2006, who formally characterized these nanodots and introduced the term "carbon dots" after synthesizing them via laser ablation (Xiao & Sun, 2018; Sun et al., 2006). Over the past two decades, CNDs have attracted significant attention due to their versatile applications in areas such as fluorescent sensing, photocatalysis, biomedicine, and optoelectronic devices (He et al., 2022; Karpushkin et al., 2024). Researchers have developed numerous synthesis methods and investigated their photoluminescence mechanisms, which are attributed to surface state, carbon core state, molecule state, and their synergistic effects (He et al., 2022). Despite extensive research, some fundamental aspects of CNDs, including their precise definition and the origin of their emissive properties, remain debated (De Boëver et al., 2022). Recent studies on CND fractionalization and assembly have expanded their potential as building blocks for larger carbon-derived materials (Zhou et al., 2022).

8.1.1 Classification of Carbon Nanodots (CNDs)

CNDs are often used interchangeably with terms like carbon nanoclusters, C-dots, and carbon dots in scientific literature (Debnath et al., 2024). Structurally, these encompass a broad family of nanomaterials exhibiting diverse structures and optical characteristics. CNDs are generally quasi-spherical in shape with either graphitic, amorphous, or even C_3N_4 crystalline cores (Sciortino et al., 2018) and can be classified as graphene-based CNDs, quantum-based CNDs (Lagos et al., 2023; Zeng et al., 2021).

8.1.1.1 Graphene-Based CNDs

The graphene quantum dots (GQDs) are nanoscale fragments of graphene characterized by one or a few layers of graphene sheets, typically with lateral dimensions under 20 nm and thicknesses corresponding to fewer than five graphene layers (~2.5 nm) (Mansuriya & Altintas, 2021; Xia et al., 2019). These dots exhibit pronounced anisotropy and contain well-defined graphene lattices, with various chemical functionalities located at the edges and within interlayer defects (Xia et al., 2019). Two key photophysical phenomena—quantum confinement and edge effects—arise from their structural features, governing their optical and electronic behaviour. Notably, the quantum confinement in GQDs is not limited to particle size but also relates to the isolation of π-conjugated domains by defects on the graphene plane, which serve as fluorescent centres (Xia et al., 2019).

8.1.1.2 Quantum-Based CNDs

Such quasi-spherical nanoparticles typically range from 2 to 10 nm, composed of multilayer graphitic lattices exhibiting good crystallinity and interlayer stacking. Often known as carbon quantum dots (CQDs), their surfaces carry abundant functional groups, notably oxygen-containing moieties, which enhance water solubility (Bartkowski et al., 2024; Xiao et al., 2025). Similar to graphene quantum dots (GQDs), CQDs display quantum confinement effects linked to both particle size and π-conjugated domains but differ from GQDs in their precursor aggregation pathways during synthesis (Xiao et al., 2025). CQDs also function effectively as electron donors or acceptors, with luminescence governed by interactions between the sp^2 carbon core and surface states (Bartkowski et al., 2024; Xiao et al., 2025).

8.1.1.3 Polymer-Based CNDs

Structurally, polymer-based carbon nanodots consist of sub-20 nm carbon cores embedded within highly crosslinked polymer networks or partially graphitized regions, surrounded by shells enriched in diverse functional groups and polymer chains (Liu et al., 2020b; Ru et al., 2022). They possess hybrid polymer/carbon architectures and are therefore often called carbonized polymer dots (CPDs), where

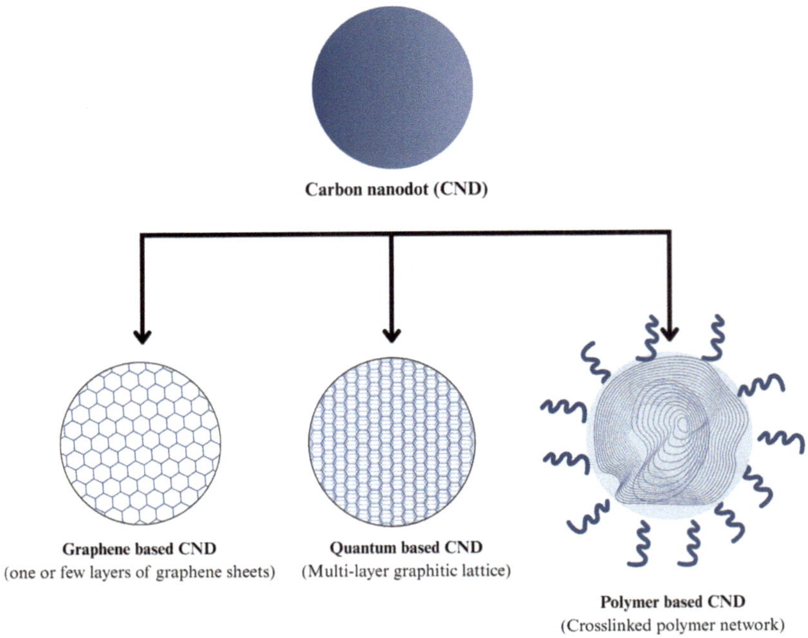

Fig. 8.1 Classification scheme of carbon nanodots

the polymeric component contributes abundant surface functionalities, polydispersity, and a highly interconnected framework formed through dehydration and carbonization processes (Ru et al., 2022). Unlike the other classes, CPDs exhibit photoluminescence primarily governed by molecular-state emissions and crosslink-enhanced emission (CEE), which can arise from both covalent bonding and noncovalent interactions, including supramolecular forces, and ionic (Ru et al., 2022). The carbon cores of CPDs often feature small subdomains composed of conjugated π-structures, contributing to their unique optical properties (Ru et al., 2022) (Fig. 8.1).

In addition to the commonly studied types, carbon nanodots have been further categorized into specialized subtypes such as hollow carbon dots (H-CDs) and heteroatom-doped CNDs (Ht-CNDs), which include nitrogen-doped CNDs (N-CNDs), boron/nitrogen co-doped CDs (BN-CNDs), and sulphur or selenium-doped CNDs (S-CNDs or Se-CNDs). Other structurally or functionally modified variants have also been reported (Bartkowski et al., 2024). Graphitic carbon nitride quantum dots (g-CNQDs) have emerged as a distinct subclass different from GQDs since they consist of networks of tri-s-triazine units connected via amino linkages, forming a poly(tri-s-triazine) framework interspersed with periodic vacancies, which imparts unique structural and electronic characteristics. (Qureshi et al., 2024).

8.1.2 Unique Features of Carbon Nanodots

8.1.2.1 Optical Features

A prominent characteristic of CNDs lies in their intense and tunable photoluminescence (PL), which extends across a broad spectral range from the ultraviolet to the near-infrared region (Aditya Khatokar et al., 2021; Azami et al., 2023; Liu, 2020; Papaioannou et al., 2018) and tunable fluorescence across the visible spectrum (Wang et al., 2020). Notably, numerous CNDs display excitation-dependent PL behaviour, a phenomenon frequently linked to factors such as particle size variations, surface defect states, and diverse surface chemistries (Liu, 2020; Papaioannou et al., 2018), with oxygen-related defects responsible for blue/green emissions and nitrogen-related defects for red emissions (Wang et al., 2020). CNDs demonstrate nanoparticle-like behaviour and molecular fluorophore characteristics, exhibiting concentration-dependent emission kinetics and quenching effects (Almoalim et al., 2023). Furthermore, Papaioannou et al. (2018) illustrated how solvent polarity and its interaction with surface functionalities influence the PL properties of glucose-derived CNDs, as demonstrated in Fig. 8.2, which depicts alterations in the PL map above and below 320 nm.

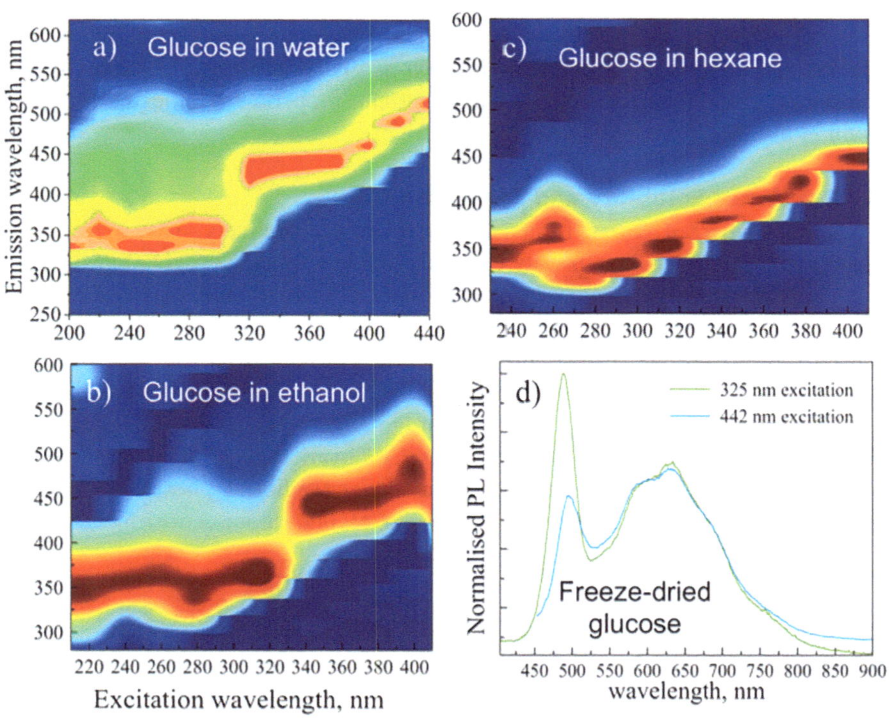

Fig. 8.2 PL maps glucose CNDs in (**a**) water (relative polarity 1), (**b**) ethanol (relative polarity 0.654), and (**c**) hexane (relative polarity 0.009). & (**d**) Normalized PL data of a freeze-dried glucose sample with two excitation wavelengths (325 nm and 442 nm) (Papaioannou et al., 2018)

8.1.2.2 Biocompatibility and Low Toxicity

CNDs have emerged as promising materials for biomedical applications due to their exceptional properties, including biocompatibility and low toxicity (Dhamodharan et al., 2022). Predominantly composed of carbon and free from heavy metals, CNDs demonstrate excellent cellular uptake, produce minimal reactive oxygen species, and exhibit low cytotoxicity, even when used at higher concentrations. Moreover, surface modifications such as PEGylation or glycosylation have been reported to significantly improve their biological compatibility and prolong their circulation time within living systems (Arcudi et al., 2019; Chan et al., 2021; Mao et al., 2016).

8.1.2.3 Surface Functionalization

Surface engineering plays a pivotal role in enhancing the functional versatility of CNDs. Their surfaces are rich in reactive functional groups, enabling facile conjugation with diverse entities, including synthetic polymers and various biomolecules (Arcudi et al., 2019; Đorđević et al., 2019; Swift et al., 2019; Xiao & Sun, 2018). Through surface modifications, researchers can tailor properties such as solubility, adjusting between hydrophilicity and hydrophobicity, as well as fine-tune the optical characteristics and specificity for targeted applications (Arcudi et al., 2019; Xiao & Sun, 2018; Goel et al., 2020). Additionally, surface passivation can significantly influence the electronic structure of CNDs, leading to shifts in their absorption spectra and alterations in fluorescence pattern. Recent findings also highlight how the strategic design of surface ligands, such as glycan moieties, can substantially affect both biological interactions and electronic properties (Arcudi et al., 2019; Swift et al., 2019).

8.1.2.4 Water Solubility and Bioavailability

CNDs have emerged as promising nanocarriers for drug delivery, offering enhanced water solubility and bioavailability of poorly soluble drugs. Nitrogen-doped CNDs coupled with paclitaxel demonstrated excellent cell viability and improved anticancer activity compared to the drug alone (Gomez et al., 2018). Similarly, curcumin-loaded CNDs exhibited increased bioavailability, sustained release, and enhanced antitumor effects (Arvapalli et al., 2020). Cyclodextrin-based nanosponges (CDNSs) represent another innovative platform for drug delivery, addressing issues of inadequate solubility and poor dissolution rates (Allahyari et al., 2019). Furthermore, silica-confined CNDs have achieved ultralong and efficient phosphorescence in aqueous solutions, overcoming the challenge of phosphorescence quenching in water (Liang et al., 2020).

8.2 Methods of Preparation

The synthesis of CNDs has evolved significantly since their initial discovery through arc discharge techniques and subsequent exploration using laser ablation (Sun et al., 2006). Early methods predominantly relied on top-down strategies, wherein bulk carbon materials were fragmented into nanoscale particles under high-energy

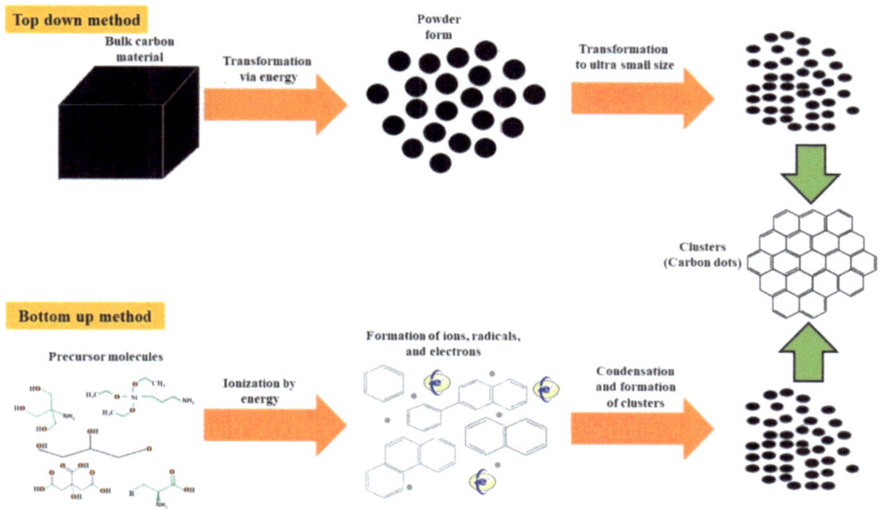

Fig. 8.3 General methods for the synthesis of carbon nanodots. Bottom-up approach: CNDs are synthesized from smaller carbon units by applying energy (electrochemical/chemical, thermal, laser, etc.). Top-down approach: CDs are synthesized by the transformation of larger carbon structures into ultra-small fragments via applying energy (thermal, mechanical, chemical, ultrasonic, etc.) (Sharma & Das, 2019)

conditions. A pivotal advancement occurred in 2013 when Zhu et al. reported the hydrothermal synthesis of CNDs from citric acid and ethylenediamine, signifying a shift from using bulk inorganic carbon sources to small organic molecules as precursors (Zhu et al., 2013). This breakthrough laid the groundwork for bottom-up approaches, which now form a major pathway for CND production. Broadly, methods for CND synthesis can be categorized into top-down and bottom-up techniques (Shi et al., 2022) (Fig. 8.3).

8.2.1 Top-Down Approaches

In top-down approaches, bulk carbon materials are reduced to nanoscale particles through various targeted physical and chemical processes (Li et al., 2023). This approach includes the laser ablation method, the arc-discharge method, the electrochemical method, and ultrasonic treatment. These methods can produce CNDs from various carbon precursors, influencing the size, morphology, and core structure of the resulting nandots (Shi et al., 2022; Goel et al., 2022a). Top-down approaches offer advantages such as scalability and control over particle size but may have limitations in terms of uniformity and reproducibility (Abid et al., 2022; Goel et al., 2022b).

8.2.1.1 Laser Ablation Method

Laser ablation has emerged as a versatile and promising approach for synthesizing CNDs, offering significant advantages in terms of operational simplicity and cost-effectiveness (Cutroneo et al., 2024). This technique generally involves three fundamental steps: initially, the carbon target absorbs the high energy delivered by a pulsed laser; subsequently, the photoelectric effect triggers electron emission from the material, often through thermionic mechanisms; and finally, intense electrostatic repulsion between positive ions and the solid lattice leads to the ejection and fragmentation of carbon atoms into nanometric domains (Fauziyah, 2022). Recent studies emphasize that laser-based synthesis can proceed without the need for complex vacuum systems and can be performed using relatively simple and accessible experimental setups, making it suitable not only for scalable production but also for fundamental research into nanomaterial properties (Cutroneo et al., 2024). Laser ablation techniques enable versatile synthesis of carbon nanodots with tunable properties. Double-pulse femtosecond lasers can produce ultrasmall CNDs (~1 nm) rich in surface groups for sensing and catalysis (Nguyen et al., 2020), while continuous-wave blue lasers yield ~10 nm luminescent CNDs from charcoal for biomedical applications (Cutroneo et al., 2024). Infrared laser ablation of graphite or diamond in liquids generates size-dependent CNDs useful in biomedicine (Dudek et al., 2017).

8.2.1.2 Arc-Discharge Method

Arc discharge is a well-established method for generating carbon nanomaterials, including CNDs. It involves the creation of an electric arc between two electrodes, commonly graphite rods, leading to their vaporization and the formation of carbon-rich soot containing various nanoparticles (Banger et al., 2023). The intense thermal energy produced during the discharge causes carbon to evaporate from the anode and subsequently condense either onto the cathode or within the surrounding environment (Timerkaev et al., 2019). Advances such as pulsed arc discharge have enhanced the precision of this technique, allowing improved control over the ablation process and achieving typical material ablation rates around 1 mg s^{-1}. Moreover, factors like electrical current, applied voltage, and the composition of the electrodes can be finely tuned to optimize the required carbon nanostructure (Corbella et al., 2019).

8.2.1.3 Electrochemical Method

Electrochemical synthesis represents a promising top-down approach for producing CNDs, wherein carbon structures are etched or fragmented into nanoscale materials. Despite its potential, relatively few studies have focused on electrochemical oxidation of small molecules as precursors for CND synthesis (Sharma & Das, 2019). The first synthesis of CDs from carbon nanotubes via electrochemical oxidation was carried out by J. Zhou et al. (2007), establishing a foundation for this technique. Control over the particle size can be achieved by maintaining the current density within the range of 20–180 mA cm^{-2}, highlighting the importance of precise operational parameters for consistent nanodot dimensions (Zaidi et al., 2022).

Typically, electrochemical etching employs conductive carbon-based materials such as graphite rods or carbon fibres as sacrificial electrodes, valued for their natural abundance and low cost. Furthermore, the sizes and compositions of the resulting CDs can be effectively tuned by adjusting key parameters, including pH, electrolyte concentration and composition, as well as the electrolysis mode, whether potentiostatic, galvanostatic, or cyclic voltammetry (Qi et al., 2016).

8.2.1.4 Ultrasonic Treatment

Ultrasonic synthesis of CNDs operates through alternating high- and low-pressure waves that generate cavitation bubbles in liquid media. The collapse of these bubbles produces intense localized energy and strong hydrodynamic shear forces, effectively breaking down macroscopic carbon materials into nanoscale CDs (Cui et al., 2021). Notably, Zhuo et al. (2012) were the first to demonstrate the ultrasonic exfoliation of graphene to produce graphene quantum dots (GQDs), highlighting the versatility of this technique for nanocarbon synthesis (Cui et al., 2021). Parameters such as ultrasonic power, reaction time, and the ratio of carbon precursors to solvents can be precisely adjusted to tailor the structural and optical properties of the resulting CDs. While frequencies in the range of 100–400 kHz are generally recognized as effective for synthesizing nanoparticles with smaller dimensions, current reports on CD synthesis predominantly employ ultrasonic frequencies around 20 kHz, and the specific influence of higher ultrasonic frequencies on CD formation remains an area open for future exploration (Kumar et al., 2020).

8.2.2 Bottom-up Approaches

CNDs can be synthesized using bottom-up approaches, which involve assembling atomic or molecular components to form nanoparticles (University of Piteşti, Romania et al., 2020). This method allows for better control over particle size and can produce CNDs with narrow size distributions and high fluorescence quantum yields (Arcudi et al., 2016). The optical properties of bottom-up synthesized CNDs are largely influenced by molecular-like fluorophores formed as intermediates during the carbonization process (Nguyen et al., 2023). While bottom-up approaches offer advantages in size control and optical properties, they may face challenges in achieving quantum size effects and controlling surface composition (Arcudi et al., 2016; Nguyen et al., 2023). This approach includes hydrothermal synthesis, solvothermal synthesis, microwave synthesis, and the pyrolysis or carbonization method.

8.2.2.1 Hydrothermal Synthesis

The hydrothermal process for bottom-up synthesis of CNDs involves the fusion of organic precursors through condensation and carbonization (Nguyen et al., 2023). The approach is widely adopted by researchers as an economical, environmentally friendly, and cost-effective technique for producing carbon dots from precursors such as saccharides, amines, organic acids, and their derivatives (Sharma & Das,

2019). The formation occurs in four steps: aggregation of small molecules, dense core formation with an extended shell, shell collapse, and core aromatization (Rigodanza et al., 2021). Synthesis parameters like temperature and reaction time significantly affect CND properties, including yield, size, and quantum yield (Quaid et al., 2022). Intermediate fluorophores formed during the process play a crucial role in determining the optical properties of CNDs (D.-K. Nguyen et al., 2023). Zhang et al. were the first to produce fluorescent carbon dots using a small organic molecule, L-ascorbic acid, through a hydrothermal synthesis route. In their method, an aqueous solution of ascorbic acid was maintained at 180 °C for 4 h inside an autoclave, followed by purification of the resulting solution via dialysis employing a membrane with a molecular weight cutoff of 8000–14,000 Da. The synthesized carbon dots exhibited an average diameter of approximately 2 nm (Sharma & Das, 2019; Zhang et al., 2010).

8.2.2.2 Solvothermal Method

The solvothermal method is a bottom-up approach for synthesizing CNDs with controllable properties and fewer defects (Liu et al., 2020a). The process allows for tuning of various properties, including emission colour, electrochemistry, and chirality (Đorđević et al., 2019) and control over particle size, morphology, and functionalization by adjusting reaction parameters such as temperature, time, and solvent choice (Huo et al., 2023). Solvent selection plays a crucial role in determining the optical properties of CDs, enabling the synthesis of particles with emissions ranging from blue to near-infrared and quantum yields up to 54% (Ding et al., 2018). Different solvents, such as water, ethanol, DMF, and acetic acid, affect the formation of functional groups and graphitic nitrogen content, leading to distinct fluorescence characteristics (Huo et al., 2023). Organic solvents containing double bonds, benzene rings, or double hydrophilic groups can be directly carbonized to form GQDs without catalysts or additional precursors (Liu et al., 2020a). The solvothermal method offers advantages such as low reaction temperatures, environmental friendliness, and high yields (Huo et al., 2023).

8.2.2.3 Microwave Synthesis

Microwave-assisted synthesis offers a rapid and efficient route for producing CNDs, often completing the entire process in a single step within minutes (Soledad-Flores et al., 2024). This technique significantly reduces reaction times while improving uniformity and control over particle characteristics. Its efficiency arises from the direct molecular excitation of polar reagents, which undergo rotational and vibrational transitions under electromagnetic irradiation in the range of 0.3 to 300 GHz, with 2.45 GHz being the frequency most utilized for synthesis purposes. The molecular friction resulting from this excitation generates internal heat, thereby accelerating reaction kinetics and enhancing process efficiency (Soledad-Flores et al., 2024). Despite notable advantages such as speed, cost-effectiveness, and environmental sustainability, microwave synthesis can still suffer from limitations, including poor size control and relatively low yields (Kong et al., 2024). To address these issues, hybrid approaches like microwave-assisted hydrothermal carbonization have been explored to improve yield and achieve better control over the properties of the resulting C-Dots (Kong et al., 2024).

8.2.2.4 Pyrolysis/Carbonization Method

Pyrolysis refers to the irreversible thermal degradation of organic substances carried out under an inert atmosphere. This process induces both physical and chemical transformations, leading to the formation of a carbon-rich solid residue. Carbonization is a thermal process wherein organic materials are subjected to extended pyrolysis under an inert atmosphere, resulting in solid residues enriched in carbon content. Typically, pyrolysis is conducted at elevated temperatures and under carefully regulated pressure conditions (Sharma & Das, 2019). CNDs can be synthesized through the pyrolysis of various organic precursors, offering a versatile and eco-friendly approach. Biomass sources like cellulose nanocrystals from oil palm empty fruit bunches have been successfully used to produce CNDs with fluorescent properties (Anindita et al., 2024; Marpongahtun et al., 2018). The pyrolysis temperature significantly affects CND characteristics, including size and optical properties (Marpongahtun et al., 2018). Laser pyrolysis of volatile organic compounds like toluene and pyridine can yield monodisperse CNDs with tunable photoluminescence (Mas et al., 2020). The synthesis of carbon nanomaterials through pyrolysis is influenced by various factors, including temperature, catalyst, and precursor type (Zahid et al., 2018).

8.3 Carbon Nanodots in Drug Delivery: The Next-Generation Nanotherapeutics

The scientific community has become more interested in carbon-based nanomaterials in recent centuries, foremost in the context of the treatment of various diseases such as cancer, microbial infections, neurological disorders, etc. Because of their cost-effective synthesis and production costs, high solubility and stability, superior biocompatibility, low toxicity, and degradability, researchers have recently shown a preference for CNDs (Rai et al., 2024). Because CNDs may integrate many functionalities on a single platform, they can perform multiple tasks at once, unlike typical nano-agents. The possible uses of various CNDs in drug delivery science are compiled in this part, utilizing their versatility to carry out various therapeutic functions (Alibrahem et al., 2025; Sameer et al., 2023).

8.3.1 Graphene Nanodots

Graphene-based carbon nanodots (G-CNDs), also touted as graphene quantum dots (GQDs) or just graphene dots (GDs), are a distinct type of carbon-based nanodots derived from graphene with a size of 2–10 nm. Owing to their modest toxicity, durability, effective surface tailorability, and size-tunable discharge in comparison to other conventional CDs, GDs have caught the attention of an array of researchers in biomedical uses. More significantly, as drug delivery nanocargos, G-CNDs constitute a new paradigm in the healthcare sector owing to their predictable drug release profiles at the cellular level. GDs exhibit strong photoluminescence, biological compatibility, and high electron mobility, making them attractive for tailored drug delivery, and thus, investigating such innovative nanomaterials puts forward an appealing means to combat several ailments (Ding et al., 2017; Dong et al., 2018; Giordano et al., 2020). In this instance, aiming to strike at detrimental microbial biofilms, an improved bacterial cellulose-based GDs inclusive nanocomposite hydrogel (BC-GDs) was engineered by Zmejkoski et al. for photodynamic intervention leveraging blue and green photons. The methodology seeks to alleviate concerns with invasive therapies and against the resistant microorganisms. Both blue & green BC-GDs revealed lethal singlet oxygen species that diminished bacterial-biological films following 30 min of illumination. Notably, green BC-GDs displayed considerable antibiotic action against *S. aureus* and *E. coli,* while the blue BC-GDs revealed antibiofilm performance against *P. aeruginosa*. Additional research examined the knack of fabricated GD-hybrids to accelerate wound repair events when exposed to bright light, demonstrating enhanced migration capacity and the activation of MMP9 and eNOS genes in human gingival fibroblasts. These unique photoactive BC-GDs hydrogel mixtures hold plenty of promise as efficient tools for combating infections and wound recovery & repair (Zmejkoski et al., 2025). Another study, which was published by Milenković et al., examines the antibacterial capacity of GDs together with Au-NPs. These nanostructures are developed in situ by irradiating G-CNDs with chloroauric acid, following the characterization of the product for its safety. On top of that, MTT experiments found that cell survival of >80% for most samples, demonstrating their non-toxic nature, while haemolytic screenings demonstrated outstanding biocompatibility with haemolysis of <2%. The results demonstrated the antibacterial effects of these materials against the MRSA strains. These findings show that Au-NPs play a major role in GDs' antibacterial promise, suggesting their use in biomedicine (Milenković et al., 2025). Furthermore, urolithin B-loaded GDs-modified MSNs were fabricated by Nezhadhossein & colleagues for potential anti-cancer delivery following their evaluation against MCF-7 breast cancer cells. Urolithin B was loaded onto the resultant nanocarriers after the GDs and MSNs were covalently attached. The diameter of the particles had been ~224 nm, while the surface bears a negative charge of ~22.6 mV, indicating appreciable stability. Having an IC_{50} of 83.19 μg/mL, the CNDs showed notable lethality and adeptly triggered cell death in a dose-responsive manner. With a dramatic increase in TNF, caspase 9, and p21 per the gene expression data. Additionally, the antioxidant capacity was validated by free radical scavenging in the ABTS assay (~76% inhibition) and in the DPPH assay (~33.4% inhibition) at 2 mg/mL. Such findings highlight the potential of GD-driven nanoplatforms as a promising candidate for targeted cancer therapy, combining effective cytotoxicity and antioxidant properties (Nezhadhossein et al., 2025).

The discomfort that patients with dry eye disease endure is severe. However, an abnormal deposition of ROS on the cornea's surface is the root cause of DED. The scientific literature has reported nitrogen-doped GDs (N-GDs), which are well-known for their ability to scavenge reactive oxygen species (ROS), as a treatment for DED (Hu et al., 2025; Huang et al., 2021). In this regard, Wu et al. investigated the utilization of N-GDs to improve treatment for DED by countering ROS and remodelling intracellular antioxidant pathways. Moreover, RGDS/N-GDs were designed by modifying N-GDs with RGDS peptides, and both CNDs were administered as ocular drops in the DED mouse model. These N-GDs effectively neutralized free radicals and activated the intracellular Nrf2 pathway, exhibiting strong antioxidant qualities. RGDS/N-GDs eye drops significantly lowered the discomfort of the DED mouse model, highlighting how altering GDs by nitrogen doping and RGDS peptide tethering enhances their antioxidant effects, therefore revealing a new therapeutic avenue for DED treatment (Wu et al., 2024). A separate research effort published by Najafi et al. revealed a cutting-edge hydrogel composite, developed employing agarose, GDs, and α-Fe_2O_3 that affords a unique pH-responsive quercetin-encapsulated controlled release system. The DLS technique verified the existence of nanostructure within the composite, highlighting a spherical morphology with an average particle size of ~280 nm while the zeta potential measurements showed a positive surface charge of ~52.8 mV. Further, the experimental assays utilizing the HepG2 cell line explored putative adverse impacts, thereby demonstrating the superior therapeutic effectiveness of GDs-loaded NPs. These outcomes underline the captivating prospects of GDs as a viable cancer treatment approach (Najafi, Khoddam, et al., 2024a). It is noteworthy that chemotherapy is still the sole treatment available in many infectious diseases like leishmaniasis, where no vaccination is there, highlighting the critical need for safer and more efficient therapies. There are encouraging opportunities to address such issues with nanomaterials, including CNDs, and particularly the GDs. To facilitate targeted drug delivery to leishmania-infected macrophages, Madhukar et al. describe a unique amine-functionalized GDs coupled with amphotericin B as an innovative, safer and more efficient therapy for leishmaniasis. Biochemical investigations, however, concluded improved antileishmanial effectiveness of the developed CD-vehicle, which resulted in quite elevated macrophage infiltration. In contrast with the standard amphotericin B, the GD-driven approach appeared safer and showed a more modest cytotoxicity with ~four-fold increased capacity towards intracellular amastigotes and ~ two-fold increased efficacy against promastigotes, giving certainty and trust regarding its said application and therefore, boosting the potential of GD-made formulations in nanomedicine (Madhukar et al., 2025). Key attributes of other GDs described in current research for their use in drug delivery are outlined in Table 8.1.

8.3.2 Quantum Nanodots

Fluorescence-based NPs featuring top-notch optical qualities, particularly high luminosity and photostability, are recognized as quantum-based carbon nanodots (Q-CNDs), carbon quantum dots (CQDs), or merely quantum dots (QDs). By modulating reaction variables, including pH, reaction temperature, and time, and

Table 8.1 Applications of different graphene nanodots explored in drug delivery science

S. No.	Graphene dots	Active(s)	Significance	Reference
1.	PAA-starch-GDs	Quercetin	The outcomes lend credence to the developed GD-nanocarrier's potential as a targeted and effective drug delivery system for the brain tumours	Ordoukhani et al. (2025)
2.	Pyrenebutyric-modified hyaluronic acid-GDs	TAK-632	GD-based dual drug delivery system significantly inhibited the proliferation of cancer cells in both in vitro and in vivo models	Lin et al. (2024)
3.	Folic acid-magnetic GDs	Curcumin	QDs exhibited a pH-dependent drug release in tandem with folic acid-guided precise targeting in tumours	Seyyedi Zadeh et al. (2023)
4.	N-doped GDs	–	GDs showed superior photonic and structural attributes with one log lower MIC when doped with nitrogen, against B. Subtilis.	Dwitya et al. (2023)
5.	GDs	Doxorubicin, deoxyadenosine	By lowering the translocation-free energy accessing the biomembrane, QDs of a suitable diameter may aid in the drug delivery process	Xue et al. (2019)
6.	Milk-derived GDs	Berberine hydrochloride	Theranostic QD-complex had shown a strong cytotoxic effect on many malignant cell line models, including breast cancer cells and cervical cancer cell lines	Thakur et al. (2016)
7.	Folate/GDs	Doxorubicin	HeLa cells were effectively inhibited by multimodal GD-based nanocarriers	Huang et al. (2014)

altering functional units, these QDs' molecular makeup can be tweaked to be put in a range of applications in biomedical science. More significantly, with appropriate modifications like surface alteration, doping, and encasing, these NPs' substantial area and capacity to encapsulate therapeutic compounds have demonstrated promise as drug delivery nanocargos (Khan et al., 2023; Sivasankarapillai et al., 2020). QDs could efficiently target distinct biological tissues or cells by combining their surface with precise ligands, thereby lowering off-target impacts and propelling effectiveness in the intended therapy. Moreover, QDs have illustrated a great promise in improving therapeutic effectiveness and thereby reducing overall toxicity when utilized as carriers for chemotherapeutic medications, phototherapy agents, and other pharmacological payloads (Lohani et al., 2023; Soumya et al., 2023). Photothermal ablation entwined with chemotherapy, for instance, displays notable promise in managing the malignancy, for which a sound, feasible, yet secure photothermal reagent must exist in this combinatorial setup. In this regard, by encapsulating QDs, a reliable photothermal agent and doxorubicin as a potent active, Liu and coworkers intuitively established a folic acid-targeted liposomal multimodal nanovehicle. Further, indocyanine green was incorporated in the lipid bilayer to improve the photothermal effects and aid the liposomal cleavage. This forged multifunctional nanoplatform was set off by the NIR pulse and implied a controlled payload release alongside demonstrating exceptional performance, including superior biological compatibility, improved tumour-specific targeting and precise cytotoxic ability, in both in vitro and in vivo research, offering a robust design and development of synergistic chemo-photothermal combination therapy (Liu et al., 2025). The complexity of designing and delivering drugs by probes makes dynamic tracking of cancer cells with theranostic properties hard to do (Goel et al., 2024). In another attempt, Dutta et al. put together doxorubicin-loaded fluorescent photon-responsive polyphenolic QDs based on phloroglucinol for superior membrane targeting while enabling combinatorial cancer therapy. Considering higher doxorubicin accumulation in the tumours, the green-emissive QDs produced more ROS, showing an improved cytotoxic level at both pH 5.0 and pH 6.5, along with an outstanding drug loading efficiency. Furthermore, the complex demonstrated superior membrane-targeting capabilities with reduced tumour cell photobleaching capacity. Pursuant to the in vivo research, the combination reveals stronger inhibitory effects, greater tumour susceptibility, and no systemic toxicity, hinting this could possibly be applied as a "smart" probe for multimodal tumour targeting (Dutta et al., 2022). Likewise, gaps in the focusing of cytotoxic agents throughout breast cancer therapy typically led to catastrophic implications. Tiwari et al. offered an alternative method to address such concerns by integrating dacarbazine into fucose-based QDs. Further, exosomes that have been pulled from breast tumours were then used to cap these QDs. In vitro studies revealed that QD-based exosomes preserved the morphological and pharmacological characteristics while facilitating the delivery of drugs to tumours, which led to superior MMP depolarization, ROS build-up, and inciting apoptosis, all of which ended up causing cells to die. The in vivo data further revealed optimized pharmacokinetic biography, continued release variables, and HSPG receptor-driven cell intake with exosome-mediated tumour accumulation. However, the overall findings strongly support the idea for the development of QD-mediated physiologically brought in nanocarriers for the purpose of combating breast cancer (Tiwari et al., 2024).

Plants contain several phytoconstituents, and among them is quercetin, a potent flavonoid with potential medicinal use that metabolically affects tumour cells and halts the advancement of a few kinds of malignancy. However, in part due to its insufficient solubility and its very hydrophobic nature, there exist difficulties in its delivery to the tumour site. Given that nanocomposite hydrogels are easily programmed and versatile, they can be used for an assortment of drug delivery applications for molecules like quercetin. They enable prolonged and strictly controlled drug release, making them valuable for therapies to combat cancer. A novel approach to enhance the bioavailability and persistence of quercetin reported by Najafi et al. used the hazelnut oil to formulate w/o/w nanocarrier by mixing it with

Cu-containing QDs to derive a pH-sensitive nanocomposite comprising carboxymethyl cellulose (CMC) and starch. This strategy aims to improve drug delivery while depicting an impressive drug loading and entrapment efficiencies ~47% and ~ 86%, respectively, placing it within the most favourable known works for similar systems. The developed nanosystem showed remarkable cellular damage over malignant cells, reducing U87-MG cell viability to below 50% yet sustaining strong cell survival of more than 92% for healthy L929 cells. These results demonstrate improved antitumor efficacy with fewer side effects, highlighting the QD-build nanocomposite's promise as a drug delivery strategy (Najafi et al., 2024b). Another work published by Jiao and coworkers, presented QDs as a novel carbon-based nanomaterial for its potential applications in cancer treatment by their in vitro and in vivo evaluation on the osteosarcoma cells. 143B cells cultured with QDs had demonstrated a high level of cytotoxicity, had dramatically reduced MMP and raised the intracellular ROS levels, thereby upregulating the production of apoptotic proteins. In the mouse model, the tumour weight inhibition was ~30% while the tumour volume inhibition rate was ~38%, with major organs showing no signs of inflammation, and all biosafety indices fell within reference values, suggesting a unique & promising approach for innovative cancer treatments (Jiao et al., 2020). Conversely, Perini et al. had carried out a comprehensive investigation that looked at the many applications of QDs in the management of glioblastoma. These particles' exceptional ability to cross the blood-brain barrier offers a significant benefit in the management of brain tumours. Additionally, surface-functionalized QDs restored transmembrane mobility and cellular intake in the 3D spheroid glioblastoma model. They further partnered with drugs such as doxorubicin and temozolomide at subtherapeutic levels. Overall, the study's findings showed a new strategy labelled photothermal interventions, in which QDs transformed near-IR light into thermal energy to increase membrane flexibility and improve chemotherapy's effects. This combination resulted in a notable reduction in cancer growth and sustenance, indicating that QDs could improve patient comfort and quality of life (Perini et al., 2023). A few further instances demonstrating QD's potential for efficient drug delivery across a variety of disease states are included in Table 8.2.

8.3.3 Polymeric Nanodots

Another kind of carbon-intensive nanomaterial is polymer-based CNDs (P-CNDs), also known as polymeric dots (Pdots), which have been created from conjugated polymer chains. When viewed alongside other CNDs, they usually feature a bigger dimension range of about 100 nm and may have been created using a variety of techniques, including emulsion polymerization and nanoprecipitation. Pdots are appealing for physiological imaging and biosensing applications by virtue of their excellent photostability, strong and programmable fluorescence, and outstanding biocompatibility (Debnath et al., 2025; Nicosia et al., 2023). In addition, a variety of groups, including carboxyl, hydroxyl, or amine groups, can be tailored on their exterior to be used for bioconjugation or to improve their safety. Different polymers

Table 8.2 Different quantum nanodots reported in the literature for drug delivery applications

S. No.	Quantum dots	Active(s)	Significance	Reference
1.	Folic acid/QDs	Doxorubicin	Folic acid-conjugated synthetic QDs demonstrated a pH-responsive drug release profile	Ghahremani et al. (2025)
2.	PAA-PEG-QDs	Quercetin	The built QDs illustrated persistent and selective drug release under acidic conditions at quite high encapsulation efficiency, stability & cytotoxicity than the free drug	Pourmadadi et al. (2024)
3.	QD-chitosan gel	Cytarabine	The QD-chitosan gel efficiently relieves the burst drug release driven by pH-dependent hydrolysis of the amido bond	Sheng et al. (2020)
4.	Carrageenan-QDs, Pullulan-QDs	–	Pullulan-QDs outperformed anticancer activity, while the carrageenan-QD set up confirmed the precedence as an antiviral agent	Emam & Ahmed (2021)
5.	Chitosan/gelatin/QDs	Curcumin	The formed nanocomposite material can be regarded as a biocompatible nanocarrier; meanwhile, it was pointed out that improved cell killing compared to the free drug in cancer cells	Ostovar et al. (2023)
6.	GDs	5-fluorouracil	The made QD-nanoconjugate came up with a decline in toxic effects for normal cells while having an anti-cancer impact like that of the free drug over lung & breast cancer cells	Cutrim et al. (2021)
7.	Cathepsin D-responsive peptide/GDs	Doxorubicin	QDs' distinctive fluorescence appeared as probes for anticipated control of drug distribution and release, in addition to drug-induced cancer cell apoptosis	Ding et al. (2017)
8.	RGD-QDs	Doxorubicin	The designed conjugates track drug distribution and release processes in glioma, alongside dramatically boosted the potency of actives	Dong et al. (2018)
9.	N-doped QDs	Methotrexate	Overall findings corroborate N-QDs as nanocarriers that prolong the cytotoxic effects of their payload for substantial cell death in malignant cells	Khodadadei et al. (2017)
10.	Nuclear localization signal peptide-QDs	Doxorubicin	The prepared QD-framework had been researched in an A549 xenograft nude mouse model, which highlighted an improved tumour suppression compared to the free drug	Yang et al. (2016)

have different chemistry, such as their solubility and hydrophobicity, and by selecting different polymers, including polyethylene glycol (PEG), polyvinylpyrrolidone (PVP), polyacrylamide (PAA), polyvinyl alcohol (PVA), and polystyrene (PS), the physicochemical attributes of Pdots can be tailored for specific applications, including drug delivery science. Nowadays, they have been widely applied in various fields, such as drug sensing, biomarker detection, and cell biological imaging. Today, as a novel treatment method, Pdots-driven nanomaterials offer a feasible direction and approach for anti-tumour therapy. For instance, photodynamic therapy has been reported as an intriguing alternative to the treatment for malignant tumours (Liu et al., 2020b; Pandey et al., 2025). Nevertheless, there are still hurdles in devising a vehicle that allows photosensitizers to be released continuously for long-term treatment. Recently, Wang et al. have come up with a peroxide-responsive injectable hydrogel formulation fabricated over PVA-based polythiophene quaternary ammonium Pdots as a photosensitizer and covalently crosslinked by another polymer. The hydrogel progressively degrades and releases Pdots when stimulated with hydrogen peroxide. The in vitro analysis notably revealed Pdots-mediated singlet oxygen production capacity, while the in vivo data depicted at least 7 days of retention of Pdots after a single dose, which established an effective tumour cell suppression with no signs of biotoxicity. The developed Pdots enabled the composite local drug delivery system to suggest a unique strategy for long-term photodynamic therapy for cancer therapy (J. Wang et al., 2024). Another emerging methodology, the lysosomal membrane permeabilization (LMP) technique, is gaining popularity as a promising chemotherapeutic approach. However, the LMP-induced non-specific cytotoxicity is still problematic, though, and it's yet unknown if lysosomal damage has an impact on the lysosomal-dependent autophagy process. In an attempt, Lei & coworkers presented a lysosomal-focused nano-strategy based on carbonized Pdots, following covalent attachment with cathepsin D-specific substrate peptide (CPds-PP) that induces LMP and disrupts autophagy degradation to mediate selective cytotoxicity to breast cancer cells. Importantly, at a specific pH, CPds-PP were produced with pH-dependent protonation and cathepsin D-responsive surface positive charge accumulation properties. Further, spatially and temporally regulated lysosomal-specific photodamage greatly increased the selective cytotoxicity of CPds-PP-induced LMP on adhering tumour cells and tumour spheres. This approach shows potential for the use of lysosome-targeting nanocarriers in drug delivery systems and anti-cancer pro-drugs (Lei et al., 2024).

Plexiform neurofibromas, a category of non-intrusive but complex tumours that typically make resection troublesome due to their hypervascularization, have been shown an opportunity to turn into malignant growths, advancing the risk of fatality. Customized procedures, like chemo-photothermal image-guided interventions, provide more efficient tailored choices when surgical removal is not feasible. Under imaging guidance, CDs may be employed as theranostic agents to regionally radiate heat tumours by virtue of their red-to-near-infrared illumination and efficient photothermal conversion in the physiologically transparent window. In this regard, Macchiaiolo et al. etched a drug-loaded CNDs-based nanoplatform with the functionalization of an N-doped graphitic core with PEG 2000 chains bearing biotin; this combination is focused as a tumour-targeting agent. Additionally, the PEGylated shell was supplemented with sirolimus as an mTOR inhibitor to target neurofibromas. Because of its near-infrared sensitivity, the drug was delivered in a regulated, localized way upon laser stimulation. Both in vitro and in 3D neurofibromas models, replicating the intricate tumour microenvironment, these NPs have strong colloidal and fluorescent stability, effective photothermal conversion, and selective tumour targeting, making them a promising theranostic agent for precise neurofibromas treatment (Macchiaiolo et al., 2025). Another distinct exploratory work focusing on Pdots is the nanodots-dispersed pH-responsive hydrogel that combines PAA and PEG polymers, encased with quercetin, was described by Pourmadadi et al., having the intent of enhancing quercetin's performance and bioavailability. The Pdots- driven nanoplatform's shape and dimensions were tweaked using the twofold emulsion technique. Remarkably higher drug loading and encapsulation efficiencies (~50% and ~ 90%, respectively) were attained via intermolecular interactions associated with PAA and PEG polymers and the developed CNDs. Indeed, research findings showed that the drug-loaded nanocarrier manifested far greater lethality than the free drug and increased the apoptotic cell destruction owing to the nanocomposite's parts linked up to inhibit cancerous cells from survival (Pourmadadi et al., 2024). A few more reports demonstrating Pdots' use for efficient drug delivery are illustrated in Table 8.3.

8.3.4 Other Carbon Nanodots

Prospective nanosystems for personalized therapies rely on CNDs and other ultra-small fluorescent NPs. Their ideal configuration enables them to seep into complex tissue habitats, enabling drug delivery, imaging, and monitoring. Surface-engineered CDs, for instance, when doped with heteroatoms, have enormous potential for drug molecule nanocarriers, antioxidant capabilities, and fluorescence detection. Likewise, Cillari & colleagues adopted an exclusive nanosystem by conjugating N, S-doped CDs with indoximod in an intuitive yet efficient manner. Following its administration, CDs significantly reduced the in vivo dissemination of breast cancer, consequently reducing IDO expression while attracting T, NKT, and NK cells. The study thus showcases the avenues for combining indoximod-CDs mediated immunotherapy with inherent pharmacological properties, offering a workable strategy for pinpointing breast cancer treatment (Cillari et al., 2025). A prevalent cancer risk that could elevate the likelihood of death from the disease is thrombosis. So, adopting heparinized CNDs as a basis, Duan et al. designed a doxorubicin-loaded hyaluronic acid-responsive drug delivery system that certainly boosted its plasma tolerability. With hyaluronic acid and an acidic environment as simultaneous stimuli, the dual-responsive system revealed ~66% release rate together with efficiently targeting the malignant cells, therefore preventing their proliferation and spread. Additionally, CNDs demonstrated outstanding drug tracking and cell imaging capabilities while revealing no adverse side effects on healthy cells. The fluorescent CNDs-based

Table 8.3 Several graphene nanodot systems used in drug delivery applications

S. No.	Polymeric nanodots	Active(s)	Significance	Reference
1.	CD/β-cyclodextrin	Pentacene	An innovative, multifunctional nanosystem for site-specific photothermal therapy and bioimaging in malignant cells	Maugeri et al. (2025)
2.	CD/ε-polylysine	Cd31	The devised CND-framework endures exemplary wound repair qualities, extensive antibacterial capacity, and excellent biological suitability	Mou et al. (2022)
3.	CD/PP-α	–	The fabricated CNDs had shown strong clinical promise for PDT-based therapy in bettering tumour inhibition efficacy and having minimal toxicity	Sajjad et al. (2022)
4.	CD/PEI-TCPP	–	The CNDs have proven dynamic control and a ratiometric pH responsiveness with pH-dependent production of ROS species in cancerous tissues	Hu et al. (2021)
5.	CD/PEI	Doxorubicin	CDs successfully suppress tumours by increasing the apoptotic incidence and the more substantial concentration of active compounds in hepatic cancer cells	Hailing et al. (2020)
6.	CD/PEG	Doxorubicin	The made CD-hybrids suggested dual chemo- and photothermal curative behaviour, contributing to the inhibition of ~80% of the MCF-7 cells	Ardekani et al. (2017)
7.	CD/DNA-hybrid hydrogel	Doxorubicin	CD-made hydrogel observed to be stable at physiological pH, while a sustained drug release in the tumour milieu	Singh et al. (2017)
8.	CD/chitosan-based hydrogel	5-fluorouracil	The designed nanotheranostic framework displayed continuous fluorescence and pH-sensitive drug dispersion coupled with an upsurge in the pro-apoptotic gene expression in lung cancer	Sachdev et al. (2016)

delivery system offers potential uses in the fields of evident drug tracking and tumour therapy (Duan et al., 2020).

The extensive build-up of β-amyloid (Aβ) peptides as clusters happened to be one of the main hallmarks of the brain in Alzheimer's disease. While Aβ peptide clumps it aggravate neuropathic and cognitive decline, and many attempts have already been made to limit Aβ self-assemblage as a possible therapeutic approach (Manchanda et al., 2023). Nevertheless, it is predestined that CNDs have been a desirable photonic nanoagent for in vivo photomodulation given their exceptional features when viewed alongside molecular dyes. They might additionally be able to build up ROS levels employing an electron transfer mechanism to denature target biotoxins, such as Aβ-complexes. In this instance, Chung et al. unveiled an aptamer-functionalized Aβ-specific red-light-responsive Pdots and their curative benefits as a light-powered nanomodulator that may reduce detrimental Aβ aggregation, both in vitro and in vivo spatiotemporally. Additionally, these nanostructures, under red radiation exposure, inhibited the formation of Aβ assemblies rich in β-sheets and attenuated the associated cytotoxicity. The in vivo CNDs-mediated photomodulation paradigm results in a notable decrease in the Aβ deposition, pointing out the potential therapeutic utility of CDs for ameliorating light-induced Aβ self-assembly in tandem with neurotoxicity (Chung et al., 2020). Additionally, CDs can be engineered to be intrinsically nano-cytotoxic to cancer cells, avoiding multidrug susceptibility associated with conventional drugs. In this regard, a chitosan-adorned folic acid-derived doxorubicin-loaded CDs were devised by Ghosh et al., which tend to be ideally suited for imaging-guided therapy for cancer due to their exceptional stability and amazing fluorescence properties. Their remarkable cytocompatibility with healthy cells, significant toxicity toward cancer cells, and pH-responsive attributes make them outstanding options for targeted cytotoxicity. Notably, upregulation of folate-receptor-mediated endocytosis drives cell uptake that underlines the exceptional lethality and selectivity of forged CDs toward cancerous tissues. The published findings imply that such CD-nanoprobes are outstanding tools for image-guided chemotherapy and pH-responsive drug release (Ghosh et al., 2025). A different investigation undertaken by Musso et al. presented a novel instance of 5-fluorouracil entrapped glucose-based CNDs, utilizing the green chemistry principles that showcased a high photothermal conversion efficiency, good photoluminescence quantum yield, and low cytotoxicity. While interacting with the overexpressed GLUT2, the developed nanocarriers revealed a specific GLUT2-mediated cell internalization. The amazing cell-targeting capacity and outstanding light-triggered cell-damaging activity of these CNDs were further validated by in vitro analysis. The developed CNDs are implied to be a revolutionary strategy in devising new nanostructures for cancer targeting owing to their sustainable synthesis and biocompatibility alongside an efficient and targeted cytotoxic effect (Musso et al., 2025).

8.4 Toxicological Landscape of Carbon Nanodots

Better functionality, an exceptional cargo-loading ability, and stronger light-responsive traits, including photoluminescence and photothermal effects, have been rendered likely by the emergence of carbon-derived nanomaterials, all of which far better line up with the peculiar delivery rate for targeted interventions. In addition, the continued advancement of nanotechnology across all fields, and particularly the development of nanomedicines, continues to emerge. Given the advancements gained thus far, there is still so much room for nanotoxicology to take root that it furnishes the rationale for the usage of nanomaterials in drug delivery (Chan et al., 2021). Accordingly, CNDs, which typically are seen as innocuous, are less cytotoxic; yet, depending upon their dosage, dimension, specific carbon-donor, mode of administration, laboratory setup, cell variety, and CND's exterior moieties, they may occasionally display lethality (Sivasankarapillai et al., 2020). However, the association among supramolecular ties within the bio-nano layout, while the resulting bio-effects remain the primary obstacle in toxicological research, provided the chemistry and distinctive features of nanomaterials outlined here, plus how they behave pertaining to bodily systems occurring in varying levels of complexity (Makhado et al., 2024). Natural precursors, including bamboo, the stem, and the roots, were used to fabricate many heteroatom-doped CNDs. Likewise, in the MTT assay, hayata and pineapple roots show over 80% cellular survival towards HacaT, MCF-7, and HT-29 cell lines at 500 μg/ml, implying that the latter are naturally compatible yet not significantly deleterious (Chen et al., 2021). In addition, Deng et al. demonstrated that CNDs made from broccoli water extract were more biologically compatible, less lethal and less genotoxic in the zebrafish model than CNDs made from organic solvents. It is noteworthy that CNDs' genotoxicity and cytotoxicity have a direct correlation with dimension and a clear association with their exposure. For instance, compared to larger CNDs, smaller sizes showed greater brain infiltration and a longer clearance time (Deng et al., 2023). Likewise, Havrdová et al. mention that CNDs up to 400 μg/ml did not drastically impact cell viability or genomic framework. Nevertheless, greater doses might trigger alterations in the architecture of cells and cell death (Havrdová et al., 2021). According to studies undertaken by Li et al., fluorescent nanomaterials smaller than 5 nm possess the ability to penetrate the blood-brain barrier and concentrate in different human organs. Further studies showcased that in vitro tests of saliva, gastric, and duodenal bile juice quenched and cleared CNDs. However, it has been reported that the naturally derived CDs exhibited a smaller size than CNDs derived from synthetic precursors, and hence, further detailed study about brain accumulation of these CNDs, along with the clearance mechanism from the brain, is required to evaluate their further safe use as the nanovehicles (Li et al., 2018). Nevertheless, despite all these challenges, researchers are still making significant progress. Since carbon nanostructures are at the forefront of nanotechnology owing to a multitude of cutting-edge events and their intriguing features that may find practical use, it is important to investigate their safety to both people and the surroundings. To put it simply, empirical studies on the cell toxicity, genetic damage, biological dispersion, and the fate of CNDs are indispensable, particularly if they are to be used for drug delivery (Parvin et al., 2024; Saifi et al., 2018). Moreover, the physicochemical functions of CNDs improve, and their toxicity is decreased by mitigating their surface characteristics with different dopants and/or functional moieties and making CNDs from reliable and organic precursors; however, quite a bit of work awaits to create comprehensive safety and regulatory standards for using CNDs in drug delivery applications for commercial use.

8.5 Major Stumbling Blocks

CNDs have garnered significant attention as nanocarriers in drug delivery owing to their biocompatibility, tunable photoluminescence, and versatile surface chemistry (Kaurav et al., 2023; Sameer et al., 2023). However, several critical barriers hinder their clinical translation and broader use. Toxicity concerns, poor biodegradability, and uncertainties regarding long-term biosafety remain significant challenges, with comprehensive preclinical studies recommended to evaluate their pharmacokinetics and immune interactions before clinical use (Debnath et al., 2025; Qi et al., 2023; Verma et al., 2019). The scalability of CND production is also impeded by heterogeneity in synthesis methods, leading to variable size, surface properties, and photoluminescence, which complicate reproducibility and quality control (Mohammed et al., 2023; Roszkowski & Durczynska, 2024). Bottom-up synthesis routes, while cost-effective, often yield products with impurities, structural defects, and inconsistent physicochemical properties, partly due to insufficient purification and incomplete characterization (Pundi & Chang, 2022; Verma et al., 2019). Further technical challenges include low drug loading capacity, poor stability, and difficulties in achieving precise control over drug release kinetics and tumour penetration (Mohan, 2023; Roszkowski & Durczynska, 2024). Additionally, regulatory hurdles exist given the necessity for thorough toxicological assessments and standardized characterisation methods to establish robust structure–property relationships (Pundi & Chang, 2022). Despite these challenges, ongoing research into surface engineering, hybrid nanostructures, and stimulus-responsive systems aims to enhance biocompatibility, targeting efficiency, and therapeutic efficacy of CND-based drug delivery systems across diverse applications, including oncology, ocular therapy, and infectious diseases (Bagheri et al., 2022; Kaurav et al., 2023; Roszkowski & Durczynska, 2024).

8.6 Conclusion and Future Perspectives

The convergence of fundamental studies with practical applications in the arena of Carbon nanodots (CNDs) is a groundbreaking development in nanotechnology that might revolutionize an array of pathologies. Carbon nanodots (CNDs) are emerging as a powerful tool in nanotechnology, with the potential to transform diagnosis and treatment across many diseases. Their high biocompatibility, stability, and unique optical and electrical properties make them highly attractive for use in imaging, drug delivery, photodynamic therapy, and other biomedical applications.

Although laboratory studies have shown impressive results, the main challenge is turning these findings into large-scale, affordable, and commercially viable products. Success will depend on strong collaboration between researchers, industry, and regulatory bodies to develop reliable synthesis methods, ensure safety, and meet quality standards. If these challenges are addressed, CNDs could significantly reduce reliance on toxic dyes and heavy-metal-based agents, offering safer and more sustainable options. Their adoption could bring major health, economic, and societal benefits, driving innovation not only in healthcare but also in other industries.

With ongoing research and coordinated efforts, CNDs are well-positioned to lead a new era of nanomaterial-based solutions that are safe, accessible, and impactful.

References

Abid, N., Khan, A. M., Shujait, S., Chaudhary, K., Ikram, M., Imran, M., Haider, J., Khan, M., Khan, Q., & Maqbool, M. (2022). Synthesis of nanomaterials using various top-down and bottom-up approaches, influencing factors, advantages, and disadvantages: A review. *Advances in Colloid and Interface Science, 300*, 102597. https://doi.org/10.1016/j.cis.2021.102597

Aditya Khatokar, J., Vinay, N., Sanjay, B., Bhargava, S., Sudhir Bale, A., Kolekar, T. R., Singh, S., Umarani, S., & Huddar, S. A. (2021). Carbon nanodots: Chemiluminescence, fluorescence and photoluminescence properties. *Materials Today: Proceedings, 43*, 3928–3931. https://doi.org/10.1016/j.matpr.2021.02.582

Alibrahem, W., Helu, N. K., Oláh, C., & Prokisch, J. (2025). Potential of carbon Nanodots (CNDs) in cancer treatment. *Nanomaterials (Basel, Switzerland), 15*(7), 560. https://doi.org/10.3390/nano15070560

Allahyari, S., Trotta, F., Valizadeh, H., Jelvehgari, M., & Zakeri-Milani, P. (2019). Cyclodextrin-based nanosponges as promising carriers for active agents. *Expert Opinion on Drug Delivery, 16*(5), 467–479. https://doi.org/10.1080/17425247.2019.1591365

Almoalim, H., Alkhaldi, M. Y., Alenaizan, A., & Jalilov, A. S. (2023). Electron donor, acceptor, and concentration-dependent quenching of Photoluminescent carbon Nanodots. *The Journal of Physical Chemistry C, 127*(26), 12640–12647. https://doi.org/10.1021/acs.jpcc.3c02460

Anindita, F., Naani, L., Permana, D., Karelius, K., & Kedang, Y. I. (2024). Study on utilizing oil palm empty fruit bunch waste into carbon Nanodots. *Journal of The Indonesian Society of Integrated Chemistry, 16*(2), 140–146. https://doi.org/10.22437/jisic.v16i2.38583

Arcudi, F., Đorđević, L., & Prato, M. (2016). Synthesis, separation, and characterization of small and highly fluorescent nitrogen-doped carbon NanoDots. *Angewandte Chemie International Edition, 55*(6), 2107–2112. https://doi.org/10.1002/anie.201510158

Arcudi, F., Đorđević, L., & Prato, M. (2019). Design, synthesis, and functionalization strategies of tailored carbon Nanodots. *Accounts of Chemical Research, 52*(8), 2070–2079. https://doi.org/10.1021/acs.accounts.9b00249

Ardekani, S. M., Dehghani, A., Hassan, M., Kianinia, M., Aharonovich, I., & Gomes, V. G. (2017). Two-photon excitation triggers combined chemo-photothermal therapy via doped carbon nanohybrid dots for effective breast cancer treatment. *Chemical Engineering Journal, 330*, 651–662.

Arvapalli, D. M., Sheardy, A. T., Allado, K., Chevva, H., Yin, Z., & Wei, J. (2020). Design of curcumin loaded carbon nanodots delivery system: Enhanced bioavailability, release kinetics, and anticancer activity. *ACS Applied Bio Materials, 3*(12), 8776–8785. https://doi.org/10.1021/acsabm.0c01144

Azami, M., Wei, J., Valizadehderakhshan, M., Jayapalan, A., Ayodele, O. O., & Nowlin, K. (2023). Effect of doping heteroatoms on the optical behaviors and radical scavenging properties of carbon Nanodots. *The Journal of Physical Chemistry C, 127*(15), 7360–7370. https://doi.org/10.1021/acs.jpcc.3c00953

Bagheri, B., Surwase, S. S., Lee, S. S., Park, H., Faraji Rad, Z., Trevaskis, N. L., & Kim, Y.-C. (2022). Carbon-based nanostructures for cancer therapy and drug delivery applications. *Journal of Materials Chemistry B, 10*(48), 9944–9967. https://doi.org/10.1039/D2TB01741E

Banger, A., Gautam, S., Jadoun, S., Jangid, N. K., Srivastava, A., Pulidindi, I. N., Dwivedi, J., & Srivastava, M. (2023). Synthetic methods and applications of carbon nanodots. *Catalysts, 13*(5), 858. https://doi.org/10.3390/catal13050858

Bartkowski, M., Zhou, Y., Nabil Amin Mustafa, M., Eustace, A. J., & Giordani, S. (2024). Carbon dots: Bioimaging and anticancer drug delivery. *Chemistry—A European Journal, 30*(19), e202303982. https://doi.org/10.1002/chem.202303982

Chan, M.-H., Chen, B.-G., Ngo, L. T., Huang, W.-T., Li, C.-H., Liu, R.-S., & Hsiao, M. (2021). Natural carbon nanodots: toxicity assessment and theranostic biological application. *Pharmaceutics, 13*(11), 1874. https://doi.org/10.3390/pharmaceutics13111874

Chen, Y.-Y., Jiang, W.-P., Chen, H.-L., Huang, H.-C., Huang, G.-J., Chiang, H.-M., Chang, C.-C., Huang, C.-L., & Juang, T.-Y. (2021). Cytotoxicity and cell imaging of six types of carbon nanodots prepared through carbonization and hydrothermal processing of natural plant materials. *RSC Advances, 11*(27), 16661–16674. https://doi.org/10.1039/d1ra01318a

Chung, Y. J., Lee, C. H., Lim, J., Jang, J., Kang, H., & Park, C. B. (2020). Photomodulating carbon dots for spatiotemporal suppression of Alzheimer's β-amyloid aggregation. *ACS Nano, 14*(12), 16973–16983. https://doi.org/10.1021/acsnano.0c06078

Cillari, R., Acúrcio, R. C., Barateiro, A., Florindo, H. F., Mauro, N., & Cavallaro, G. (2025). Harnessing sulfur-doped carbon nanodots conjugated with IDO inhibitors act as a dual-mode breast cancer immunotherapy. *Journal of Controlled Release, 381*, 113575. https://doi.org/10.1016/j.jconrel.2025.02.071

Cohen, E. N., Kondiah, P. P. D., Choonara, Y. E., Du Toit, L. C., & Pillay, V. (2020). Carbon dots as nanotherapeutics for biomedical application. *Current Pharmaceutical Design, 26*(19), 2207–2221. https://doi.org/10.2174/1381612826666200402102308

Corbella, C., Portal, S., Zolotukhin, D. B., Martinez, L., Lin, L., Kundrapu, M. N., & Keidar, M. (2019). Pulsed anodic arc discharge for the synthesis of carbon nanomaterials. *Plasma Sources Science and Technology, 28*(4), 045016. https://doi.org/10.1088/1361-6595/ab123c

Cui, L., Ren, X., Sun, M., Liu, H., & Xia, L. (2021). Carbon dots: Synthesis, properties and applications. *Nanomaterials, 11*(12), 3419. https://doi.org/10.3390/nano11123419

Cutrim, E. S. M., Vale, A. A. M., Manzani, D., Barud, H. S., Rodríguez-Castellón, E., Santos, A. P. S. A., & Alcântara, A. C. S. (2021). Preparation, characterization and in vitro anticancer performance of nanoconjugate based on carbon quantum dots and 5-fluorouracil. *Materials Science & Engineering. C, Materials for Biological Applications, 120*, 111781. https://doi.org/10.1016/j.msec.2020.111781

Cutroneo, M., Malinsky, P., Slepicka, P., & Torrisi, L. (2024). Blue laser for production of carbon dots. *Polymers, 16*(19), 2801. https://doi.org/10.3390/polym16192801

De Boëver, R., Town, J. R., Li, X., & Claverie, J. P. (2022). Carbon dots for carbon dummies: The quantum and the molecular questions among some others. *Chemistry—A European Journal, 28*(47), e202200748. https://doi.org/10.1002/chem.202200748

Debnath, R., Ikbal, A. M. A., Ravi, N. K., Kargarzadeh, H., Palit, P., & Thomas, S. (2025). Carbon nanodots-based polymer nanocomposite: A potential drug delivery armament of phytopharmaceuticals. *Polymers, 17*(3), 365. https://doi.org/10.3390/polym17030365

Debnath, R., Singh, A., Saini, A., Seni, K., Sharma, A., Bisht, D. S., Sharma, K., Goel, H., Chawla, V., & Chawla, P. A. (2024). Cutting edge Nanoplatforms with smart bio-sensing applications: Paving the way for sustainable green approaches. *Current Nanomaterials*. https://doi.org/10.2174/0124054615342051241002061227

Deng, W., Zang, C., Li, Q., Sun, B., Mei, X., Bai, L., Shang, X., Deng, Y., Xiao, Y., Ghiladi, R. A., Lorimer, G. H., Zhang, X., & Wang, J. (2023). Hydrothermally derived green carbon dots from broccoli water extracts: Decreased toxicity, enhanced free-radical scavenging, and anti-inflammatory performance. *ACS Biomaterials Science & Engineering, 9*(3), 1307–1319. https://doi.org/10.1021/acsbiomaterials.2c01537

Dhamodharan, D., Byun, H.-S., Varsha Shree, M., Veeman, D., Natrayan, L., & Stalin, B. (2022). Carbon nanodots: Synthesis, mechanisms for bio-electrical applications. *Journal of Industrial and Engineering Chemistry, 110*, 68–83. https://doi.org/10.1016/j.jiec.2022.03.014

Ding, H., Wei, J., Zhang, P., Zhou, Z., Gao, Q., & Xiong, H. (2018). Solvent-controlled synthesis of highly luminescent carbon dots with a wide color gamut and narrowed emission peak widths. *Small, 14*(22), 1800612. https://doi.org/10.1002/smll.201800612

Ding, H., Zhang, F., Zhao, C., Lv, Y., Ma, G., Wei, W., & Tian, Z. (2017). Beyond a carrier: Graphene quantum dots as a probe for programmatically monitoring anti-cancer drug delivery, release, and response. *ACS Applied Materials & Interfaces, 9*(33), 27396–27401. https://doi.org/10.1021/acsami.7b08824

Dong, J., Wang, K., Sun, L., Sun, B., Yang, M., Chen, H., Wang, Y., Sun, J., & Dong, L. (2018). Application of graphene quantum dots for simultaneous fluorescence imaging and tumor-targeted drug delivery. *Sensors and Actuators B: Chemical, 256*, 616–623. https://doi.org/10.1016/j.snb.2017.09.200

Đorđević, L., Arcudi, F., & Prato, M. (2019). Preparation, functionalization and characterization of engineered carbon nanodots. *Nature Protocols, 14*(10), 2931–2953. https://doi.org/10.1038/s41596-019-0207-x

Duan, Q., Ma, L., Zhang, B., Zhang, Y., Li, X., Wang, T., Zhang, W., Li, Y., & Sang, S. (2020). Construction and application of targeted drug delivery system based on hyaluronic acid and heparin functionalised carbon dots. *Colloids and Surfaces. B, Biointerfaces, 188*, 110768. https://doi.org/10.1016/j.colsurfb.2019.110768

Dudek, M., Rosowski, A., Koperkiewicz, A., Grobelny, J., Wach, R., Sharp, M., French, P., Janasz, L., & Kozanecki, M. (2017). Carbon nanoparticles fabricated by infrared laser ablation of graphite and polycrystalline diamond targets: Carbon nanoparticles fabricated by infrared laser ablation of graphite. *Physica Status Solidi (a), 214*(2), 1600318. https://doi.org/10.1002/pssa.201600318

Dutta, S. D., Hexiu, J., Kim, J., Sarkar, S., Mondal, J., An, J. M., Lee, Y., Moniruzzaman, M., & Lim, K.-T. (2022). Two-photon excitable membrane targeting polyphenolic carbon dots for long-term imaging and pH-responsive chemotherapeutic drug delivery for synergistic tumor therapy. *Biomaterials Science, 10*(7), 1680–1696. https://doi.org/10.1039/D1BM01832A

Dwitya, S. S., Hsueh, Y. H., Wang, S. S., & Lin, K. S. (2023). Ultrafine nitrogen-doped graphene quantum dot structure and antibacterial activities against Bacillus subtilis 3610. *Materials Chemistry and Physics, 295*, 127135.

Emam, H. E., & Ahmed, H. B. (2021). Antitumor/antiviral carbon quantum dots based on carrageenan and pullulan. *International Journal of Biological Macromolecules, 170*, 688–700. https://doi.org/10.1016/j.ijbiomac.2020.12.151

Fauziyah, N. A. (2022). A brief review of C-dots preparation using top-down and bottom-up approaches. *International Journal of Eco-Innovation in Science and Engineering, 3*(01), 25–29. https://doi.org/10.33005/ijeise.v3i01.54

Giordano, P., Valentina, P., Gabriele, C., Marcello, D., Jacopo, G., Aniello, P., Monica, R., Daniela, F., Chiara, P., Federica, T., Wanda, L., Ornella, P., Marco, D. S., & Massimiliano, P. (2020). Graphene quantum dots' surface chemistry modulates the sensitivity of glioblastoma cells to chemotherapeutics. *International Journal of Molecular Sciences, 21*(17), 6301. https://doi.org/10.3390/ijms21176301

Ghahremani, H., Jafar Molaei, M., & Salimi, E. (2025). Dextran-derived carbon quantum dots for drug delivery and food preservation. *Diamond and Related Materials, 158*, 112615.

Ghosh, T., Mandal, M., Shee, M., Das, T. K., Mandal, M., Banerji, P., & Das, N. C. (2025). Fabrication of folic acid-derived carbon dot-conjugated chitosan nanospheres as theragnostic agents for pH-responsive anticancer drug delivery. *ACS Applied Bio Materials, 8*(4), 3096–3110. https://doi.org/10.1021/acsabm.4c01962

Gomez, I. J., Arnaiz, B., Cacioppo, M., Arcudi, F., & Prato, M. (2018). Nitrogen-doped carbon nanodots for bioimaging and delivery of paclitaxel. *Journal of Materials Chemistry B, 6*(35), 5540–5548. https://doi.org/10.1039/C8TB01796D

Goel, H., Arora, P., Riya, Chawla, V., Singla, R., & Chawla, P. A. (2024). Stratagems for breast cancer management in the era of nanotheranostics. In M. A. Barkat, F. J. Ahmad, M. A. Rahman, & M. A. Ansari (Eds.), *Nanotheranostics for diagnosis and therapy*. Springer. https://doi.org/10.1007/978-981-97-3115-2_6

Goel, H., Saini, K., Razdan, K., Khurana, R. K., Elkordy, A. A., & Singh, K. K. (2022a). Chapter 3—In vitro physicochemical characterization of nanocarriers: A road to optimization. In P. Kesharwani & K. K. Singh (Eds.), *Nanoparticle therapeutics* (pp. 133–179). Academic.

Goel, H., Siddiqui, L., Mahtab, A., & Talegaonkar, S. (2022b). Fabrication design, process technologies, and convolutions in the scale-up of nanotherapeutic delivery systems. In *Nanoparticle therapeutics* (pp. 47–131). Elsevier.

Goel, H., Razdan, K., Singla, R., Talegaonkar, S., et al (2020). Engineered site-specific vesicular systems for colonic delivery: Trends and implications. *Current Pharmaceutical Design, 26*(42), 5441–5455. https://doi.org/10.2174/1381612826666200813132301

Hailing, Y., Xiufang, L., Lili, W., Baoqiang, L., Kaichen, H., Yongquan, H., Qianqian, Z., Chaoming, M., Xiaoshuai, R., Rui, Z., Hui, L., Pengfei, P., & Hong, S. (2020). Doxorubicin-loaded fluorescent carbon dots with PEI passivation as a drug delivery system for cancer therapy. *Nanoscale, 12*(33), 17222–17237. https://doi.org/10.1039/d0nr01236j

Havrdová, M., Urbančič, I., Bartoň Tománková, K., Malina, L., Štrancar, J., & Bourlinos, A. B. (2021). Self-targeting of carbon dots into the cell nucleus: Diverse mechanisms of toxicity in NIH/3T3 and L929 cells. *International Journal of Molecular Sciences, 22*(11), 5608. https://doi.org/10.3390/ijms22115608

He, C., Xu, P., Zhang, X., & Long, W. (2022). The synthetic strategies, photoluminescence mechanisms and promising applications of carbon dots: Current state and future perspective. *Carbon, 186*, 91–127. https://doi.org/10.1016/j.carbon.2021.10.002

Hu, X., Zhang, Q., Dai, X., Sun, J., & Gao, F. (2021). Dual-emission carbonized polymer dots for Ratiometric pH sensing, pH-dependent generation of singlet oxygen, and imaging-guided dynamics monitoring of photodynamic therapy. *ACS Applied Bio Materials, 4*(10), 7663–7672. https://doi.org/10.1021/acsabm.1c00892

Hu, R., Shi, J., Xie, C.-M., & Yao, X.-L. (2025). Dry eye disease: Oxidative stress on ocular surface and cutting-edge antioxidants. *Global Challenges (Hoboken, NJ), 9*(7), e00068. https://doi.org/10.1002/gch2.202500068

Huang, L., Song, J., & Chen, B. (2014). A novel targeting drug carrier to deliver chemical bonded and physical entrapped anti-tumor drugs. *International Journal of Pharmaceutics, 466*(1–2), 52–57. https://doi.org/10.1016/j.ijpharm.2014.03.006

Huang, L., Gao, H., Wang, Z., Zhong, Y., Hao, L., & Du, Z. (2021). Combination Nanotherapeutics for dry eye disease treatment in a rabbit model. *International Journal of Nanomedicine, 16*, 3613–3631. https://doi.org/10.2147/IJN.S301717

Huo, Y., Xiu, S., Meng, L.-Y., & Quan, B. (2023). Solvothermal synthesis and applications of micro/nano carbons: A review. *Chemical Engineering Journal, 451*, 138572. https://doi.org/10.1016/j.cej.2022.138572

Jiao, Y., Guo, Y., Fan, Y., Wang, R., Li, X., Wu, H., Meng, Z., Yang, X., Cui, Y., Liu, H., Pan, L., Maimaitijuma, T., Zhang, J., Wang, Y., Cao, Y., & Zhang, T. (2020). Triggering of apoptosis in osteosarcoma 143B cell line by carbon quantum dots via the mitochondrial apoptotic signal pathway. *BioMed Research International, 2020*, 2846297. https://doi.org/10.1155/2020/2846297

Karpushkin, E. A., Kharochkina, E. S., Lopatina, L. I., & Sergeev, V. G. (2024). Carbon Nanodots: Preparation, properties, application (A review). *Журнал Общей Химии, 94*(1), 136–164. https://doi.org/10.31857/S0044460X24010122

Kaurav, H., Verma, D., Bansal, A., Kapoor, D. N., & Sheth, S. (2023). Progress in drug delivery and diagnostic applications of carbon dots: A systematic review. *Frontiers in Chemistry, 11*, 1227843. https://doi.org/10.3389/fchem.2023.1227843

Khan, M. S., Sheikh, A., Abourehab, M. A. S., Gupta, N., & Kesharwani, P. (2023). Understanding the theranostic potential of quantum dots in cancer management. *Materials Today Communications, 36*, 106424. https://doi.org/10.1016/j.mtcomm.2023.106424

Khodadadei, F., Safarian, S., & Ghanbari, N. (2017). Methotrexate-loaded nitrogen-doped graphene quantum dots nanocarriers as an efficient anticancer drug delivery system. *Materials Science & Engineering. C, Materials for Biological Applications, 79*, 280–285. https://doi.org/10.1016/j.msec.2017.05.049

Kong, J., Wei, Y., Zhou, F., Shi, L., Zhao, S., Wan, M., & Zhang, X. (2024). Carbon quantum dots: Properties, preparation, and applications. *Molecules, 29*(9), 2002. https://doi.org/10.3390/molecules29092002

Kumar, R., Kumar, V. B., & Gedanken, A. (2020). Sonochemical synthesis of carbon dots, mechanism, effect of parameters, and catalytic, energy, biomedical and tissue engineering applications. *Ultrasonics Sonochemistry, 64*, 105009. https://doi.org/10.1016/j.ultsonch.2020.105009

Lagos, K. J., García, D., Cuadrado, C. F., De Souza, L. M., Mezzacappo, N. F., Da Silva, A. P., Inada, N., Bagnato, V., & Romero, M. P. (2023). Carbon dots: Types, preparation, and their boosted antibacterial activity by photoactivation. Current status and future perspectives. *Wiley Interdisciplinary Reviews. Nanomedicine and Nanobiotechnology, 15*(4), e1887. https://doi.org/10.1002/wnan.1887

Lei, Y., Tian, B.-H., Li, X.-X., Sun, M.-Y., Guo, X.-L., Wang, Y.-D., Zhou, H.-Q., Ma, R.-S., & Liang, H.-X. (2024). Surface charge accumulation of functionalized carbonized polymer dots selectively induces lysosomal membrane permeabilization of breast cancer cells. *Chemical Engineering Journal, 494*, 152710. https://doi.org/10.1016/j.cej.2024.152710

Li, S., Jiang, C., Wang, H., Cong, S., & Tan, M. (2018). Fluorescent nanoparticles present in Coca-Cola and Pepsi-cola: Physiochemical properties, cytotoxicity, biodistribution and digestion studies. *Nanotoxicology, 12*(1), 49–62. https://doi.org/10.1080/17435390.2017.1418443

Li, X., Yu, L., He, M., Chen, C., Yu, Z., Jiang, S., Wang, Y., Li, L., Li, B., Wang, G., Shen, A., & Fan, J. (2023). Review on carbon dots: Synthesis and application in biology field. *BMEMat, 1*(4), e12045. https://doi.org/10.1002/bmm2.12045

Lin, J., Liu, R., Li, X., Zhao, Y., Kong, L., Li, Y., Zhu, S., & Wang, L. (2024). Cobalt-doped carbon nanodots as an interfacial modifier for efficient solar-driven water splitting. *Applied Catalysis B: Environmental, 343*, 123544.

Liang, Y.-C., Gou, S.-S., Liu, K.-K., Wu, W.-J., Guo, C.-Z., Lu, S.-Y., Zang, J.-H., Wu, X.-Y., Lou, Q., Dong, L., Gao, Y.-F., & Shan, C.-X. (2020). Ultralong and efficient phosphorescence from silica confined carbon nanodots in aqueous solution. *Nano Today, 34*, 100900. https://doi.org/10.1016/j.nantod.2020.100900

Liu, C., Chang, Q., Fan, X., Meng, N., Lu, J., Shu, Q., Xie, Y., Celia, C., Wei, G., & Deng, X. (2025). Rational construction of CQDs-based targeted multifunctional nanoplatform for synergistic chemo-photothermal tumor therapy. *Journal of Colloid and Interface Science, 677*, 79–90. https://doi.org/10.1016/j.jcis.2024.08.037

Liu, H., Lv, X., Li, C., Qian, Y., Wang, X., Hu, L., Wang, Y., Lin, W., & Wang, H. (2020a). Direct carbonization of organic solvents toward graphene quantum dots. *Nanoscale, 12*(20), 10956–10963. https://doi.org/10.1039/D0NR01903H

Liu, J., Li, R., & Yang, B. (2020b). Carbon dots: A new type of carbon-based nanomaterial with wide applications. *ACS Central Science, 6*(12), 2179–2195. https://doi.org/10.1021/acscentsci.0c01306

Liu, M. (2020). Optical properties of carbon dots: A review. *Nano, 1*(1), 1–12. https://doi.org/10.37256/nat.112020124.1-12

Lohani, A., Durgapal, S., & Morganti, P. (2023). 9 – Quantum dots: An emerging implication of nanotechnology in cancer diagnosis and therapy. In M. Henini & M. O. Rodrigues (Eds.), *Quantum materials, devices, and applications* (pp. 243–262). Elsevier. https://doi.org/10.1016/B978-0-12-820566-2.00008-9

Macchiaiolo, M., Cillari, R., Gonfiantini, M. V., Cacchione, A., Mastronuzzi, A., Cavallaro, G., Vecchio, D., & Mauro, N. (2025). Reduction of tumor burden via precise photothermal release of sirolimus by carbon nanodots in 3D plexiform neurofibroma-in-a-dish models. *Surfaces and Interfaces, 59*, 105978. https://doi.org/10.1016/j.surfin.2025.105978

Madhukar, P., Kesarwani, R., Pandey, S., Singh, V. K., Rao Gedda, M., Singh, O. P., Shaz, M., Kumar, R., & Sundar, S. (2025). Amine-functionalized graphene quantum dots conjugated with amphotericin B: Synthesis, characterization, and in vitro evaluation for visceral Leishmaniasis treatment. *ACS Omega, 10*(25), 26627–26638. https://doi.org/10.1021/acsomega.5c00879

Makhado, B. P., Oladipo, A. O., Gumbi, N. N., De Kock, L. A., Andraos, C., Gulumian, M., & Nxumalo, E. N. (2024). Unravelling the toxicity of carbon nanomaterials—From cellular interactions to mechanistic understanding. *Toxicology In Vitro, 100*, 105898. https://doi.org/10.1016/j.tiv.2024.105898

Manchanda, N., Aggarwal, A., Setya, S., & Talegaonkar, S. (2023). Digital intervention for the management of Alzheimer's disease. *Current Alzheimer Research.* https://doi.org/10.2174/1567205020666230206124155

Mansuriya, B. D., & Altintas, Z. (2021). Carbon dots: Classification, properties, synthesis, characterization, and applications in health care—An updated review (2018–2021). *Nanomaterials, 11*(10), 2525. https://doi.org/10.3390/nano11102525

Mao, Q.-X., Han, L., Shu, Y., Chen, X.-W., & Wang, J.-H. (2016). Improving the biocompatibility of carbon nanodots for cell imaging. *Talanta, 161*, 54–61. https://doi.org/10.1016/j.talanta.2016.08.031

Marpongahtun, G., S., Muis, Y., Andriayani, Novita, T., & Piliang, A. F. (2018). Synthesis of carbon Nanodots from cellulose nanocrystals oil palm empty fruit by pyrolysis method. *Journal of Physics: Conference Series, 1120*, 012071. https://doi.org/10.1088/1742-6596/1120/1/012071

Mas, N., Hueso, J. L., Martinez, G., Madrid, A., Mallada, R., Ortega-Liebana, M. C., Bueno-Alejo, C., & Santamaria, J. (2020). Laser-driven direct synthesis of carbon nanodots and application as sensitizers for visible-light photocatalysis. *Carbon, 156*, 453–462. https://doi.org/10.1016/j.carbon.2019.09.073

Maugeri, L., Fangano, G., Butera, E., Forte, G., Bonacci, P. G., Musso, N., Ruffino, F., Ferreri, L., Consoli, G. M. L., & Petralia, S. (2025). Emissive Pentacene-loaded βcyclodextrin-derived C-Nanodots exhibit red-light triggered Photothermal effect. *Pharmaceutics., 17*(5), 543. https://doi.org/10.3390/pharmaceutics17050543

Milenković, M., Jovanović, S., Marković, Z., Ciasca, G., Di Santo, R., Mead, J. L., Mojsin, M., Dojčinović, B., Milivojević, D., & Marković, B. T. (2025). Graphene quantum dots enhanced with gold nanoparticles for advanced antibacterial applications. *Journal of Drug Delivery Science and Technology, 111*, 107207. https://doi.org/10.1016/j.jddst.2025.107207

Mohammed, J., Desu, P. K., Namratha, J. R., & Rao, G. K. (2023). Applications of carbon dots (CDs) in drug delivery. *Advances in Pharmacology and Pharmacy, 11*(1), 36–45. https://doi.org/10.13189/app.2023.110104

Mohan, V. (2023). Recent advances of carbon dots for treatment of Alzheimer's and Parkinson's diseases. *International Journal of Advanced Nano Computing and Analytics., 2*(1), 39–59. https://doi.org/10.61797/ijanca.v2i1.240

Mou, Z., Gao, Z., & Hu, Y. (2022). Orange emissive carbon dots for fluorescent determination of hypoxanthine in fish. *Spectrochimica Acta. Part A, Molecular and Biomolecular Spectroscopy, 269*, 120734. https://doi.org/10.1016/j.saa.2021.120734

Musso, N., Bonacci, P. G., Letizia Consoli, G. M., Maugeri, L., Terrana, M., Lanzanò, L., Longo, E., Buscarino, G., Consoli, A., & Petralia, S. (2025). Biofriendly glucose-derived carbon nanodots: GLUT2-mediated cell internalization for an efficient targeted drug delivery and light-triggered cancer cell damage. *Journal of Colloid and Interface Science, 696*, 137873. https://doi.org/10.1016/j.jcis.2025.137873

Najafi, M., Khoddam, Z., Masnavi, M., Pourmadadi, M., & Abdouss, M. (2024a). Physicochemical and in vitro characterization of agarose based nanocarriers incorporated with graphene quantum dots/ α-Fe2O3 for targeted drug delivery of quercetin to liver cancer treatment. *Materials Chemistry and Physics, 320*, 129333. https://doi.org/10.1016/j.matchemphys.2024.129333

Najafi, M., Pourmadadi, M., Abdous, M., Rahdar, A., & Pandey, S. (2024b). Formulation of double nanoemulsions based on pH-sensitive carboxymethyl cellulose /starch copper doped carbon quantum dots for quercetin controlled release. *Journal of Molecular Liquids, 400*, 124543. https://doi.org/10.1016/j.molliq.2024.124543

Nezhadhossein, B., Afshar, A. S., Homayouni Tabrizi, M., Nematpour, F. S., & Mohtashami, M. (2025). Efficacy of urolithin B-loaded mesoporous silica nanoparticles modified with graphene quantum dots against MCF-7 breast cancer cells. *Scientific Reports, 15*(1), 25473. https://doi.org/10.1038/s41598-025-10904-3

Nguyen, D.-K., Le, T.-S., Le, Q.-T., & Mai, X.-D. (2023). The roles of intermediate fluorophores on the optical properties of bottom-up synthesized carbon nanodots. *HPU2 Journal of Science:*

Natural Sciences and Technology, 2(2), 68–82. https://doi.org/10.56764/hpu2.jos.2023.2.2.68-82

Nguyen, V., Zhao, N., Yan, L., Zhong, P., Nguyen, V. C., & Le, P. H. (2020). Double-pulse femtosecond laser ablation for synthesis of ultrasmall carbon nanodots. *Materials Research Express, 7*(1), 015606. https://doi.org/10.1088/2053-1591/ab6124

Nicosia, A., Mineo, P., Micali, N., & Villari, V. (2023). Dopamine-coated carbon nanodots: A supramolecular approach to polydopamine composite. *International Journal of Molecular Sciences, 24*(20), 15384. https://doi.org/10.3390/ijms242015384

Ordoukhani, P., Pourmadadi, M., & Abdouss, M. (2025). Development and assessment of polyacrylamide-starch hydrogel nanocomposites with graphene quantum dots for targeted quercetin delivery to brain tumors. *Carbohydrate Polymer Technologies and Applications, 10*, 100848.

Ostovar, S., Pourmadadi, M., Shamsabadipour, A., & Mashayekh, P. (2023). Nanocomposite of chitosan/gelatin/carbon quantum dots as a biocompatible and efficient nanocarrier for improving the curcumin delivery restrictions to treat brain cancer. *International Journal of Biological Macromolecules, 242*(Pt 3), 124986. https://doi.org/10.1016/j.ijbiomac.2023.124986

Pandey, M., Pradhan, N., Abu Elella, M. H., Makhado, E., Samy, M., Botlhoko, O. J., & Allouss, D. (2025). Frontiers in polymer/carbon dots nanocomposites: Recent advances in synthesis and diverse applications. *Surfaces and Interfaces, 60*, 106101. https://doi.org/10.1016/j.surfin.2025.106101

Papaioannou, N., Marinovic, A., Yoshizawa, N., Goode, A. E., Fay, M., Khlobystov, A., Titirici, M.-M., & Sapelkin, A. (2018). Structure and solvents effects on the optical properties of sugar-derived carbon nanodots. *Scientific Reports, 8*(1), 6559. https://doi.org/10.1038/s41598-018-25012-8

Parvin, N., Kumar, V., Joo, S. W., & Mandal, T. K. (2024). Emerging trends in nanomedicine: Carbon-based nanomaterials for healthcare. *Nanomaterials (Basel, Switzerland), 14*(13), 1085. https://doi.org/10.3390/nano14131085

Perini, G., Palmieri, V., Friggeri, G., Augello, A., De Spirito, M., & Papi, M. (2023). Carboxylated graphene quantum dots-mediated photothermal therapy enhances drug-membrane permeability, ROS production, and the immune system recruitment on 3D glioblastoma models. *Cancer Nanotechnology, 14*(1), 13. https://doi.org/10.1186/s12645-023-00168-9

Pourmadadi, M., Tajiki, A., Abdouss, M., Beig Mohammadi, A., Kharaba, Z., Rahdar, A., & Díez-Pascual, A. M. (2024). Novel carbon quantum dots incorporated polyacrylic acid/polyethylene glycol pH-sensitive nanoplatform for drug delivery. *Inorganic Chemistry Communications, 159*, 111814. https://doi.org/10.1016/j.inoche.2023.111814

Pundi, A., & Chang, C.-J. (2022). Recent advances in synthesis, modification, characterization, and applications of carbon dots. *Polymers, 14*(11), 2153. https://doi.org/10.3390/polym14112153

Qi, B.-P., Bao, L., Zhang, Z.-L., & Pang, D.-W. (2016). Electrochemical methods to study Photoluminescent carbon Nanodots: Preparation, photoluminescence mechanism and sensing. *ACS Applied Materials & Interfaces, 8*(42), 28372–28382. https://doi.org/10.1021/acsami.5b11551

Qi, J., Zhang, R., Liu, X., Liu, Y., Zhang, Q., Cheng, H., Li, R., Wang, L., Wu, X., & Li, B. (2023). Carbon dots as advanced drug-delivery nanoplatforms for antiinflammatory, antibacterial, and anticancer applications: A review. *ACS Applied Nano Materials, 6*(11), 9071–9084. https://doi.org/10.1021/acsanm.3c01207

Quaid, T., Ghalandari, V., & Reza, T. (2022). Effect of synthesis process, synthesis temperature, and reaction time on chemical, morphological, and quantum properties of carbon dots derived from loblolly pine. *Biomass, 2*(4), 250–263. https://doi.org/10.3390/biomass2040017

Qureshi, Z. A., Dabash, H., Ponnamma, D., & Abbas, M. K. G. (2024). Carbon dots as versatile nanomaterials in sensing and imaging: Efficiency and beyond. *Heliyon, 10*(11), e31634. https://doi.org/10.1016/j.heliyon.2024.e31634

Rai, M., Ingle, A. P., Törős, G., & Prokisch, J. (2024). Assessing the efficacy of carbon nanodots derived from curcumin on infectious diseases. *Expert Review of Anti-Infective Therapy, 22*(12), 1107–1121. https://doi.org/10.1080/14787210.2024.2409401

Rigodanza, F., Burian, M., Arcudi, F., Đorđević, L., Amenitsch, H., & Prato, M. (2021). Snapshots into carbon dots formation through a combined spectroscopic approach. *Nature Communications, 12*(1), 2640. https://doi.org/10.1038/s41467-021-22902-w

Roszkowski, S., & Durczynska, Z. (2024). Advantages and limitations of nanostructures for biomedical applications. *Advances in Clinical and Experimental Medicine, 34*(3), 447–456. https://doi.org/10.17219/acem/186846

Ru, Y., Waterhouse, G. I. N., & Lu, S. (2022). Aggregation in carbon dots: Special issue: Emerging investigators. *Aggregate, 3*(6), e296. https://doi.org/10.1002/agt2.296

Sachdev, A., Matai, I., & Gopinath, P. (2016). Carbon dots incorporated polymeric hydrogels as multifunctional platform for imaging and induction of apoptosis in lung cancer cells. *Colloids and Surfaces. B, Biointerfaces, 141*, 242–252. https://doi.org/10.1016/j.colsurfb.2016.01.043

Saifi, M. A., Khan, W., & Godugu, C. (2018). Cytotoxicity of nanomaterials: Using nanotoxicology to address the safety concerns of nanoparticles. *Pharmaceutical Nanotechnology, 6*(1), 3–16. https://doi.org/10.2174/2211738505666171023152928

Sajjad, F., Han, Y., Bao, L., Yan, Y., Shea, D. O., Wang, L., & Chen, Z. (2022). The improvement of biocompatibility by incorporating porphyrins into carbon dots with photodynamic effects and pH sensitivities. *Journal of Biomaterials Applications, 36*(8), 1378–1389. https://doi.org/10.1177/08853282211050449

Sameer, M., Arif, Y., Aqil, A., Nadaf, A., Rafiya, K., Hasan, N., Kesharwani, P., & Ahmad, F. J. (2023). Carbon nanodots as a remedial nanovesicles for drug delivery. *European Polymer Journal, 200*, 112515. https://doi.org/10.1016/j.eurpolymj.2023.112515

Sciortino, A., Cannizzo, A., & Messina, F. (2018). Carbon Nanodots: A review—From the current understanding of the fundamental Photophysics to the full control of the optical response. *C, 4*(4), 67. https://doi.org/10.3390/c4040067

Seyyedi Zadeh, E., Ghanbari, N., Salehi, Z., Derakhti, S., Amoabediny, G., Akbari, M., & Tokmedash, M. A. (2023). Smart pH-responsive magnetic graphene quantum dots nanocarriers for anticancer drug delivery of curcumin. *Materials Chemistry and Physics, 297*, 127336.

Sharma, A., & Das, J. (2019). Small molecules derived carbon dots: Synthesis and applications in sensing, catalysis, imaging, and biomedicine. *Journal of Nanobiotechnology, 17*(1), 92. https://doi.org/10.1186/s12951-019-0525-8

Sheng, Y., Dai, W., Gao, J., Li, H., Tan, W., Wang, J., Deng, L., & Kong, Y. (2020). pH-sensitive drug delivery based on chitosan wrapped graphene quantum dots with enhanced fluorescent stability. *Materials Science & Engineering. C, Materials for Biological Applications, 112*, 110888. https://doi.org/10.1016/j.msec.2020.110888

Shi, W., Han, Q., Wu, J., Ji, C., Zhou, Y., Li, S., Gao, L., Leblanc, R. M., & Peng, Z. (2022). Synthesis mechanisms, structural models, and Photothermal therapy applications of top-down carbon dots from carbon powder, graphite, graphene, and carbon nanotubes. *International Journal of Molecular Sciences, 23*(3), 1456. https://doi.org/10.3390/ijms23031456

Singh, S., Mishra, A., Kumari, R., Sinha, K. K., Singh, M. K., & Das, P. (2017). Carbon dots assisted formation of DNA hydrogel for sustained release of drug. *Carbon, 114*, 169–176. https://doi.org/10.1016/j.carbon.2017.12.020

Sivasankarapillai, V. S., Kirthi, A. V., Akksadha, M., Indu, S., Dharshini, U. D., Pushpamalar, J., & Karthik, L. (2020). Recent advancements in the applications of carbon nanodots: Exploring the rising star of nanotechnology. *Nanoscale Advances, 2*(5), 1760–1773. https://doi.org/10.1039/C9NA00794F

Soledad-Flores, O., Bailón-Ruiz, S. J., & Román-Velázquez, F. (2024). Rapid synthesis of non-toxic, water-stable carbon dots using microwave irradiation. *The Micropaleontologist, 4*(4), 659–669. https://doi.org/10.3390/micro4040040

Soumya, K., More, N., Choppadandi, M., Aishwarya, D A., Singh, G., & Kapusetti, G. (2023). A comprehensive review on carbon quantum dots as an effective photosensitizer and drug delivery system for cancer treatment. *Biomedical Technology, 4*, 11–20. https://doi.org/10.1016/j.bmt.2023.01.005

Sun, Y.-P., Zhou, B., Lin, Y., Wang, W., Fernando, K. A. S., Pathak, P., Meziani, M. J., Harruff, B. A., Wang, X., Wang, H., Luo, P. G., Yang, H., Kose, M. E., Chen, B., Veca, L. M., & Xie, S.-Y. (2006). Quantum-sized carbon dots for bright and colorful photoluminescence. *Journal of the American Chemical Society, 128*(24), 7756–7757. https://doi.org/10.1021/ja062677d

Swift, T. A., Oliver, T. A. A., Galan, M. C., & Whitney, H. M. (2019). Functional nanomaterials to augment photosynthesis: Evidence and considerations for their responsible use in agricultural applications. *Interface Focus, 9*(1), 20180048. https://doi.org/10.1098/rsfs.2018.0048

Thakur, M., Mewada, A., Pandey, S., Bhori, M., Singh, K., Sharon, M., & Sharon, M. (2016). Milk-derived multi-fluorescent graphene quantum dot-based cancer theranostic system. *Materials Science & Engineering. C, Materials for Biological Applications, 67*, 468–477. https://doi.org/10.1016/j.msec.2016.05.007

Timerkaev, B. A., Shakirov, B. R., Galieva, A. L., Vafina, A. G., Kaleeva, A. A., & Yalaltdinova, G. D. (2019). Synthesis of carbon nanostructures in electric discharge. *Journal of Physics: Conference Series, 1328*(1), 012039. https://doi.org/10.1088/1742-6596/1328/1/012039

Tiwari, P., Shukla, R. P., Yadav, K., Singh, N., Marwaha, D., Gautam, S., Bakshi, A. K., Rai, N., Kumar, A., Sharma, D., & Mishra, P. R. (2024). Dacarbazine-primed carbon quantum dots coated with breast cancer cell-derived exosomes for improved breast cancer therapy. *Journal of Controlled Release: Official Journal of the Controlled Release Society, 365*, 43–59. https://doi.org/10.1016/j.jconrel.2023.11.005

Modan, E. M., & Plăiaşu, A. G. (2020). Advantages and disadvantages of chemical methods in the elaboration of nanomaterials. The Annals of "Dunarea de Jos" University of Galati, Fascicle IX: Metallurgy and Materials Science, 43(1), 53–60. https://doi.org/10.35219/mms.2020.1.08

Verma, N. C., Yadav, A., & Nandi, C. K. (2019). Paving the path to the future of carbogenic nanodots. *Nature Communications, 10*(1), 2391. https://doi.org/10.1038/s41467-019-10394-8

Wang, J., Liang, K., Li, J., Zhang, Y.-X., Xue, X.-K., Chen, T.-J., Hao, Y.-L., Wu, J.-S., & Ge, J.-C. (2024). H_2O_2-responsive injectable polymer dots hydrogel for long-term photodynamic therapy of tumors. *Chinese Journal of Polymer Science, 42*(11), 1690–1698. https://doi.org/10.1007/s10118-024-3155-z

Wang, L., Zhang, X., Yang, K., Wang, L., & Lee, C.-S. (2020). Oxygen/nitrogen-related surface states controlled carbon nanodots with tunable full-color luminescence: Mechanism and bio-imaging. *Carbon, 160*, 298–306. https://doi.org/10.1016/j.carbon.2020.01.029

Wu, Z., Xia, W., Ou, L., Zheng, L., Hou, B., Pan, T., Sun, W., Koole, L. H., Shao, Y., & Qi, L. (2024). Utilization of nitrogen-doped graphene quantum dots to neutralize ROS and modulate intracellular antioxidant pathways to improve dry eye disease therapy. *International Journal of Nanomedicine, 19*, 2691–2708. https://doi.org/10.2147/IJN.S445398

Xia, C., Zhu, S., Feng, T., Yang, M., & Yang, B. (2019). Evolution and synthesis of carbon dots: From carbon dots to carbonized polymer dots. *Advanced Science, 6*(23), 1901316. https://doi.org/10.1002/advs.201901316

Xiao, L., & Sun, H. (2018). Novel properties and applications of carbon nanodots. *Nanoscale Horizons, 3*(6), 565–597. https://doi.org/10.1039/C8NH00106E

Xiao, Y., Wang, Z., Fu, J., Zhang, J., He, Q., Lu, H., Zhou, Q., & Wang, H. (2025). Recent advances in the synthesis, characterization, and application of carbon dots in the field of wastewater treatment: A comprehensive review. *Water, 17*(2), 210. https://doi.org/10.3390/w17020210

Xu, X., Ray, R., Gu, Y., Ploehn, H. J., Gearheart, L., Raker, K., & Scrivens, W. A. (2004). Electrophoretic analysis and purification of fluorescent single-walled carbon nanotube fragments. *Journal of the American Chemical Society, 126*(40), 12736–12737. https://doi.org/10.1021/ja040082h

Xue, Z., Sun, Q., Zhang, L., Kang, Z., Liang, L., Wang, Q., & Shen, J. W. (2019). Graphene quantum dot assisted translocation of drugs into a cell membrane. *Nanoscale, 11*(10), 4503–4514. https://doi.org/10.1039/c8nr10091h

Yang, L., Wang, Z., Wang, J., Jiang, W., Jiang, X., Bai, Z., He, Y., Jiang, J., Wang, D., & Yang, L. (2016). Doxorubicin conjugated functionalizable carbon dots for nucleus targeted delivery and enhanced therapeutic efficacy. *Nanoscale, 8*(12), 6801–6809. https://doi.org/10.1039/c6nr00247a

Zahid, M. U., Pervaiz, E., Hussain, A., Shahzad, M. I., & Niazi, M. B. K. (2018). Synthesis of carbon nanomaterials from different pyrolysis techniques: A review. *Materials Research Express, 5*(5), 052002. https://doi.org/10.1088/2053-1591/aac05b

Zaidi, Z., Maiti, N., Ali, M. I., Sharma, G., Moin, S., Padhy, H., Balaji, G. L., & Sundramurthy, V. P. (2022). Fabrication, characteristics, and therapeutic applications of carbon-based Nanodots. *Journal of Nanomaterials, 2022*(1), 8031495. https://doi.org/10.1155/2022/8031495

Zeng, Q., Feng, T., Tao, S., Zhu, S., & Yang, B. (2021). Precursor-dependent structural diversity in luminescent carbonized polymer dots (CPDs): The nomenclature. *Light: Science & Applications, 10*(1), 142. https://doi.org/10.1038/s41377-021-00579-6

Zhang, B., Liu, C., & Liu, Y. (2010). A novel one-step approach to synthesize fluorescent carbon nanoparticles. *European Journal of Inorganic Chemistry, 2010*(28), 4411–4414. https://doi.org/10.1002/ejic.201000622

Zhou, J., Booker, C., Li, R., Zhou, X., Sham, T.-K., Sun, X., & Ding, Z. (2007). An electrochemical avenue to blue luminescent nanocrystals from multiwalled carbon nanotubes (MWCNTs). *Journal of the American Chemical Society, 129*(4), 744–745. https://doi.org/10.1021/ja0669070

Zhou, Y., Zhang, W., & Leblanc, R. M. (2022). Structure–property–activity relationships in carbon dots. *The Journal of Physical Chemistry B, 126*(51), 10777–10796. https://doi.org/10.1021/acs.jpcb.2c06856

Zhu, S., Meng, Q., Wang, L., Zhang, J., Song, Y., Jin, H., Zhang, K., Sun, H., Wang, H., & Yang, B. (2013). Highly photoluminescent carbon dots for multicolor patterning, sensors, and bioimaging. *Angewandte Chemie, 125*(14), 4045–4049. https://doi.org/10.1002/ange.201300519

Zhuo, S., Shao, M., & Lee, S.-T. (2012). Upconversion and downconversion fluorescent graphene quantum dots: Ultrasonic preparation and photocatalysis. *ACS Nano, 6*(2), 1059–1064. https://doi.org/10.1021/nn2040395

Zmejkoski, D. Z., Zdravković, N. M., Mitić, D. D., Marković, Z. M., Budimir Filimonović, M. D., Milivojević, D. D., & Todorović Marković, B. M. (2025). Graphene quantum dots in bacterial cellulose hydrogels for visible light-activated antibiofilm and angiogenesis in infection management. *International Journal of Molecular Sciences, 26*(3), Article 3. https://doi.org/10.3390/ijms26031053

Carbon Nanodots in the Therapy of Infectious Diseases

9

Rahul Bhagat, Arun Ghuge, Harris Panakkal, Avinash Ingle, Mahendra Rai, and Indarchand Gupta

Abstract

Carbon dots are being considered to be novel nanomaterials showing promising antimicrobial applications against bacterial, fungal and viral pathogens. While interacting with microorganisms, they disrupt the cell membrane, induce oxidative stress and interfere with other microbial cellular processes. They also show the synergistic effects when used in combination with antibiotics and antiviral agents. It enhances their efficacy and shows promising results against multidrug resistance. The synergistic effects appear to be promising and novel strategies to disarm the drug-resistant microbes. They hold the potential to combat the broad spectrum of microbes, but on the other hand, toxicity remains a crucial factor for their translation into clinical practice. Although many studies advocate low toxicity of CNDs, concerns regarding their long-term effects and biocompatibility require attention. Therefore, optimisation and modifications of CNDs to minimize the potential harmful effects will ensure their safe applications. Considering their potential, the present chapter aims to discuss the antimicrobial potential of CNDs against microbes, including those that have developed resistance. The chapter further discusses their synergistic effects in combination with antibiotics and

R. Bhagat · H. Panakkal · I. Gupta (✉)
Department of Biotechnology, Government Institute of Science, Chhatrapati Sambhajinagar, Maharashtra, India

A. Ghuge
Department of Forensic Biology, Government Institute of Forensic Science, Nagpur, Maharashtra, India

A. Ingle
Biotechnology Centre, Department of Agricultural Botany, Dr. Panjabrao Deshmukh Agricultural University, Akola, Maharashtra, India

M. Rai
Department of Chemistry and Pharmacy, The Federal University of Piaui (UFPI), Teresina, Brazil

M. Rai et al. (eds.), *Carbon Nanodots*, https://doi.org/10.1007/978-981-95-4705-0_9

antiviral agent, applications in photodynamic therapy and diagnosis, along with their potential harmful effects.

Keywords

Carbon Nanodots · Synergistic effects · MDR · Antimicrobial effects · Toxicity

9.1 Introduction

Infectious diseases are defined as illnesses or disorders commonly caused by various pathogenic microorganisms, like bacteria, viruses, parasites, and fungi (Prabhu, 2023). Moreover, these can be spread directly or indirectly (vector-borne) from one individual to another. Although groundbreaking research and recent advances in terms of the discovery of several antimicrobial agents have been made for the treatment of infectious diseases over the last century, millions of deaths are still happening around the globe due to diseases like pneumonia, flu, tuberculosis, HIV/AIDS, malaria, etc. According to the estimation of the World Health Organization (WHO) and the Centre for Infectious Disease Research (CIDR), the mortality rates of many infectious diseases may have worsened over the past few decades. Therefore, such infectious diseases are a major burden for public healthcare systems. Moreover, concerns about such diseases have been further exacerbated by the emergence of new and more virulent strains of multidrug-resistant (MDR) pathogens (Tripathi et al., 2019; Alara & Alara, 2024; Mondol et al., 2025). It has been proposed that the rise in MDR by pathogens is mainly due to the continuous decline in the discovery and development of new antibiotics. Moreover, according to Dong et al. (2020), by 2050, the acceleration of antibiotic resistance could result in 300 million additional deaths and incur an additional US$100 trillion in costs.

These concerns created pressure for the search for innovative treatment strategies for infectious diseases. In this context, nanotechnology has emerged as a promising frontier in medicine. To date, several nanomaterials are being used in the treatment of various infectious diseases; however, among these nanomaterials, carbon nanodots (CNDs) have attracted a great deal of attention due to their unique physicochemical properties, including small size, biocompatibility, low toxicity, tunable surface functionalities, and intrinsic fluorescence. CNDs are a new type of novel and emerging bioactive nanomaterials of less than 10 nm in size. Based on their unique features, CNDs have been demonstrated in a wide range of applications in biomedicine. Initially, their role was investigated only for bioimaging and sensing applications; however, recently, they have been demonstrated as promising antimicrobial and antiviral agents, opening new avenues for their application in the diagnosis, treatment, and prevention of infectious diseases Guo et al. (2022).

This chapter aims to provide an in-depth overview of recent advances in the use of CNDs in the treatment of infectious diseases. Moreover, their mechanisms of antimicrobial action and roles in combating multidrug-resistant pathogens have also been discussed. In addition, focus has also been given to its toxicological concerns and clinical potential in infectious diseases.

9.2 Applications of CNDs Against Bacterial, Fungal, and Viral Pathogens

CNDs are carbon-based fluorescent nanomaterials having antibacterial, antifungal, and antiviral applications because of their small size, high surface area, biocompatibility, surface functionality, photoluminescence, and reactive oxygen species. Significantly, CNDs, because of their ultra-small particle size, exhibit the characteristic of rapid renal metabolism (in vivo), which reduces the risk of long-term toxicity. CNDs apply antibacterial, antifungal, and antiviral effects (Fig. 9.1) through various modes of action.

They can generate reactive oxygen species (ROS) such as singlet ions or superoxide anion by UV-Visible light irradiation, which ultimately leads to the damage of bacterial membranes and nucleic acid material, along with proteins. Positively charged CNDs interact with negatively charged membranes and cause bacterial cell lysis. CNDs have been used in delivering anti-inflammatory drugs directly to the site of inflammation to improve their efficacy and reduce side effects.

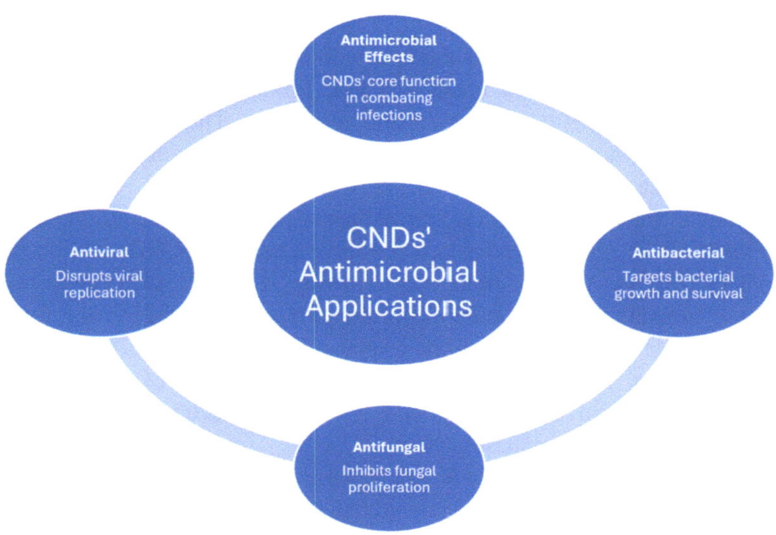

Fig. 9.1 Antimicrobial applications of CNDs

9.2.1 Antibacterial Applications

Generally, pathogenic or opportunistic pathogens are responsible for bacterial infectious diseases and are cured by antibiotic treatment. Presently, the mismanagement of antibiotics is a serious concern for multidrug resistance. The United States Centers for Disease Control and Prevention (CDC) reported that methicillin-resistant *Staphylococcus aureus* (MRSA), vancomycin-resistant *Enterococcus faecium* (VRE), and drug-resistant *Streptococcus pneumoniae* have been identified as serious public health threats. That's why the use of obsolete antibiotics for many diseases is no longer adequate for treatment and wound healing (Russell et al., 2020; Durão et al., 2018). Therefore, there is an urgent requirement for an alternative solution to combat multidrug-resistance bacterial infection and reduce the side effects.

In bacterial infectious diseases, carbon dots show high bactericidal properties by damaging the bacterial membrane upon photoactivation (Sun et al., 2024). As per the literature, hydrothermally synthesized ~6 nm CNDs with zinc ions, high fluorescence and biocompatibility are used for biofilm inhibition and broad-spectrum killing of bacteria. These ~6 nm CNDs penetrated both cell walls and biofilms of *S. aureus* and *Pseudomonas aeruginosa*, inhibited the growth and biofilm formation without toxic effect on mammalian cells. The green hydrothermal method produced Zn Cdots with ROS antioxidant activity, showing potent activity against *S. aureus* and *S. marcescens*, outperforming the raw biomass extract. The qCQDs showed antibacterial activity against Gram-positive and Gram-negative bacteria. Zhao et al. (2021) reported that the qCQD reduced the death of rats and promoted the healing of infected wounds and recovery.

9.2.2 Antifungal Applications

Fungi are the major concern in pathogenicity, the decay of agricultural material and food. Carbon nanodots are a useful material for antifungal applications across biomedicine, environmental protection, and agriculture. Their essential characteristic properties, such as surface chemistry, capacity for ROS induction, and imaging properties, make them suitable agents and drug carriers. Among all the nanomaterials, the carbon nanodots showed promising results in antifungal activity.

Sturabotti et al. (2024) synthesised the three classes of CNDs and carried out in-depth in vitro and in vivo biological screening against *C. albicans* to test their antifungal, antiadhesion, and antibiofilm formation activities. Moreover, found that positively charged nanoparticles were responsible for inhibiting the adhesion and biofilm formation, interacting with the cellular membrane of *C. albicans*, and increasing the survival of *G. mellonella* infected larvae after the injection with positively charged nanoparticles.

Fungal keratitis is an infectious eye disease that can lead to blindness. The conventional antifungal drugs have few limitations because of ocular barrier permeability. The carbon dot can penetrate the corneal barrier by opening the tight junctions, allowing it to reach the lesion site and effectively kill the fungi. Chen et al. (2024) demonstrated that in vitro and in vivo studies exhibited good biocompatibility and antifungal activity, significantly improving the therapeutic effect in a mouse model of fungal keratitis. Kim et al. (2022) reported the antifungal activities by the disk diffusion test against five fungal strains. The CNDs–ZnO showed antifungal activities against *Saccharomyces cerevisiae*, *Rhodotorula mucilaginosa*, *Aspergillus flavus*, *A. niger*, and *A. terreus*.

9.2.3 Antiviral Applications

A virus is a microscopic infectious agent that can replicate only inside a host organism. They are the inanimate particles on the Earth. A virus can infect the host by completing its attachment, penetration, replication, and budding. CNDs are the molecules responsible for stopping the viral infection at attachment, penetration, replication and budding stages (Huseen, 2020; Hussen et al., 2022; Hasan et al., 2022). Du et al. (2016) reported that the multiplication of pseudorabies virus (PRV) and porcine reproductive and respiratory syndrome virus (PRRSV) was inhibited in the porcine kidney (PK-15) cells and monkey kidney (MARC-145) cells, respectively, treated with CNDs. The inhibition was being carried out by the activation of interferon IFN-α, as Type I interferons (IFNs), including IFN-α and IFN-β, are the best known antiviral innate immune molecules (Gomez et al., 2005).

CNDs functionalised with ethylenediamine/citric acid as precursor and post-modification of CNDs with boronic acid have been reported to prevent the replication of human coronavirus. The functionalised CNDs were found to prevent the HCoV-229E entry. Herein, the surface functional groups of CNDs interacted with HCoV-229E entry receptors, thereby inhibiting their entry into the cell (Łoczechin et al., 2019; Kwee et al., 2022).

The antimicrobial applications of CNDs, their mechanism of action, and notable outcomes of various studies are summarised in Table 9.1.

9.3 Synergistic Effects of CNDs with Antibiotics and Antiviral Agents

Along with the antimicrobial applications, the CNDs also show synergistic effects when used in combination with antibiotics and antiviral agents.

9.3.1 Synergistic Effects with Antibiotics

The effect of carbon dots against various microorganisms is reported by several researchers (Lin et al., 2022; Shamkhali et al., 2025). Moreover, the carbon dots have been used in drug delivery systems. The use of antibiotics as frontline treatment for microbial infections is common practice. However, uncontrolled and prolonged use, and improper dosage could lead to the development of antibiotic resistance.

Table 9.1 Types of carbon nanodots and their effect on types of pathogens

Pathogen type	CNDs form	Key mechanisms	Notable outcome	Reference
Bacteria	CND–ZnO nanocomposite (from *Sargassum horneri*)	Membrane damage + ROS	Broad-spectrum (e.g., *B. cereus*, *S. aureus*, *E. coli*); low toxicity in zebrafish models	Kim et al. (2022)
	Kanamycin-derived CNDs (KCDs) in photo-activated hydrogel	ROS generation + biofilm disruption	Effective infected wound healing under light activation	Wang et al. (2023)
Fungi	Ultrasmall positively charged CNDs (stage-melting method)	Opening corneal tight junctions & membrane killing	Cured fungal keratitis in the mouse model	Chen et al. (2024)
	Surface-tailored CNDs (positive charge)	Inhibit adhesion + biofilm, internalization	90% survival in *C. albicans*-infected larvae	Sturabotti et al. (2024)
	N-doped CQDs (microwave-assisted) on cellulosic material	Adhesion, membrane penetration, ROS, Fe^{3+} chelation	Prevents fungal degradation of wood	Li et al. (2018), and Travlou et al. (2018)
Viruses	Polyamine-modified CNDs	Bind viral envelopes, block entry	Active in vitro against flaviviruses, WSSV	Huang et al. (2020)
	Herpes-simplex virus entry-inhibitor CNDs	Entry blockade	High efficacy against HSV-1 in cell culture	Barras et al. (2016)

Antimicrobial resistance is a crucial challenge around the globe, causing higher mortality and economic burden. Therefore, the discovery of new antimicrobials, controlled use, and combined use of multiple drugs are some of the alternative strategies employed to combat drug-resistant pathogens. Still, various compounds such as CDs, antimicrobial polymers, and antimicrobial peptides have been tested and proven effective against bacterial infections. The CDs have emerged as a promising antimicrobial agent due to their prominent advantages like low-cost production, greater biocompatibility, surface functionality, excellent physicochemical and photoluminescence properties, and reactive oxygen species (ROS) generation capability (Wang et al., 2025). The combined use of antibiotics and carbon dots is a more promising approach to control microbial pathogens. The synergistic effects of the antibiotics and carbon nanomaterials can increase therapeutic outcomes, impede resistance development, minimize required drug doses, and magnify the antimicrobial potency against diverse pathogens.

Nevertheless, the bactericidal effect of CDs can be exerted through multiple mechanisms, including disruption of bacterial membrane, by interfering and destabilizing proteins and enzymes, DNA cutting and binding, photodynamic effect, which increases generation of ROS and phototherapy, where an increase in ambient temperature causes biochemical damage (Ghirardello et al., 2021; Wang et al., 2025). The synergistic effects of the antibiotic and CDs lead to efficient membrane disruption as CDs destabilise the membrane, which causes elevated penetration of antibiotic. The limited effectiveness of antibiotics against biofilms is primarily due to protective mechanisms possessed by bacteria within biofilms. On the contrary, the combination of CDs with antibiotics significantly inhibits biofilm formation. The CDs can disrupt the biofilm matrix effectively, hence inhibiting biofilm growth. The CDs promote the synthesis of ROS, and such functionalised CDs can sensitise bacteria to antibiotics. These carbon-based nanomaterials can act as a delivery system to transport antibiotics directly to infection sites, augmenting the local concentration of antibiotics, thus increasing overall antibacterial efficacy (Wang et al., 2025). Additionally, Tiwari et al. (2021) investigated the synergistic effect of the carbon dots and tetracycline against *E. coli* and reported the considerable improvement of the antibacterial efficiency over separate treatment of either tetracycline or C-dot. The results indicate the significant inhibition of bacterial growth with six times lower tetracycline (0.25 μg/ml), along with 0.75 μg/mL C-dot concentration, than the MIC of tetracycline. A recent study by Shamkhali et al. (2025) demonstrated that application of CDs in conjunction with meropenem antibiotic resulted in stronger biocompatibility, more stability and solubility, and improved permeability of the meropenem drug as compared with the drug administered independently. Furthermore, CDs conjugated with ampicillin antibiotic have shown significant potency against *E. coli* cells. The minimum amount of AMP (14 μg mL − 1) required to prevent the growth of *E. coli* cells in the case of CDs-AMP conjugate over to free AMP (25 μg mL − 1), thus specifying the synergistic benefits of the CDs-AMP combinations (Jijie et al., 2018). The levofloxacin-based CDs could be a strong agent to control drug-resistant bacteria and to overcome the restrictions of antibiotics in treatment (Wu et al., 2022). However, Pajavand et al. (2023) investigated the effect of CDs alone and a combination of CDs plus ciprofloxacin on biofilm development. Authors reported the inhibition of biofilm-forming isolates of *P. aeruginosa* when a combination of CDs and ciprofloxacin was used, signifying the synergistic effect. Qie et al. (2022) prepared chiral biomolecule functionalized CDs (BAPTCDs) using a one-step solvothermal method. These BAPTCDs were further tested against *E. coli* and *S. aureus,* and it was found that they specifically bind with the test bacteria, showing synergistic antibacterial action with an efficiency of over 99%. The excellent photothermal properties of these CDs are known to cause bacterial cell wall breakage via a rapid temperature rise. Meanwhile, Rodriguez et al. (2025) studied the photoactivated bactericidal role of CDs and their combination with antibiotics against drug-resistant bacteria *E. faecalis* and *L. monocytogenes.* They tested the effect of CDs along with different antibiotics, including erythromycin, streptomycin and tetracycline. The findings reveal that the combination of tested antibiotics and CDs did not unexpectedly improve the comprehensive antibacterial potency of photoactive CDs against *E. faecalis*. In contrast, the CDs plus streptomycin combination considerably lessens the antibacterial potential of CDs. Henceforth, the use of different synergistic antimicrobial strategies leverages the synergistic connections among various antimicrobial agents, including CDs and antibiotics, causing improved efficiency of antibiotics even at lower amounts.

9.3.2 Synergistic Effects with Antiviral Agents

As mentioned earlier, viruses are one of the most life-threatening pathogens, causing more than 30% mortality from infectious diseases (Hansen et al., 2016). The whole world experienced the severity of the recent pandemic COVID-19, caused by SARS-CoV-2, thus infecting more than 417 million people and resulting in over six million deaths (Chen et al., 2022). Henceforth, there is an utmost need and demand for antiviral compounds to combat viral infections. The designing and development of new antiviral drugs is a time-consuming process and requires various phases before their therapeutic approval for human use. Moreover, synthesis of a new class of highly flexible drugs, long-term usable with virostatic and/or virucidal potential to supplement current pharmacological strategies against viral infection and transmission, is highly desirable. Antiviral agents are being effectively used to control viral infections, and their availability is much less compared to antibacterial agents. The main challenges associated with developing antiviral compounds are that, at higher concentrations, they may show toxicity to host cells and also targeting the virus life cycle is a very complex process. Therefore, searching for new antiviral compounds with higher efficiency and lower toxicity to the host is highly inevitable. In this regard, many plant-based metabolites, new lead molecules and nanomaterials like CDs serve as promising antiviral candidates. The nanomaterials possess unique properties like shape, size, surface areas to volume ratios, and surface electric charge, etc., responsible for their widespread usage in diverse sectors. Among the numerous nanoscale materials, CDs have been studied due to their beneficial properties and applications. The CDs have shown remarkable antiviral activity due to multiple modes of action, and they are water-soluble and noncytotoxic (Innocenzi et al., 2025). CDs can inactivate viral pathogens by joining carbon nanostructures to viral proteins, thus inhibiting virus multiplication. The combination of CDs and other antiviral compounds has shown synergistic effects with improved antiviral efficacy. Thus, indicating that the joint effect of the antiviral molecules and CDs is better than the sum of their independent effects. However, this synergistic effect results in increased antiviral activity, lowered toxicity to the host, and lower dosages of individual counterparts (da Silva Júnior et al., 2021).

The carbon quantum dots synthesised using curcumin (Cur-CQDs) have shown the effect on the viral infection cycle through binding with specific sites on the E protein of Japanese encephalitis virus (JEV). Further, the interaction of CDs with E protein prevents entry and fusion of the virus to host cells. Such inhibitory effect of CDs can be further increased by conjoining them with other antiviral drugs (Chen et al., 2022). Moreover, the efficiency of CDs in synergistic action is based on their size and shape, biocompatibility, functionalization, and surface properties. The selection of an antiviral agent that complements the mode of action of CDs and targets various life stages of the virus is crucial for proper antiviral action. The CDs conjugated with boronic acid or amine could potentially be used to treat herpes simplex virus type 1 (HSV-1) infection, as they impede the virus from entering host cells. These nanomaterials were shown to interact with HSV-1 in a dose-dependent manner with EC50 = 80 ng mL − 1 (Barras et al., 2016). Furthermore, carbon quantum dots (CQD) functionalized with 4-aminophenylboronic acid showed virucidal effect on human coronavirus with an EC50 lowered to 5.2 ± 0.7 μg/mL (Łoczechin et al., 2019). The study by Du et al. in 2018 emphasized a one-step synthesis of cationic carbon dots (CCM-CDs) using curcumin, resulting in uniform and stable nanomaterials with significant antiviral action. Hitherto, Huang et al. (2019) prepared CDs (BZM-CDs) using a series of benzoxazine monomers and tested their in vitro virus-blocking activity against flaviviruses like Zika, Dengue viruses, and Japanese encephalitis, and non-enveloped viruses such as adenovirus-correlated virus and porcine parvovirus. It was reported that BZM-CDs instantly bind to the virion surface and destroy the virus-cell interconnection. Nevertheless, the synergistic effect of CDs with traditional antiviral compounds and antibiotics embraces the potential for overcoming challenges such as limited bioavailability, antimicrobial resistance, and limited efficacy of current treatment practices.

9.4 Mechanisms of Antimicrobial Activity of CNDs

As discussed earlier, CND possesses a great antimicrobial potential. Although CNDs exhibit a broad spectrum of antimicrobial properties against various types of microbes, including bacteria, fungi, drug-resistant bacteria, multidrug-resistant bacteria, and viruses, their mechanisms of antimicrobial action differ for each of these microbes (Rai et al., 2024).

As far as bacteria are concerned, various hypothetical mechanisms for their antibacterial actions have been proposed (see Fig. 9.2).

It is well-known that the cell wall is the outermost layer of bacteria, which plays a crucial role in maintaining the cell shape. It also provides mechanical strength to the cell, inhibits damage, helps in the movement and growth of the cell, and also provides sensitivity to antibiotics. The cell wall in the case of both Gram-positive and Gram-negative bacteria usually possesses a negative charge due to the presence of different components in their cell wall (pepticoglycan and acidic polysaccharides like teichoic acid in the case of Gram-positive bacteria and peptidoglycan, porin, and lipopolysaccharide in the case of Gram-negative bacteria). However, damage to the bacterial cell wall can lead to cell disruption and death. It has been reported that CNDs generally possess a positive surface charge and are attracted towards negatively charged bacteria through electrostatic interactions. This interaction can lead to the induction of reactive oxygen species (ROS) by bacteria, causing mechanical damage to the cell membrane, destroying the integrity of the cell membrane, leading to cytoplasmic leakage, and cell apoptosis. In addition, CNDs can affect the cell wall synthesis and cell membrane electron transport. Moreover, the entry of CNDs into

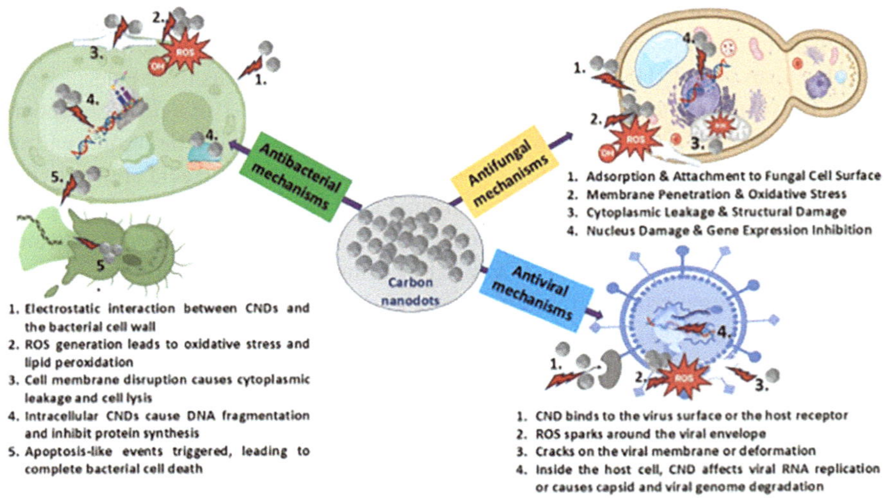

Fig. 9.2 Mechanism of antimicrobial action of CNDs

the cell can affect protein and DNA synthesis; sometimes, their interaction with DNA can cause DNA fragmentation (Bing et al., 2016).

In the case of fungi, similar mechanisms of action have been proposed, which are proposed for bacteria (see Fig. 9.2). It has been proposed that CNDs showed promising antifungal activity mainly due to their ability to disrupt fungal cell membranes and induce oxidative stress when interacting with fungal cell membranes. The unique features of CNDs, such as their small size, high surface energy, and functional groups present on their surface (in the case of engineered CNDs), enable their interaction with and penetration into fungal cells. This interaction then led to membrane damage, leakage of cellular contents, and ultimately, cell death (Sturabotti et al., 2024). Moreover, some of the studies demonstrated that sometimes CNDs can induce ROS within the cell and generate various free radicals, which also contribute to their antifungal effects by damaging various cellular components like lipids, proteins, and DNA (Liu et al., 2023). In addition, Li et al. (2018) studied the antifungal efficacy of biodegradable CDs against two plant pathogenic fungi, i.e., *Rhizoctonia solani* and *Pyricularia grisea*, and the results recorded showed that the synthesized CDs can directly enter the fungal nucleus and destroy it by inhibiting the growth of fungal cells. Moreover, in the case of *R. solani*, it was observed that CDs can significantly inhibit the expression of the non-ribosomal peptide synthase gene.

As far as the antiviral mechanisms of action are concerned, only a few studies are available. According to these studies, CNDs are reported to produce ROS that induce oxidative stress and hamper the viral functions. CNDs designed through the functionalization of species antiviral molecules on their surface can attach to the virions, leading to inhibition of their entry into the cells, or they can also attach to the cell receptor, hampering virus entry. Moreover, it was proposed that after binding to the virus, CNDs can destroy the surface membrane and capsid structure (Innocenzi et al., 2025).

9.5 CNDs' Role in Combating Multidrug-Resistant Pathogens

Management of MDR pathogens is an ever-challenging task. However, CNDs are also reported to combat the MDR bacteria, fungi, and viruses.

9.5.1 MDR Bacteria

MDR bacterial infections pose a serious global threat, with pathogens such as *S. aureus* (MRSA), *P. aeruginosa*, *Acinetobacter baumannii*, and *Klebsiella pneumoniae* evading most antibiotics. CNDs have emerged as a promising alternative, showing broad-spectrum bactericidal activity even against these "superbugs" (Lin et al., 2022). In vitro, various CND formulations exhibit potent anti-MRSA effects. For example, protein-derived CNDs achieved a low minimum inhibitory concentration (MIC) of ~37.5 μg/mL against MRSA (Lin et al., 2022). Nitrogen-doped CNDs synthesized from amino acids effectively killed MRSA and other bacteria without inducing any detectable resistance, even after 30 days of repeated exposure (Lin et al., 2022). Significantly, these nanodots retained high efficacy in vivo. In a rat wound model co-infected with MRSA and *E. coli*, CND treatment cleared the infection as effectively as the antibiotic levofloxacin, with minimal toxicity (Lin et al., 2022). Similarly, glucose-derived CNDs used on MRSA-infected wounds in rats produced healing outcomes equivalent to vancomycin yet showed negligible side effects in major organs (Lin et al., 2022). Such results underscore that CNDs can match conventional antibiotics in treating MRSA infections, a notable feat given MRSA's notorious drug resistance.

Beyond Gram-positive MRSA, CNDs have also demonstrated strong activity against multidrug-resistant (MDR) Gram-negative bacteria. A broad-spectrum CND, derived from positively charged glutathione, exhibited high bactericidal activity not only against *S. aureus* but also against *P. aeruginosa* and *E. coli*, with no resistance observed during prolonged testing (Lin et al., 2022). Another study reported "red" carbon dots (R-CDs) that could penetrate biofilms of carbapenem-resistant *A. baumannii*, effectively disrupting established biofilms and eradicating the bacteria (Lin et al., 2022). In a mouse wound infection, these R-CDs successfully treated MDR *A. baumannii* and MRSA wounds, and importantly, no microbial resistance was observed with repeated R-CD use (Lin et al., 2022). MDR *K. pneumoniae*, which often resists even last-resort drugs, has also been susceptible to CND-based approaches. For instance, hydrophobic carbon dots embedded in a polymer matrix were shown to produce bactericidal reactive oxygen species under visible light, killing *K. pneumoniae* (as well as *E. coli* and *S. aureus*) on contact surfaces without harming mammalian cells (Lin et al., 2022). This points to potential uses of CND coatings or wound dressings to control hospital-acquired infections by pathogens like *K. pneumoniae*.

Notably, CNDs can be engineered to augment or replace existing antibiotics. Levofloxacin-derived carbon dots (LCDs) preserved the drug's antibacterial motifs and exhibited an MIC of only 0.125 μg/mL against *S. aureus*, a potency higher than free levofloxacin (Lin et al., 2022). These LCDs maintained broad-spectrum activity against MRSA, *Enterococcus faecalis*, *Listeria monocytogenes*, *P. aeruginosa,* and other difficult pathogens, all while showing a low tendency to induce resistance (Lin et al., 2022). Several CND systems have also demonstrated in vivo efficacy superior to standard antibiotics. Quaternary-ammonium-functionalized carbon dots (QCQDs) achieved admirable clearance of MRSA pneumonia in mice, significantly reducing lung infection and inflammation (Lin et al., 2022). In fact, mice treated with QCQDs showed regression of MRSA-induced pneumonia on par with vancomycin therapy (Lin et al., 2022). Likewise, in a murine pneumonia model, treatment with levofloxacin-derived CNDs resulted in better infection control than levofloxacin itself, without any damage to normal tissues (Lin et al., 2022). These animal studies highlight the clinical relevance of CNDs: they not only combat MDR bacteria effectively but can do so with equal or improved outcomes compared to traditional antibiotics.

9.5.2 MDR Fungi

Carbon nanodots have also shown promise against drug-resistant fungal pathogens. In particular, *Candida auris*—an emerging MDR yeast that is resistant to multiple antifungal classes—has been a recent target. Nitrogen-doped CNDs engineered from a medicinal plant source exhibited potent fungicidal activity against several clinical isolates of *C. auris*, demonstrating significantly enhanced killing of this pathogen (Vijeata et al., 2025). CNDs can also impede fungal biofilms, which are a major resistance mechanism in Candida infections. By tailoring surface functional groups, researchers have tuned carbon dots to disrupt *C. albicans* biofilm formation and adhesion (Sturabotti et al., 2024). Notably, positively charged CNDs were able to not only inhibit *C. albicans* biofilm growth in vitro but also interact with fungal cell membranes and improve survival in an infected Galleria (wax moth) larvae model (Sturabotti et al., 2024). These findings suggest that CNDs can be designed to overcome fungal defences (like biofilms) and could serve as a new class of antifungals against organisms like *C. auris*, which are becoming increasingly resistant to all available drugs. This is of high clinical importance, as *C. auris* outbreaks are rising and current antifungal options are limited.

9.5.3 MDR Viruses

Although viruses are not typically labelled "MDR," many (such as HIV and hepatitis viruses) can evolve resistance to antiviral drugs, and novel viruses like coronaviruses

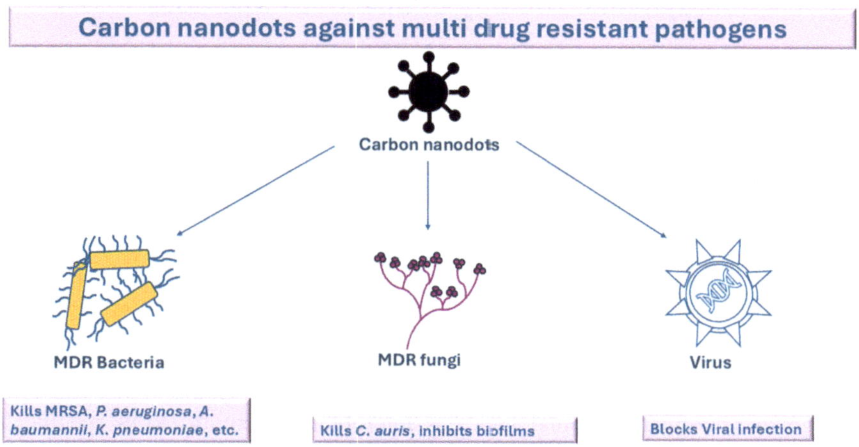

Fig. 9.3 Antimicrobial activity of carbon nanodots (CNDs) against multidrug-resistant (MDR) pathogens. CNDs exhibit broad-spectrum effects by eliminating MDR bacteria (e.g., MRSA, *P. aeruginosa*, *A. baumannii*, *K. pneumoniae*), inhibiting MDR fungi like *C. auris* and fungal biofilms, and blocking viral infections through virion inactivation or entry inhibition

present huge treatment challenges. Here, carbon nanodots have demonstrated broad antiviral capabilities that could bypass conventional resistance issues. CNDs often work by binding to viral particles or host receptors, thereby blocking the virus's life cycle. For example, boronic-acid functionalized carbon dots were shown to prevent HIV-1 infection in cell culture by attaching to the HIV envelope glycoprotein (gp120) and inhibiting the virus's entry into host cells (Kotta et al., 2020). Carbon dots have also been effective against coronaviruses: one report found that amine/boronic-acid modified CNDs completely inhibited the entry and replication of human coronavirus HCoV-229E in vitro (Kotta et al., 2020). During the COVID-19 pandemic, researchers applied CND technology to combat SARS-CoV-2. In one study, converting a porphyrin molecule into carbon dots drastically increased its antiviral performance, significantly disrupting SARS-CoV-2 propagation in vitro compared to the parent compound (Fibriani et al., 2023). In another case, curcumin-derived carbon dots displayed potent anti-enterovirus activity in vivo: when newborn mice were challenged with a lethal dose of enterovirus 71 (EV71), treatment with these carbon dots reduced viral mortality and even prevented paralysis in the infected mice (Kotta et al., 2020). Such examples illustrate the wide-ranging antiviral action of CNDs, from HIV to novel coronaviruses, achieved through mechanisms that viruses are less able to evade. This broad-spectrum activity is particularly valuable for viruses that rapidly develop drug resistance or those for which existing therapies are inadequate (Fig. 9.3).

9.6 Photodynamic Therapy of Infectious Microbes

In addition to the discussion above, bacterial infectious diseases are certainly systemic. They are treated by using various antibiotics and their misuse leads to the development of MDR. Therefore, it will be important to look for an alternative to get the desired effects and reduce or avoid the side effects. In such a situation, the photodynamic therapy seems attractive as they have comparative advantages over traditional therapies (see Fig. 9.4).

This therapy is used to treat diseases in their early or advanced stage. Use of CNDs for photodynamic therapy shows a potential approach to fight against infectious microbes, specifically in the scenario of rising antibiotic resistance. CNDs serve as efficient photosensitizers in photodynamic therapy. Upon activation by light, it induces the generation of reactive oxygen species (ROS). ROS can cause damage to microbial cells and thus has microbicidal activity. ROS specifically damage the bacterial cell membrane, leading to the leakage of cytoplasmic content, protein denaturation, DNA damage and lipid peroxidation (Paramanantham et al., 2019; Sun et al., 2024). This mechanism offers a broad-spectrum antibacterial activity with low risk of development of resistance and seems to be an important alternative to traditional antibiotics. Lin and Baba (2023) reported the enhancement of photodynamic activity of Hemin-modified CNDs by improving electron-hole separation efficiency. It provides oxygen self-supply, which is crucial in hypoxic environments typical of bacterial infections. As per Knoblauch and Geddes (2020), CNDs are an inexpensive and eco-friendly solution for antimicrobial photodynamic therapy. However, their clinical applications are still in nascent stages (Lin & Baba, 2023), posing a huge scope for further necessary research to optimize their use.

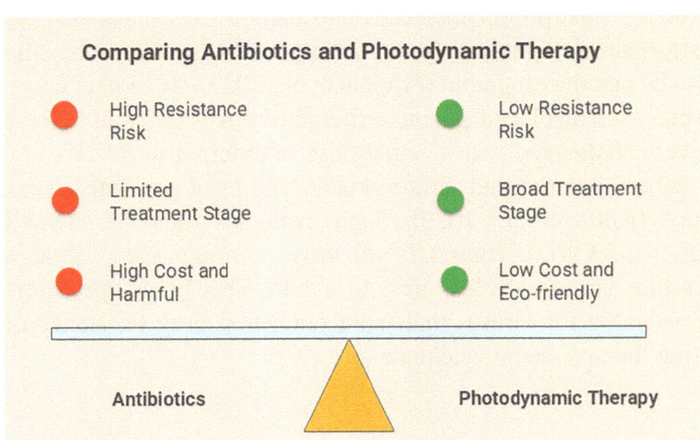

Fig. 9.4 Comparison of traditional antibiotics and photodynamic therapy

9.7 Early Detection of Microbes Using CNDs

The early diagnosis of microbes, especially to find the root cause of infectious disease, is extremely important for timely treatment. CNDs have emerged as promising candidate in the field of biosensors owing to their unique characteristics. Use of CNDs in such applications is reported to have advantages in the form of high sensitivity, cost-effectiveness, and rapidity. They appear to be valuable tools for early detection of various pathogens, including bacteria and viruses. Diagnosis of bloodstream infection is usually a time-consuming process that involves blood culturing. Blood culturing is the current standard for the diagnosis of bloodstream infection. However, the detection time for blood culture is about 1–5 days or even more. Therefore, there is a need for early and rapid diagnosis. Accordingly, Kuang et al. (2023) have synthesised cefminox sodium carbon nanodots (CS-CDs) by a rapid microwave method. It was later investigated for the detection of bacterial infection. CS-CDs were found to detect *E.coli* within a span of approximately 6 h, detecting their concentration ranging from 0.5×10^6 to 1×10^9 CFU/mL with a detection limit (LOD) of 3.7×10^5 CFU/mL. Such observations suggest the high potential of CNDs for the detection of bacterial infections.

The CNDs can employ the fluorescence-based detection of pathogens. They can be engineered to respond to specific pH changes or other biochemical signals, enabling the turn-off-on mode of pathogen detection (Cao et al., 2022). As per the study performed by Cao et al. (2022), CNDs are highly sensitive and selective. This property allows the detection of the African swine fever virus. It can detect the virus with a detection limit of 15.21 copies μL^{-1}. The CNDs' surface can be modified to enhance their binding affinity for targeting microbes. This mechanism improves the detection accuracy of CNDs for multiple pathogens simultaneously (Anand et al., 2020; Zor et al., 2022). However, due to such a promising outcome, there is a pressing need to explore the CNDs for the detection of major viral infectious agents.

9.8 Toxicity Concerns and Clinical Potential of CNDs in Infectious Diseases

CND poses significant toxicity concerns, exerting adverse effects on human and environmental health. The CND enters the human body via respiration and can pass into the bloodstream. Thus, they can potentially affect various organs (Qamar et al., 2024). CND can infiltrate the cell membrane and disturb the normal cell metabolism. It can result in the accumulation of reactive oxygen species (ROS) in the cytoplasm. The accumulation of ROS induces oxidative stress, which may lead to cell death (Paula et al., 2016; Qamar et al., 2024). CNDs can also penetrate the nucleus and thus interact with the cellular DNA. It can cause DNA damage and thus interfere with the complete metabolism of the cell (Ray et al., 2020).

When released into the environment, it may show harmful effects on different trophic levels. For example, CND has been reported to inhibit the growth of aquatic plants and show acute toxicity in *Artemia franciscana*, an aquatic organism (Kim et al., 2016). Structurally, CNDs are stable and hence, they can persist in the environment for a long period. Therefore, they have the potential for bioaccumulation and biomagnification in food chains (Kim et al., 2016; Prokisch 2024). While the toxicity of CNDs is a significant concern, it is important to consider the potential for safe design and application. Advances in nanotechnology could lead to the development of safer CNDs with reduced toxicity profiles. Additionally, ongoing research into the mechanisms of CND toxicity and their environmental fate can inform regulatory frameworks and safety guidelines, ensuring that the benefits of these materials are realized without compromising health and environmental safety.

9.9 Conclusion

CNDs are considered a new member of the nanomaterials family and hence they attract considerable attention over other types of nanomaterials due to their unique properties, which mainly include photoluminescence, easy surface modification, easy preparation, low toxicity, low side effects, and lower probability to develop resistance. Recently, much attention has been paid by the research community to the applications of carbon dots in various sectors. This nanomaterial exhibits remarkable properties, including physicochemical, thermal, and optical. Nowadays, the use of non-antibiotic antimicrobial agents, such as CNDs, is used as an effective alternative against microbial infections. This carbon-based nanomaterial is used in the medicinal field due to its significant antimicrobial and antiviral potential. Moreover, it has also been shown to be effective in combating multiple drug resistance in various pathogens. Across bacteria, fungi, and viruses, carbon nanodots have proven effective against some of the most drug-resistant pathogens of clinical concern. They offer several advantages: facile tuning of surface chemistry, inherent antimicrobial. Equally important, CNDs have shown a marked ability to circumvent resistance—bacteria exposed to certain CNDs did not develop new resistance, even over weeks of pressure. While these findings are largely preclinical, they highlight a new therapeutic strategy with real potential. Ongoing in vivo successes (e.g., in animal models of MRSA pneumonia and candidiasis) strengthen the case for translating CNDs into clinical use. Therefore, CNDs represent a cutting-edge nanomedicine approach that could bolster our arsenal against MDR infections, addressing a critical need in the treatment of infectious diseases. Due to their unique properties, CNDs are being researched to explore their potential in photodynamic therapy for combating infectious microbes, primarily bacteria and viruses. CNDs have been pioneering the diagnosis of various diseases with high sensitivity and specificity. However, along with their applications, the toxicity assessment of CNDs is also an important factor that needs to be considered before any therapeutic applications. Their continued development and optimization may pave the way for safer and more effective treatments where traditional antibiotics or antifungals have failed.

Authors' Contributions Collaborators AG, RB, and HP contributed to the applications of CNDs against various pathogens, their synergistic effects with antibiotics and antiviral agents, and their role in combating multidrug-resistant pathogens, respectively. AI contributed to the introduction and mechanism of antimicrobial activity of CNDs. All co-authors have drawn figures and tables in their respective contributed sections. MR conceptualized the chapter, integrated scientific content, and finalized the conclusion. IG outlined the chapter framework, contributed to writing about the toxicity of CNDs, and assisted MR in manuscript preparation for final approval from all authors.

References

Alara, J. A., & Alara, O. R. (2024). An overview of the global alarming increase of multiple drug resistant: A major challenge in clinical diagnosis. *Infectious Disorders – Drug Targets, 24*(3), e250723219043. https://doi.org/10.2174/1871526523666230725103902

Anand, A., Manavalan, G., Mandal, R. P., Chang, H.-T., Chiou, Y.-R., & Huang, C.-C. (2020). Carbon dots for bacterial detection and antibacterial applications-a minireview. *Current Pharmaceutical Design, 25*(46), 4848–4860. https://doi.org/10.2174/1381612825666191216150948

Barras, A., Pagneux, Q., Sane, F., Wang, Q., Boukherroub, R., Hober, D., & Szunerits, S. (2016). High efficiency of functional carbon nanodots as entry inhibitors of herpes simplex virus type. *ACS Applied Materials & Interfaces, 8*(15), 9004–9013.

Bing, W., Sun, H., Yan, Z., Ren, J., & Qu, X. (2016). Programmed bacteria death induced by carbon dots with different surface charge. *Small, 12*(34), 4713–4718. https://doi.org/10.1002/smll.201600294

Cao, G., Qiu, Y., Long, K., Xiong, Y., Shi, M., Li, Y., Nie, F., Huo, D., & Hou, C. (2022). Carbon nanodots combined with loop-mediated isothermal amplification (LAMP) for detection of African swine fever virus (ASFV). *Microchimica Acta, 189*(9), 342. https://doi.org/10.1007/s00604-022-05390-7

Chen, H. H., Lin, C. J., Anand, A., Lin, H. J., Lin, H. Y., Mao, J. Y., Wang, P. H., Tseng, Y. J., Tzou, W. S., Huang, C. C., & Wang, R. Y. L. (2022). Development of antiviral carbon quantum dots that target the Japanese encephalitis virus envelope protein. *Journal of Biological Chemistry, 298*(6), 101957. https://doi.org/10.1016/j.jbc.2022.101957

Chen, H., Geng, X., Ning, Q., Shi, L., Zhang, N., He, S. Zhao, M., Zhang, J., Li, Z., Shi, J., & Li, J. (2024). Biophilic positive carbon dot exerts antifungal activity and augments corneal permeation for fungal keratitis. *Nano Letters, 24*(13), 4044–4053. https://doi.org/10.1021/acs.nanolett.4c01042

da Silva Júnior, A. H., Macuvele, D. L. P., Riella, H. G., Soares, C., & Padoin, N. (2021). Are carbon dots effective for ion sensing and antiviral applications? A state-of-the-art description from synthesis methods to cost evaluation. *Journal of Materials Research and Technology, 12*, 688–716. https://doi.org/10.1016/j.jmrt.2021.02.069

Dong, X., Liang, W., Meziani, M. J., Sun, Y. P., & Yang, L. (2020). Carbon dots as potent antimicrobial agents. *Theranostics, 10*(2), 671–686. https://doi.org/10.7150/thno.39863

Du, T., Liang, J. G., Dong, N., Liu, L., Fang, L. R., Xiao, S. B., et al. (2016). Carbon dots as inhibitors of virus by activation of type I interferon response. *Carbon, 110*, 278–285. https://doi.org/10.1016/j.carbon.2016.09.008

Du, T., Dong, N., Fang, L. R., Lu, J., Bi, J., Xiao, S. B., & Han, H. Y. (2018). Multisite inhibitors for enteric coronavirus: Antiviral cationic carbon dots based on curcumin. *ACS Applied Nano Materials, 1*(10), 5451–5459.

Durão, P., Balbontín, R., & Gordo, I. (2018). Evolutionary mechanisms shaping the maintenance of antibiotic resistance. *Trends in Microbiology, 26*(8), 677–691. https://doi.org/10.1016/j.tim.2018.01.005

Fibriani, A., Taharuddin, A. A. P., Yamahoki, N., Stephanie, R., Laurelia, J., Agustiyanti, D. F., & Giri-Rachman, E. A. (2023). Porphyrin-derived carbon dots for an enhanced antiviral activity targeting the CTD of SARS-CoV-2 nucleocapsid. *Journal of Genetic Engineering and Biotechnology, 21*(1), 93. https://doi.org/10.1186/s43141-023-00548-z

Ghirardello, M., Ramos-Soriano, J., & Galan, M. C. (2021). Carbon dots as an emergent class of antimicrobial agents. *Nanomaterials (Basel), 11*(8), 1877. https://doi.org/10.3390/nano11081877

Gomez, C. R., Boehmer, E. D., & Kovacs, E. J. (2005). The aging innate immune system. *Current Opinion in Immunology, 17*(5), 457–462. https://doi.org/10.1016/j.coi.2005.07.013

Guo, B., Liu, G., Hu, C., Lei, B., & Liu, Y. (2022). The structural characteristics and mechanisms of antimicrobial carbon dots: A mini review. *Materials Advances, 3*, 7726–7741. https://doi.org/10.1039/D2MA00625A

Hansen, V., Oren, E., Dennis, L. K., & Brown, H. E. (2016). Infectious disease mortality trends in the United States, 1980–2014. *JAMA, 316*(20), 2149–2151.

Hasan, A. H., Hussen, N. H., Shakya, S., Jamalis, J., Pratama, M. R. F., Chander, S., Kharkwal, H., & Murugesan, S. (2022). In silico discovery of multi-targeting inhibitors for the COVID-19 treatment by molecular docking, molecular dynamics simulation studies, and ADMET predictions. *Structural Chemistry, 33*(5), 1645–1665. https://doi.org/10.1007/s11224-022-01996-y

Huang, S., Gu, J., Ye, J., Fang, B., Wan, S., & Wang, C. (2019). Benzoxazine monomer derived carbon dots as a broad-spectrum agent to block. *Journal of Colloid and Interface Science, 542*, 198–206.

Huang, H. T., Lin, H. J., Huang, H. J., Huang, C. C., Lin, J. H., & Chen, L. L. (2020). Synthesis and evaluation of polyamine carbon quantum dots (CQDs) in Litopenaeus vannamei as a therapeutic agent against WSSV. *Scientific Reports, 10*(1), 7343. https://doi.org/10.1038/s41598-020-64325-5

Huseen, N. H. A. (2020). Docking study of naringin binding with COVID-19 main protease enzyme. *Iraqi Journal of Pharmaceutical Sciences, 29*(2), 231–238. https://doi.org/10.31351/vol29iss2pp231-238

Hussen, N. H., Hasan, A. H., Jamalis, J., Shakya, S., Chander, S., Kharkwal, H., Murugesan, S., Bastikar, V. A., & Gupta, P. P. (2022). Potential inhibitory activity of phytoconstituents against black fungus: In silico ADMET, molecular docking and MD simulation studies. *Computational Toxicology, 24*, 100247. https://doi.org/10.1016/j.comtox.2022.100247

Innocenzi, P., De Forni, D., & Lori, F. (2025). Antiviral activity of carbon dots: Strategies and mechanisms of action. *Small Structures, 6*, 2400401. https://doi.org/10.1002/sstr.202400401

Jijie, R., Barras, A., Bouckaert, J., Dumitrascu, N., Szunerits, S., & Boukherroub, R. (2018). Enhanced antibacterial activity of carbon dots functionalized with ampicillin combined with visible light triggered photodynamic effects. *Colloids and Surfaces B: Biointerfaces, 170*, 347–354. https://doi.org/10.1016/j.colsurfb.2018.06.040

Kim, T. H., White, A. R., Sirdaarta, J. P., Ji, W., Cock, I. E., St John, J., Boyd, S. E., Brown, C. L., & Li, Q. (2016). Yellow-emitting carbon Nanodots and their flexible and transparent films for White LEDs. ACS Applied Materials & Interfaces, 8(48), 33102–33111. https://doi.org/10.1021/acsami.6b12113

Kim, K. W., Chung, D., Jung, S.-H., Kwon, Y. M., Kim, J. Y. H., & Baek, K. (2022). Antimicrobial effect of carbon nanodots–ZnO nanocomposite synthesized using *Sargassum horneri*. *Journal of Marine Science and Engineering, 10*(10), 1546. https://doi.org/10.3390/jmse10101546

Knoblauch, R., & Geddes, C. D. (2020). Carbon nanodots in photodynamic antimicrobial therapy: A review. *Materials, 13*(18), 4004. https://doi.org/10.3390/MA13184004

Kotta, S., Aldawsari, H. M., Badr-Eldin, S. M., Alhakamy, N. A., Md, S., Nair, A. B., & Deb, P. K. (2020). Exploring the potential of carbon dots to combat COVID-19. *Frontiers in Molecular Biosciences, 7*, 616575. https://doi.org/10.3389/fmolb.2020.616575

Kuang, Y., Song, M., Zhou, X., Mi, J., Zhang, Z., Liu, G., Shen, Z., Liu, Z., Chen, C., Wu, M. X., Zhao, Y., Yang, B., & Jiang, Y. (2023). Cefminox sodium carbon nanodots for treatment and bacterial detection of bloodstream infection. *Chemical Engineering Journal, 470*, 143988. https://doi.org/10.1016/j.cej.2023.143988

Kwee, Y., Zhou, Y., Fahmi, M. Z., Sharon, M., & Kristanti, A. N. (2022). Progress on applying carbon dots for inhibition of RNA virus infection. *Nano, 6*(4), 436–450. https://doi.org/10.7150/ntno.73918

Li, H., Huang, J., Song, Y., Zhang, M., Wang, H., Lu, F., Huang, H., Liu, Y., Dai, X., Gu, Z., Yang, Z., Zhou, R., & Kang, Z. (2018). Degradable carbon dots with broad-spectrum antibacterial activity. *ACS Applied Materials & Interfaces, 10*(32), 26936–26946. https://doi.org/10.1021/acsami.8b08832

Lin, H., & Baba, Y. (2023). Specific chemiluminescence imaging and enhanced photodynamic therapy of bacterial infections by hemin-modified carbon dots. *Small, 19*(31), e2207868. https://doi.org/10.1002/smll.202207868

Lin, F., Wang, Z., & Wu, F.-G. (2022). Carbon dots for killing microorganisms: An update since 2019. *Pharmaceuticals, 15*(10), 1236. https://doi.org/10.3390/ph15101236

Liu, Y., Li, W., Wu, K., Lei, B., Chen, J., Zhang, X., Lei, H., Duan, X., & Huang, R. (2023). Antifungal molecular details of MNQ-derived novel carbon dots against *Penicillium digitatum. Food Chemistry, 413*, 135687.

Łoczechin, A., Séron, K., Barras, A., Giovanelli, E., Belouzard, S., Chen, Y. T., Metzler-Nolte, N., Boukherroub, R., Dubuisson, J., & Szunerits, S. (2019). Functional carbon quantum dots as medical countermeasures to human coronavirus. *ACS Applied Materials & Interfaces, 11*(46), 42964–42974.

Mondol, S. M., Islam, M. R., Mia, M. E., Hassan, M. H., Farhad, F., Akter, K., Shakil, S. K., Islam, I., Rakhi, N. N., Mustary, J. F., Amiruzzaman, & Rahaman, M. M. (2025). Molecular and genomic insights into multidrug-resistant (MDR) and extensively drug-resistant (XDR) Pseudomonas aeruginosa causing burn wound infections in Bangladesh. *Scientific Reports, 15*, 25445. https://doi.org/10.1038/s41598-025-11614-6

Pajavand, H., Mobarez, A. M., Barati, A., Nikkhah, M., Delnavazi, M. R., Abiri, R., Alvandi, A. H., & Karimiravesh, R. (2023). Evaluation of combined carbon dots and ciprofloxacin on the expression level of pslA, pelA, and ppyR genes and biofilm production in ciprofloxacin-resistant *Pseudomonas aeruginosa* isolates from burn wound infection in Iran. *Journal of Global Antimicrobial Resistance, 35*, 289–296. https://doi.org/10.1016/j.jgar.2023.10.005

Paramanantham, P., Anju, V. T., Dyavaiah, M., & Siddhardha, B. (2019). Applications of carbon-based nanomaterials for antimicrobial photodynamic therapy, Springer, . pp. 237–259., doi: https://doi.org/10.1007/978-3-030-16534-5_12,

Paula, A. J., Padovani, G. C., Durán, N., & Souza Filho, A. G. (2016). Nanotoxicology of carbon-based nanomaterials. In *Nanomaterials for biomedicine* (pp. 105–137). Springer. https://doi.org/10.1007/978-3-319-25907-9_7

Prabhu, S. R. (2023). Infectious and communicable diseases: An overview. In Textbook of general pathology for dental students. Springer. https://doi.org/10.1007/978-3-031-31244-1_9

Prokisch, J., Nguyen, D. H. H., Muthu, A., Ferroudj, A., Singh, A., Agrawal, S., Rajput, V. D., Ghazaryan, K., El-Ramady, H., & Rai, M. (2024). Carbon Nanodot–microbe–plant nexus in agroecosystem and antimicrobial applications. *Nanomaterials, 14*(15), 1249. https://doi.org/10.3390/nano14151249

Qamar, W., Gulia, S., Athar, M., Ahmad, R., Imam, M. T., Chandra, P., Singh, B. P., Haque, R., Hassan, M. I., & Rahman, S. (2024). An insight into impact of nanomaterials toxicity on human health. *PeerJ, 12*, e17807. https://doi.org/10.7717/peerj.17807

Qie, X., Zan, M., Gui, P., Chen, H., Wang, J., Lin, K., Mei, Q., Ge, M., Zhang, Z., Tang, Y., Dong, W. F., & Yizhi, S. (2022). Design, synthesis, and application of carbon dots with synergistic antibacterial activity. *Frontiers in Bioengineering and Biotechnology, 10*, 894100. https://doi.org/10.3389/fbioe.2022.894100

Rai, M., Ingle, A. P., Törős, G., & Prokisch, J. (2024). Assessing the efficacy of carbon nanodots derived from curcumin on infectious diseases. *Expert Review of Anti-Infective Therapy, 22*(12), 1107–1121. https://doi.org/10.1080/14787210.2024.2409401

Ray, S. S., Gusain, R., & Kumar, N. (2020). Toxicity of carbon nanomaterials. In *Carbon nanomaterials for Agri-food and environmental applications* (pp. 365–385). Elsevier. https://doi.org/10.1016/B978-0-12-821959-1.00016-7

Rodriguez, C. E., Adcock, A. F., Dong, X., et al. (2025). Photoactivated antimicrobial activity of carbon dots in combination with different co-agents and mechanistic implications. *Journal of Nanoparticle Research, 27*, 169. https://doi.org/10.1007/s11051-025-06364-5

Russell, S. P., Neary, C., Abd Elwahab, S., Powell, J., O'Connell, N., Power, L., Tormey, S., Merrigan, B. A., & Lowery, A. J. (2020). Breast infections—Microbiology and treatment in an era of antibiotic resistance. *The Surgeon, 18*(1), 1–7. https://doi.org/10.1016/j.surge.2019.03.003

Shamkhali, L., Mobarez, A. M., Siadat, S. D., & Pajavand, H. (2025). Synergistic antibacterial effects of carbon dots derived from *Lactobacillus acidophilus* alone and in combination against carbapenem-resistant *Klebsiella pneumoniae. Journal of Infection and Public Health, 18*(5), 102724. https://doi.org/10.1016/j.jiph.2025.102724

Sturabotti, E., Camilli, A., Moldoveanu, V. G., Bonincontro, G., Simonetti, G., Valletta, A., Serangeli, I., Miranda, E., Amato, F., Marrani, A. G., Migneco, L. M., Sennato, S., Simonis, B., Vetica, F., & Leonelli, F. (2024). Targeting the antifungal activity of carbon dots against *Candida albicans* biofilm formation by tailoring their surface functional groups. *Chemistry–A European Journal, 30*, e202303631. https://doi.org/10.1002/chem.202303631

Sun, L., Zhao, Y., Peng, H., Zhang, Y., Li, W., Wang, X., & Liu, J. (2024). Carbon dots as a novel photosensitizer for photodynamic therapy of cancer and bacterial infectious diseases: Recent advances. *Journal of Nanobiotechnology, 22*, 210. https://doi.org/10.1186/s12951-024-02479-4

Tiwari, D. K., Jha, G., Tiwari, M., Kerkar, S., Das, S., & Gobre, V. V. (2021). Synergistic antibacterial potential and cell surface topology study of carbon Nanodots and tetracycline against E. Coli. Frontiers in Bioengineering and Biotechnology, 9, 626276. https://doi.org/10.3389/fbioe.2021.626276

Travlou, N. A., Ginnakoudakis, D. A., Algarra, M., Labella, A. M., Rodríguez-Castellón, E., & Bandosz, T. J. (2018). S- and N-doped carbon quantum dots: Surface chemistry dependent antibacterial activity. *Carbon, 135*, 104–111. https://doi.org/10.1016/j.carbon.2018.03.026

Tripathi, L. P., Chen, Y. A., Mizuguchi, K., & Morita, E. (2019). Network-based analysis of host-pathogen interactions. In S. Ranganathan, M. Gribskov, K. Nakai, & C. Schönbach (Eds.), *Encyclopedia of bioinformatics and computational biology* (Vol. 3, pp. 932–937). Elsevier.

Vijeata, A., Ghawri, H., Chaudhary, G. R., Raj, K., & Chaudhary, S. (2025). Engineering fungicidal nitrogen-doped carbon dots from *trillium govanianum* rhizomes: A theoretical and experimental approach against *Candida auris. Applied Surface Science, 709*, 163789. https://doi.org/10.1016/j.apsusc.2025.163789

Wang, Z., Zhang, P., Yin, C., Li, Y., Liao, Z., Yang, C., Liu, H., Wang, W., Fan, C., Sun, D., & Cheng, L. (2023). Antibiotic-derived carbon-nanodot-decorated hydrogel for reactive oxygen species-enhanced anti-infection through biofilm damage. *Advanced Functional Materials, 33*(13), 2300341. https://doi.org/10.1002/adfm.202300341

Wang, S., Wang, D., Wang, G., Zhang, M., Sun, Y., & Ding, J. (2025). Antibacterial carbon dots. *Materials Today Bio, 30*, 101383. https://doi.org/10.1016/j.mtbio.2024.101383

Wu, L. N., Yang, Y. J., Huang, L. X., Zhong, Y., Chen, Y., Gao, Y. R., Lin, L. Q., Lei, Y. Y., & Liu, A. L. (2022). Levofloxacin-based carbon dots to enhance antibacterial activities and combat antibiotic resistance. *Carbon, 186*, 452–464. https://doi.org/10.1016/j.carbon.2021.10.020

Zhao, C., Wang, X., Yu, L., Wu, L., Hao, X., Liu, Q., Lin, L., Huang, Z., Ruan, Z., Weng, S., Liu, A., & Lin, X. (2022). Quaternized carbon quantum dots with broad-spectrum antibacterial activity for the treatment of wounds infected with mixed bacteria. *Acta Biomaterialia, 138*, 528–544. https://doi.org/10.1016/j.actbio.2021.11.010

Zor, E., Mollarasouli, F., Karadurmus, L., Ozcelikay, G., & Ozkan, S. A. (2022). Carbon dots in the detection of pathogenic bacteria and viruses. *Critical Reviews In Analytical Chemistry, 54*(2), 219–246. https://doi.org/10.1080/10408347.2022.2072168

Emerging Role of Carbon Nanodots in Cancer Therapy

10

Jigar Nandha, Rashmin Patel, and Mrunali Patel

Abstract

Recent advancements in theranostic properties of carbon nanodots have positioned them as a promising tool in cancer examination. Their distinctive physicochemical properties offer diverse benefits, making them a compelling option for potential theranostic applications. The benefits of carbon nanodots include biocompatible behavior, surface conjugation and functionalization abilities, and adjustable size, making them a targeted nanoplatform for drug delivery and bioimaging of cancerous tumor cells. However, several challenges must be addressed to realize their full potential, such as understanding pharmacokinetics and biodistribution patterns comprehensively, addressing potential safety and toxicity concerns and developing robust synthesis methods to improve their therapeutic responses. This chapter aims to provide a succinct review of research on carbon nanodots within the fields of medicinal and biomedical sciences, emphasizing the available opportunities and current obstacles in utilizing them as a novel nanoplatform for cancer treatment employing therapeutic and diagnostic imaging.

Keywords

Carbon nanodots · Theranostic applications · Surface functionalization · Safety considerations

J. Nandha
Department of Regulatory Affairs, Alembic Pharmaceuticals Limited, Vadodara, Gujarat, India

R. Patel · M. Patel (✉)
Ramanbhai Patel College of Pharmacy, Charotar University of Science and Technology, Changa, Petlad, Anand, Gujarat, India
e-mail: mrunalipatel.ph@charusat.ac.in

M. Rai et al. (eds.), *Carbon Nanodots*, https://doi.org/10.1007/978-981-95-4705-0_10

10.1 Introduction

Cancer represents a significant global health challenge, characterized by high incidence and mortality rates. Despite the evolution in therapeutic options, the five-year survival rate for cancer patients remains relatively low. According to the Global Cancer Observatory (GLOBOCAN), nearly 20 million new cancer cases and approximately 9.7 million cancer-related deaths were recorded in 2022. Among various types, lung cancer in men and breast cancer in women stand out as the most frequently diagnosed and foremost causes of cancer mortality (Bray et al., 2024). Smoking is a major contributor to lung cancer deaths, accounting for about 81%, emphasizing the urgent need for innovative technologies and treatment options aimed at effectively eradicating cancer (Siegel et al., 2023). Recent trends highlight the growing interest in integrating diagnostic and therapeutic approaches on a single nanoplatform for cancer treatment (Bhattacharya et al., 2023). This integration enhances specificity and sensitivity in cancer treatment while also enabling the tracking of tumor locations and metastases, which is vital for improving therapeutic outcomes.

A wide variety of nanotherapeutic agents are accessible based on a few imaging methods, which include magnetic resonance with fluorescence, photoacoustic with fluorescence, positron emission computed tomography with magnetic resonance, and magnetic resonance with positron emission computed tomography with fluorescence (Oraevsky, 2018). Additionally, several combined treatment approaches, like radiotherapy with chemotherapy and photothermal therapy paired with gene therapy, are routinely employed in cancer care (de Santana et al., 2023; Guo et al., 2022; Rallis et al., 2021; Shu et al., 2023; Tang et al., 2024). Each of these combination methods has distinct characteristics and limitations (Unnikrishnan et al., 2021). Inorganic nanoparticles have demonstrated unique physicochemical properties that make them suitable for use in combined diagnostic and therapeutic applications. However, safety concerns, particularly related to heavy metals used in these nanoparticles, pose significant challenges and limit their broader application as theranostic platforms (Amaldoss et al., 2022). To address these issues, there is a critical need for the development of safe, biocompatible, and innovative materials that exhibit dual functionalities while ensuring acceptable long-term toxicity profiles. Over the past decade, carbon nanodots have gained considerable attention across various fields due to their unique optical, mechanical, electrical, and thermal properties.

Carbon nanodots are defined as a new category of quasi-spherical particles based on carbon nanomaterials with a size of below 10 nm (Li et al., 2021). Due to their diverse composition and adjustable structural behaviors, carbon nanodots show extraordinary features such as tremendous chemical stability, good biocompatibility, water solubility, ease of modification, lower toxicity, and modifiable fluorescence with higher quantum yield. It is fabricated by the two types of synthetic approaches, bottom-up and top-down (Banger et al., 2023; Ozyurt et al., 2023), as detailed in Table 10.1. This progress in nanomaterial research opens promising avenues for

Table 10.1 Summary of synthesis methods for carbon nanodots

Sr. no.	Synthesis method	Brief description
1	*Top-down method*	Involves breaking large carbon structures using high energy, leading to higher cost and higher energy consumption.
	(a) Arc discharge method	It is used for the synthesis of carbon nanodots from basic graphite rods or carbonaceous sources through an oxidation process.
	(b) Laser ablation method	The process involves the use of a high-energy laser beam on bulk carbonaceous materials (target material), followed by cooling and particle nucleation to form carbon nanodots.
	(c) Ultrasonic method	The method involves the use of high energy produced by an ultrasonic process to break down the larger-sized carbon-based nanomaterials, like carbon nanotubes and graphite, into carbon nanodots.
	(d) Electrochemical method	The process involves the preparation of carbon nanodots by an electrochemical oxidation and reduction process at room temperature and pressure.
2	*Bottom-up approach*	Involves building carbon nanodots from molecular or atomic precursors under controlled conditions, offering easy instrumentation and a cost-effective approach.
	(a) Thermal method	The process involves the preparation of carbon nanodots by thermal breakdown, which leads to the carbonization of carbonaceous materials (large-sized) at high temperatures.
	(b) Microwave-assisted method	The process involves the use of a broad spectrum of microwaves, usually in the frequency range of 300 MHz to 300 GHz (1 m to 1 mm), to create a homogenous temperature to form carbon nanodots.
	(c) Hydrothermal method	The process is eco-friendly and nontoxic, involves the use of organic compounds (often water soluble) as precursors, and the reaction is carried out in a hydrothermal reactor, which holds higher temperature and pressure. The method can produce carbon nanodots from glucose, citric acid, chitosan, and protein.
	(d) Template method	The method is very little utilized. It involves the use of structured templates (like mesoporous silica nanoparticles or polymers), followed by two processing steps: (a) calcining and (b) etching steps to produce carbon nanodots.

more effective cancer therapies and diagnostics, paving the way for improved patient outcomes in the fight against this pervasive disease.

Carbon nanodots are emerging as exceptional candidates for cancer imaging due to their unique properties, as illustrated in Fig. 10.1. They have been extensively evaluated for diagnosing cancerous tumors through various advanced imaging techniques, including magnetic resonance, photoacoustic imaging, and fluorescence imaging (Das et al., 2023). In addition to their imaging capabilities, some carbon nanodots possess inherent anticancer properties that make them suitable for use in photothermal and photodynamic therapies. Photodynamic therapy involves the use of photosensitizing agents that, when activated by light (from lasers or other sources), generate reactive oxygen species (ROS). These species can effectively

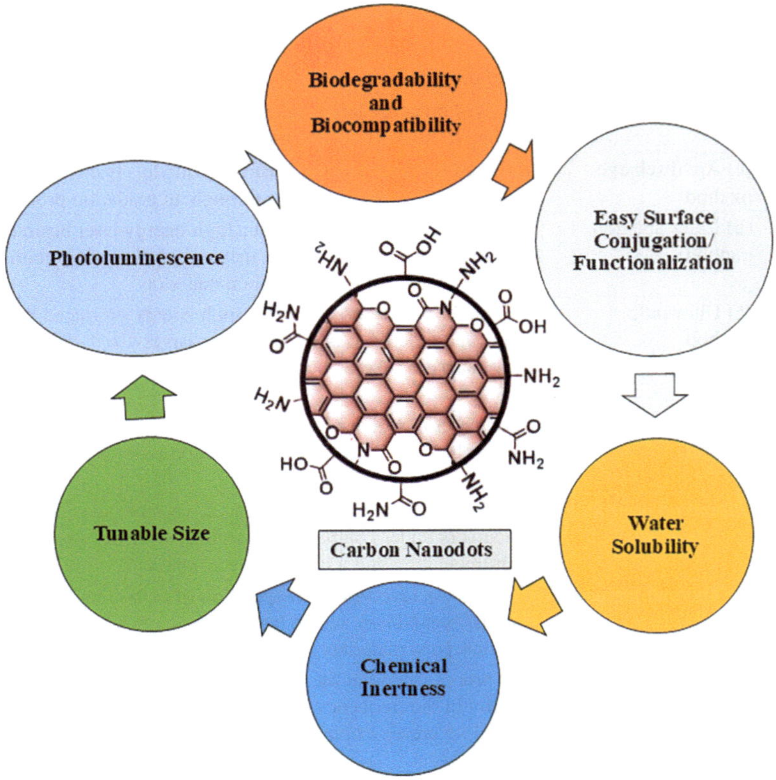

Fig. 10.1 Properties of carbon nanodots

induce oxidative damage to cancerous cells by interacting with biological macromolecules within them (Correia et al., 2021). Photothermal therapy is generally linked with the photothermal agents that absorb specific wavelengths of photon energy, subsequently releasing heat in the tumor region. This localized heating can lead to irreversible damage to tumor cells. Both therapies offer the advantage of targeting only the tumor area while sparing surrounding healthy tissues. This results in comparatively lower systemic toxicity than traditional chemotherapy and radiotherapy approaches (Zhao et al., 2021). Given these attributes, carbon nanodots with their unique phototheranostic properties hold significant promise in the treatment of cancerous tumors (Ray et al., 2022). The primary aim of the chapter is to provide a comprehensive overview of the emerging role of carbon nanodots in cancer therapy. This chapter seeks to elucidate the dual applications of carbon nanodots as both imaging/diagnostic tools and therapeutic agents in the fight against cancer (Bhattacharya et al., 2024).

10.2 Mechanism of Action

The mechanism of action of carbon nanodots as a theranostic platform with different techniques/delivery systems is described below:

10.2.1 Photodynamic Therapy

Due to the excellent photosensitizing properties, carbon dots have been utilized in photosensitizer-mediated photodynamic therapy (PDT). Acting as photosensitizing agents, carbon nanodots play a significant role in generating large quantities of ROS under light stimulation that eventually leads to cell death. This ROS production reinforces the therapeutic effectiveness of carbon nanodots in clinical contexts, offering improved outcomes and reduced chances of drug resistance. The therapeutic mechanism involves the excitation of the photosensitizing agent from its ground state to an excited state when exposed to specific wavelengths of light. Two primary photochemical pathways are involved. In the first pathway (Type I), the excited photosensitizing agent undergoes electron transfer with intracellular molecules, such as DNA, proteins, and lipids, resulting in the formation of free radicals. These radicals then react with molecular oxygen to generate hydrogen peroxide (H_2O_2), hydroxyl radicals ($\cdot OH$), and superoxide anions (O_2^-). In the second pathway (Type II), energy transfer occurs from the excited photosensitizing agent directly to oxygen molecules via intersystem crossing, leading to the formation of singlet oxygen ($^1O_2$). Both pathways often proceed simultaneously, contributing to the generation of various ROS types, thereby amplifying the therapeutic impact of PDT, and supporting its development as an advanced treatment modality (Sun et al., 2024).

10.2.2 Chemotherapy and Gene Delivery

Carbon nanodots can act as multifunctional platforms for both therapy and diagnostics (theranostics) in cancer management. When therapeutic compounds are integrated with carbon nanodots, either through π–π stacking, covalent bonding, nonspecific surface adsorption, or hydrophobic interactions, the water solubility of poorly soluble drugs like paclitaxel, doxorubicin, and oxaliplatin can be significantly improved. These nanoconjugates can accumulate preferentially at tumor sites via the enhanced permeability and retention (EPR) effect, and active targeting can also be achieved by attaching ligands such as antibodies, aptamers, peptides, hyaluronic acid, and folic acid (Adrita et al., 2020; Kong et al., 2018; Pardo et al., 2018). Additionally, carbon nanodots are explored for gene delivery applications. One study demonstrated that DNA could be electrostatically bound to positively charged carbon nanodots, functionalized with polyethyleneimine, enabling successful gene transfection within a few hours. In both drug and gene delivery applications, the intrinsic fluorescence of carbon nanodots across different excitation wavelengths supports their role as valuable diagnostic probes (Adrita et al., 2020; Othman et al., 2024).

10.2.3 Photothermal Therapy

Carbon nanodots also serve as components in photothermal therapy, functioning within hybrid delivery systems aimed at both treating and diagnosing cancer. This therapy involves the selective absorption of near-infrared (NIR) light by carbon-based nanomaterials, which then convert the light energy into heat. The localized heat generation results in the targeted destruction of cancer cells via hyperthermia. These multifunctional nanoplatforms enhance the effectiveness of cancer treatments through synergistic mechanisms and exemplify the broad therapeutic potential of carbon nanodots in oncological applications (Adrita et al., 2020; Bao et al., 2018; Qi et al., 2023).

10.3 Applications in Cancer Therapy

10.3.1 As Diagnostic Tools

Carbon nanodots are increasingly recognized as a detection tool in cancer treatment. These nanoparticles, characterized by their heterogeneous structure and distinct optical and chemical properties, have emerged as a pivotal advancement in biolabeling technologies. Their versatility spans from energy applications to therapeutic uses, offering high sensitivity and specificity. In comparison to conventional organic dye-based fluorescent probes, carbon nanodots demonstrate superior water solubility, biocompatibility, minimal toxicity, and straightforward synthesis processes, factors that significantly enhance their suitability in cancer detection. One of their key features is the ability to activate multiple fluorescence emissions under a single light source with minimal spectral overlap, facilitating the simultaneous detection of multiple targets. This dual functionality, as both diagnostic (via optical traits) and therapeutic agents, enhances their utility in oncology (Singh et al., 2021). Their strong tissue penetration (up to 1 cm) and low background interference also position them as valuable tools in detecting lymph node metastases (Bentolila et al., 2009).

The carbon nanodots also serve as effective fluorescent probes in imaging applications. Despite their potential, challenges remain, particularly regarding in vivo toxicity. However, integrating metal ions such as gadolinium (Gd) with it has been shown to mitigate toxicity and prevent leakage (Jiang et al., 2021). Cancer cells often display altered iron metabolism. Thus, many iron ion (Fe^{3+}) sensors based on carbon nanodots have been developed, owing to disruptions in Fe^{3+} homeostasis as implicated in oncogenesis (Bhattacharya et al., 2023). Additionally, one study demonstrated that biomarker glutathione-modified carbon dots could detect cancer progression by correlating glutathione (GSH) levels with fluorescence intensity (Traverso et al., 2013). GSH plays a crucial role as an antioxidant that protects the cells from oxidative stress, and its depletion is associated with tumor development (Valenti et al., 2023). The fluorescence enhancement of dual GSH and Fe^{3+} doped carbon nanodots allows for differentiation between malignant and normal cells, offering a reliable imaging tool (Gao et al., 2018). Green fluorescent carbon nanodots-based probes have been designed for visual cancer diagnostics (Hettiarachchi et al., 2019; Tiron et al., 2020; Zhao et al., 2017). Further, the functionalization of carbon nanodots with folic acid has enabled receptor-specific imaging, particularly targeting folate-overexpressing cancer cells such as A549 human lung adenocarcinoma cells. The conjugation specifically distinguished between cancerous and normal healthy cells and showed outstanding biocompatibility (Saljoughi et al., 2020; Yang et al., 2022). Furthermore, researchers like Wu et al. have developed signal amplification systems using functionalized carbon nanodots for the sensitive and selective detection of specific cancers, such as cervical and breast cancers (Wu et al., 2013). These were utilized in electrochemiluminescent sensors, which, due to the high surface area of carbon nanodots, offered label-free, highly accurate detection. Compared to organic dyes, carbon nanodots are easier to synthesize, more hydrophilic, and exhibit low toxicity, making them superior candidates for bioimaging applications (Singh et al., 2021).

10.3.2 As Bioimaging Agents

Carbon nanodots are increasingly used as contrast agents for in vivo optical imaging because of their photoluminescent properties when exposed to light (Devi et al., 2022). They tend to build up more in cancer tissues than in normal ones, allowing clearer and real-time imaging as illustrated in Fig. 10.2. Advances in nanotech have improved their targeting ability, especially through interactions with biomolecules (Tiernan et al., 2013). Ideal carbon nanodots for imaging should have strong light emission, low toxicity, work well in the body, and be safely cleared by the kidneys.

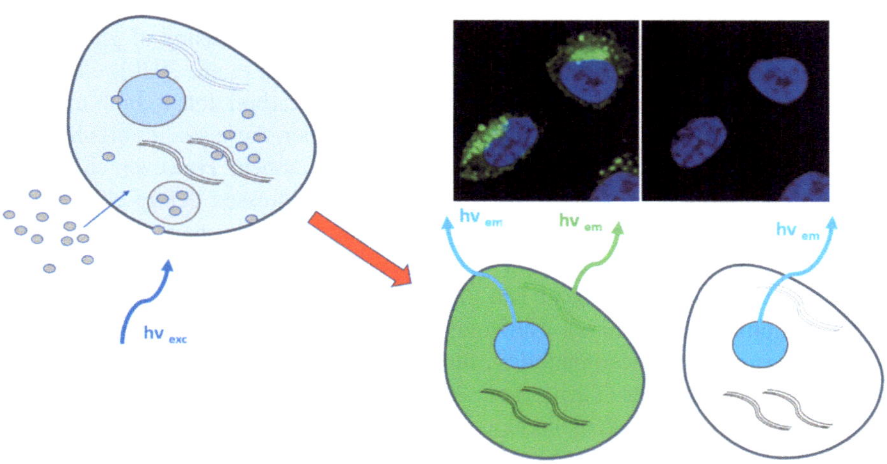

Fig. 10.2 Schematic illustration of the cellular internalization pathway of carbon nanodots for bioimaging applications in cancer diagnosis. (Reproduced from Nocito et al. (2021) under the terms of the Creative Commons Attribution License (CC BY), with permission from MDPI)

These nanoprobes are suitable for both in vitro and in vivo systems (Belkahla et al., 2019; Choppadandi et al., 2021). While early studies showed these dots spreading evenly across tissues, later work confirmed they tend to concentrate more in tumors due to the enhanced permeability and retention (EPR) effect (Sengar et al., 2022). Some researchers have taken this further by designing carbon nanodots that target specific parts of cancer cells. One study focused on directing them to mitochondria and used fluorescent dots combined with magnetic silica particles for enhanced imaging. They added a magnetic field to guide the dots more precisely to tumor cells, leading to higher uptake in both cancer and normal cells like A549 and HFF (human foreskin fibroblast) lines (Zhang et al., 2015). These distinctive characteristics of carbon nanodots, particularly their selective uptake at tumor sites, form a crucial foundation for future advancements in targeted cancer therapies (Zhang et al., 2021). Zwitterionic carbon nanodots have shown promise in cancer bioimaging due to their ability to interact with the tumor microenvironment. These dots can easily bond with biomolecules through carboxylic groups. When exposed to tumor conditions, they gain a positive surface charge, improving their uptake by cancer cells. For instance, Sri et al. used these dots for imaging cancer cells, while Kong et al. attached them to the AS1411 aptamer to detect nucleolin-positive MCF-7 (Michigan Cancer Foundation-7) cells (Sri et al., 2018; Kong et al., 2020). Researchers have also explored pH-responsive carbon nanodots, which turn fluorescent in acidic tumor environments but remain inactive at normal pH, allowing precise imaging (Phuong et al., 2020). Some studies combined carbon dots with agents like titanium dioxide, enabling both imaging and treatment by triggering cell death. Qian et al. developed a system where carbon dots linked with aconitic acid responded to folic acid, aiding in the imaging of various cancer cells such as A549, HeLa (Henrietta Lacks), and SMMC-7721 (Shanghai Medical Clinic 7721) based on folate receptor levels (Qian et al., 2018). Similarly, Das and his team created a probe using carbon dots and gold nanoparticles to detect high glutathione levels in cancer cells (Das et al., 2016). Li et al. designed dots for targeted sensing of glutathione, using fluorescence changes to distinguish cancer cells (Li et al., 2020). Gao et al. went further by combining doxorubicin with modified carbon dots to both treat and image HeLa cells effectively (Gao et al., 2017). Together, these studies highlight the potential of carbon nanodots in cancer diagnosis and treatment. However, challenges like toxicity and complex synthesis still need to be addressed before clinical use. Ongoing research has shown promising variations of carbon nanodots with self-targeting capabilities. For example, Zheng et al. developed self-targeted carbon nanodots using L-aspartic acid and D-glucose via pyrolysis, which specifically targets glioma (brain cancer). The study revealed that this novel carbon dot type possesses excellent biocompatibility, tunable emission, and remarkable selectivity for C6 brain cancer cells (Zheng et al., 2015). Li et al. also constructed self-targeted large amino acid-mimicking carbon nanodots directed at brain tumors for both targeted imaging and drug delivery in brain tumors (Li et al., 2020). Despite their potential, further refinement is essential to minimize toxicity and enhance clinical applicability (Shen et al., 2022).

10.3.3 As Drug Delivery Carriers

Carbon nanodots are gaining attention as promising drug delivery vehicles because of their unique set of advantages. They are easy to produce, allow precise control over particle size, exhibit minimal toxicity, and are compatible with biological systems. One of their most valuable features is their ability to support sustained drug release. These properties make them a favorable alternative to conventional drug delivery systems, which often face limitations such as toxicity, instability, or complex synthesis processes. In addition to their biocompatibility, carbon nanodots possess fluorescent properties that span multiple colors (Singh et al., 2018; Zuo et al., 2015). This enables them to serve dual roles, not just as drug carriers but also as imaging agents, as shown in Fig. 10.3. Their surface chemistry is highly adaptable, which means they can be modified with different functional groups to increase drug-loading efficiency and target specificity while keeping toxicity low. While gold nanoparticles have been widely used in imaging, photothermal therapy, and biosensing due to their optical qualities, their potential for clinical use is hindered by concerns about toxicity (Alkilany & Murphy, 2010; Medici et al., 2021). Carbon nanodots, on the other hand, provide a more biocompatible alternative. Their surface can be easily functionalized to attach various drug molecules and targeting ligands, unlocking numerous possibilities for precision drug delivery (Sameer et al., 2023). A comparative overview of commonly explored nanocarrier systems in cancer therapy

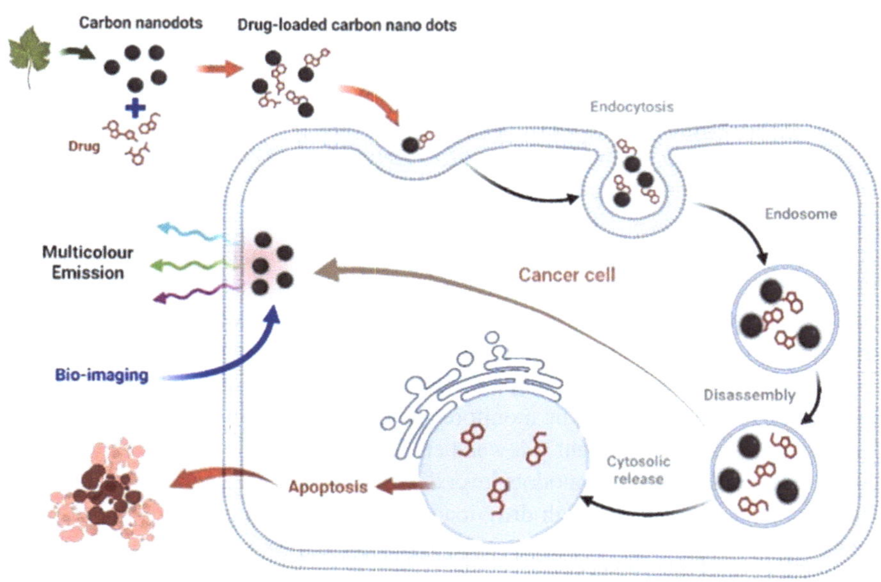

Fig. 10.3 Depiction of the mechanism by which carbon nanodots function as a theranostic agent in cancer treatment, integrating diagnostic imaging and therapeutic action. (Reproduced from Malhan et al. (2024) with permission from Elsevier. © 2024 Elsevier B.V)

is presented in Table 10.2, highlighting the advantages of carbon nanodots in terms of compatibility, imaging ability, and drug delivery potential.

Seçme and İlhan investigated the potential biomedical application of natural herbal substances when conjugated with metal nanoparticles in ovarian cancer. Researchers have used silver as a metal moiety for the fabrication of nanoparticles (AgNPs), followed by the surface functionalized with thymol carbon nanodot (ThCND). Results of the cell line Study using OVCAR-3 cells indicated that carbon nanodots complex (ThCND-AgNPs) showed tremendous apoptotic, antiproliferative and antiinvasive effects in ovarian cancer (Seçme & İlhan, 2025). Tang et al. described a smart drug delivery platform using carbon nanodots in combination with doxorubicin. This system was designed to allow real-time observation of drug release via a mechanism called fluorescence resonance energy transfer (FRET) (Tang et al., 2013). In this model, doxorubicin molecules were connected to the carbon nanodots through π–π interactions. The presence of an acidic environment, such as that found in cancerous tissues (around pH 5.0), triggered a strong fluorescence response, indicating drug release. In contrast, at near-neutral pH ($\approx$ 7.4), which corresponds to normal tissues, the release remained minimal as indicated by low fluorescence intensity. This pH-sensitive behavior helps protect healthy normal cells while ensuring targeted drug delivery at tumor sites. Bayda et al. fabricated another innovative system where carbon nanodots with carboxyl-rich surfaces were combined with doxorubicin through electrostatic interactions involving amine and carboxyl groups (Bayda et al., 2017). The observed pH difference across cancerous cells and normal healthy cells triggered the release of the drug from the complex. The results of the experiments demonstrated that this nanoplatform was more cytotoxic to breast and colon cancer cell lines than the free drug alone. Moreover, an in vivo study over 2 months revealed that the nanocomplex was well tolerated by mice, with no signs of illness and significant tumor shrinkage compared to untreated groups or those given free doxorubicin. However, the authors emphasized the need to further refine the synthesis process to ensure that no harmful byproducts are generated in the final formulation (Cailotto et al., 2018).

In another approach, Zheng et al. created a multifunctional nanomedicine by attaching an oxaliplatin derivative to fluorescent carbon nanodots (Zheng et al., 2014). The two components were joined through a condensation reaction between carboxyl and amino groups. Once inside tumor cells through endocytosis, where the intracellular environment is highly reductive, this bond was broken, converting the prodrug form back into active oxaliplatin. Because the nanodots are fluorescent, their distribution in the body could be monitored in real-time, enabling better control over dosing and timing of treatment. Mewada et al. contributed to the field by designing folic acid-modified carbon nanodots loaded with doxorubicin (Mewada et al., 2014). This formulation showed a high drug-loading capacity (around 86%) and selective uptake by cancer cells due to the presence of folic acid, which targets folate receptors commonly overexpressed in tumors. The presence of folate receptors on cancer cells enables carbon nanodots to selectively target these cells, highlighting their potential as a valuable tool in biomedical research, as shown in Fig. 10.4. The platform also showed strong biocompatibility and greater cytotoxicity compared to the free drug,

Table 10.2 Comparative evaluation of carbon nanodots and common nanocarriers in cancer applications

Nanocarrier system	Biological compatibility	Intrinsic imaging ability	Drug carrying efficiency	Toxicological characteristics	Stage of development
Carbon nanodots	Highly compatible with cells and tissues	Naturally fluorescent across multiple wavelengths	Supports diverse drug attachment due to a modifiable surface	Shows low toxicity in lab and animal models with minimal immune response	Under preclinical investigation with promising theranostic potential
Gold nanoparticles	Moderate compatibility, influenced by dose	Excellent optical performance for diagnostics and therapy	Limited drug loading due to the smaller surface area	Risk of accumulation and toxicity at higher concentrations	Evaluated in early-stage clinical trials with some safety concerns
Liposomal formulations	Made from biologically safe lipids	Requires additional agents for imaging	Effective for drug encapsulation and controlled release	Well-tolerated with multiple approved formulations	Approved for clinical use for several cancer therapies
Polymeric nanocarriers	Compatibility depends on polymer structure	Moderate imaging capabilities if modified	High drug payload flexibility based on polymer design	Biocompatibility varies, but is generally acceptable	Widely studied with several candidates in trials
Quantum dots	Variable biocompatibility depending on composition	Very strong fluorescence for deep tissue imaging	Limited loading capacity for therapeutic agents	Often exhibits toxicity due to heavy metal content	Mostly limited to research settings due to safety risks
Dendrimers	Designed for safe interaction with cells	Requires external modification for imaging	High due to the multibranched structure and surface reactivity	Safety depends on surface group chemistry and size	Under early research and development for therapeutic use

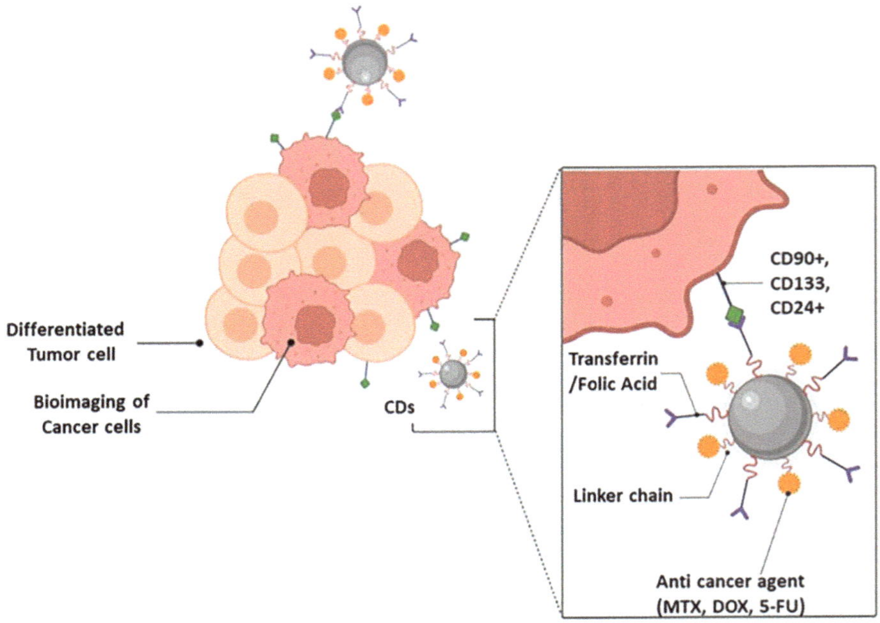

Fig. 10.4 Illustration of the role of carbon nanodots as a theranostic agent in cancer management, highlighting their combined diagnostic and therapeutic potential. (Reproduced from Kaurav et al. (2023) under the terms of the Creative Commons Attribution License (CC BY), with permission from Frontiers, Journal, Copyright © 2023)

along with the added benefit of enabling cancer cell imaging. Wang et al. tested a similar strategy using carbon nanodots derived from beer waste, which were functionalized with doxorubicin. This system demonstrated extended drug release in MCF-7 breast cancer cells (Wang et al., 2015). The combination of therapeutic effects and built-in imaging capability showed that the complex was nontoxic, highly biocompatible, and suitable for biomedical use. Zeng et al. also worked on developing green-emitting carbon nanodots that were non-covalently bonded with doxorubicin using amine and carboxyl groups. Their results, validated through both in vitro and in vivo studies, indicated that this system was stable, selectively accumulated in tumor cells, and exhibited reduced toxicity (Zeng et al., 2016).

Taken together, these studies highlight the potential of carbon nanodots as multifunctional nanocarriers for targeted cancer therapy. Their surface chemistry supports the attachment of a variety of therapeutic agents and targeting molecules, while their optical properties make it possible to track drug distribution in real time. This dual functionality enhances treatment precision, reduces side effects, and opens new avenues in cancer nanomedicine (Mishra et al., 2018).

10.4 Safety Considerations and Challenges in Clinical Translation

Carbon nanodots have emerged as valuable tools in biomedical research, particularly in applications like imaging, sensing, and diagnostics (Lodha et al., 2024; Qureshi et al., 2024). Recent advancements in the synthesis and functionalization of these nanomaterials have significantly expanded their role beyond basic imaging to include complex theranostic applications and optical uses such as photocatalysis and environmental sensing. One of the reasons for their growing popularity in medicine is the ability to conjugate carbon nanodots with various functional molecules (Lodha et al., 2024). This adaptability makes them highly effective for both drug delivery and bioimaging across a wide range of medical conditions. As carbon is the primary building block of these nanostructures, they inherently possess several beneficial properties. These include high biocompatibility, excellent stability, ease of functionalization thanks to abundant surface groups, and an environmentally friendly profile (Hao et al., 2021; Lu et al., 2021; Ma et al., 2019).

Multiple in vitro studies using diverse cell lines have consistently reported that carbon nanodots exhibit minimal cytotoxicity, even when applied at relatively high concentrations (Sun et al., 2016). This suggests they are generally safe for use in cellular environments. Supporting this, in vivo experiments have demonstrated favorable safety outcomes as well. These studies indicate that carbon nanodots are typically eliminated from the body through natural routes such as renal or hepatobiliary excretion, reinforcing their biocompatibility and low risk of long-term accumulation in tissues (Liu et al., 2018). Because of their fluorescent properties, carbon nanodots serve as effective probes in biological systems. Their brightness, stability, and nontoxic nature enable precise tracking and imaging within living organisms. The safety of fluorescent carbon nanodots has also been specifically evaluated in vitro using NIH/3T3 mouse fibroblast cells (Bhattacharya et al., 2023). These assessments considered the potential cytotoxic effects, how the particles distribute within cells, factors influenced by their surface charge, and chemical characteristics. Further in vivo investigations, including histological examinations of vital organs such as the lungs, kidneys, bladder, brain, spleen, heart, and testicles in rat models, have shown no significant inflammatory or adverse responses. These findings collectively support the conclusion that carbon nanodots are safe for biomedical use. Their favorable safety profile and functional versatility strongly support their use as theranostic agents, integrating both diagnostic and therapeutic capabilities in a single platform (Xu et al., 2023).

Despite these encouraging preclinical outcomes, several hurdles still hinder the successful clinical translation of carbon nanodots. The absence of multi-model investigations presents a major hurdle to the clinical applications of such nanoplatforms in human volunteers. One of the key challenges includes a lack of thorough understanding of the pharmacokinetic behavior, like metabolism and elimination mechanisms, as this addresses their concerns regarding the toxicity profile in biological environments (Bhattacharya et al., 2023). Additionally, considering the variations in size, surface chemistry, and functionalization methodology can ensure coherent biological behavior and enhanced therapeutic outcomes. By focusing on robust toxicological assessments, improving the scalability of production, and implementing standardized characterization techniques, we can pave the way for carbon nanodots to achieve practical and impactful applications in clinical therapy.

Carbon nanodots are commonly conjugated for targeting cancerous disease, but the materials utilized for fabrication can cause some level of toxicity, which makes it difficult for efficient therapy. The surface conjugation of anticancer molecules with carbon nanodots results in enhanced cytotoxic potential against tumor cells, because of the photothermal properties of carbon nanodots, which ultimately raises the ROS alongside the effects of the parent moiety. While this synergistic effect can be beneficial, the increase in ROS poses a higher risk to normal healthy cells due to a lack of target specificity (Butowska et al., 2023; Feng et al., 2022; Vargason et al., 2021). However, it is important to consider that beyond target specificity, the toxicity profile of carbon nanodots is also tremendously dependent on the surface modifications. In an experimental study conducted by Kuznietsova and his research team, in vitro studies revealed the absence of toxicity in the A549 cell line. However, significant kidney and other organ-related toxicities were reported in vivo in mice after multiple dosing of a specific type of carbon dot called CD_3011 (surface enriched with phenol and carboxyl groups containing nitrogen) (Kuznietsova et al., 2023). Similarly, another type enriched with aniline and toluidine groups (CDN19) showed initial adverse effects, such as mortality rates and changes in body weight, although the mice eventually recovered without any noticeable long-term toxicity. These investigations underscore the importance of thorough investigations into potential long-term toxicity, particularly with repeated or high-dose administrations, in drug delivery applications.

On a more positive note, an investigation conducted by Wibrianto et al. reported a low level of toxicity of naproxen-loaded boron-based carbon nanodots following administration (20 mg/kg) in female mice (Wibrianto et al., 2024). Additionally, fabrication of biocompatible and fluorescent carbon nanodots from banana biomass highlights a promising avenue for disease detection, biosensing and imaging applications (Bartkowski et al., 2024). Using banana biomass as a natural and renewable resource not only demonstrated an eco-friendly approach but also facilitated the fabrication of fluorescent carbon dots with a higher level of biocompatibility and limited toxicity, making them ideal for biomedical applications (Soumya et al., 2023). Similarly, Rodríguez-Varillas et al. produced eco-friendly carbon nanodots from tomato juice, benefiting from its inherent antioxidant properties (Rodríguez-Varillas et al., 2022). In vitro studies displayed cytotoxic potential of these carbon nanodots against A549 cells due to inherent antioxidant properties, while demonstrating a low level of toxicity to normal cells after 72 h observation. Overall, the results suggested that the carbon nanodots fabricated from tomato offer promising opportunities in biomedical applications with commendable biocompatibility. Researchers are currently exploring alternative raw materials like biowaste for the synthesis of carbon nanodots, addressing safety concerns and enhancing the sustainability of long-term applications. The hydrothermal carbonization approach has recently garnered special attention for its simplicity and minimal environmental effect. These newly developed approach offers tremendous advantages over conventional quantum dots, including easy accessibility, sufficient biocompatibility, reduced toxicity concerns, and excellent biodegradability. While the regulatory processes related to its biomedical use remain insufficient to fully recognize carbon nanodots for promising clinical applications, their excellent imaging properties position them as the safest substitute nanoplatform for theranostic applications, i.e., therapeutic and biological imaging purposes. As investigations continue to strengthen their safety profiles and address regulatory concerns, carbon nanodots hold great potential in biomedicine, particularly in cancer therapy and imaging, paving the way for their enduring acceptance in clinical settings (Bhattacharya et al., 2023).

10.5 Advancing Frontiers and Future Opportunities

The future of carbon nanodots in the biomedical area is full of potential, especially when it comes to their innovative theranostic applications in cancer therapy and other illnesses. These remarkable nanoplatforms offer a range of benefits such as cost effectiveness, minimal environmental impact, reduced toxicity, higher level of cell penetration, improved water solubility, good biocompatibility, high accuracy in imaging via fluorescence probe and relatively easy surface modification/conjugation. Their versatility opens the door to diverse applications, from drug delivery to advanced chemical sensing. One of the advanced aspects of this technology is the role of aptamers, synthetic nucleic acid ligands that can bind to different targets such as amino acids, active drug molecules and proteins (Davodabadi et al., 2023). Conjugation of such aptamers with nanomaterials like carbon nanodots improves their use as promising tools in the biomedical field and bioanalysis. Looking ahead, future research on carbon nanodots is mainly focused on large-scale production with controllable particle sizes, rapid purification and efficient surface conjugation and functionalization. A key goal is to explore how their unique parameters can combine with the other available technologies, like instrumental analysis and immunoassays, to drive enhanced outcomes in medical applications. However, challenges in terms of harmful biological as well as pharmacological consequences associated with crucial physicochemical properties of carbon nanodots need to be considered. Ensuring their compatibility with biological environments is a significant challenge for their use in biomedical applications. For effective clinical use, several parameters need to be evaluated, including cellular level toxicity, organ accumulation, immunogenic responses, potential agglomeration, and suitability in biological media. Hence, the development of a supplementary biological model is vital for better clarity of biological interactions of nanomaterials used in the fabrication of carbon nanodots and their theranostic clinical applications. Future research studies should prioritize in vivo monitoring and evaluation of various fundamental interaction parameters.

Carbon nanodots generally exhibit lower toxicity compared to other nanomaterials, encouraging innovation focused on surface conjugation and functionalization. By conjugation with biocompatible macromolecules such as polyethylene glycol, chitosan, or gelatin, any hydrophobic interactions can be counteracted that could compromise their safety in biological systems. Future innovations must focus on fabricating nanotheranostic platforms utilizing biodegradable, biorenewable, and biocompatible materials derived from animal byproducts or plant oils. Various nanocarrier systems, including liposomes, nanoparticles, and drug conjugates, have been approved for their clinical usage across various diseases. Thus, refining approved drug formulations through surface conjugation can enhance their diagnostic capabilities. Moreover, careful consideration of the physicochemical properties of starting materials and nanomaterials throughout the processing steps is crucial for establishing their safety profiles, regardless of the fabrication method. This will constitute a fundamental step in risk characterization and allow a detailed evaluation of any possible risks associated with nanoformulations. Despite their multiple potential applications, it is imperative to address potential challenges related to toxicity, long-term effects, and biocompatibility through careful investigations. Lastly, optimizing the fabrication process for scalability, reproducibility, and cost-effectiveness is critical for driving the translation of these nanomaterials into clinical practice. The prospect of carbon nanodots is promising, and with continued innovation, they may transform the landscape of modern cancer therapy.

10.6 Conclusion

Carbon nanodots are revolutionizing cancer therapy with their unique properties that facilitate easy attachment to anticancer agents, selective ligands, and various therapeutic molecules. Research indicates that these nanodots surpass other nanoplatforms, displaying impressive dual capabilities for therapy and imaging that enhance safety in clinical applications. Based on several in vitro studies using cell lines and in vivo studies using animal models, it can be inferred that carbon nanodots are multifunctional and exhibit low toxicity, but challenges like controlling size, optimizing surface functionalization, and refining synthesis remain. Their significant optical properties enable effective drug delivery and bioimaging through photothermal and photodynamic therapies, specifically targeting cancerous cells. Carbon nanodots can serve multiple roles, from acting as carriers for medicinal agents to generating fluorescence and optical probes for diagnostics. However, each application requires careful evaluation of the toxicity patterns and safety profiles. Moving forward, more in-depth research is required to ensure their safety for humans and the environment, addressing the challenges linked to toxicity and pharmacological issues. As the new avenues in carbon nanodots are explored, focusing on synthesis variations and surface modifications will help mitigate risks and improve their applications in cancer therapy and diagnostics. With ongoing innovations, the carbon nanodots hold tremendous promise for unlocking new potentials in imaging and cancer theranostics.

References

Adrita, S. H., Tasnim, K. N., Ryu, J. H., & Sharker, S. M (2020). Nanotheranostic carbon dots as an emerging platform for cancer therapy. *Journal of Nanotheranostics, 1*(1), 58–77. https://doi.org/10.3390/JNT1010006

Alkilany, A. M., & Murphy, C. J. (2010). Toxicity and cellular uptake of gold nanoparticles: What we have learned so far? *Journal of Nanoparticle Research, 12*(7), 2313. https://doi.org/10.1007/S11051-010-9911-8

Amaldoss, M. J. N., Yang, J. L., Koshy, P., Unnikrishnan, A., & Sorrell, C. C. (2022). Inorganic nanoparticle-based advanced cancer therapies: Promising combination strategies. *Drug Discovery Today, 27*(12), 103386. https://doi.org/10.1016/J.DRUDIS.2022.103386

Banger, A., Gautam, S., Jadoun, S., Jangid, N. K., Srivastava, A., Pulidindi, I. N., Dwivedi, J., & Srivastava, M. (2023). Synthetic methods and applications of carbon nanodots. *Catalysts, 13*(5), 858. https://doi.org/10.3390/CATAL13050858

Bao, X., Yuan, Y., Chen, J., Zhang, B., Li, D., Zhou, D., Jing, P., Xu, G., Wang, Y., Holá, K., Shen, D., Wu, C., Song, L., Liu, C., Zbořil, R., & Qu, S. (2018). In vivo theranostics with near-infrared-emitting carbon dots—Highly efficient photothermal therapy based on passive targeting after intravenous administration. *Light: Science & Applications, 7*(1), 1–11. https://doi.org/10.1038/s41377-018-0090-1

Bartkowski, M., Zhou, Y., Mustafa, M. N. A., Eustace, A. J., & Giordani, S. (2024). CARBON DOTS: Bioimaging and anticancer drug delivery. *Chemistry – A European Journal, 30*(19), e202303982. https://doi.org/10.1002/CHEM 202303982

Bayda, S., Hadla, M., Palazzolo, S., Kumar, V., Caligiuri, I., Ambrosi, E., Pontoglio, E., Agostini, M., Tuccinardi, T., Benedetti, A., Riello, P., Canzonieri, V., Corona, G., Toffoli, G., & Rizzolio, F. (2017). Bottom-up synthesis of carbon nanoparticles with higher doxorubicin efficacy. *Journal of Controlled Release, 248*, 144–152. https://doi.org/10.1016/J.JCONREL.2017.01.022

Belkahla, H., Boudjemaa, R., Caorsi, V., Pineau, D., Curcio, A., Lomas, J. S., Decorse, P., Chevillot-Biraud, A., Azaïs, T., Wilhelm, C., Randriamahazaka, H., & Hémadi, M. (2019). Carbon dots, a powerful non-toxic support for bioimaging by fluorescence nanoscopy and eradication of bacteria by photothermia. Nanoscale Advances, 1(7), 2571–2579. https://doi.org/10.1039/C9NA00140A

Bentolila, L. A., Ebenstein, Y., & Weiss, S. (2009). Quantum dots for in vivo small-animal imaging. *Journal of Nuclear Medicine, 50*(4), 493–496. https://doi.org/10.2967/JNUMED.108.053561

Bhattacharya, T., Shin, G. H., & Kim, J. T. (2023). Carbon dots: Opportunities and challenges in cancer therapy. *Pharmaceutics, 15*(3), 1019. https://doi.org/10.3390/PHARMACEUTICS15031019

Bhattacharya, T., Preetam, S., Mukherjee, S., Kar, S., Roy, D. S., Singh, H., Ghose, A., Das, T., & Mohapatra, G. (2024). Anticancer activity of quantum size carbon dots: Opportunities and challenges. *Discover Nano, 19*(1), 1–33. https://doi.org/10.1186/S11671-024-04069-7

Bray, F., Laversanne, M., Sung, H., Ferlay, J., Siegel, R. L., Soerjomataram, I., & Jemal, A. (2024). Global cancer statistics 2022: GLOBOCAN estimates of incidence and mortality worldwide for 36 cancers in 185 countries. *CA: a Cancer Journal for Clinicians, 74*(3), 229–263. https://doi.org/10.3322/CAAC.21834

Butowska, K., Han, X., Gong, N., El-Mayta, R., Haley, R. M., Xue, L., Zhong, W., Guo, W., Wang, K., & Mitchell, M. J. (2023). Doxorubicin-conjugated siRNA lipid nanoparticles for combination cancer therapy. *Acta Pharmaceutica Sinica B, 13*(4), 1429–1437. https://doi.org/10.1016/J.APSB.2022.07.011

Cailotto, S., Amadio, E., Facchin, M., Selva, M., Pontoglio, E., Rizzolio, F., Riello, P., Toffoli, G., Benedetti, A., & Perosa, A. (2018). Carbon dots from sugars and ascorbic acid: Role of the precursors on morphology, properties, toxicity, and drug uptake. *ACS Medicinal Chemistry Letters, 9*(8), 832–837. https://doi.org/10.1021/ACSMEDCHEMLETT.8B00240/SUPPL_FILE/ML8B00240_SI_001.PDF

Choppadandi, M., Guduru, A. T., Gondaliya, P., Arya, N., Kalia, K., Kumar, H., & Kapusetti, G. (2021). Structural features regulated photoluminescence intensity and cell internalization of carbon and graphene quantum dots for bioimaging. Materials Science and Engineering: C, 129, 112366. https://doi.org/10.1016/J.MSEC.2021.112366

Correia, J. H., Rodrigues, J. A., Pimenta, S., Dong, T., & Yang, Z. (2021). Photodynamic therapy review: Principles, photosensitizers, applications, and future directions. *Pharmaceutics, 13*(9), 1332. https://doi.org/10.3390/PHARMACEUTICS13091332

Das, S., Mondal, S., & Ghosh, D. (2023). Carbon quantum dots in bioimaging and biomedicines. *Frontiers in Bioengineering and Biotechnology, 11*, 1333752. https://doi.org/10.3389/FBIOE.2023.1333752/BIBTEX

Das, K., Sarkar, S., & Das, P. K. (2016). Fluorescent Indicator Displacement Assay: Ultrasensitive Detection of Glutathione and Selective Cancer Cell Imaging. ACS Applied Materials and Interfaces, 8(39), 25691–25701. https://doi.org/10.1021/ACSAMI.6B06353/SUPPL_FILE/AM6B06353_SI_001.PDF

Davodabadi, F., Mirinejad, S., Fathi-Karkan, S., Majidpour, M., Ajalli, N., Sheervalilou, R., Sargazi, S., Rozmus, D., Rahdar, A., & Diez-Pascual, A. M. (2023). Aptamer-functionalized quantum dots as theranostic nanotools against cancer and bacterial infections: A comprehensive overview of recent trends. *Biotechnology Progress, 39*(5). https://doi.org/10.1002/BTPR.3366

de Santana, W. M. O. S., Surur, A. K., Momesso, V. M., Lopes, P. M., Santilli, C. V., & Fontana, C. R. (2023). Nanocarriers for photodynamic-gene therapy. *Photodiagnosis and Photodynamic Therapy, 43*, 103644. https://doi.org/10.1016/J.PDPDT.2023.103644

Devi, S., Kumar, M., Tiwari, A., Tiwari, V., Kaushik, D., Verma, R., Bhatt, S., Sahoo, B. M., Bhattacharya, T.,Alshehri, S., Ghoneim, M. M., Babalghith, A. O., & Batiha, G. E. S. (2022). Quantum dots: An emerging approach for cancer therapy. *Frontiers in Materials, 8*, 798440. https://doi.org/10.3389/FMATS.2021.798440/XML/NLM

Feng, S., Lu, J., Wang, K., Di, D., Shi, Z., Zhao, Q., & Wang, S. (2022). Advances in smart mesoporous carbon nanoplatforms for photothermal–enhanced synergistic cancer therapy. *Chemical Engineering Journal, 435*, 134886. https://doi.org/10.1016/J.CEJ.2022.134886

Gao, G., Jiang, Y. W., Jia, H. R., Yang, J., & Wu, F. G. (2018). On-off-on fluorescent nanosensor for Fe3+ detection and cancer/normal cell differentiation via silicon-doped carbon quantum dots. *Carbon, 134*, 232–243. https://doi.org/10.1016/J.CARBON.2018.02.063

Gao, N., Yang, W., Nie, H., Gong, Y., Jing, J., Gao, L., & Zhang, X. (2017). Turn-on theranostic fluorescent nanoprobe by electrostatic self-assembly of carbon dots with doxorubicin for targeted cancer cell imaging, in vivo hyaluronidase analysis, and targeted drug delivery. Biosensors and Bioelectronics, 96, 300–307. https://doi.org/10.1016/J.BIOS.2017.05.019

Guo, S., Song, Z., Ji, D. K., Reina, G., Fauny, J. D., Nishina, Y., Ménard-Moyon, C., & Bianco, A. (2022). Combined photothermal and photodynamic therapy for cancer treatment using a multifunctional graphene oxide. *Pharmaceutics, 14*(7), 1365. https://doi.org/10.3390/PHARMACEUTICS14071365/S1

Hao, X., Huang, L., Zhao, C., Chen, S., Lin, W., Lin, Y., Zhang, L., Sun, A., Miao, C., Lin, X., Chen, M., & Weng, S. (2021). Antibacterial activity of positively charged carbon quantum dots without detectable resistance for wound healing with mixed bacteria infection. *Materials Science & Engineering C, Materials for Biological Applications, 123*, 111971. https://doi.org/10.1016/J.MSEC.2021.111971

Hettiarachchi, S. D., Graham, R. M., Mintz, K. J., Zhou, Y., Vanni, S., Peng, Z., & Leblanc, R. M. (2019). Triple conjugated carbon dots as a nano-drug delivery model for glioblastoma brain tumors. *Nanoscale, 11*(13), 6192–6205. https://doi.org/10.1039/C8NR08970A

Jiang, Q., Liu, L., Li, Q., Cao, Y., Chen, D., Du, Q., Yang, X., Huang, D., Pei, R., Chen, X., & Huang, G. (2021). NIR-laser-triggered gadolinium-doped carbon dots for magnetic resonance imaging, drug delivery and combined photothermal chemotherapy for triple negative breast cancer. *Journal of Nanobiotechnology, 19*(1), 1–15. https://doi.org/10.1186/S12951-021-00811-W/TABLES/1

Kaurav, H., Verma, D., Bansal, A., Kapoor, D. N., & Sheth, S. (2023). Progress in drug delivery and diagnostic applications of carbon dots: A systematic review. *Frontiers in Chemistry, 11*, 1227843. https://doi.org/10.3389/FCHEM.2023.1227843/XML/NLM

Kong, T., Hao, L., Wei, Y., Cai, X., & Zhu, B. (2018). Doxorubicin conjugated carbon dots as a drug delivery system for human breast cancer therapy. *Cell Proliferation, 51*(5), e12488. https://doi.org/10.1111/CPR.12488

Kong, T., Zhou, R., Zhang, Y., Hao, L., Cai, X., & Zhu, B. (2020). AS1411 aptamer modified carbon dots via polyethylenimine-assisted strategy for efficient targeted cancer cell imaging. Cell Proliferation, 53(1). https://doi.org/10.1111/CPR.12713

Kuznietsova, H., Géloën, A., Dziubenko, N., Zaderko, A., Alekseev, S., Lysenko, V., & Skryshevsky, V. (2023). In vitro and in vivo toxicity of carbon dots with different chemical compositions. *Discover Nano, 18*(1), 1–17. https://doi.org/10.1186/S11671-023-03891-9/FIGURES/10

Li, S., Su, W., Wu, H., Yuan, T., Yuan, C., Liu, J., Deng, G., Gao, X., Chen, Z., Bao, Y., Yuan, F., Zhou, S., Tan, H., Li, Y., Li, X., Fan, L., Zhu, J., Chen, A. T., Liu, F., et al. (2020). Targeted tumour theranostics in mice via carbon quantum dots structurally mimicking large amino acids. *Nature Biomedical Engineering, 4*(7), 704–716. https://doi.org/10.1038/s41551-020-0540-y

Li, S., Li, L., Tu, H., Zhang, H., Silvester, D. S., Banks, C. E., Zou, G., Hou, H., & Ji, X. (2021). The development of carbon dots: From the perspective of materials chemistry. *Materials Today, 51*, 188–207. https://doi.org/10.1016/J.MATTOD.2021.07.028

Liu, J. J., Li, D., Zhang, K., Yang, M., Sun, H., & Yang, B. (2018). One-step hydrothermal synthesis of nitrogen-doped conjugated carbonized polymer dots with 31% efficient red emission for in vivo imaging. *Small, 14*(15), 1703919. https://doi.org/10.1002/SMLL.201703919

Lodha, S. R., Merchant, J. G., Pillai, A. J., Gore, A. H., Patil, P. O., Nangare, S. N., Kalyankar, G. G., Shah, S. A., Shah, D. R., & Patole, S. P. (2024). Carbon dot-based fluorescent sensors for pharmaceutical detection: Current innovations, challenges, and future prospects. *Heliyon, 10*(24), e41020. https://doi.org/10.1016/J.HELIYON.2024.E41020

Lu, J., Wang, K., Lei, W., Mao, Y., Di, D., Zhao, Q., & Wang, S. (2021). Polydopamine-carbon dots functionalized hollow carbon nanoplatform for fluorescence-imaging and photothermal-enhanced thermochemotherapy. *Materials Science and Engineering: C, 122*, 111908. https://doi.org/10.1016/J.MSEC.2021.111908

Ma, H., Li, S., Zhang, H., Wei, Y., & Jiang, L. (2019). Fabrication of polydopamine-based layer-by-layer nanocomposites for combined pH-sensitive chemotherapy and photothermal therapy. *Colloids and Surfaces A: Physicochemical and Engineering Aspects, 561*, 332–340. https://doi.org/10.1016/J.COLSURFA.2018.10.072

Malhan, A., Guleria, M., Das, U., Singh, S., Prajapati, B. G., Mohite, P., Bhattacharya, S., Chidrawar, V. R., Puri, A., & Datta, D. (2024). Navigating the future of cancer management through carbon nanodots: A review. *Nano-Structures & Nano-Objects, 39*, 101217. https://doi.org/10.1016/J.NANOSO.2024.101217

Medici, S., Peana, M., Coradduzza, D., & Zoroddu, M. A. (2021). Gold nanoparticles and cancer: Detection, diagnosis and therapy. *Seminars in Cancer Biology, 76*, 27–37. https://doi.org/10.1016/J.SEMCANCER.2021.06.017

Mewada, A., Pandey, S., Thakur, M., Jadhav, D., & Sharon, M. (2014). Swarming carbon dots for folic acid mediated delivery of doxorubicin and biological imaging. *Journal of Materials Chemistry B, 2*(6), 698–705. https://doi.org/10.1039/C3TB21436B

Mishra, V., Patil, A., Thakur, S., & Kesharwani, P. (2018). Carbon dots: Emerging theranostic nanoarchitectures. *Drug Discovery Today, 23*(6), 1219–1232. https://doi.org/10.1016/J.DRUDIS.2018.01.006

Nocito, G., Calabrese, G., Forte, S., Petralia, S., Puglisi, C., Campolo, M., Esposito, E., & Conoci, S. (2021). Carbon dots as promising tools for cancer diagnosis and therapy. *Cancers, 13*(9), 1991. https://doi.org/10.3390/CANCERS13091991

Oraevsky, A. A. (2018). Gold and silver nanoparticles as contrast agents for optoacoustic tomography. In *Photoacoustic imaging and spectroscopy* (pp. 373–386). https://doi.org/10.1201/9781420059922-40

Othman, H. O., Anwer, E. T., Ali, D. S., Hassan, R. O., Mahmood, E. E., Ahmed, R. A., Muhammad, R. F., & Smaoui, S. (2024). Recent advances in carbon quantum dots for gene delivery: A comprehensive review. *Journal of Cellular Physiology*. https://doi.org/10.1002/JCP.31236

Ozyurt, D., Al Kobaisi, M., Hocking, R. K., & Fox, B. (2023). Properties, synthesis, and applications of carbon dots: A review. *Carbon Trends, 12*, 100276. https://doi.org/10.1016/J.CARTRE.2023.100276

Pardo, J., Peng, Z., & Leblanc, R. M. (2018). Cancer targeting and drug delivery using carbon-based quantum dots and nanotubes. *Molecules, 23*(2), 378. https://doi.org/10.3390/MOLECULES23020378

Phuong, P. T. M., Won, H. J., Robby, A. I., Kim, S. G., Im, G. B., Bhang, S. H., Lee, G., & Park, S. Y. (2020). NIRvis-Induced pH-Sensitive TiO2Immobilized Carbon Dot for Controllable Membrane-Nuclei Targeting and Photothermal Therapy of Cancer Cells. ACS Applied Materials and Interfaces, 12(34), 37929–37942. https://doi.org/10.1021/ACSAMI.0C11979/SUPPL_FILE/AM0C11979_SI_001.PDF

Qi, K., Sun, B., Liu, S.-Y., & Zhang, M. (2023). Research progress on carbon materials in tumor photothermal therapy. *Biomedicine & Pharmacotherapy, 165*, 115070. https://doi.org/10.1016/J.BIOPHA.2023.115070

Qian, J., Quan, F., Zhao, F., Wu, C., Wang, Z., & Zhou, L. (2018). Aconitic acid derived carbon dots: Conjugated interaction for the detection of folic acid and fluorescence targeted imaging of folate receptor overexpressed cancer cells. Sensors and Actuators B: Chemical, 262, 444–451. https://doi.org/10.1016/J.SNB.2018.01.227

Qureshi, Z. A., Dabash, H., Ponnamma, D., & Abbas, M. K. G. (2024). Carbon dots as versatile nanomaterials in sensing and imaging: Efficiency and beyond. *Heliyon, 10*(11), e31634. https://doi.org/10.1016/J.HELIYON.2024.E31634

Rallis, K. S., Yau, T. H. L., & Sideris, M. (2021). Chemoradiotherapy in cancer treatment: Rationale and clinical applications. *Anticancer Research, 41*(1), 1–7. https://doi.org/10.21873/ANTICANRES.14746

Ray, P., Moitra, P., & Pan, D. (2022). Emerging theranostic applications of carbon dots and its variants. *View, 3*(2), 20200089. https://doi.org/10.1002/VIW.20200089

Rodríguez-Varillas, S., Fontanil, T., Obaya, Á. J., Fernández-González, A., Murru, C., & Badía-Laíño, R. (2022). Biocompatibility and antioxidant capabilities of carbon dots obtained from tomato (Solanum lycopersicum). *Applied Sciences, 12*(2), 773. https://doi.org/10.3390/APP12020773

Saljoughi, H., Khakbaz, F., & Mahani, M. (2020). Synthesis of folic acid conjugated photoluminescent carbon quantum dots with ultrahigh quantum yield for targeted cancer cell fluorescence imaging. *Photodiagnosis and Photodynamic Therapy, 30*, 101687. https://doi.org/10.1016/J.PDPDT.2020.101687

Sameer, M., Arif, Y., Aqil, A., Nadaf, A., Rafiya, K., Hasan, N., Kesharwani, P., & Ahmad, F. J. (2023). Carbon nanodots as a remedial nanovesicles for drug delivery. *European Polymer Journal, 200*, 112515. https://doi.org/10.1016/J.EURPOLYMJ.2023.112515

Seçme, M., & İlhan, H. (2025). Synthesis and characterization of thymol carbon nanodot functionalized silver nanoparticles (ThCND-AgNPs) and evaluation of their antiproliferative, anti-invasive, and apoptotic effects on OVCAR-3 ovarian cancer cells. *Microscopy Research and Technique, 88*(3), 668–677. https://doi.org/10.1002/JEMT.24724

Sengar, P., Chauhan, K., & Hirata, G. A. (2022). Progress on carbon dots and hydroxyapatite based biocompatible luminescent nanomaterials for cancer theranostics. *Translational Oncology, 24*, 101482. https://doi.org/10.1016/J.TRANON.2022.101482

Shen, C. L., Liu, H. R., Lou, Q., Wang, F., Liu, K. K., Dong, L., & Shan, C. X. (2022). Recent progress of carbon dots in targeted bioimaging and cancer therapy. *Theranostics, 12*(6), 2860. https://doi.org/10.7150/THNO.70721

Shu, Y., Linghu, X., Zhao, Y., Chen, Z., Zhang, J., Shan, D., Liu, W., Di, M., & Wang, B. (2023). Photodynamic and photothermal therapy-driven synergistic cancer treatment assisted by zeolitic imidazolate framework-8: A review. *Journal of Drug Delivery Science and Technology, 81*, 104272. https://doi.org/10.1016/J.JDDST.2023.104272

Siegel, R. L., Miller, K. D., Wagle, N. S., & Jemal, A. (2023). Cancer statistics, 2023. *CA: a Cancer Journal for Clinicians, 73*(1), 17–48. https://doi.org/10.3322/CAAC.21763

Singh, I., Arora, R., Dhiman, H., & Pahwa, R. (2018). Carbon quantum dots: Synthesis, characterization and biomedical applications. *Turkish Journal of Pharmaceutical Sciences, 15*(2), 219. https://doi.org/10.4274/TJPS.63497

Singh, G., Kaur, H., Sharma, A., Singh, J., Alajangi, H. K., Kumar, S., Singla, N., Kaur, I. P., & Barnwal, R. P. (2021). Carbon based nanodots in early diagnosis of cancer. *Frontiers in Chemistry, 9*, 669169. https://doi.org/10.3389/fchem.2021.669169

Soumya, K., More, N., Choppadandi, M., Aishwarya, D. A., Singh, G., & Kapusetti, G. (2023). A comprehensive review on carbon quantum dots as an effective photosensitizer and drug delivery system for cancer treatment. *Biomedical Technology, 4*, 11–20. https://doi.org/10.1016/J.BMT.2023.01.005

Sri, S., Kumar, R., Panda, A. K., & Solanki, P. R. (2018) Highly Biocompatible, Fluorescence, and Zwitterionic Carbon Dots as a Novel Approach for Bioimaging Applications in Cancerous Cells. ACS Applied Materials and Interfaces, 10(44), 37835–37845. https://doi.org/10.1021/ACSAMI.8B13217/SUPPL_FILE/AM8B13217_SI_001.PDF

Sun, S., Zhang, L., Jiang, K., Wu, A., & Lin, H. (2016). Toward high-efficient red emissive carbon dots: Facile preparation, unique properties, and applications as multifunctional theranostic agents. *Chemistry of Materials, 28*(23), 8659–8668. https://doi.org/10.1021/ACS.CHEMMATER.6B03695/SUPPL_FILE/CM6B03695_SI_001.PDF

Sun, L., Zhao, Y., Peng, H., Zhou, J., Zhang, Q., Yan, J., Liu, Y., Guo, S., Wu, X., & Li, B. (2024). Carbon dots as a novel photosensitizer for photodynamic therapy of cancer and bacterial infectious diseases: Recent advances. *Journal of Nanobiotechnology, 22*(1), 1–24. https://doi.org/10.1186/S12951-024-02479-4

Tang, J., Kong, B., Wu, H., Xu, M., Wang, Y., Wang, Y., Zhao, D., & Zheng, G. (2013). Carbon nanodots featuring efficient FRET for real-time monitoring of drug delivery and two-photon imaging. *Advanced Materials (Deerfield Beach, Fla.), 25*(45), 6569–6574. https://doi.org/10.1002/ADMA.201303124

Tang, F., Ding, A., Xu, Y., Ye, Y., Li, L., Xie, R., & Huang, W. (2024). Gene and photothermal combination therapy: Principle, materials, and amplified anticancer intervention. *Small, 20*(6), 2307078. https://doi.org/10.1002/SMLL.202307078

Tiernan, J. P., Perry, S. L., Verghese, E. T., West, N. P., Yeluri, S., Jayne, D. G., & Hughes, T. A. (2013). Carcinoembryonic antigen is the preferred biomarker for in vivo colorectal cancer targeting. *British Journal of Cancer, 108*(3), 662–667. https://doi.org/10.1038/BJC.2012.605

Tiron, C. E., Luta, G., Butura, M., Zugun-Eloae, F., Stan, C. S., Coroaba, A., Ursu, E. L., Stanciu, G. D., & Tiron, A. (2020). NHF-derived carbon dots: Prevalidation approach in breast cancer treatment. *Scientific Reports, 10*(1), 1–10. https://doi.org/10.1038/s41598-020-69670-z

Traverso, N., Ricciarelli, R., Nitti, M., Marengo, B., Furfaro, A. L., Pronzato, M. A., Marinari, U. M., & Domenicotti, C. (2013). Role of glutathione in cancer progression and chemoresistance. *Oxidative Medicine and Cellular Longevity, 2013*. https://doi.org/10.1155/2013/972913

Unnikrishnan, B. S., Sen, A., Preethi, G. U., Joseph, M. M., Maya, S., Shiji, R., Anusree, K. S., & Sreelekha, T. T. (2021). Folic acid-appended galactoxyloglucan-capped iron oxide nanoparticles as a biocompatible nanotheranostic agent for tumor-targeted delivery of doxorubicin. *International Journal of Biological Macromolecules, 168*, 130–142. https://doi.org/10.1016/J.IJBIOMAC.2020.11.205

Valenti, G. E., Tasso, B., Traverso, N., Domenicotti, C., & Marengo, B. (2023). Glutathione in cancer progression and chemoresistance: An update. *Redox Experimental Medicine, 2023*(1). https://doi.org/10.1530/REM-22-0023

Vargason, A. M., Anselmo, A. C., & Mitragotri, S. (2021). The evolution of commercial drug delivery technologies. *Nature Biomedical Engineering, 5*(9), 951–967. https://doi.org/10.1038/s41551-021-00698-w

Wang, Z., Liao, H., Wu, H., Wang, B., Zhao, H., & Tan, M. (2015). Fluorescent carbon dots from beer for breast cancer cell imaging and drug delivery. *Analytical Methods, 7*(20), 8911–8917. https://doi.org/10.1039/C5AY01978H

Wibrianto, A., Putri, F. S. D., Nisa, U. K., Mahyahani, N., Sugito, S. F. A., Wardana, A. P., Sakti, S. C. W., Chang, J. Y., & Fahmi, M. Z. (2024). Strategic assessment of boron-enriched carbon dots/naproxen: Diagnostic, toxicity, and in vivo therapeutic evaluation. *Molecular Pharmaceutics, 21*(2), 801–812. https://doi.org/10.1021/ACS.MOLPHARMACEUT.3C00919/SUPPL_FILE/MP3C00919_SI_001.PDF

Wu, L., Wang, J., Ren, J., Lia, W., & Qu, X. (2013). Highly sensitive electrochemiluminescent cytosensing using carbon nanodot@Ag hybrid material and graphene for dual signal amplification. *Chemical Communications, 49*(50), 5675–5677. https://doi.org/10.1039/C3CC42637H

Xu, H., Zhang, Y., Zhang, H., Zhang, Y., Xu, Q., Lu, J., Feng, S., Luo, X., Wang, S., & Zhao, Q. (2023). Smart polydopamine-based nanoplatforms for biomedical applications: State-of-art and further perspectives. *Coordination Chemistry Reviews, 488*, 215153. https://doi.org/10.1016/J.CCR.2023.215153

Yang, J., Cheng, Q., Zhu, C. C., Huang, H. Y., Su, J. B., Ni, W., Ruan, B., Ma, N., & Tsai, F. C. (2022). Folic acid conjugated carbon dots doped metal-organic framework materials for cell imaging. *Materials Today Chemistry, 26*, 101202. https://doi.org/10.1016/J.MTCHEM.2022.101202

Zeng, Q., Shao, D., He, X., Ren, Z., Ji, W., Shan, C., Qu, S., Li, J., Chen, L., & Li, Q. (2016). Carbon dots as a trackable drug delivery carrier for localized cancer therapy in vivo. *Journal of Materials Chemistry B, 4*(30), 5119–5126. https://doi.org/10.1039/C6TB01259K

Zhang, Y., Shen, Y., Teng, X., Yan, M., Bi, H., & Morais, P. C. (2015). Mitochondria-targeting nanoplatform with fluorescent carbon dots for long time imaging and magnetic field-enhanced cellular uptake. *ACS Applied Materials and Interfaces, 7*(19), 10201–10212. https://doi.org/10.1021/ACSAMI.5B00405/SUPPL_FILE/AM5B00405_SI_001.PDF

Zhang, S. L., Xu, J. J., Shen, Y., Wu, T., Wang, Y., Zhao, X., & Chen, H. Y. (2021). Nucleolin-targeted ratiometric fluorescent carbon dots with a remarkably large emission wavelength shift for precise imaging of cathepsin B in living cancer cells. Analytical Chemistry, 93(8), 4042–4050. https://doi.org/10.1021/ACS.ANALCHEM.0C05046/SUPPL_FILE/AC0C05046_SI_001.PDF

Zhao, X., Zhang, J., Shi, L., Xian, M., Dong, C., & Shuang, S. (2017). Folic acid-conjugated carbon dots as green fluorescent probes based on cellular targeting imaging for recognizing cancer cells. *RSC Advances, 7*(67), 42159–42167. https://doi.org/10.1039/C7RA07002K

Zhao, L., Zhang, X., Wang, X., Guan, X., Zhang, W., & Ma, J. (2021). Recent advances in selective photothermal therapy of tumor. *Journal of Nanobiotechnology, 19*(1), 1–15. https://doi.org/10.1186/S12951-021-01080-3/FIGURES/7

Zheng, M., Liu, S., Li, J., Qu, D., Zhao, H., Guan, X., Hu, X., Xie, Z., Jing, X., & Sun, Z. (2014). Integrating oxaliplatin with highly luminescent carbon dots: An unprecedented theranostic agent for personalized medicine. *Advanced Materials, 26*(21), 3554–3560. https://doi.org/10.1002/ADMA.201306192

Zheng, M., Ruan, S., Liu, S., Sun, T., Qu, D., Zhao, H., Xie, Z., Gao, H., Jing, X., & Sun, Z. (2015). Self-targeting fluorescent carbon dots for diagnosis of brain cancer cells. *ACS Nano, 9*(11), 11455–11461. https://doi.org/10.1021/ACSNANO.5B05575/SUPPL_FILE/NN5B05575_SI_001.PDF

Zuo, J., Jiang, T., Zhao, X., Xiong, X., Xiao, S., & Zhu, Z. (2015). Preparation and application of fluorescent carbon dots. *Journal of Nanomaterials, 2015*(1), 787862. https://doi.org/10.1155/2015/787862

Part III

Carbon Nanodots in Food and Agriculture

Carbon Dots in Food and Beverages: Formation, Benefits, Risks, and Toxicity 11

Duyen H. H. Nguyen and József Prokisch

Abstract

Carbon dots (CDs) are fluorescent nanoparticles formed during food processing through high-temperature reactions such as the Maillard reaction, caramelization, and pyrolysis. These particles, generally less than 10 nm in size, exhibit unique optical and chemical properties, making them suitable for applications in biosensing and food safety testing. Despite their promising potential, concerns regarding their toxicity and long-term effects on human health persist, necessitating further research. The integration of CDs in food applications holds significant promise, but comprehensive validation studies are essential to ensure their safety and efficacy. Continued exploration of novel synthesis techniques and functionalization strategies will likely lead to further advancements in their use within the food industry.

D. H. H. Nguyen (✉)
Institute of Animal Science, Faculty of Agricultural and Food Sciences and Environmental Management, Biotechnology and Nature Conservation, University of Debrecen, Debrecen, Hungary

Institute of Life Sciences, Vietnam Academy of Science and Technology, Ho Chi Minh City, Vietnam

Doctoral School of Nutrition and Food Science, University of Debrecen, Debrecen, Hungary
e-mail: nguyen.huu.huong.duyen@agr.unideb.hu

J. Prokisch
Institute of Animal Science, Faculty of Agricultural and Food Sciences and Environmental Management, Biotechnology and Nature Conservation, University of Debrecen, Debrecen, Hungary

M. Rai et al. (eds.), *Carbon Nanodots*, https://doi.org/10.1007/978-981-95-4705-0_11

Keywords

Carbon nanodots · Food applications · Sensing · Food safety · Mitigating strategies

11.1 Introduction

Carbon dots (CDs) are a fascinating class of carbon-based nanomaterials that have gained considerable attention over the past decade due to their unique optical properties, excellent water solubility, tunable photoluminescence, and remarkable biocompatibility. Typically less than 10 nm in size, CDs are composed primarily of carbon, along with other heteroatoms such as hydrogen, oxygen, and nitrogen, which contribute to their surface functionality and diverse applications (Wang et al., 2020). Initially discovered as synthetic nanomaterials, CDs are now known to also form naturally during high-temperature food processing, raising questions about their formation, functionality, and potential implications for food safety and human health. In 2012, CDs were first reported in food caramels such as bread and caramelized sugars (Sk et al., 2012).

During food processing, particularly under thermal treatments such as roasting, grilling, baking, frying, and boiling, complex chemical reactions—including Maillard reactions and caramelization—can lead to the generation of CDs (H. Wang et al., 2020). These nanoparticles have been identified in a variety of processed food items, ranging from grilled meats (Li et al., 2017) and roasted coffee (Nguyen et al., 2024b) to caramelized sugars (Sk et al., 2012), and even in orange juice (Sahu et al., 2012). The composition, size, and surface chemistry of these food-derived CDs are influenced by several factors, including the type of raw material, heating temperature, duration, and moisture content. As such, even subtle changes in cooking conditions may significantly alter the physicochemical characteristics and biological interactions of CDs.

One of the most intriguing aspects of CDs is their intrinsic fluorescence, which allows them to be used in a wide range of applications, including biosensing (Buiculescu et al., 2016), food quality monitoring (Nguyen et al., 2024a), and biomedical imaging (Mancini et al., 2023). In the context of food science, their ability to act as optical sensors for detecting spoilage (Zhao et al., 2022), contaminants (Xu et al., 2025), or harmful additives (Su et al., 2022) offers great promise. In addition, their relatively low cytotoxicity and good biocompatibility, particularly in comparison to other engineered nanomaterials, make them attractive candidates for future integration into food packaging, nutraceuticals, and smart labeling systems (Manzoor et al., 2023; Wang et al., 2020).

However, despite their appealing properties and functional versatility, the presence of CDs in the food supply also raises important safety concerns (Nguyen et al., 2024a). While some studies have reported low toxicity and minimal adverse effects at relevant concentrations, others have indicated that CDs formed under extreme thermal conditions may exhibit increased cytotoxicity, oxidative stress potential, or interactions with cellular components (Li et al., 2017). Moreover, current toxicological assessments are still limited, with most studies focusing on short-term in vitro experiments or small animal models. The lack of comprehensive long-term in vivo studies, particularly those mimicking realistic human exposure scenarios, presents a significant gap in our understanding of the health impacts associated with chronic ingestion of CDs through the diet.

Given these dual perspectives—potential benefits and safety uncertainties—it is essential to systematically investigate the origin, behavior, and biological effects of CDs in food systems. This requires an interdisciplinary approach combining food chemistry, nanotechnology, toxicology, and materials science.

In this chapter, we aim to provide a comprehensive overview of carbon dots originating from food and beverages. We begin by discussing the synthesis and formation pathways of CDs during food processing, including both naturally occurring and deliberately synthesized examples. Next, we highlight the emerging applications and functional advantages of CDs in food science and technology. This is followed by an evaluation of the associated risks and toxicological profiles, based on the latest findings from both in vitro and in vivo studies. Finally, we explore risk mitigation strategies and propose future research directions needed to ensure the safe use of CDs in the food industry while leveraging their beneficial attributes.

11.2 Synthesis and Sources of Carbon Dots in Food and Beverages

11.2.1 Natural Formation During Food Processing

Carbon dots (CDs) can naturally form during food processing, especially through high-temperature reactions that vary based on the food precursor and processing temperature. Their presence in food is primarily linked to thermal chemical processes, such as the Maillard reaction, caramelization, and pyrolysis. These processes not only enhance the color, flavor, and texture of foods but also create fluorescent carbon-based nanoparticles with unique optical and chemical properties.

11.2.1.1 Mechanisms of Carbon Dot Formation in Food Processing

The formation of CDs in foods is mostly related to the degradation of food components, such as carbohydrates, proteins, and lipids, at high temperature (Wang et al., 2020). The polymerization of these precursors under specific conditions can lead to the formation of highly soluble and fluorescent CDs (Shi et al., 2014). The longer heat treatments, the smaller CDs with higher yield and stronger fluorescence intensity (Nguyen et al., 2024a). The presence of heteroatoms, such as nitrogen, can further enhance the optical properties and yield of the CDs (Shi et al., 2014). There are three key reactions related to this formation, including the following:

- *Maillard reaction*: This reaction occurs when reducing sugars interact with amino acids while processing foods at high temperatures. The Maillard reaction is crucial for developing unique flavors in heat-processed foods such as bread, grilled meats, and fried potatoes by creating a range of aromatic compounds and a brown coloration. Recently, the formation of CDs alongside other Maillard reaction products has been reported (Li et al., 2018; Nguyen et al., 2024c). Additionally, CDs have been noted to form from by-products, specifically melanoidins (Li et al., 2019).
- *Caramelization*: The thermal decomposition of sugars at high temperatures results in complex aromatic and colored compounds, leading to the formation of CDs. In 2012, CDs were first reported in caramelized sugars, opening a new journey of studying CDs in foods (Sk et al., 2012). The extent of CDs generation depends on temperature, pH, and sugar type (Nguyen et al., 2024c).
- *Pyrolysis*: As high temperatures close to the formation of CDs, pyrolysis, the high-temperature decomposition of organic matter occurs in the absence of oxygen, leading to the direct conversion of biomolecules into carbonaceous nanomaterials, including CDs. Pyrolysis-driven CD formation is relevant in charred or heavily roasted foods (Li et al., 2017; Song et al., 2018; Wang et al., 2020).

These dry heat methods typically involve higher temperatures (200–350 °C), resulting in variations in CD size, fluorescence intensity, and chemical composition depending on the food precursor and processing conditions. Additionally, wet heat methods such as brewing can also lead to CD formation, albeit at lower temperatures compared to dry heat methods, highlighting the diverse pathways through which CDs can emerge in processed foods. Furthermore, some studies reported that chemical treatment can lead to the formation of CDs like the presence of tetraethylenepentamine (TEPA) in fruit affected enhancing carbon–nitrogen (C–N) groups of CDs synthesis (De Yro et al., 2019) or the presence of sodium hydroxide (NaOH) in the immersion solution n significantly impacted CDs formation in pretzel production (Semsey et al., 2024). Depending on the food sources and the process methods, the properties and size of CDs could be varied (Table 11.1).

The process of creating CDs in processed foods was outlined by Wang et al. (2020) (Fig. 11.1). It begins with the pyrolysis of food components such as proteins, carbohydrates, or lipids. This is followed by polymerization, which produces a fluorescent polymer chain, ultimately leading to the formation of a CD core that exhibits strong fluorescence.

11.2.1.2 Occurrence of Carbon Dots in Common Foods

Various cooking methods, such as frying, roasting, grilling, and baking, can create CDs, as shown in Table 11.1. Additionally, naturally occurring CDs have been found in several foods that undergo thermal processing:

- *Roasted coffee*: The Maillard reaction and caramelization during roasting result in the formation of CDs. These nanoparticles contribute to coffee's unique

Table 11.1 Reported carbon dots in processed foods

Food source	Processing method	Temperature range	Particle size range	Fluorescence properties	Chemical structure	References
Food model, sandwich leftover	Hydrothermal carbonization	–	–	Lifetime increased with temperature	Pyridinic N and graphitic N increased; amino N and pyrrolic N decreased	Ahn et al. (2019)
Kangkung (*Ipomoea aquatica*)—roots, stems, leaves	Frying and roasting	250 °C, 88 °C	1–10 nm	Red (674–677 nm) and green (500–510 nm) for frying; green (511–519 nm) for roasting	Carbon–carbon double bond (C=C), carbonyl group (C=O), carbon–hydrogen bond (C–H), hydroxyl group (O–H) functional groups	Dwandaru et al. (2019)
Hamburger	Grilling (pyrolysis)	220 °C, 260 °C, 300 °C	Decreased with temperature increase	Quantum yield: 23.25% at 260 °C	–	Li et al. (2017)
Arabica coffee beans	Brewing	Over 90 °C	3.2 ± 0.9 nm	Excitation wavelength at 370 nm	–	Nguyen et al. (2024b)
Pretzels	Baking	200 °C	3.5 nm	Excitation wavelength at 380 nm	–	Semsey et al. (2024)
Bread (outer brown part), jaggery, sugar caramel, corn flakes, biscuits	Heating (caramelization)	180 °C, 200 °C	4–30 nm (varied by food source)	Quantum yields: 1.2% (bread), 0.55% (jaggery), 0.63% (sugar caramel).	Coated with hydrophilic carbohydrate units	Sk et al. (2012)
Chicken	Roasting	200 °C, 250 °C, 300 °C	17.1–2.1 nm (depend on temperature)	Quantum yields: 6.71% (200 °C), 12.85% (250 °C), 17.46% (300 °C)	–	Song et al. (2018)
Lamb	Baking	200 °C, 300 °C, 350 °C	–	Quantum yield increased with temperature	Surface functional groups changed with temperature	Wang et al. (2020)

(continued)

Table 11.1 (continued)

Food source	Processing method	Temperature range	Particle size range	Fluorescence properties	Chemical structure	References
Breadcrumbs	Frying	180 °C	–	–	–	Yin et al. (2014)
Calamansi waste, pineapple waste	Hydrothermal treatment	–	–	Strong blue fluorescence at 365 nm	Carbon–nitrogen (C–N) functional groups	De Yro et al. (2019)
Mature vinegar	Isolated from Chinese mature vinegar	–	0.5–2.5 nm	Quantum yields: 5.71%	–OH;C=O; –COOH	Cao et al. (2017)
Pizza	Isolated from pizza	–	~3.33 nm	Quantum yields: 2.14%	C=O; –OH; –NH2; COOH	Cong et al. (2019)
Croaker	Isolated from canned yellow croaker	–	1.8–5.8 nm	Quantum yields: 9.7%	C–O–C; C=O; –OH;CO–NH	Li et al. (2018)
Pike Eel	Grilling	160 °C, 200 °C, 230 °C, 260 °C, 300 °C	1.75–4.2 nm	Quantum yields:12.86–80.16%	C=O; C=C; C-O; –OH	Bi et al. (2018)

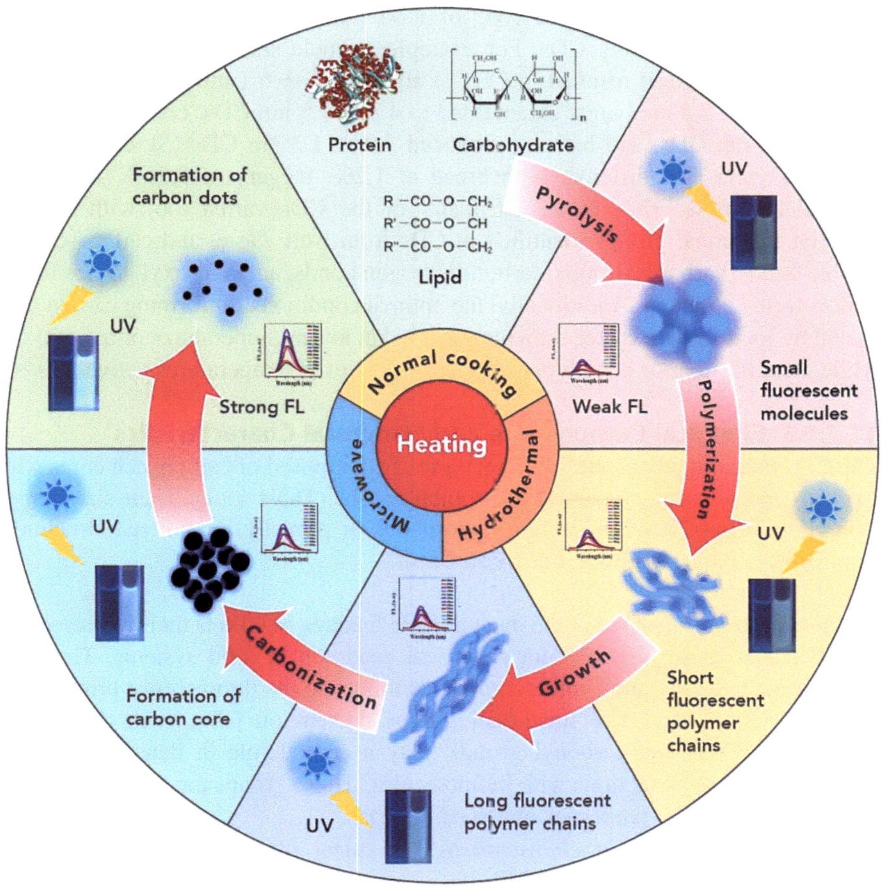

Fig. 11.1 The formation mechanism of carbon dots (CDs) in processed foods. (Reprinted from Wang et al., 2020)

fluorescence properties and may influence its antioxidant capacity (Nguyen et al., 2024b).

- *Grilled meats*: The thermal breakdown of proteins and fats at high temperatures leads to the generation of CDs, particularly in charred portions (Li et al., 2017; Song et al., 2018). These nanoparticles may coexist with other heat-induced compounds, some of which are linked to health concerns.
- *Baked goods*: CDs form as a by-product of the Maillard reaction and caramelization in bread crusts, cookies, and other baked products (Semsey et al., 2024; Sk et al., 2012). Their formation depends on baking temperature, sugar content, and protein composition.

As shown in Fig. 11.1, the type of food material significantly impacts the properties of the resulting CDs. For example, particle sizes varied depending on the food source: bread resulted in CDs of about 27.5 ± 6.1 nm, jaggery produced CDs of 20.3 ± 7.5 nm, sugar caramel led to 4.3 ± 1.5 nm CDs, cornflakes yielded 10.5 ± 2.8 nm CDs, and biscuits produced 3.9 ± 1.3 nm CDs (Sk et al., 2012). Quantum yield also differed, with bread at 1.2%, jaggery at 0.55%, and sugar caramel at 0.63%. The surface chemistry of the CDs varied too, with carbon-nitrogen functional groups identified in CDs from fruit waste, and carbon–carbon double bonds, carbonyl groups, carbon-hydrogen bonds, and hydroxyl groups found in dots from kangkung. Additionally, the optimal conditions for forming carbon dots varied by food source; coffee brewing occurred at lower temperatures of over 90 °C, while meat roasting or baking processes typically used temperatures above 200 °C.

11.2.1.3 Chemical Composition and Structural Characteristics

The CDs found in processed foods are typically composed of carbon-rich cores with oxygen- and nitrogen-containing functional groups (Table 11.1). Their size ranges from 1 to 10 nm, and they exhibit amorphous or partially graphitized structures. Several key characteristics define these CDs:

- *Optical properties*: One of the most notable features of CDs is their fluorescence, which can cause difficulty for analytical detection in food systems. The size, surface chemistry, and synthesis methods influence the fluorescence properties.
- *Surface chemistry*: The functional groups present on CDs, such as carbon-nitrogen (C-N) and carbonyl groups, play a crucial role in determining their solubility and interactions with biomolecules. These groups are also responsible for the fluorescent properties observed in CDs.
- *Temperature-dependent characteristics*: In meat products and baked goods subjected to dry heat (200–350 °C), higher temperatures consistently yield smaller particles and higher fluorescence quantum yields. For example, CDs formed from roasted chicken decrease in size from 17.1 nm at 200 °C to 2.1 nm at 300 °C, with quantum yields rising from 6.71% to 17.46% (Song et al., 2018). Grilling hamburgers shows similar temperature-dependent reductions in particle size and modifications in morphology (Li et al., 2017), while coffee brewing at over 90 °C produces dots averaging 3.2 nm (Nguyen et al., 2024b). In baked goods, particle sizes range widely—from 3.9 nm in biscuits to 27.5 nm in bread (Sk et al., 2012)—and vary along with differences in fluorescence and surface chemistry.

Food-derived CDs commonly reported with key findings are summarized in Table 11.1. A decrease in particle size was observed with increasing temperature, with sizes ranging from 1 to 20 nm. Quantum yield also varied based on the precursors (food varieties), heat treatment temperature, and processing method.

11.2.2 Synthetic Incorporation in Food Applications

Beyond their natural occurrence, CDs are also intentionally synthesized for applications in the food industry. Due to their unique properties and ease of synthesis, CDs have been explored as food additives; sensors for traceability food compositions, detecting additives residuals and foodborne disease, or smart packaging materials (Moradi et al., 2023; Shi et al., 2019; Zhao et al., 2023). These CDs can be produced from environmentally friendly sources, including food, food by-products, and microorganisms (Moradi et al., 2023).

Various synthesis techniques allow for the controlled production of CDs with tailored properties for specific food-related functions. Each technique offers different advantages regarding scalability, purity, and physicochemical properties (Chen et al., 2023; Shi et al., 2019). Three primary synthesis techniques are commonly applied to obtain CDs:

- *Hydrothermal synthesis*: Organic precursors (e.g., citric acid, glucose, or natural sources) are heated in water under high pressure, promoting carbonization and nanoparticle formation (De Yro et al., 2019; Sa et al., 2017). The resulting CDs typically have dimensions ranging from 10 to 22 nm and exhibit strong photoluminescence in the violet-blue region (Tomskaya et al., 2018). The luminescence properties of CDs can be tuned by varying the excitation wavelength and are influenced by particle size and surface functional groups (Smagulova et al., 2017). Nitrogen doping, achieved by adding ammonia during synthesis, can enhance the electronic and transport properties of CDs (Tomskaya et al., 2018).
- *Solvothermal synthesis*: Similar to hydrothermal synthesis, but organic solvents replace water, allowing for the tuning of properties of CDs (e.g., hydrophobicity, fluorescence). For example, CDs with high photoluminescence and quantum yields have been produced from orange juice (Sahu et al., 2012) and food waste (Park et al., 2014), exhibiting excellent biocompatibility and imaging capabilities. Banana juice has also been used to create green luminescent CDs with pH-dependent properties (De & Karak, 2013).
- *Microwave-assisted synthesis*: Precursors are rapidly heated using microwave irradiation, enabling fast and energy-efficient CD production with controlled particle size and surface properties. Various food-derived precursors, including lemon juice (MoziOğlu, 2022), carbohydrates (Wang et al., 2011b), eggshell membrane (Wang et al., 2012), and denatured proteins (Liu et al., 2016), have been used to synthesize CDs using microwaves. Additionally, CDs produced from lemon juice exhibit antimicrobial effects against various pathogens (MoziOğlu, 2022). The microwave-assisted approach offers advantages over traditional hydrothermal methods, including shorter reaction times and simpler procedures (MoziOğlu, 2022; Wang et al., 2011b). Additionally, highly crystalline CDs with UV emission and quantum confinement have been synthesized from fresh tomatoes using microwave-assisted pyrolysis (Liu et al., 2017).

Each method impacts the yield, optical behavior, and functional properties of CDs, making the choice of synthesis technique crucial for specific food applications. However, the quantum yield of CDs may vary depending on the synthesis method and precursor used (MoziOğlu, 2022; Wang et al., 2012). Besides, various functionalization strategies are employed to modify the properties of CDs depending on the application purpose:

- *Surface modification*: Functional groups (e.g., carboxyl, amine) are introduced to improve dispersibility and stability in food matrices.
- *Elemental doping*: Nitrogen (N), sulfur (S), and phosphorus (P) doping alter fluorescence and antioxidant activity, making CDs more suitable for sensing or preservation.
- *Encapsulation*: CDs are embedded within biopolymers (e.g., starch, chitosan) to ensure controlled release and minimize potential toxicity in food applications.

The integration of synthetic CDs in food applications holds *promising potential* but requires further studies on *bioavailability, interaction with food components, and long-term safety assessments* (Nguyen et al., 2024a)

11.3 Benefits of Carbon Dots in Food and Beverages

11.3.1 Functional Properties

Besides the fluorescence ability of carbon dots, which can turn on-off depending on the interacting compound, carbon dots (CDs) exhibit a range of functional properties like antioxidant activity or antimicrobial properties that make them valuable in various applications, particularly in the food and beverage industry, illustrated in Fig. 11.2.

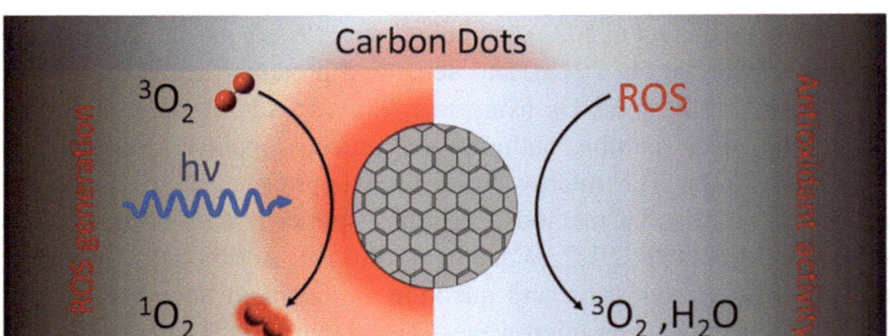

Fig. 11.2 Carbon dots as antioxidant material. (Reprinted from Innocenzi & Stagi, 2023)

11.3.1.1 Antioxidant Activity

Studies indicate that CDs show significant scavenging activities against various radicals, such as DPPH,·OH, and O_2·-(Rodríguez-Varillas et al., 2022; Zhao et al., 2022). Additionally, CDs demonstrate effective antioxidant activities that help mitigate lipid oxidation, which can negatively impact the texture, color, and overall appearance of food products. By reducing these oxidative processes, CDs improve the stability and shelf life of food items, ensuring consistent quality for consumers. When added to frying oils, CDs notably decreased lipid and protein oxidation in fried meatballs during refrigerated storage, surpassing the performance of some conventional antioxidants (Zhao et al., 2022).

Moreover, CDs derived from natural sources like tomatoes exhibit higher antioxidant power than previously reported CDs (Rodríguez-Varillas et al., 2022). The integration of CDs into food packaging materials can enhance mechanical strength, UV barrier properties, and water barrier properties, while also providing antioxidant and antibacterial functionalities (Zhao et al., 2022). When incorporated into polymers like polyvinyl alcohol or gelatin/chitosan, CDs impart antioxidant and antimicrobial properties, extending the shelf life of various foods (Fu et al., 2022; Zhao et al., 2022). These nanoparticles can be synthesized from environmentally sustainable sources, including microorganisms and food by-products (Moradi et al., 2023). Some CDs can function as both oxidants and antioxidants depending on their structure and exposure to light, emphasizing the importance of understanding structure–property relationships for specific applications (Innocenzi & Stagi, 2023).

11.3.1.2 Antimicrobial Properties

Carbon dots (CDs) are reported to inhibit microbial growth through mechanisms such as reactive oxygen species (ROS) generation, positioning them as promising agents for extending the shelf life of food products and beverages while maintaining quality and safety standards (Chaudhary et al., 2025; Lin et al., 2022; Mancini et al., 2023). One of the primary modes of action is the generation of reactive oxygen species (ROS), which play a crucial role in inhibiting the growth of bacteria and fungi. The interaction of CDs with microbial cells increases the levels of endogenous ROS, leading to oxidative stress and subsequent cellular damage, ultimately impairing microbial viability (Collins et al., 2025; Mancini et al., 2023). Research has demonstrated that incorporating CDs into food packaging materials significantly enhances their antimicrobial properties. Studies reveal that CDs can reduce the viability of foodborne pathogens, such as *Escherichia coli* and *Listeria monocytogenes,* when integrated into biopolymer-based films (Collins et al., 2025; Sheikh et al., 2023). Additionally, the capacity of CDs to decrease oxygen transmission rates within packaging materials further inhibits microbial growth by creating an oxygen-free environment, which is essential for prolonging the freshness of food (Dong et al., 2017). These qualities have garnered attention in an era where approximately one-third of global food production is wasted due to spoilage. While CDs show promise as natural antimicrobial agents, their interactions with food components and microbial resistance patterns require further investigation.

11.3.2 Technological Applications

Beyond their biological benefits, CDs exhibit unique physicochemical properties that enable their application in food technology. Their fluorescent nature, antioxidant and antimicrobial effects, and ability to function as biosensors make them suitable for innovative food applications like monitoring nutrients, enhancing food packaging, detecting contaminants, or detecting foodborne pathogens (Fig. 11.3). In food safety, CDs have been utilized as fluorescent probes for detecting contaminants such as metal ions, pesticides, and banned additives (Shi et al., 2019). Furthermore, CDs have shown potential in developing sustainable technologies for rapid detection of food contaminants and nutrients (Manzoor et al., 2023). However, challenges remain in improving CD synthesis procedures and developing more specific sensing probes for complex food samples (X. Shi et al., 2019).

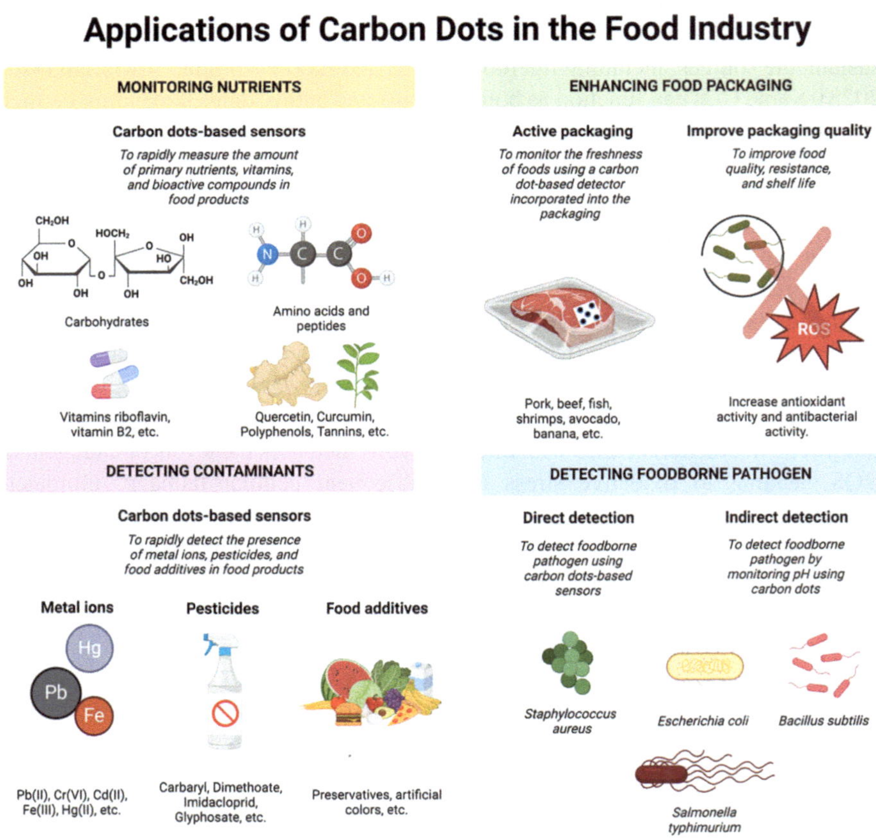

Fig. 11.3 Primary applications of carbon dots (CDs) in the food industry. (Reprinted from Nguyen et al., 2024a)

11.3.2.1 Detecting Contaminants

Monitoring contaminants in food products is crucial for ensuring public health, with regulations mandating effective detection methods. Carbon dots (CDs) have emerged as promising nanomaterials for identifying various food contaminants due to their unique fluorescence properties and surface functionalization capabilities. The integration of CD-based fluorescent sensors offers a noninvasive, rapid, and cost-effective method for detecting food contaminants, reducing reliance on complex analytical techniques.

Heavy Metal Detection

CDs are effective in detecting heavy metals such as mercury (Hg^{2+}) (Wise et al., 2022), lead (Pb^{2+}) (Levallois et al., 2018), cadmium (Cd^{2+}) (Yu et al., 2017), and arsenic (As^{3+}). Their strong binding interactions with these metal ions result in measurable fluorescence changes, facilitating rapid and sensitive detection. For instance, CD-based probes have been utilized to detect mercury in seafood (He et al., 2020) and lead in water and beverages (Olorunyomi et al., 2022).

Pesticide and Toxin Detection

Functionalized CDs can identify organophosphate pesticides like malathion and chlorpyrifos in food samples (Lin et al., 2018) or residue carbaryl in water (Xu et al., 2025). Additionally, they can detect mycotoxins such as aflatoxins and ochratoxins through fluorescence changes upon interaction with these compounds. For example, a CD-based sensor can detect aflatoxin B in coix seed (Li et al., 2024). CD-based sensors have been successfully applied to detect pesticide residues in fruits, vegetables, and water samples, demonstrating their potential for food safety and environmental monitoring applications (Lin et al., 2018; Wu et al., 2017).

Food Additive Detection

CD-based sensors can also detect several food preservatives such as nitrite (Liu et al., 2024), potassium sorbate (Jia et al., 2019), and sulfite (Wang et al., 2023). Furthermore, various artificial colors included in foods can be detected with CDs, including sunset yellow (Su et al., 2022) and indigo carmine (Albalawi et al., 2023).

11.3.2.2 Detecting Foodborne Pathogens

Foodborne pathogens, including *Salmonella*, *Escherichia coli*, and *Listeria monocytogenes*, are major contributors to foodborne illnesses, leading to significant public health concerns worldwide. Although reliable, conventional detection methods such as culture-based techniques and polymerase chain reaction (PCR) assays often require prolonged incubation times or sophisticated equipment (Zhao et al., 2014). In contrast, CD-based biosensors offer rapid, highly sensitive, and specific pathogen detection. CDs can be functionalized with various surface groups, enabling conjugation with biomolecules like aptamers for selective recognition of target bacteria (Sargazi et al., 2022). Amphiphilic CDs have demonstrated the ability to bind to bacterial cells, allowing detection through fluorescence spectroscopy and microscopy (Nandi et al., 2015). The antimicrobial properties of CDs are attributed to their surface functional groups, which generate free radicals (Maruthapandi et al., 2021). CD-based sensors have been used to detect *Escherichia coli* (Zhao et al., 2022), *Staphylococcus aureus* (Gao et al., 2024), *Listeria* and *Salmonella* (Dong et al., 2021).

11.3.2.3 Monitoring Food Nutrients

Accurate monitoring of nutrients like vitamins, amino acids, glucose, and minerals is vital for food quality and health. CDs are promising for nutrient sensing due to their photoluminescence, *biocompatibility, and easy surface functionalization. These fluorescent nanoprobes detect* micronutrients through changes in fluorescence. For example, silver-doped carbon dots (CDs) can detect glucose (Haiyang et al., 2024), and zinc and chlorine co-doped carbon dots (CDs) are used to sense vitamin B_2 (riboflavin) in beverages (Meng and Wu, 2024). CDs doped with nitrogen show potential for detecting quercetin in red wines (Sasikumar et al., 2023) and polyphenols in green tea (Wei et al., 2020). Compared to traditional methods, CD-based sensors are rapid, cost-effective, and portable, making them ideal for real-time monitoring in industrial and household settings.

11.3.2.4 Enhancing Food Packaging

CDs can be integrated into smart packaging to enhance food safety, quality monitoring, and shelf-life extension. For example, CDs combined with antimicrobial bacterial cellulose membranes can significantly reduce the development of *Escherichia coli* in ground beef (Ghorbani et al., 2024). Furthermore, CDs enhance UV-blocking properties, antioxidant activity, and antimicrobial efficacy against various pathogens (Ezati et al., 2022; Moradi et al., 2023). These nanoparticles can generate reactive oxygen species under light irradiation, inhibiting microbial growth and extending food shelf life (Ezati et al., 2022).

11.4 Risks and Toxicity of Carbon Dots in Food and Beverages

11.4.1 Risks

The use of carbon dots (CDs) in enhancing the efficiency of food and beverage processing is an innovative approach. The implementation of the more recent advances could result in more competitive low-cost processes for the sector. Nevertheless, complex food safety laws and regulations must be navigated carefully to protect public health. These regulatory constraints extend through all food and packaging aspects of food production processes (Pan et al., 2019). Legislation and regulation could anticipate technical development that, when addressed to the prospective nature of rule-making, may prevent or at least reduce the number of bottlenecks (Majid et al., 2024; Moradi et al., 2023).

The current landscape of regulatory frameworks presents significant challenges, particularly in addressing the intricate nuances of defining "product-specific" categories for this new class of materials. Adding to these complexities are the methodologies used for detection and analysis. Although there are established protocols for evaluating the safety of nanoparticles, industry stakeholders and researchers have yet to reach a consensus on standardizing testing procedures for nanomaterials (Engelmann & Gaymard, 2017). As discussions around the industrial use of untested materials continue to grow, recent talks have emerged about the potential for grouping similar nanomaterials based on their analogous technological foundations. Moreover, the advancements and possible health implications of CDs differ considerably from those observed in previous nanomaterial studies. This divergence arises from the unique chemical composition, varying mechanisms of toxicity, and distinctive bioavailability exhibited by CDs, presenting substantial challenges for stakeholders advocating for the integration of CD-based materials into practical applications. Consequently, there is an urgent need for relevant parties to navigate the complexities of comprehensive validation studies that are both scientifically rigorous and suitable for comparative analysis with peer-reviewed research. Given these challenges, it is vital for leading technology coherence bodies to allocate the necessary resources to bridge the gaps in our understanding of CDs' toxicity. This will enable a more informed approach to their use in various applications.

11.4.2 Toxicological Concerns

Despite the promising applications of carbon dots (CDs) in food and beverages, their safety profile remains a critical concern. The potential toxicity of CDs is influenced by their size, surface chemistry, and concentration, as well as the route of exposure (ingestion, inhalation, or dermal contact). Key toxicological concerns include cytotoxicity, genotoxicity, bioaccumulation risks, and physicochemical influences on toxicity (Fig. 11.4).

11.4.2.1 Cytotoxicity and Genotoxicity

The toxicity of carbon dots (CDs) is influenced by factors such as precursor materials, surface passivation agents, and size. While CDs generally exhibit low toxicity, higher concentrations and smaller sizes can increase toxicity, leading to oxidative stress and cell damage. Smaller CDs (10–20 nm) tend to be more toxic than larger ones (40–100 nm), and higher concentrations generally increase toxicity (Fan et al., 2019). The choice of surface passivation molecules and functionalization can significantly affect CD cytotoxicity (Wang et al., 2011a). Positive charge, nitrogen content, and the use of high molecular weight polyamines in synthesis contribute to increased toxicity, while PEG decoration can reduce it (Fan et al., 2019).

Studies indicate that CDs do not cause significant genotoxic effects at moderate concentrations. However, higher doses can lead to significant effects. In vitro studies on cancer cell lines demonstrated that CDs can cause DNA damage and cell cycle arrest, even at low concentrations (Havrdová et al., 2022; Şimşek et al., 2020). Prolonged exposure to certain types of CDs in mice resulted in lethality and organ

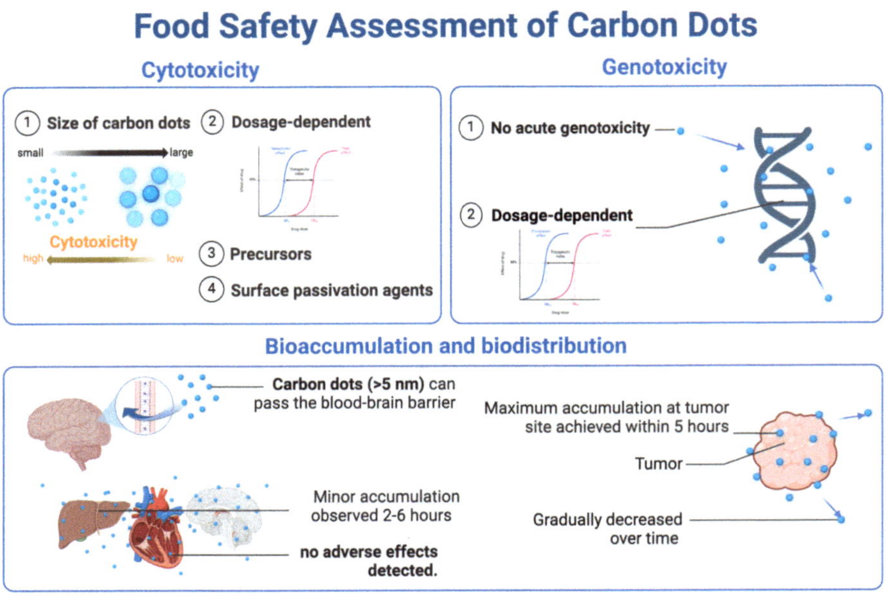

Fig. 11.4 Food safety assessment of carbon dots (CDs) including cytotoxicity, genotoxicity, bioaccumulation and biodistribution. (Reprinted from Nguyen et al., 2024a)

injuries, particularly those with oxygen- and nitrogen-containing functional groups (Kuznietsova et al., 2023). Overall, CDs show low genotoxicity in various tests.

11.4.2.2 Bioaccumulation and Biodistribution

Carbon dots (CDs) show minor bioaccumulation in organs, but further study is needed to understand in vivo biodistributions. Studies have shown that CDs can penetrate the blood–brain barrier (BBB) and accumulate in tumors, demonstrating potential for targeted drug delivery (Liao et al., 2021; Zhang et al., 2021). CDs exhibit unique properties such as biocompatibility, small size, and tunable optical characteristics, making them suitable for bioimaging and drug delivery systems (Singh & Singh, 2023). However, concerns regarding their toxicity and long-term effects on the central nervous system persist, necessitating further investigation. Research has demonstrated that CDs can cross the blood-brain barrier, accumulate in the brain, liver, and heart, and cause neurodegenerative effects (Li et al., 2018). While CDs show promise for treating neurodegenerative diseases and brain tumors, more research is needed to fully understand their in vivo biodistribution, clearance mechanisms, and potential health impacts (Liao et al., 2021; Zhang et al., 2021).

11.5 Strategies for Mitigating the Risk of Carbon Dots

There are four primary strategies to reduce the risk of carbon dots (CDs): (1) selecting appropriate precursors to synthesize CDs, (2) optimizing synthesis methods, (3) modifying the surface agents, and (4) developing detection protocols of CDs in foods (Fig. 11.5).

11.5.1 Selecting the Appropriate Precursors

The choice of precursors is crucial in controlling the properties and reducing the toxicity of CDs (Nguyen et al., 2024a). Using safe, nontoxic materials can improve biocompatibility and customize CDs for various applications, including in the food industry (Hola et al., 2014; Peng et al., 2016; Yao et al., 2019; Zeng et al., 2021).

11.5.2 Optimizing Synthesis Methods

Optimizing synthesis methods for CDs can significantly impact their properties, performance, and toxicity. Different methods, such as hydrothermal, microwave-assisted, and solvothermal techniques, influence the size, morphology, surface chemistry, and optical properties of CDs (Wang et al., 2014). Lowering synthesis temperatures and using assist solvents can obtain stronger fluorescent CDs with

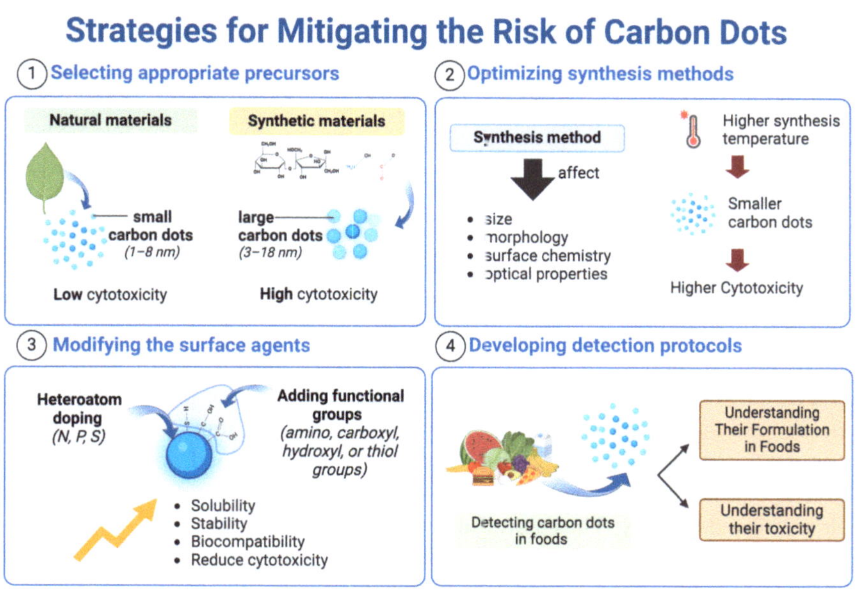

Fig. 11.5 Recommended strategies for mitigating the risk of carbon dots (CDs). (Reprinted from Nguyen et al., 2024a)

smaller sizes without increasing toxicity (Nguyen et al., 2024a, b, c). Surface passivation and functionalization during synthesis can improve stability, dispersibility, and biocompatibility (Atabaev, 2018). Choosing the right synthesis method also affects the scalability and reproducibility of CD production (De & Karak, 2013).

11.5.3 Surface Modification and Functionalization

Surface modification and functionalization of CDs are crucial for enhancing their stability, biocompatibility, and functionality. Introducing functional groups like amino, carboxyl, and hydroxyl groups improves solubility, stability, and reactivity and enables further conjugation to biomolecules and targeting ligands (Park et al., 2014). Doping CDs with heteroatoms like nitrogen, sulfur, or phosphorus enhances photoluminescence and catalytic activity (Wang et al., 2023). Encapsulation with polymers or silica provides additional protection and controlled release capabilities (Sawant et al., 2016).

11.5.4 Development of Specific Protocols to Quantify the Carbon Dots Content

Developing quantitative methods for measuring carbon dots is crucial for accurate dosing. Spectroscopic techniques like fluorescence and UV-Vis spectroscopy, electrochemical methods, and advanced imaging techniques such as AFM, SEM, TEM, and Raman microscopy are being used to quantify CDs based on their emission intensity, absorbance, and morphology (Li et al., 2012; Nguyen et al., 2024a, b, c; Pan et al., 2020; Saraswat and Yadav, 2020).

11.6 Conclusion

In conclusion, carbon dots (CDs) are fluorescent nanoparticles formed during food processing through high-temperature reactions such as the Maillard reaction, caramelization, and pyrolysis. Typically less than 10 nm in size, these particles exhibit unique optical and chemical properties, making them suitable for applications in biosensing and food safety testing. Despite their promising potential, concerns about their toxicity and long-term effects on human health persist, necessitating further research. The integration of CDs in food applications holds significant promise, but comprehensive validation studies are essential to ensure their safety and efficacy. The careful selection of precursors, optimization of synthesis methods, and meticulous surface modification have led to CDs with enhanced properties, stability, and biocompatibility, making them suitable for use in the food industry, biomedicine, and beyond. Establishing specific protocols to quantify CDs accurately ensures reliable dosing and application. Future exploration of novel synthesis techniques, functionalization strategies, and application domains will reveal more groundbreaking CD uses.

References

Ahn, J., Song, Y., Kwon, J. E., Woo, J., & Kim, H. (2019). Characterization of food waste-driven carbon dot focusing on chemical structural, electron relaxation behavior and Fe3+ selective sensing. *Data in Brief, 25*, 104038. https://doi.org/10.1016/j.dib.2019.104038

Albalawi, M. A., Gomaa, H., El Hamd, M. A., Abourehab, M. A. S., & Abdel-Lateef, M. A. (2023). Detection of indigo carmine dye in juices via application of photoluminescent europium-doped carbon dots from tannic acid. *Luminescence, 38*(2), 92–98. https://doi.org/10.1002/bio.4417

Atabaev, T. (2018). Doped carbon dots for sensing and bioimaging applications: A Minireview. *Nanomaterials, 8*(5), 342. https://doi.org/10.3390/nano8050342

Bi, J., Li, Y., Wang, H., Song, Y., Cong, S., Yu, C., Zhu, B.-W., & Tan, M. (2018). Presence and formation mechanism of foodborne carbonaceous nanostructures from roasted pike eel (Muraenesox cinereus). *Journal of Agricultural and Food Chemistry, 66*(11), 2862–2869. https://doi.org/10.1021/acs.jafc.7b02303

Cao, L., Song, X., Song, Y., Bi, J., Cong, S., Yu, C., & Tan, M. (2017). Fluorescent nanoparticles from mature vinegar: Their properties and interaction with dopamine. *Food & Function, 8*(12), 4744–4751. https://doi.org/10.1039/C7FO01475A

Buiculescu, R., Stefanakis, D., Androulidaki, M., Ghanotakis, D., & Chaniotakis, N. A. (2016). Controlling carbon nanodot fluorescence for optical biosensing. *The Analyst, 141*(13), 4170–4180. https://doi.org/10.1039/C6AN00783J

Chaudhary, A., Siddiqui, M. W., Gaikwad, K. K., Ahmad, M. S., Homa, F., Singh, D. R., Lata, D., Alam, T., Saurabh, A., Deshmukh, R., & Mousavi Khaneghah, A. (2025). Novel applications of carbon dots (CDs) for enhanced preservation of postharvest fruit and vegetables. *Food Control, 172*, 111182. https://doi.org/10.1016/j.foodcont.2025.111182

Chen, L., Wang, C., Liu, C., & Chen, S. (2023). Facile access to fabricate carbon dots and perspective of large-scale applications. *Small, 19*(31), 2206671. https://doi.org/10.1002/smll.202206671

Collins, J., Yang, L., Dong, X., & Sun, Y.-P. (2025). Antimicrobial properties of carbon "quantum" dots for food safety applications. *Journal of Nanoparticle Research, 27*(2), 35. https://doi.org/10.1007/s11051-025-06239-9

Cong, S., Wang, N., Wang, K., Wu, Y., Li, D., Song, Y., Prakash, S., & Tan, M. (2019). Fluorescent nanoparticles in the popular pizza: Properties, biodistribution and cytotoxicity. *Food & Function, 10*(5), 2408–2416. https://doi.org/10.1039/C8FO01944D

De, B., & Karak, N. (2013). A green and facile approach for the synthesis of water soluble fluorescent carbon dots from banana juice. *RSC Advances, 3*(22), 8286. https://doi.org/10.1039/c3ra00088e

De Yro, P. A. N., Quaichon, G. M. O., Cruz, R. A. T., Emolaga, C. S., Que, M. C. O., Magdaluyo, E. R., & Basilia, B. A. (2019). Hydrothermal synthesis of carbon quantum dots from biowaste for bio-imaging. *AIP Conference Proceedings*, 020007. https://doi.org/10.1063/1.5094310

Dong, X., Awak, M. A., Tomlinson, N., Tang, Y., Sun, Y.-P., & Yang, L. (2017). Antibacterial effects of carbon dots in combination with other antimicrobial reagents. *PLoS One, 12*(9), e0185324. https://doi.org/10.1371/journal.pone.0185324

Dong, X., Wang, P., Darby, J. P., Tang, Y., Overton, C. M., Kathariou, S., Sun, Y.-P., & Yang, L. (2021). Photoactivated carbon dots for inactivation of foodborne pathogens *listeria* and salmonella. *Applied and Environmental Microbiology, 87*(23), e01042–e01021. https://doi.org/10.1128/AEM.01042-21

Dwandaru, W. S. B., Fauzi, F., Sari, D. S., Sari, E. K., Santoso, I., & Suhendar, H. (2019). Optical properties comparison of carbon Nanodots synthesized from Kangkung (Ipomoea aquatica)

with deep frying and roasting techniques. *Jurnal Penelitian Fisika Dan Aplikasinya (JPFA), 9*(2), 123–131. https://doi.org/10.26740/jpfa.v9n2.p123-131

Engelmann, W., & Gaymard, S. (2017). Consumers' health and safety in relation to the use of nanotechnology in food: Challenges and perspectives from the humanities and social sciences.

Ezati, P., Rhim, J.-W., Molaei, R., & Rezaei, Z. (2022). Carbon quantum dots-based antifungal coating film for active packaging application of avocado. *Food Packaging and Shelf Life, 33*, 100878. https://doi.org/10.1016/j.fpsl.2022.100878

Fan, J., Claudel, M., Ronzani, C., Arezki, Y., Lebeau, L., & Pons, F. (2019). Physicochemical characteristics that affect carbon dot safety: Lessons from a comprehensive study on a nanoparticle library. *International Journal of Pharmaceutics, 569*, 118521. https://doi.org/10.1016/j.ijpharm.2019.118521

Fu, B., Liu, Q., Liu, M., Chen, X., Lin, H., Zheng, Z., Zhu, J., Dai, C., Dong, X., & Yang, D.-P. (2022). Carbon dots enhanced gelatin/chitosan bio-nanocomposite packaging film for perishable foods. *Chinese Chemical Letters, 33*(10), 4577–4582. https://doi.org/10.1016/j.cclet.2022.03.048

Gao, X., Zhang, H., Liu, L., Jia, M., Li, X., & Li, J. (2024). Nano-biosensor based on manganese dioxide nanosheets and carbon dots for dual-mode determination of Staphylococcus aureus. *Food Chemistry, 432*, 137144. https://doi.org/10.1016/j.foodchem.2023.137144

Ghorbani, M., Moradi, M., Tajik, H., Molaei, R., & Alizadeh, A. (2024). Carbon dots embedded bacterial cellulose membrane as active packaging: Toxicity, in vitro release and application in minced beef packaging. *Food Chemistry, 433*, 137311. https://doi.org/10.1016/j.foodchem.2023.137311

Haiyang, L., Guantong, L., Nan, Z., Zhanye, Y., Xinge, J., Bing, Z., & Tian, Y. (2024). Ag-carbon dots with peroxidase-like activity for colorimetric and SERS dual mode detection of glucose and glutathione. *Talanta, 273*, 125898. https://doi.org/10.1016/j.talanta.2024.125898

Havrdová, M., Urbančič, I., Tománková, K. B., Malina, L., Poláková, K., Štrancar, J., & Bourlinos, A. B. (2022). Intracellular trafficking of cationic carbon dots in cancer cell lines MCF-7 and HeLa—Time lapse microscopy, concentration-dependent uptake, viability, DNA damage, and cell cycle profile. *International Journal of Molecular Sciences, 23*(3), 1077. https://doi.org/10.3390/ijms23031077

He, X., Han, Y., Luo, X., Yang, W., Li, C., Tang, W., Yue, T., & Li, Z. (2020). Terbium (III)-referenced N-doped carbon dots for ratiometric fluorescent sensing of mercury (II) in seafood. *Food Chemistry, 320*, 126624. https://doi.org/10.1016/j.foodchem.2020.126624

Hola, K., Zhang, Y., Wang, Y., Giannelis, E. P., Zboril, R., & Rogach, A. L. (2014). Carbon dots—Emerging light emitters for bioimaging, cancer therapy and optoelectronics. *Nano Today, 9*(5), 590–603. https://doi.org/10.1016/j.nantod.2014.09.004

Innocenzi, P., & Stagi, L. (2023). Carbon dots as oxidant-antioxidant nanomaterials, understanding the structure-properties relationship. A critical review. *Nano Today, 50*, 101837. https://doi.org/10.1016/j.nantod.2023.101837

Jia, Y., Hu, Y., Li, Y., Zeng, Q., Jiang, X., & Cheng, Z. (2019). Boron doped carbon dots as a multifunctional fluorescent probe for sorbate and vitamin B12. *Microchimica Acta, 186*(2), 84. https://doi.org/10.1007/s00604-018-3196-5

Kuznietsova, H., Géloën, A., Dziubenko, N., Zaderko, A., Alekseev, S., Lysenko, V., & Skryshevsky, V. (2023). In vitro and in vivo toxicity of carbon dots with different chemical compositions. *Discover Nano, 18*(1), 111. https://doi.org/10.1186/s11671-023-03891-9

Levallois, P., Barn, P., Valcke, M., Gauvin, D., & Kosatsky, T. (2018). Public health consequences of Lead in drinking water. *Current Environmental Health Reports, 5*(2), 255–262. https://doi.org/10.1007/s40572-018-0193-0

Li, G., Xu, J., & Xu, K. (2023). Physiological functions of carbon dots and their applications in agriculture: A review. *Nanomaterials, 13*(19), 2684. https://doi.org/10.3390/nano13192684

Li, H., Kang, Z., Liu, Y., & Lee, S.-T. (2012). Carbon nanodots: Synthesis, properties and applications. *Journal of Materials Chemistry, 22*(46), 24230. https://doi.org/10.1039/c2jm34690g

Li, Y., Bi, J., Liu, S., Wang, H., Yu, C., Li, D., Zhu, B.-W., & Tan, M. (2017). Presence and formation of fluorescence carbon dots in a grilled hamburger. *Food & Function, 8*(7), 2558–2565. https://doi.org/10.1039/C7FO00675F

Li, D., Na, X., Wang, H., Xie, Y., Cong, S., Song, Y., Xu, X., Zhu, B.-W., & Tan, M. (2018). Fluorescent carbon dots derived from Maillard reaction products: Their properties, biodistribution, cytotoxicity, and antioxidant activity. *Journal of Agricultural and Food Chemistry, 66*(6), 1569–1575. https://doi.org/10.1021/acs.jafc.7b05643

Li, D., Xie, Y., Na, X., Li, Y., Dai, C., Li, Y., & Tan, M. (2019). Insights into melanoidin conversion into fluorescent nanoparticles in the Maillard reaction. *Food & Function, 10*(7), 4414–4422. https://doi.org/10.1039/C9FO00383E

Liao, J., Yao, Y., Lee, C.-H., Wu, Y., & Li, P. (2021). In vivo biodistribution, clearance, and biocompatibility of multiple carbon dots containing nanoparticles for biomedical application. *Pharmaceutics, 13*(11), 1872. https://doi.org/10.3390/pharmaceutics13111872

Lin, B., Yan, Y., Guo, M., Cao, Y., Yu, Y., Zhang, T., Huang, Y., & Wu, D. (2018). Modification-free carbon dots as turn-on fluorescence probe for detection of organophosphorus pesticides. *Food Chemistry, 245*, 1176–1182. https://doi.org/10.1016/j.foodchem.2017.11.038

Lin, F., Wang, Z., & Wu, F.-G. (2022). Carbon dots for killing microorganisms: An update since 2019. *Pharmaceuticals, 15*(10), 1236. https://doi.org/10.3390/ph15101236

Liu, Y., Huang, Z., Wang, X., Hao, Y., Yang, J., Wang, H., & Qu, S. (2025). Recent advances in highly luminescent carbon dots. *Advanced Functional Materials, 35*(17), 2420587. https://doi.org/10.1002/adfm.202420587

Liu, X., Li, T., Hou, Y., Wu, Q., Yi, J., & Zhang, G. (2016). Microwave synthesis of carbon dots with multi-response using denatured proteins as carbon source. *RSC Advances, 6*(14), 11711–11718. https://doi.org/10.1039/C5RA23081K

Liu, W., Li, C., Sun, X., Pan, W., Yu, G., & Wang, J. (2017). Highly crystalline carbon dots from fresh tomato: UV emission and quantum confinement. *Nanotechnology, 28*(48), 485705. https://doi.org/10.1088/1361-6528/aa900b

Majid, A., Ahmad, K., Tan, L., Niaz, W., Na, W., Huiru, L., & Wang, J. (2024). The advanced role of carbon quantum dots in Nano-food science: Applications, bibliographic analysis, safety concerns, and perspectives. *C, 11*(1), 1. https://doi.org/10.3390/c11010001

Mancini, F., Menichetti, A., Degli Esposti, L., Montesi, M., Panseri, S., Bassi, G., Montalti, M., Lazzarini, L., Adamiano, A., & Iafisco, M. (2023). Fluorescent carbon dots from food industry by-products for cell imaging. *Journal of Functional Biomaterials, 14*(2), 90. https://doi.org/10.3390/jfb14020090

Manzoor, S., Dar, A. H., Dash, K. K., Pandey, V. K., Srivastava, S., Bashir, I., & Khan, S. A. (2023). Carbon dots applications for development of sustainable technologies for food safety: A comprehensive review. *Applied Food Research, 3*(1), 100263. https://doi.org/10.1016/j.afres.2023.100263

Maruthapandi, M., Saravanan, A., Das, P., Luong, J. H. T., & Gedanken, A. (2021). Microbial inhibition and biosensing with multifunctional carbon dots: Progress and perspectives. *Biotechnology Advances, 53*, 107843. https://doi.org/10.1016/j.biotechadv.2021.107843

Meng, L., & Wu, H. (2024). Preparation of carbon quantum dots and their application in the detection of vitamin B2. *RSC Advances, 14*(22), 15499–15506. https://doi.org/10.1039/D4RA01388C

Moradi, M., Molaei, R., Kousheh, S. A., Guimarães, J. T., & McClements, D. J. (2023). Carbon dots synthesized from microorganisms and food by-products: Active and smart food packaging applications. *Critical Reviews in Food Science and Nutrition, 63*(14), 1943–1959. https://doi.org/10.1080/10408398.2021.2015283

MoziOğlu, E. (2022). Microwave-based synthesis of carbon dots from lemon juice for biotechnological applications. *Osmaniye Korkut Ata Üniversitesi Fen Bilimleri Enstitüsü Dergisi, 5*(2), 600–611. https://doi.org/10.47495/okufbed.1024008

Nandi, S., Ritenberg, M., & Jelinek, R. (2015). Bacterial detection with amphiphilic carbon dots. *The Analyst, 140*(12), 4232–4237. https://doi.org/10.1039/C5AN00471C

Nguyen, D., El-Ramady, H., & Prokisch, J. (2024a). Food safety aspects of carbon dots: A review. *Environmental Chemistry Letters*. https://doi.org/10.1007/s10311-024-01779-3
Nguyen, D., Muthu, A., El-Ramady, H., Béni, Á., & Prokisch, J. (2024b). Detection and formation of fluorescent carbon nanodots in coffee brews and its relationship with other compositions. *Journal of Food Composition and Analysis, 132*, 106347. https://doi.org/10.1016/j.jfca.2024.106347
Nguyen, D., Muthu, A., El-Ramady, H., Daróczi, L., Nagy, L., Kéki, S., Béni, Á., Csarnovics, I., & Prokisch, J. (2024c). Optimization of extraction conditions to synthesize green carbon nanodots using the Maillard reaction. *Materials Advances*. https://doi.org/10.1039/D4MA00037D
Olorunyomi, J. F., White, J. F., Gengenbach, T. R., Caruso, R. A., & Doherty, C. M. (2022). Fabrication of a reusable carbon dot/gold nanoparticle/metal–organic framework film for fluorescence detection of Lead ions in water. *ACS Applied Materials & Interfaces, 14*(31), 35755–35768. https://doi.org/10.1021/acsami.2c09122
Pan, M., Yin, Z., Liu, K., Du, X., Liu, H., & Wang, S. (2019). Carbon-based nanomaterials in sensors for food safety. *Nanomaterials, 9*(9), 1330. https://doi.org/10.3390/nano9091330
Peng, H., Li, Y., Jiang, C., Luo, C., Qi, R., Huang, R., Duan, C.-G., & Travas-Sejdic, J. (2016). Tuning the properties of luminescent nitrogen-doped carbon dots by reaction precursors. *Carbon, 100*, 386–394. https://doi.org/10.1016/j.carbon.2016.01.029
Park, S. Y., Lee, H. U., Park, E. S., Lee, S. C., Lee, J.-W., Jeong, S. W., Kim, C. H., Lee, Y.-C., Huh, Y. S., & Lee, J. (2014). Photoluminescent green carbon Nanodots from food-waste-derived sources: Large-scale synthesis, properties, and biomedical applications. *ACS Applied Materials & Interfaces, 6*(5), 3365–3370. https://doi.org/10.1021/am500159p
Rodríguez-Varillas, S., Fontanil, T., Obaya, Á. J., Fernández-González, A., Murru, C., & Badía-Laíño, R. (2022). Biocompatibility and antioxidant capabilities of carbon dots obtained from tomato (Solanum lycopersicum). *Applied Sciences, 12*(2), 773. https://doi.org/10.3390/app12020773
Sa, S., Mn, E., Ae, T., & An, K. (2017). Synthesis of carbon dots with tunable luminescence. *Journal of Material Science & Engineering, 06*(05). https://doi.org/10.4172/2169-0022.1000376
Sahu, S., Behera, B., Maiti, T. K., & Mohapatra, S. (2012). Simple one-step synthesis of highly luminescent carbon dots from orange juice: Application as excellent bio-imaging agents. *Chemical Communications, 48*(70), 8835. https://doi.org/10.1039/c2cc33796g
Sargazi, S., Er, S., Mobashar, A., Gelen, S. S., Rahdar, A., Ebrahimi, N., Hosseinikhah, S. M., Bilal, M., & Kyzas, G. Z. (2022). Aptamer-conjugated carbon-based nanomaterials for cancer and bacteria theranostics: A review. *Chemico-Biological Interactions, 361*, 109964. https://doi.org/10.1016/j.cbi.2022.109964
Saraswat, V., & Yadav, M. (2020). Carbon dots as green corrosion inhibitor for mild steel in HCl solution. *ChemistrySelect, 5*(25), 7347–7357. https://doi.org/10.1002/slct.202000625
Sasikumar, K., Rajamanikandan, R., & Ju, H. (2023). Nitrogen- and Sulfur-Codoped strong green fluorescent carbon dots for the highly specific quantification of quercetin in food samples. *Materials, 16*(24), 7686. https://doi.org/10.3390/ma16247686
Sawant, V. J., Bamane, S. R., Kanase, D. G., Patil, S. B., & Ghosh, J. (2016). Encapsulation of curcumin over carbon dot coated TiO 2 nanoparticles for pH sensitive enhancement of anticancer and anti-psoriatic potential. *RSC Advances, 6*(71), 66745–66755. https://doi.org/10.1039/C6RA13851A
Semsey, D., Nguyen, D., Törős, G., Muthu, A., Labidi, S., El-Ramady, H., Béni, Á., Rai, M., & József, P. (2024). Analysis of fluorescent carbon Nanodot formation during pretzel production. *Nanomaterials, 14*(13), 1142. https://doi.org/10.3390/nano14131142
Sheikh, M. A., Chandok, R. S., & Abida, K. (2023). High energy density storage, antifungal activity and enhanced bioimaging by green self-doped heteroatom carbon dots. *Discover Nano, 18*(1), 132. https://doi.org/10.1186/s11671-023-03910-9

Shi, Q.-Q., Li, Y.-H., Xu, Y., Wang, Y., Yin, X.-B., He, X.-W., & Zhang, Y.-K. (2014). High-yield and high-solubility nitrogen-doped carbon dots: Formation, fluorescence mechanism and imaging application. *RSC Advances, 4*(4), 1563–1566. https://doi.org/10.1039/C3RA45762A

Shi, X., Wei, W., Fu, Z., Gao, W., Zhang, C., Zhao, Q., Deng, F., & Lu, X. (2019). Review on carbon dots in food safety applications. *Talanta 194*, 809–821. https://doi.org/10.1016/j.talanta.2018.11.005

Şimşek, S., Şüküroğlu, A. A., Yetkin, D., Özbek, B., Battal, D., & Genç, R. (2020). DNA-damage and cell cycle arrest initiated anti-cancer potency of super tiny carbon dots on MCF7 cell line. *Scientific Reports, 10*(1), 13880. https://doi.org/10.1038/s41598-020-70796-3

Singh, N., & Singh, S. (2023). A review on carbon dots use in neurological disorder. *Journal of Survey in Fisheries Sciences.* https://doi.org/10.53555/sfs.v8i3.1397

Sk, M. P., Jaiswal, A., Paul, A., Ghosh, S. S., & Chattopadhyay, A. (2012). Presence of amorphous carbon nanoparticles in food caramels. *Scientific Reports, 2*(1), 383. https://doi.org/10.1038/srep00383

Song, X., Wang, H., Zhang, R., Yu, C., & Tan, M. (2018). Bio-distribution and interaction with dopamine of fluorescent nanodots from roasted chicken. *Food & Function, 9*(12), 6227–6235. https://doi.org/10.1039/C8FO01159A

Su, K., Xiang, G., Jin, X., Wang, X., Jiang, X., He, L., Zhao, W., Sun, Y., & Cui, C. (2022). Gram-scale synthesis of nitrogen-doped carbon dots from locusts for selective determination of sunset yellow in food samples. *Luminescence, 37*(1), 118–126. https://doi.org/10.1002/bio.4152

Tomskaya, A. E., Egorova, M. N., Kapitonov, A. N., Nikolaev, D. V., Popov, V. I., Fedorov, A. L., & Smagulova, S. A. (2018). Synthesis of luminescent N-doped carbon dots by hydrothermal treatment. *Physica Status Solidi (b), 255*(1), 1700222. https://doi.org/10.1002/pssb.201700222

Wang, Y., Anilkumar, P., Cao, L., Liu, J.-H., Luo, P. G., Tackett, K. N., Sahu, S., Wang, P., Wang, X., & Sun, Y.-P. (2011a). Carbon dots of different composition and surface functionalization: Cytotoxicity issues relevant to fluorescence cell imaging. *Experimental Biology and Medicine, 236*(11), 1231–1238. https://doi.org/10.1258/ebm.2011.011132

Wang, X., Qu, K., Xu, B., Ren, J., & Qu, X. (2011b). Microwave assisted one-step green synthesis of cell-permeable multicolor photoluminescent carbon dots without surface passivation reagents. *Journal of Materials Chemistry, 21*(8), 2445. https://doi.org/10.1039/c0jm02963g

Wang, Q., Liu, X., Zhang, L., & Lv, Y. (2012). Microwave-assisted synthesis of carbon nanodots through an eggshell membrane and their fluorescent application. *The Analyst, 137*(22), 5392. https://doi.org/10.1039/c2an36059d

Wang, L., Yin, Y., Jain, A., & Zhou, H. S. (2014). Aqueous phase synthesis of highly luminescent, nitrogen-doped carbon dots and their application as bioimaging agents. *Langmuir, 30*(47), 14270–14275. https://doi.org/10.1021/la5031813

Wang, H., Su, W., & Tan, M. (2020). Endogenous fluorescence carbon dots derived from food items. *The Innovation, 1*(1), 100009. https://doi.org/10.1016/j.xinn.2020.04.009

Wang, B. J., Xu, Z. Y., Sun, Z., Li, Z. Q., Luo, Y. H., Luo, H. Q., & Li, N. B. (2023). A wide-range ratiometric sensor mediating fluorescence and scattering based on carbon dots/metal–organic framework composites for the detection of bisulfite/sulfite in sugar. *Analytical and Bioanalytical Chemistry, 415*(19), 4639–4647. https://doi.org/10.1007/s00216-023-04763-y

Wei, Q., Liu, T., Pu, H., & Sun, D. (2020). Development of a fluorescent microwave-assisted synthesized carbon dots/cu 2+ probe for rapid detection of tea polyphenols. *Journal of Food Process Engineering, 43*(7), e13419. https://doi.org/10.1111/jfpe.13419

Wise, J. P., Young, J. L., Cai, J., & Cai, L. (2022). Current understanding of hexavalent chromium [Cr(VI)] neurotoxicity and new perspectives. *Environment International, 158*, 106877. https://doi.org/10.1016/j.envint.2021.106877

Wu, X., Song, Y., Yan, X., Zhu, C., Ma, Y., Du, D., & Lin, Y. (2017). Carbon quantum dots as fluorescence resonance energy transfer sensors for organophosphate pesticides determination. *Biosensors and Bioelectronics, 94*, 292–297. https://doi.org/10.1016/j.bios.2017.03.010

Xu, Y., Liu, Y., Feng, L., Sun, X., Wang, M., Xia, Y., Yang, L., & Yuwang, J. (2025). Raspberry-derived carbon dots for specific detection of intracellular copper ions. *Analytical Methods, 17*(16), Article 16. https://doi.org/10.1039/D5AY00225G

Yao, B., Huang, H., Liu, Y., & Kang, Z. (2019). Carbon dots: A small conundrum. *Trends in Chemistry, 1*(2), 235–246. https://doi.org/10.1016/j.trechm.2019.02.003

Yin, J., Yuan, S., Guo, Y., Yu, H., Cheng, Y., Xie, Y., Qian, H., & Yao, W. (2024). Carbon dots in fried breadcrumbs: In vitro toxicity to normal rat kidney cells and its control method. *Process Biochemistry, 139*, 125–136. https://doi.org/10.1016/j.procbio.2024.01.026

Yu, G., Zheng, W., Wang, W., Dai, F., Zhang, Z., Yuan, Y., & Wang, Q. (2017). Health risk assessment of Chinese consumers to cadmium via dietary intake. *Journal of Trace Elements in Medicine and Biology, 44*, 137–145. https://doi.org/10.1016/j.jtemb.2017.07.003

Zeng, Q., Feng, T., Tao, S., Zhu, S., & Yang, B. (2021). Precursor-dependent structural diversity in luminescent carbonized polymer dots (CPDs): The nomenclature. *Light: Science & Applications, 10*(1), 142. https://doi.org/10.1038/s41377-021-00579-6

Zhang, W., Sigdel, G., Mintz, K. J., Seven, E. S., Zhou, Y., Wang, C., & Leblanc, R. M. (2021). Carbon dots: A future blood–brain barrier penetrating nanomedicine and drug nanocarrier. *International Journal of Nanomedicine, 16*, 5003–5016. https://doi.org/10.2147/IJN.S318732

Zhao, X., Lin, C.-W., Wang, J., & Oh, D. H. (2014). Advances in rapid detection methods for foodborne pathogens. *Journal of Microbiology and Biotechnology, 24*(3), 297–312. https://doi.org/10.4014/jmb.1310.10013

Zhao, L., Zhang, M., Wang, H., & Devahastin, S. (2022). Effect of addition of carbon dots to the frying oils on oxidative stabilities and quality changes of fried meatballs during refrigerated storage. *Meat Science, 185*, 108715. https://doi.org/10.1016/j.meatsci.2021.108715

Zhao, L., Zhang, M., Mujumdar, A. S., & Wang, H. (2023). Application of carbon dots in food preservation: A critical review for packaging enhancers and food preservatives. *Critical Reviews in Food Science and Nutrition, 63*(24), 6738–6756. https://doi.org/10.1080/10408398.2022.2039896

Carbon Nanodots in Food Safety and Preservation

12

A. Ramalakshmi, M. Balakrishnan, C. Devadharshini, and D. Akila

Abstract

Carbon nanodots (CNDs) have emerged as multifunctional nanomaterials with significant promise to improve food safety and preservation measures. CNDs have a broad range of applications in active and intelligent food packaging systems due to their tiny particle size, high surface area-to-volume ratio, outstanding photoluminescence, and surface functionalisation capabilities. They exhibit high antibacterial action by breaking microbial cell membranes and producing oxidative stress via the production of reactive oxygen species, which inhibit foodborne pathogens. The incorporation of CNDs into packaging materials increases barrier characteristics by decreasing oxygen transfer and moisture permeability, consequently minimising oxidative deterioration and microbiological contamination. CNDs also modulate enzymatic activities such as polyphenol oxidase and peroxidase, effectively delaying enzymatic browning and softening in fresh produce. Their luminous features allow for real-time monitoring of food freshness by detecting environmental changes such as pH shifts and the presence of spoilage-associated volatile gases. CND-based edible coatings give direct antimicrobial protection to food surfaces while keeping sensory qualities, including colour, texture, and scent. The capacity to synthesise CNDs from renewable sources promotes sustainable packaging activities that coincide with current industry trends. This chapter thoroughly discusses the functional processes, uses, and future prospects of carbon nanodots in food

A. Ramalakshmi (✉) · M. Balakrishnan (✉) · C. Devadharshini · D. Akila
Department of Agricultural Microbiology, Tamil Nadu Agricultural University, Coimbatore, Tamil Nadu, India

Department of Food Process Engineering, Tamil Nadu Agricultural University, Coimbatore, Tamil Nadu, India
e-mail: ramalakshmi.a@tnau.ac.in; bala_tnau@tnau.ac.in

M. Rai et al. (eds.), *Carbon Nanodots*, https://doi.org/10.1007/978-981-95-4705-0_12

preservation, emphasising their position as a transformative instrument for increasing food quality, safety, and shelf life.

Keywords

Antimicrobial · Antioxidant · Biofilm · Food packaging · Shelf life

12.1 Introduction

The global food supply is confronted with considerable threats stemming from food spoilage and contamination, primarily instigated by microbial activity and exacerbated by adverse environmental conditions. The incidence of foodborne illnesses has exhibited a persistent increase, with the World Health Organisation estimating that 420,000 fatalities occurred worldwide due to contaminated food in 2010, along with 600 million instances of foodborne diseases each year (WHO, 2015). This escalating burden is intricately connected to microbial threats that are shaped by intrinsic food characteristics, microbial load, processing methodologies, and environmental determinants such as climate change (Karanth et al., 2023).

Climate change further intensifies microbial hazards. Alterations in global temperature, precipitation patterns, and occurrences of extreme weather events affect the behaviour and survivability of foodborne pathogens (IPCC, 2014). These alterations are particularly crucial for ready-to-eat foods and bulk products, where an increase in spoilage potential and pathogen virulence has been documented (Misiou & Koutsoumanis, 2022).

The economic ramifications are similarly noteworthy. The Food and Agriculture Organisation (FAO) revealed that nearly 931 million tons of food were wasted worldwide in 2019, with microbial decay being a notable factor (FAO, 2019). Food spoilage not only incurs considerable financial losses but also aggravates resource wastage and contributes to environmental degradation through heightened greenhouse gas emissions (Karanth et al., 2023).

The limitations of conventional preservatives and packaging materials have required the development of more effective, long-term approaches to food protection. Traditional packaging often relies on synthetic polymers and chemical preservatives, which present several limitations, including poor barrier properties, low biodegradability, potential toxicity, and insufficient inhibition of microbial growth over extended periods (Babu, 2022). Because food items are highly sensitive to chemical, physical, and biological contamination along the supply chain, existing approaches for ensuring long-term safety and quality have proven inadequate, prompting research into advanced nanotechnology solutions.

To address these challenges, nanotechnology has emerged as a formidable field for food preservation. Antimicrobial nanoparticles, including silver, titanium dioxide, and zinc oxide, have been rigorously investigated for their capacity to impede microbial proliferation through various mechanisms such as the generation of reactive oxygen species, ion release, and disruption of cellular membranes (Espitia et al., 2012; Fernández et al., 2010, as cited in Nazir & Azaz Ahmad Azad, 2019). Techniques for nanoencapsulation have been developed to enhance the stability and targeted delivery of antimicrobials and antioxidants, thereby improving food shelf life and safety (Donsì et al., 2011, as cited in Nazir & Azaz Ahmad Azad, 2019).

Among recent advancements, carbon nanodots (CNDs) have attracted attention due to their antimicrobial characteristics and applications in food safety. CNDs, distinguished by their diminutive dimensions, extensive surface area, and functionalisable surfaces, have exhibited significant antimicrobial efficacy through mechanisms involving membrane disruption and the induction of oxidative stress (Nazir & Azaz Ahmad Azad, 2019). Their prospective applications in active packaging and edible coatings are currently under exploration, presenting promising opportunities for spoilage detection and microbial inhibition in food products.

Revolutionary packaging frameworks that employ nanotechnology, particularly nanosensors comprising gold nanoparticles, titanium dioxide, and zinc oxide nanocomposites, are being formulated to assess spoilage markers such as gas emissions, pH alterations, and microbial contamination (Chaudhry et al., 2008). These advancements facilitate the early detection of spoilage and furnish real-time information to stakeholders throughout the food supply chain.

Another notable benefit of CNDs in food applications is their biocompatibility and water solubility. These characteristics separate CNDs from numerous metal-based nanoparticles, allowing for their safe incorporation into edible films and coatings with minimal deleterious consequences (Wu et al., 2024). Furthermore, CNDs' inherent fluorescence enables real-time monitoring and tracking of food quality changes. CNDs can detect microbiological contamination, rotting gases, or chemical changes in packaging conditions by quenching and recovering fluorescence (Babu, 2022). Thus, the introduction of carbon nanodots provides a multi-functional and environmentally friendly alternative for improving food safety and increasing shelf life. CNDs' antibacterial, antioxidant, and enzyme-inhibitory properties, along with improved biocompatibility, solubility, and fluorescence-based monitoring, position them as a game-changing breakthrough in the next generation of active and intelligent food packaging solutions.

The primary aim of this chapter is to provide a comprehensive overview of the functional mechanisms and practical applications of carbon nanodots (CNDs) in enhancing food safety and preservation. Specifically, this chapter seeks to explore the antimicrobial, antioxidant, and enzyme-inhibitory properties of carbon nanodots and their implications for extending shelf life and inhibiting microbial contamination in various food matrices. It also critically evaluates the role of CNDs in microbial detection, and incorporates toxicological and regulatory perspectives to address concerns related to consumer safety, environmental impact, and the current legal status of CNDs in food contact materials.

12.2 Carbon Nanodots as a Multifunctional Preservative

12.2.1 Antimicrobial Strategies: Combatting Foodborne Pathogens

Carbon nanodots (CNDs) have emerged as multifunctional nanomaterials with wide-ranging antibacterial activity. Their modes of action are multifaceted, including physical, chemical, and biological disruption of microbial structures and activities. These processes are especially relevant in food safety applications, where avoiding microbial contamination and persistence is critical.

12.2.1.1 Disrupting Microbial Cell Membranes

CNDs' principal antibacterial method is to interact directly with microbial cell walls and membranes. CNDs, with their tiny size (<10 nm), spherical shape, and numerous surface functional groups, may firmly cling to bacterial surfaces (Fig. 12.1). Nitrogen doping or surface passivation with amino groups results in a positive surface charge, which enhances electrostatic interactions with negatively charged bacterial membranes (Guo et al., 2022). Experimental investigations have demonstrated that CND adhesion to bacterial cell walls causes surface roughening and microstructural damage. Li et al. (2018), as quoted in Chatzimitakos and Stalikas (2020), showed that vitamin C-derived CNDs stuck to the bacterial membrane, causing morphological changes and reduced cellular integrity. CNDs could potentially coat bacterial surfaces, effectively separating them from the external nutritional media and contributing to bacteriostatic effects. Following initial surface attachment, CNDs can enter and destabilise the lipid bilayer, resulting in physical rupture and

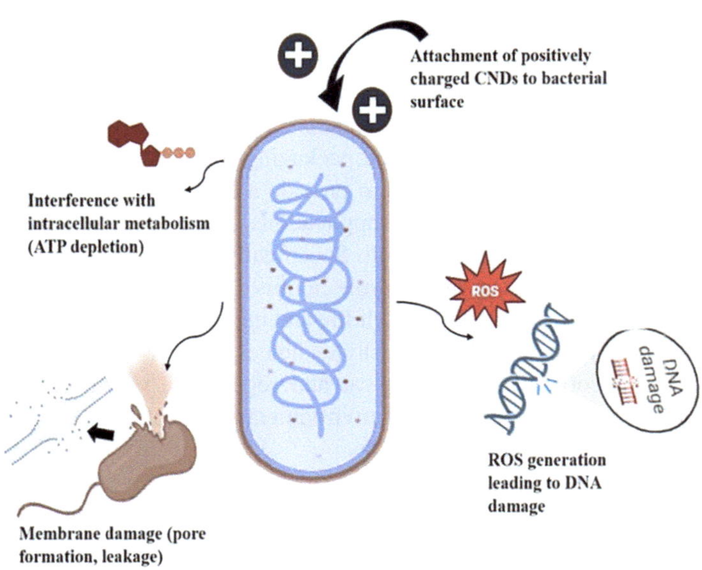

Fig. 12.1 Mechanism of antimicrobial activity of carbon quantum dots against bacterial cell

intracellular leaking. Guo et al. (2022) found that CNDs enhanced bacterial membrane permeability, causing uncontrolled efflux of essential intracellular components such as nucleic acids, proteins, and ions. This process causes the breakdown of cellular homeostasis and, eventually, microbial cell death. The significance of CND size and surface curvature on membrane rupture has been highlighted. Hui et al. (2020), referenced in Guo et al. (2022), found that carbon dots with Gaussian curvature similar to bacterial membrane curvature can improve mechanical disruption efficiency.

12.2.1.2 Induction of Oxidative Stress via Reactive Oxygen Species (ROS)

Another important antibacterial action of CNDs is the development of oxidative stress via the production of reactive oxygen species (ROS). When activated by light (e.g. visible light, UV-A), CNDs function as photosensitisers, creating ROS via two different pathways: Type I reaction: Electron transfer generates radicals such as hydroxyl radicals (-OH) and superoxide anions (-O2). Type II reaction: Energy transfer produces extremely reactive singlet oxygen (1O2). Knoblauch and Geddes (2020) found that excitation of CNDs generates electron–hole pairs (e−/h+) that combine with surrounding molecular oxygen and water, producing ROS. These ROS cause significant damage to microbial cells, targeting lipids, proteins, and nucleic acids. Knoblauch and Geddes (2020) mentioned that a study demonstrates CNDs can emit Type I and Type II ROS based on surface functionalisation and excitation circumstances.

ROS, such as -OH and 1O2, are not easily detoxified by microbial defence mechanisms, making ROS-mediated killing less susceptible to resistance development compared to traditional antibiotics (He et al., 2021; Knoblauch & Geddes, 2020). ROS produced by CNDs can impair microbial respiration by destroying critical enzymes involved in energy metabolism. Chatzimitakos et al. (2020) studied the metabolomic profile of *Escherichia coli* exposed to CNDs and identified abnormalities in membrane integrity but little influence on central metabolism. Specific modifications included stimulation of α-linolenic acid metabolism.

Furthermore, CNDs can interact directly with microbial genetic materials. CNDs can intercalate into nucleic acid structures, resulting in unwinding and distortion (Chatzimitakos et al., 2020; Li et al., 2018). These interactions impede DNA replication and transcription, further limiting microbial growth and survival.

12.2.1.3 Inhibition of Biofilm Formation and Pathogen Adhesion

CNDs have been found to reduce microbial adhesion and biofilm development, two key early phases in food surface contamination. Guo et al. (2022) ascribe surface adhesion inhibition to interference with microbial surface proteins and extracellular polysaccharide matrices required for biofilm development. CNDs with certain surface curvatures or functions can inhibit stable microbial adhesion, lowering the risk of biofilm formation on food processing equipment, packaging, and food surfaces. Mature biofilms defend bacteria from environmental stressors, sanitisers, and antibiotics. CNDs have antibiofilm properties that inhibit not only biofilm initiation but also break pre-existing biofilms. Lin et al. (2018), as reported by Chatzimitakos and Stalikas (2020), found that CNDs produced from *Lactobacillus plantarum* prevented *E. coli* biofilm development while causing little cytotoxicity to planktonic cells. CNDs interact with the structural components of the biofilm matrix, reducing mechanical stability and making bacterial colonies easier to remove. This feature is especially useful in the food business, where biofilm growth is a constant source of contamination and spoilage.

12.2.2 Anti-Oxidative Shield

Controlling oxidative degradation is critical to ensuring food product quality, safety, and shelf life. Reactive oxygen species (ROS) and reactive nitrogen species (RNS) are known to start chain reactions that cause lipid peroxidation, protein denaturation, colour degradation, and sensory loss in food items. Carbon nanodots (CNDs) have high surface area, numerous functional groups, and biocompatibility, making them promising antioxidants (Innocenzi & Stagi, 2023).

12.2.2.1 Scavenging Free Radicals and Reactive Species

CNDs' antioxidant mechanisms primarily include direct scavenging of free radicals via two major pathways: hydrogen atom transfer (HAT) and single electron transfer (SET) processes. Studies utilising DPPH tests and cyclic voltammetry have revealed that CNDs can efficiently neutralise radicals by donating hydrogen atoms or electrons, stabilising the reactive species (Zhang et al., 2017, 2018). CNDs may scavenge numerous ROS, including hydroxyl radicals (OH), superoxide anions (O_2^-), nitric oxide (NO), and DPPH radicals. Surface functional groups (e.g. hydroxyl, carboxyl, amine) boost electron density and promote radical neutralisation, resulting in high scavenging effectiveness (Wei et al., 2017; Gao et al., 2023). Doping CNDs with heteroatoms like nitrogen and sulphur improves their antioxidative properties by creating more electron-rich sites and changing surface chemistry (Zhang et al., 2018; Gao et al., 2023). In lipid-rich food matrices, oxidation causes rancidity and off-flavours. Studies have shown that CNDs can decrease lipid peroxidation and oxidative degradation in lipid models. CNDs do this by quenching lipid peroxyl radicals (LOO) during lipid oxidation.

Electron transfer interrupts radical chain reactions, stabilising lipid molecules before oxidation (Wang et al., 2017; Wei et al., 2017). Applications in meat and dairy products help prevent the formation of secondary oxidation products such as malondialdehyde (MDA), thereby preserving sensory and nutritional integrity during storage. The oxidative destruction of sensitive bioactives such as vitamins (particularly vitamin C and E), carotenoids, and flavour components results in a loss of nutritional value and consumer approval. CNDs stabilise molecules by scavenging ambient ROS, preventing degradation. Specifically, CNDs improved the preservation of β-carotene and anthocyanins under oxidative stress conditions (Zheng et al., 2018, quoted by Gao et al., 2023). The shelf life of flavours, particularly those containing unsaturated chemicals, was increased by inhibiting oxidation-induced volatilisation and degradation.

12.2.2.2 Stabilising Food Quality and Extending Shelf Life

Oxidative breakdown of pigments such as myoglobin in meats and chlorophyll in plants causes unpleasant colour changes. CNDs prevent oxidative processes and maintain the natural look of food products (Innocenzi & Stagi, 2023). CNDs inhibit oxidative cross-linking of proteins, which preserves the texture of foods like meats and dairy gels (Zhang et al., 2017). Another important benefit is the preservation of vitamins and vital fatty acids, which ensures nutritional integrity throughout the product's shelf life. CNDs' radical scavenging ability prevents the development of aldehydes, ketones, and short-chain hydrocarbons that cause rancid odours and flavours in fats and oils. In perishable food model systems, CNDs dramatically reduced peroxide and anisidine levels relative to controls (Gao et al., 2023). Furthermore, incorporating CNDs into packaging materials or directly applying them to food systems provides a powerful technique for natural food preservation that avoids the use of synthetic antioxidants like BHA and BHT, which have raised consumer health concerns.

12.2.3 Modulation of Food Enzymatic Activities

Carbon nanodots (CNDs) provide a unique technique for enzyme modulation that is important for food safety and preservation. Their capacity to inhibit important spoilage-related enzymes like polyphenol oxidase (PPO) and peroxidase, as well as control enzymatic softening pathways, makes them ideal bio-nano agents for extending shelf life without the use of synthetic chemical preservatives.

12.2.3.1 Enzymatic Regulation: Slowing Down Spoilage Processes

Enzymatic browning in fruits and vegetables is primarily caused by polyphenol oxidase (PPO) and peroxidase enzymes. These enzymes catalyse the oxidation of phenolic substrates to quinones, which polymerise to generate brown pigments, reducing product quality.

Recent research has shown that carbon nanodots can alter enzymatic activity by interacting with enzyme structural regions. Shi et al. (2017) found that highly crystalline graphite-derived carbon nanodots (GDs) can preferentially bind to enzyme surfaces, causing conformational changes that affect activity. Li et al. (2015) found that phosphate-functionalised carbon dots enhanced the activity of laccase enzymes, which are structurally similar to PPO. Laccase inhibition by phosphate-functionalised carbon dots was accomplished via surface adsorption and modification of the enzyme's active site configuration (Li et al., 2015). Given the molecular similarities between laccase and PPO, analogous interactions of CNDs with PPO are possible, resulting in the prevention of enzymatic browning in fresh produce. Beyond solid foods, enzymatic browning affects juices and other processed foods in which oxidative enzymes remain active after processing. The use of CNDs provides a means to extend freshness by suppressing oxidative processes. Shi et al. (2017) found that the surface chemistry of carbon dots, specifically the presence of oxygen-containing groups like hydroxyl and carbonyl groups, significantly affects their binding to enzyme surfaces and activity modulation. Tailoring CNDs with appropriate surface functions thus provides a tailored technique for inhibiting browning enzymes in liquid and processed food matrices without using chemical additives.

12.2.3.2 Controlling Enzymatic Softening and Ripening

In addition to browning, enzymatic softening caused by enzymes such as pectinases and cellulases leads to rapid loss of fruit and vegetable texture after harvest. Although direct investigations on pectinase inhibition by CNDs are sparse, Shi et al. (2017) established modulation principles for lipase activity, providing a strong mechanistic basis.

GD4-CND complexes have been demonstrated to affect enzyme kinetics by modifying the Michaelis–Menten parameters: Lower Km values indicate higher substrate affinity, and increased Vmax values indicate improved enzyme activity or stability. These conformational effects are caused by non-covalent interactions (hydrogen bonding, hydrophobic interactions) between CND surfaces and protein structures. Similar concepts could be used to control or limit the activity of cell-wall disintegrating enzymes, reducing tissue softening and maintaining firmness in fresh food. Shi et al. (2017) found that GD4 complexes improved the thermostability of enzymes, as shown by higher T50 values (temperature at which 50% activity is lost). This stabilisation effect might be reversed for undesirable enzymes (e.g. those that speed up spoiling) by changing CND surface chemistry to encourage deactivation or denaturation at mild storage conditions. As a result, strategically deploying CNDs with customised surface groups may decrease softening enzyme activities, increase postharvest shelf life, and minimise dependency on synthetic chemical preservatives.

12.3 Carbon Nanodots in Food Packaging: A Smart Barrier Against Contamination

12.3.1 Active Food Packaging Incorporating Carbon Nanodots

The addition of carbon nanodots (CNDs) to food packaging materials provides an excellent antibacterial barrier against foodborne microorganisms. Table 12.1 presents the list of different CND sources and their various applications in the food industry. CNDs strongly limit microbial growth when embedded in polymer matrices due to their large surface area, abundance of functional groups, and capacity to produce reactive oxygen species (ROS). Wu et al. (2024) demonstrated that corn starch-based edible films containing 5% carbon nanodots achieved a 99.9% reduction in *Staphylococcus aureus* and *Escherichia coli*, indicating strong antibacterial action against both Gram-positive and Gram-negative bacteria. The antibacterial

Table 12.1 List of CND sources, their synthesis methods and application in food industry

Source of CND	Synthesis method	Application in food industry	Reference
Avocado (*Persea americana*)	Hydrothermal (200 °C, 6 h)	Utilized in packaging films with UV-barrier, antibacterial, and antioxidant properties and extended shelf life	Ezati et al. (2022)
Tea waste and grape pomace	Hydrothermal (180 °C, 5 h)	Used as antioxidant additives in oil samples	Murru et al. (2020)
Polyethylenimine and L-cysteine	Hydrothermal (200 °C, 6 h)	Created CD infused hydrogel films to extend freshness and storage quality of bananas	Alzahrani (2024)
Potato skins (*Solanum tuberosum)*	Hydrothermal (200 °C, 6 h)	Applied in packaging films for antioxidant and antibacterial activity	Min et al. (2022)
Moringa oleifera roots (Moringa roots)	Hydrothermal (220 °C, 10 h)	Incorporated into packaging films exhibiting antioxidant and antimicrobial properties	Wang et al. (2021)
Glucose and L-tryptophan	Hydrothermal (200 °C, 8 h)	Formulated CD-based chitosan hydrogels to enhance freshness of strawberries and oranges	Zhang et al. (2023a)
Crayfish shells	Hydrothermal (200 °C, 8 h)	Developed for fluorescence detection of thiamphenicol in food samples	Chen et al. (2022)
Coffee husk waste	Hydrothermal (220 °C, 10 h)	Incorporated into CMC/CQD composite films to delay spoilage of fresh-cut apples	Yang et al. (2024)
Baker's yeast	Hydrothermal (200 °C, 7 h)	Employed as antimicrobial and antioxidant additives in food packaging	Ghorbani et al. (2022)
Radish residue (*Raphanus sativus*)	Hydrothermal (200 °C, 6 h)	Enhanced starch/chitosan CD incorporated films for preservation of salmon fillets	Wang et al. (2024)

process is essentially based on the formation of ROS under ambient light, which causes oxidative damage to microbial membranes. CNDs also impair bacterial membrane integrity by direct contact, leading to internal component leakage and, ultimately, cell death. Dong et al. (2020) found that carbon dot-treated surfaces effectively killed germs through both ROS-mediated oxidative stress and photothermal effects. The continuous antimicrobial activity of CND-based packaging reduces microbial load on food surfaces, increasing shelf life without the need for additional chemical preservatives.

Water vapour and oxygen intrusion contribute significantly to food spoilage by promoting microbial growth, oxidative rancidity, and textural degradation. CNDs, when introduced into biopolymer matrices, greatly improve barrier characteristics by making structures more compact and less permeable. Wu et al. (2024) found that maize starch films containing carbon nanodots have lower water vapour permeability compared to pure starch films. The decrease in permeability was due to the homogeneous dispersion of CNDs inside the polymer matrix, which improved the tortuousness of gas and moisture transport paths. Dong et al. (2020) found that carbon dot-modified cellulose nanofibre films had better oxygen barrier characteristics, preventing oxidative deterioration in packaged goods. CND-enhanced packaging sheets slow down spoiling mechanisms by restricting oxygen and moisture transfer, improving the stability and shelf life of a variety of foods.

Carbon nanodot-enhanced packaging sheets can alter internal gas composition, resulting in less ethylene production and slower respiration rates, delaying senescence in fresh products. Wu et al. (2024) found that food wrapped in CND-based films had reduced respiration rates and a longer shelf life compared to controls. Gaikwad et al. (2023) found that carbon dot-based films prevent microbial growth and protect against oxidative senescence by lowering ROS buildup in fruits and vegetables. Thus, CND-enhanced active packaging serves the dual purpose of limiting microbiological contamination and oxidative deterioration.

12.3.2 Edible Coatings and Surface Treatments with Carbon Nanodots

Carbon nanodots have been effectively integrated into edible coatings applied directly to food surfaces. These coatings act as functional barriers, preventing microbial contamination and oxidative degradation. Wu et al. (2024) developed edible antibacterial films using corn starch/CND composites. When applied directly to meat items, they delivered instantaneous microbial inactivation while maintaining sensory qualities. Patil et al. (2020) found that carbon dot-reinforced polyvinyl alcohol (PVA) composite coatings provide improved UV protection, potentially preserving food quality by preventing photodegradation.

CND-based edible coatings provide rapid antimicrobial protection against pathogenic bacteria. Studies show that coatings not only inhibit surface microbial colonisation but also considerably lower overall microbial burden during storage (Knoblauch & Geddes, 2020). Wu et al. (2024) found that coated beef samples had significantly lower microbial populations compared to uncoated controls. Antimicrobial effectiveness is linked to ROS production, membrane contact, and suppression of bacterial metabolic pathways (Dong et al., 2020). Maintaining sensory qualities, including colour, texture, scent, and flavour, is critical for customer approval. Riahi et al. (2022) found that CMC-based coatings containing carbon quantum dots efficiently retained the colour and texture of treated fruits. Gaikwad et al. (2023) found that edible coatings with CNDs improve firmness and minimise enzymatic browning in fresh vegetables.

12.3.3 Carbon Nanodots as Real-Time Quality Monitors

Carbon nanodots (CNDs) have demonstrated great potential as real-time sensors for detecting food deterioration using fluorescence-based processes. Their photoluminescent features allow for the sensitive detection of biochemical changes such as the release of spoilage-related gases (ammonia, hydrogen sulphide) and pH changes in the food environment. Notably, pH-responsive CNDs generated from natural materials such as tangerine peel and resazurin (Tan/Res CDs) showed a quick colorimetric and fluorescence response to biogenic amines, which are commonly created during microbial protein deterioration (Ezati et al., 2023a, b). When exposed to ammonia vapour, these Tan/Res CDs change colour from yellow to brown and decrease in fluorescence strength, enabling early detection of spoilage before sensory quality deteriorates. Similarly, lignin-derived carbon quantum dots embedded in cellulose nanofibre films have been designed to check food freshness in real time. These films function as ratiometric fluorescence sensors, detecting the buildup of spoilage-related biogenic amines (Si et al., 2022). Flexible nanocomposite sensors offer an eco-friendly and scalable solution for detecting freshness in commodities such as prawns and pork.

CND-based sensors not only detect spoiling but also provide as visual early warning systems via immediate colour changes, allowing for quick spoilage detection on-site without the need for complex equipment. Zhang et al. (2022) created a functional hydrogel sensor in which exposure to volatile amines from prawn rotting triggered the in situ production of polymer-carbon nanodots (PCNDs), resulting in an observable fluorescence-based colour alteration. Fluorescence intensity associated with trimethylamine (TMA) concentrations, providing a smartphone-based freshness monitoring system. Furthermore, intelligent packaging ideas take advantage of carbon-based nanomaterials' capacity to detect freshness in response to environmental changes such as pH and gas concentration. Luo et al. (2020) describe smart systems that use colour-changing films or coatings to provide real-time, non-invasive quality monitoring.

CND-based fluorescence sensors provide several benefits over standard analytical techniques, including high sensitivity, low cost, mobility, quick response, and superior biocompatibility. They reduce the need for sophisticated sample preparation and specialised apparatus, which is consistent with current trends in on-site, user-friendly food quality evaluation (Chelladurai et al., 2024). In the future, expanding the range of detectable spoilage indicators, enhancing the stability of carbon nanodots (CNDs) under diverse food storage conditions, and integrating these sensors into commercial packaging materialscould significantly advance global food safety monitoring.

12.4 Carbon Nanodots for the Detection of Microbes in Food

Because of their exceptional biocompatibility, low toxicity, and distinctive optical characteristics, carbon nanodots (CNDs) have become extremely effective instruments in the field of food safety. Their intrinsic fluorescence, especially when responsive to environmental stimuli, enables them to serve as effective nanosensors for microbial detection in various food systems. Real-time imaging of microbiological contamination, spoilage processes, and quality degradation is made possible by these nanoparticles' excitation-dependent or -independent emission (Wang et al., 2021). CNDs can interact with microbial by-products like ammonia and hydrogen sulphide during food degradation, leading to fluorescence changes—either amplification or quenching—based on the analyte and the CND surface. These nanodots can be incorporated into edible films or sensor strips, enabling easy detection of microbial growth and food freshness without complex equipment (Mancini et al., 2023).

Additionally, CNDs' selectivity toward particular bacterial species is increased when they are functionalised with particular targeted molecules, such as antibiotics, peptides, or antibodies. For example, using fluorescence resonance energy transfer (FRET) or photoinduced electron transfer (PET) processes, antibiotic-functionalised CNDs have shown selective detection of foodborne pathogens such as *Escherichia coli* and *Listeria monocytogenes* (Ma et al., 2024). The usefulness of CNDs for food-related applications has also been enhanced by green synthesis techniques that use biomass or microbial metabolites. High water solubility and biocompatibility are characteristics of CNDs made from lactic acid bacteria or plant extracts that are necessary for their safe incorporation into food packaging systems (Cui et al., 2021). CNDs have expanded their use in contemporary food surveillance systems by being included in multiplex biosensing platforms that detect infections and chemical residues concurrently, in addition to pH and gas detection (Han et al., 2022).

12.5 Toxicological and Regulatory Perspectives

Although carbon dots are typically thought to be less harmful than semiconductor quantum dots, their production and design can still have a variety of characteristics that might cause cytotoxicity (Hola et al., 2014). The way that CNDs interact with biological systems is greatly influenced by their size and surface characteristics. Surface functionalisation can alter how CNDs interact with proteins and cell membranes, while smaller CNDs can be more readily absorbed by cells and perhaps pass across the blood–brain barrier (Du et al., 2023a, b). CNDs can be cytotoxic, according to in vitro research employing a variety of cell lines, but this is frequently dose- and time-dependent. Additionally, in vivo research using animal models has shown that CNDs may be absorbed by a variety of organs and tissues, some of which may be able to bioaccumulate (Makhado et al., 2024). After 24 h of incubation, researchers discovered that carbon dots were dispersed throughout the cytoplasm but not in the nucleus. When carbon dots were given orally to mice in studies, they entered the brain within 2 h, as indicated by the mouse brain's enhanced fluorescence, which shows that the carbon dots had crossed the blood–brain barrier. According to a recent study, carbon dots can build up in the cytoplasm and nucleus of L929 cells, killing them (Shabbir et al., 2022).

The absence of generally recognised regulatory criteria or guidelines for CNDs at the moment may prevent their broad use (Nguyen et al., 2025). Before using CNDs in a variety of applications, a comprehensive risk evaluation is essential to determine their potential toxicity and environmental effects. This entails evaluating their environmental interactions, long-term impacts, and bioaccumulation potential (Awasthi et al., 2024). CNDs have not yet received a thorough regulatory classification under food safety guidelines. Although debates are underway, organisations such as the US FDA and the European Food Safety Authority (EFSA) have not yet formally approved CNDs for use in food contact materials. To avoid accidental exposure, future laws must provide uniform toxicological screening, biodegradability evaluation, and migration testing (Chan et al., 2021).

12.6 Future Prospects: Transforming Food Safety with Carbon Nanodots

Despite substantial advances, some critical gaps and obstacles remain in the use of carbon nanodots for food safety and preservation. Carbon nanodots' antibacterial, antioxidant, enzyme-inhibitory, and sensing properties have been proven via the current study. However, the long-term durability of CND-based films and coatings under a variety of storage circumstances, including temperature, humidity, and light exposure, needs more research to ensure consistent performance throughout the shelf life of packaged goods. More standardised techniques are needed to assess the migration behaviour of carbon nanodots from packaging materials to food items, especially under real-world storage and handling situations. Comprehensive toxicological studies are required to ensure the biocompatibility and safety of carbon nanodots, particularly for edible coatings and direct-contact applications. While laboratory studies have shown that CNDs are effective at detecting specific spoilage markers such as ammonia and biogenic amines, broader sensing capabilities that cover a wider range of microbial metabolites and environmental parameters are required to improve the reliability of real-time freshness monitoring systems. Scalability of carbon nanodot synthesis, particularly from natural and food-grade precursors, remains an important consideration for commercial viability. Current approaches frequently entail small-scale or laboratory-based production, and the transition to cost-effective, industrial-scale manufacture must be established while maintaining the functional qualities of the nanodots. The incorporation of carbon nanodots into flexible, biodegradable, and multifunctional packaging solutions that correspond with sustainability objectives is an important future area. Furthermore, compliance with regulatory requirements and the establishment of clear guidelines for the use of nanomaterials in food-contact applications are critical to encouraging commercial acceptance. Future research will necessitate multidisciplinary collaboration among materials scientists, food technologists, toxicologists, and regulatory authorities in order to fully realise carbon nanodots' promise for building safer, longer-lasting, and more sustainable food packaging solutions.

12.7 Conclusion

Carbon nanodots are a significant development in food preservation, integrating multifunctional protective qualities on a single nanoscale substrate. Their shown antibacterial, antioxidant, enzyme-regulatory, and real-time sensing characteristics allow for complete food safety, shelf-life extension, and waste reduction techniques. Carbon nanodots in active packaging materials act as excellent barriers to microbiological contamination and oxidative degradation. Edible coatings enhanced with carbon nanodots provide direct protection to food products while retaining sensory and nutritional integrity. Smart packaging solutions that make use of carbon nanodots' photoluminescent capabilities allow for early spoilage detection, which aids in speedy quality control. The capacity to generate carbon nanodots from sustainable sources and customise their surface chemistry for specific functional results demonstrates their suitability for a wide range of culinary applications. Future incorporation of carbon nanodots into commercially feasible, regulatory-compliant packaging solutions has the potential to transform food safety monitoring and preservation methods. Carbon nanodots thus represent a potential and adaptable instrument in global efforts to ensure safer, fresher, and more sustainable food systems.

References

Alzahrani, A. (2024). Fluorescent carbon dots in situ polymerized biodegradable semi-interpenetrating tough hydrogel films with antioxidant and antibacterial activity for applications in food industry. *Food Chemistry, 447*, 138905.

Awasthi, S., Srivastava, A., Kumar, D., Pandey, S. K., Mubarak, N. M., Dehghani, M. H., & Ansari, K. (2024). An insight into the toxicological impacts of carbon nanotubes (CNTs) on human health: A review. *Environmental Advances, 18*, 100601. https://doi.org/10.1016/j.envadv.2024.100601

Babu, P. J. (2022). Nanotechnology mediated intelligent and improved food packaging. *International Nano Letters, 12*(1), 1–14.

Chan, M.-H., Chen, B.-G., Ngo, L. T., Huang, W.-T., Li, C.-H., Liu, R.-S., & Hsiao, M. (2021). Natural carbon Nanodots: Toxicity assessment and Theranostic biological application. *Pharmaceutics, 13*(11), 1874. https://doi.org/10.3390/pharmaceutics13111874

Chatzimitakos, T., & Stalikas, C. (2020). Antimicrobial properties of carbon quantum dots. In *Nanotoxicity* (pp. 301–315). Elsevier.

Chatzimitakos, T. G., Kasouni, A. I., Troganis, A. N., & Stalikas, C. D. (2020). Exploring the antibacterial potential and unraveling the mechanism of action of non-doped and heteroatom-doped carbon nanodots. *Journal of Nanoparticle Research, 22*, 1–13.

Chaudhry, Q., Scotter, M., Blackburn, J., Ross, B., Boxall, A., Castle, L., et al. (2008). Applications and implications of nanotechnologies for the food sector. *Food Additives and Contaminants, 25*(3), 241–258.

Chelladurai, D., Alaguthevar, R., Murugesan, B., Subburamu, K., Khan, A., & Rhim, J. W. (2024). Carbon quantum dots: Progress toward food safety and sustainability. *Food Bioscience, 105016*.

Chen, S., Ouyang, W., Zhu, Y., He, L., Zou, L., Ao, X., et al. (2022). Facile synthesis of N, S-doped carbon quantum dots from food waste as fluorescent probe for sensitive detection of thiamphenicol and its analogues in real food samples along with an application in bioimaging. *Food, 11*(16), 2414.

Cui, L., Ren, X., Sun, M., Liu, H., & Xia, L. (2021). Carbon dots: Synthesis, properties and applications. *Nanomaterials (Basel), 11*(12), 3419. https://doi.org/10.3390/nano11123419

Dong, X., Liang, W., Meziani, M. J., Sun, Y. P., & Yang, L. (2020). Carbon dots as potent antimicrobial agents. *Theranostics, 10*(2), 671.

Donsì, F., Annunziata, M., Sessa, M., & Ferrari, G. (2011). Nanoencapsulation of essential oils to enhance their antimicrobial activity in foods. *LWT-Food Science and Technology, 44*(9), 1908–1914.

Du, Y., Li, Y., Liu, Y., Liu, N., Cheng, Y., Shi, Q., Liu, X., Tao, Z., Guo, Y., Zhang, J., Askaria, N., & Li, H. (2023a). Stalk-derived carbon dots as nanosensors for Fe3+ ions detection and biological cell imaging. *Frontiers in Bioengineering and Biotechnology, 11*. https://doi.org/10.3389/fbioe.2023.1187632

Du, X., Zhang, M., Ma, Y., Wang, X., Liu, Y., Huang, H., & Kang, Z. (2023b). Size-dependent antibacterial of carbon dots by selective absorption and differential oxidative stress of bacteria. *Journal of Colloid and Interface Science, 634*, 44–53.

Espitia, P. J. P., Soares, N. D. F. F., Coimbra, J. S. D. R., de Andrade, N. J., Cruz, R. S., & Medeiros, E. A. A. (2012). Zinc oxide nanoparticles: Synthesis, antimicrobial activity and food packaging applications. *Food and Bioprocess Technology, 5*, 1447–1464.

Ezati, P., Rhim, J. W., Molaei, R., & Rezaei, Z. (2022). Carbon quantum dots-based antifungal coating film for active packaging application of avocado. *Food Packaging and Shelf Life, 33*, 100878.

Ezati, P., Khan, A., Priyadarshi, R., Bhattacharya, T., Tammina, S. K., & Rhim, J. W. (2023a). Biopolymer-based UV protection functional films for food packaging. *Food Hydrocolloids, 142*, 108771.

Ezati, P., Khan, A., Rhim, J. W., Kim, J. T., & Molaei, R. (2023b). pH-responsive strips integrated with resazurin and carbon dots for monitoring shrimp freshness. *Colloids and Surfaces B: Biointerfaces, 221*, 113013.

FAO. (2019). *The state of food and agriculture 2019: Moving forward on food loss and waste reduction*. UN.

Fernández, A., Picouet, P., & Lloret, E. (2010). Cellulose-silver nanoparticle hybrid materials to control spoilage-related microflora in absorbent pads located in trays of fresh-cut melon. *International Journal of Food Microbiology, 142*(1–2), 222–228.

Gao, F., Liu, J., Gong, P., Yang, Y., & Jiang, Y. (2023). Carbon dots as potential antioxidants for the scavenging of multi-reactive oxygen and nitrogen species. *Chemical Engineering Journal, 462*, 142338.

Ghorbani, M., Tajik, H., Moradi, M., Molaei, R., & Alizadeh, A. (2022). One-pot microbial approach to synthesize carbon dots from baker's yeast-derived compounds for the preparation of antimicrobial membrane. *Journal of Environmental Chemical Engineering, 10*(3), 107525.

Guo, B., Liu, G., Hu, C., Lei, B., & Liu, Y. (2022). The structural characteristics and mechanisms of antimicrobial carbon dots: A mini review. *Materials Advances, 3*(21), 7726–7741.

Han, Y., Yang, W., Luo, X., He, X., Zhao, H., Tang, W., Yue, T., & Li, Z. (2022). Carbon dots based ratiometric fluorescent sensing platform for food safety. *Critical Reviews in Food Science and Nutrition, 62*(1), 244–260. https://doi.org/10.1080/10408398.2020.1814197

He, Z., Xu, Q., Newland, B., Foley, R., Lara-Sáez, I., Curtin, J. F., & Wang, W. (2021). Reactive oxygen species (ROS): Utilizing injectable antioxidative hydrogels and ROS-producing therapies to manage the double-edged sword. *Journal of Materials Chemistry B, 9*(32), 6326–6346.

Hola, K., Zhang, Y., Wang, Y., Giannelis, E. P., Zboril, R., & Rogach, A. L. (2014). Carbon dots—Emerging light emitters for bioimaging, cancer therapy and optoelectronics. *Nano Today, 9*(5), 590–603. https://doi.org/10.1016/j.nantod.2014.09.004

Hui, W., Yang, Y., Xu, Q., Gu, H., Feng, S., Su, Z., Zhang, M., Wang, J., Li, X., Fang, J., & Xia, F. (2020). Red-carbon-quantum-dot-doped SnO2 composite with enhanced electron mobility for efficient and stable perovskite solar cells. *Advanced Materials, 32*(4), 1906374.

Innocenzi, P., & Stagi, L. (2023). Carbon dots as oxidant-antioxidant nanomaterials, understanding the structure-properties relationship. A critical review. *Nano Today, 50*, 101837.

IPCC. Summary for Policy Makers. Geneva: IPCC (2014).

Karanth, S., Feng, S., Patra, D., & Pradhan, A. K. (2023). Linking microbial contamination to food spoilage and food waste: The role of smart packaging, spoilage risk assessments, and date labeling. *Frontiers in Microbiology, 14*, 1198124.

Knoblauch, R., & Geddes, C. D. (2020). Carbon nanodots in photodynamic antimicrobial therapy: A review. *Materials, 13*(18), 4004.

Li, W. Q., Wang, Z., Hao, S., Sun, L., Nisic, M., Cheng, G., Zhu, C., Wan, Y., Ha, L., & Zheng, S. Y. (2018). Mitochondria-based aircraft carrier enhances in vivo imaging of carbon quantum dots and delivery of anticancer drug. *Nanoscale, 10*(8), 3744–3752.

Li, H., Guo, S., Li, C., Huang, H., Liu, Y., & Kang, Z. (2015). Tuning laccase catalytic activity with phosphate functionalized carbon dots by visible light. *ACS Applied Materials & Interfaces, 7*(18), 10004–10012.

Lin, F., Li, C., & Chen, Z. (2018). Bacteria-derived carbon dots inhibit biofilm formation of Escherichia coli without affecting cell growth. *Frontiers in Microbiology, 9*, 259.

Luo, X., Han, Y., Chen, X., Tang, W., Yue, T., & Li, Z. (2020). Carbon dots derived fluorescent nanosensors as versatile tools for food quality and safety assessment: A review. *Trends in Food Science & Technology, 95*, 149–161.

Ma, G., Li, X., Cai, J., & Wang, X. (2024). Carbon dots-based fluorescent probe for detection of foodborne pathogens and its potential with microfluidics. *Food Chemistry, 451*, 139385. https://doi.org/10.1016/j.foodchem.2024.139385

Makhado, B. P., Oladipo, A. O., Gumbi, N. N., De Kock, L. A., Andraos, C., Gulumian, M., & Nxumalo, E. N. (2024). Unravelling the toxicity of carbon nanomaterials—From cellular interactions to mechanistic understanding. *Toxicology In Vitro, 100*, 105898. https://doi.org/10.1016/j.tiv.2024.105898

Mancini, F., Menichetti, A., Degli Esposti, L., Montesi, M., Panseri, S., Bassi, G., Montalti, M., Lazzarini, L., Adamiano, A., & Iafisco, M. (2023). Fluorescent carbon dots from food industry by-products for cell imaging. *Journal of Functional Biomaterials, 14*(2), 90. https://doi.org/10.3390/jfb14020090

Min, S., Ezati, P., & Rhim, J. W. (2022). Gelatin-based packaging material incorporated with potato skins carbon dots as functional filler. *Industrial Crops and Products, 181*, 114820.

Misiou, O., & Koutsoumanis, K. (2022). Climate change and its implications for food safety and spoilage. *Trends in Food Science & Technology, 126*, 142–152.

Murru, C., Badía-Laíño, R., & Díaz-García, M. E. (2020). Synthesis and characterization of green carbon dots for scavenging radical oxygen species in aqueous and oil samples. *Antioxidants, 9*(11), 1147.

Nazir, S., & Azaz Ahmad Azad, Z. R. (2019). Food nanotechnology: An emerging technology in food processing and preservation. In *Health and safety aspects of food processing technologies* (pp. 567–576).

Nguyen, D. H. H., El-Ramady, H., & Prokisch, J. (2025). Food safety aspects of carbon dots: A review. *Environmental Chemistry Letters, 23*(1), 337–360. https://doi.org/10.1007/s10311-024-01779-3

Patil, A. S., Waghmare, R. D., Pawar, S. P., Salunkhe, S. T., Kolekar, G. B., Sohn, D., & Gore, A. H. (2020). Photophysical insights of highly transparent, flexible and re-emissive PVA@WTR-CDs composite thin films: A next generation food packaging material for UV blocking applications. *Journal of Photochemistry and Photobiology A: Chemistry, 400*, 112647.

Riahi, Z., Rhim, J. W., Bagheri, R., Pircheraghi, G., & Lotfali, E. (2022). Carboxymethyl cellulose-based functional film integrated with chitosan-based carbon quantum dots for active food packaging applications. *Progress in Organic Coatings, 166*, 106794.

Shabbir, H., Wojtaszek, K., Rutkowski, B., Csapó, E., Bednarski, M., Adamiec, A., Głuch-Lutwin, M., Mordyl, B., Druciarek, J., Kotańska, M., Ozga, P., & Wojnicki, M. (2022). Milk-derived carbon quantum dots: Study of biological and chemical properties provides evidence of toxicity. *Molecules, 27*(24), 8728. https://doi.org/10.3390/molecules27248728

Shi, R., Li, H., Wu, E., Xiong, L., Lv, R., Guo, R., et al. (2017). Simultaneous enzymatic activity modulation and rapid determination of enzyme kinetics by highly crystalline graphite dots. *Nanoscale, 9*(24), 8410–8417.

Si, L., Shi, Z., Hou, J., Miao, C., Hou, Q., Xu, Z., & Ai, S. (2022). Lignin-derived carbon dot/cellulose nanofiber films for real-time food freshness monitoring. *ACS Applied Nano Materials, 5*(11), 16620–16632.

Wang, L., Li, B., Li, L., Xu, F., Xu, Z., Wei, D., et al. (2017). Ultrahigh-yield synthesis of N-doped carbon nanodots that down-regulate ROS in zebrafish. *Journal of Materials Chemistry B, 5*(38), 7848–7860.

Wang, Z., Liu, Q., Leng, J., Liu, H., Zhang, Y., Wang, C., et al. (2021). The green synthesis of carbon quantum dots and applications for sulcotrione detection and anti-pathogen activities. *Journal of the Saudi Chemical Society, 25*(12), 101373.

Wang, D., Wang, X., Zhou, S., Ren, L., Meng, Y., Ma, R., et al. (2024). Radish residue carbon dots-based novel starch/chitosan film with high antioxidant, biocompatibility, and antibacterial activities for salmon fillets' active packaging. *International Journal of Biological Macromolecules, 273*, 133107.

Wei, J., Sun, W., Pan, W., Yu, X., Sun, G., & Jiang, H. (2017). Comparing the effects of different oxygen-containing functional groups on sulfonamides adsorption by carbon nanotubes: Experiments and theoretical calculation. *Chemical Engineering Journal, 312*, 167–179.

WHO, U. (2015). *WHO estimates of the global burden of foodborne diseases*. World Health Organization.

Wu, Y., Zhang, J., Hu, X., Huang, X., Zhang, X., Zou, X., & Shi, J. (2024). Preparation of edible antibacterial films based on corn starch/carbon nanodots for bioactive food packaging. *Food Chemistry, 444*, 138467.

Yang, J., Li, Y., Liu, B., Wang, K., Li, H., & Peng, L. (2024). Carboxymethyl cellulose-based multifunctional film integrated with polyphenol-rich extract and carbon dots from coffee husk waste for active food packaging applications. *Food Chemistry, 448*, 139143.

Zhang, W., Zeng, Z., & Wei, J. (2017). Electrochemical study of DPPH radical scavenging for evaluating the antioxidant capacity of carbon nanodots. *The Journal of Physical Chemistry C, 121*(34), 18635–18642.

Zhang, W., Chavez, J., Zeng, Z., Bloom, B., Sheardy, A., Ji, Z., et al. (2018). Antioxidant capacity of nitrogen and sulfur codoped carbon nanodots. *ACS Applied Nano Materials, 1*(6), 2699–2708.

Zhang, Y., Dong, S., Chen, J., Gao, W., & Shi, X. (2022). Amines-mediated β-glucose pentaacetate to generate photoluminescent polymer-carbon nanodots for visual monitoring the freshness of shrimp. *Talanta, 249*, 123706.

Zhang, L., Gao, X., Feng, Y., Yan, Y., Zhu, H., Liu, S., et al. (2023a). Chitosan-based hydrogel-incorporated Trp-CDs with antibacterial properties and pH-mediated fluorescence response as a smart food preservation material. *ACS Applied Materials & Interfaces, 15*(37), 44097–44108.

Zhang, Y. F., Sun, Y. W., Liu, X. H., An, Z. X., & Yang, X. F. (2023b). The study on bactericidal effect and ultrastructural alterations of chlorocresol nanoemulsion disinfectant against *Staphylococcus aureus*. *Polish Journal of Veterinary Sciences, 26*(1), 13–20.

Zheng, M., Wang, C., Wang, Y., Wei, W., Ma, S., Sun, X., & He, J. (2018). Green synthesis of carbon dots functionalized silver nanoparticles for the colorimetric detection of phoxim. *Talanta, 185*, 309–315.

Role of Carbon Nanodots in Plant Growth and Development

13

Kajal Gautam, Mohit Bhatt, Hukum Singh, and Santan Barthwal

Abstract

Carbon nanodots (CDs), a novel class of carbon-based nanomaterials, are rapidly emerging as promising agents in sustainable agriculture due to their unique physicochemical, optical, and biological properties. With sizes typically below 10 nm, CDs offer excellent biocompatibility, tunable fluorescence, high water solubility, and rich surface functionalization, making them ideal candidates for applications ranging from plant growth promotion to stress mitigation and smart nutrient delivery. This chapter comprehensively explores the synthesis strategies of CDs, including green, top-down, and bottom-up approaches, along with their characterization and mechanisms of interaction within plant systems. It further highlights how CDs enhance seed germination, nutrient uptake, photosynthesis, and biomass production, while also functioning as efficient nanocarriers for fertilizers, agrochemicals, and hormones. The role of CDs in alleviating abiotic and biotic stresses through antioxidant activity and gene regulation is discussed in depth. Additionally, recent advances in real-time plant monitoring, nanofertilizer design, and pathogen detection are presented, showcasing the multifaceted utility of CDs in modern precision agriculture. Despite their promising benefits, challenges related to large-scale synthesis, regulatory gaps, environmental fate, and field-level validation persist. This chapter concludes by addressing these concerns and outlining future research opportunities, positioning carbon nanodots

K. Gautam
Department of Chemistry, School of Advanced Engineering, UPES, Dehradun, India

M. Bhatt
Department of Physics, School of Advanced Engineering, UPES, Dehradun, India

H. Singh (✉) · S. Barthwal
ICFRE-Forest Research Institute, P.O. New Forest, Dehradun, Uttarakhand, India

M. Rai et al. (eds.), *Carbon Nanodots*, https://doi.org/10.1007/978-981-95-4705-0_13

as next-generation tools for climate-resilient and resource-efficient agricultural systems.

Keywords

Carbon nanodots · Plant growth · Green nanotechnology · Smart delivery systems · Stress tolerance · Physiological and biochemical responses · Sustainable agriculture

13.1 Introduction

Nanotechnology, defined as the manipulation of matter at the atomic and molecular scale within the range of 1–100 nm, is emerging as a transformative approach across numerous scientific domains (Patel et al., 2021). In agriculture, it has garnered considerable attention for its ability to provide novel, precision-based solutions to the longstanding challenges of food insecurity (Khan et al., 2022a, b; Neme et al., 2021), declining crop productivity (Mali et al., 2020), and environmental degradation (Guerra et al., 2018). Enabling the engineering of materials with tailored physicochemical properties is revolutionizing how crops are grown, protected, and monitored under changing climatic and environmental conditions (Xu, 2022). In agriculture, nanomaterials (NMs) have been integrated into a wide array of crop management strategies (Agrawal et al., 2022). These materials exhibit unique characteristics, including a high surface-area-to-volume ratio, enhanced reactivity (Sajid, 2022), and quantum confinement effects (Njuguna et al., 2021), making them well suited for interacting with biological systems in beneficial ways (Mu et al., 2014). Their multifunctional potential has led to the development of smart fertilizers (Solanki et al., 2015), nano-pesticides (Tang et al., 2023), biosensors (Li et al., 2024), and controlled-release systems (Jadhav et al., 2024), improving the efficiency of resource utilization, enhancing plant health, and minimizing environmental pollution (Wahab et al., 2024). Notably, the shift toward biodegradable and biocompatible nanomaterials has aligned these innovations with global sustainability goals, ecological safety, and climate resilience objectives.

Carbon nanodots (CDs), a relatively new class of carbon-based nanoparticles derived from green synthesis and conventional methods, have gained increasing interest due to their exceptional optical, chemical, and biological properties (Qureshi et al., 2024; Gautam et al., 2025). CDs are quasi-spherical, zero-dimensional nanoparticles, typically smaller than 10 nm, composed mainly of carbon and variable proportions of other elements such as oxygen, hydrogen, and nitrogen (Yadav & Daniel, 2024). Their surfaces are richly functionalized with groups such as hydroxyl (–OH), carboxyl (–COOH), and amine ($-NH_2$), which improve their aqueous solubility and biocompatibility, and allows for diverse chemical modifications (Yan et al., 2018). Based on their structure, crystallinity, and synthesis method, CDs are broadly categorized into three types: (a) *carbon quantum dots (CQDs)* possess a crystalline core and exhibit strong quantum confinement effects, making their fluorescence highly size-dependent and tunable properties ideal for bioimaging and optical tracking (Saraswat et al., 2024), (b) *graphene quantum dots (GQDs)* are composed of single or few-layer graphene sheets, known for their pronounced edge effects, excellent electronic properties, and potential in biosensing and optoelectronic applications (Kadian et al., 2021), and (c) *carbonized polymer dots (CPDs),* are typically amorphous or partially crystalline, synthesized from organic or polymeric precursors and are valued for their ease of functionalization and diverse surface chemistry (Xia et al., 2019).

What makes CDs particularly attractive in agriculture is their exceptional photoluminescence, which enables emission across a broad spectral range from ultraviolet (UV) to near-infrared (NIR) wavelengths (Hallaji et al., 2022). This optical tunability facilitates noninvasive plant imaging (Kumar et al., 2023), molecular tagging, and real-time stress detection, providing researchers and agronomists with valuable insights into plant physiology and responses to environmental stimuli (Wang et al., 2017). Importantly, CDs are also recognized for their low cytotoxicity, distinguishing them from many metal-based nanoparticles, which may pose risks to both plants and ecosystems (Prokisch et al., 2024). When produced through green synthesis routes, utilizing renewable precursors such as fruit peels, plant biomass, agricultural residues, or food waste, CDs become environmentally sustainable. These eco-friendly synthesis strategies reduce harmful by-products and energy consumption, supporting the circular bioeconomy and contributing to long-term environmental and economic sustainability (Sah et al., 2024).

Additionally, the surface modifiability of CDs enables them to serve as effective nanocarriers for delivering agrochemicals, micronutrients, plant growth regulators, and even genetic materials (Chaudhary et al., 2024). Their small size facilitates cellular and subcellular interactions, allowing for enhanced permeability through plant cell walls and membranes (Chaudhary et al., 2024). Such interactions open doors to advanced applications in nutrient management, stress mitigation, pathogen resistance, and nano-enabled precision farming.

The multifunctional nature of CDs, which combines optical sensing, targeted delivery, and environmental compatibility, positions them as an auspicious tool in climate-smart and sustainable agriculture. Their ability to support plant growth under both optimal and stressful conditions makes them particularly valuable for addressing the increasing demands for food production in the face of climate uncertainty, land scarcity, and declining soil quality (Minhans et al., 2025).

There is a need to comprehensively explore the potential of carbon nanodots in agriculture, particularly their role in promoting plant growth, development, and resilience. This chapter delves into the synthesis techniques and classifications of CDs, followed by an examination of their unique properties and characterization methods. The subsequent sections examine how CDs influence critical physiological processes, including seed germination, photosynthesis, nutrient uptake, and stress tolerance. Furthermore, this chapter discusses the use of CDs as smart delivery vehicles, their role in defense responses, and the ecotoxicological considerations associated with their use. Understanding the recent advances and the existing challenges provides a holistic view of the emerging role of carbon nanodots in next-generation agro-nanotechnology and sustainable farming systems.

13.2 Synthesis and Characterization of Carbon Nanodots (CDs)

The synthesis of carbon nanodots (CDs) has gained momentum due to their unique properties and wide-ranging applications in biological and agricultural sciences (Maholiya et al., 2023). Numerous top-down and bottom-up approaches have been developed, enabling precise control over their size, surface functionalities, and optical characteristics (Anadebe et al., 2024). A sustainable and scalable synthesis process is crucial for aligning with the objectives of green chemistry and ensuring environmental compatibility, particularly in agricultural applications.

13.2.1 Synthesis Approaches

13.2.1.1 Top-Down Methods

Top-down strategies involve breaking down larger carbon structures into nanoscale fragments. These methods generally require harsh chemical or physical treatments and are effective in producing CDs with well-defined crystalline cores.

Arc Discharge and Laser Ablation These methods involve high-energy physical processes to cleave carbon sources such as graphite. Although these methods are effective, they often require post-treatment steps to enhance biocompatibility (Banger et al., 2023).

Electrochemical Oxidation This technique utilizes electrochemical cells to fragment graphite or other carbon materials into CDs, offering better control over surface chemistry (Wang et al., 2025).

Ultrasonic Treatment Ultrasonication is applied to carbonaceous precursors, such as soot or graphite, to induce mechanical disintegration, yielding nano-sized carbon dots (Wang et al., 2015).

13.2.1.2 Bottom-Up Methods

Bottom-up methods are more suitable for green and low-cost synthesis, as they assemble CDs from molecular precursors through chemical or thermal reactions (Abid et al., 2022).

Hydrothermal and Solvothermal Methods Organic precursors (e.g., citric acid, glucose, amino acids, or plant extracts) are subjected to high temperatures and pressures, leading to carbonization and the formation of CDs. These are widely used due to their simplicity, eco-friendliness, and suitability for bulk production (Varadharajan et al., 2024).

Microwave-Assisted Synthesis This method offers rapid and energy-efficient heating, resulting in uniform particle size and high photoluminescence. It is increasingly used in green synthesis from biowaste (Guan et al., 2014).

Thermal Pyrolysis This technique involves direct heating of organic materials to initiate carbonization. This technique is straight forward and scalable; however, control over surface functionality may require additional processing (Usman & Cheng, 2024).

Green Synthesis Leveraging natural precursors like fruit peels, leaves, flower extracts, or even agricultural waste, green synthesis methods are environmentally benign and cost-effective. They eliminate the need for toxic chemicals and contribute to sustainability goals (Abuzeid et al., 2023).

13.2.2 Characterization Techniques

Characterizing the synthesized CDs is crucial to understanding their structural, chemical, and optical properties, which determine their functionality in plant systems.

Transmission Electron Microscopy (TEM) This technique provides high-resolution images to determine the size, shape, and morphology of CDs. Typically, CDs appear as quasi-spherical dots with diameters of less than 10 nm (Liu et al., 2017).

Fourier Transform Infrared Spectroscopy (FTIR) This technique identifies surface functional groups like hydroxyl, carboxyl, and amine moieties, which are essential for solubility, reactivity, and plant interaction.

X-ray Diffraction (XRD) This determines the crystalline or amorphous nature of CDs. While carbon quantum dots show crystalline features, CPDs often appear amorphous.

Raman Spectroscopy This is used to study the graphitic nature and defects in carbon structures through D and G band intensity analysis (Dubale et al., 2014).

UV-Visible Spectroscopy This technique helps identify absorption peaks, typically indicating π–π^* transitions (aromatic C=C) and n–π^* transitions (C=O), which are reflective of the carbon core structure (Akash & Rehman, 2020).

Photoluminescence (PL) Spectroscopy It measures the emission behavior of CDs under different excitation wavelengths. CDs generally exhibit excitation-dependent fluorescence, a key property for bioimaging and sensing applications (Wang & Lu, 2022).

Zeta Potential Measurement This evaluates surface charge, colloidal stability, and dispersion behavior in aqueous environments, which are critical for applications in plant systems (Midekessa et al., 2020).

13.2.3 Importance of Controlled Synthesis

Achieving reproducible results and consistent quality in CD synthesis is vital for their integration into agriculture. Parameters such as pH, precursor concentration, temperature, and reaction time directly influence the particle size, surface groups, and fluorescence properties, all of which determine the effectiveness of CDs in interacting with plant cells (Li et al., 2020a, b). Through careful selection of methods and precursors, especially those derived from renewable biomass, CDs can be tailored to suit specific agricultural applications, such as nutrient delivery, stress detection, or growth promotion.

13.3 Mechanisms of Interaction Between Carbon Nanodots and Plants

Understanding how CDs interact with plants at the cellular and molecular levels is essential for optimizing their use in agriculture (Su et al., 2018). Once introduced into plant systems, CDs influence multiple physiological processes, depending on their size, surface charge, chemical composition, and functional groups. These properties determine their uptake, movement within plant tissues, and ultimate bioactivity (Chen et al., 2020a, b, 2021).

13.3.1 Entry Pathways of Carbon Nanodots into Plant Systems

Carbon nanodots (CDs), due to their ultrasmall size, functional surface groups, and high aqueous dispersibility, infiltrate plant tissues through multiple routes (Maholiya et al., 2023). While root uptake and foliar application are the most prominent and studied methods (Shahid et al., 2017), several additional entry mechanisms are being explored in controlled environments, in vitro cultures, and using advanced delivery techniques. Figure 13.1 illustrates the multiple entry pathways of nanoparticles into the plant system. These include seed priming, stem injection, and pollen-mediated uptake (Acharya et al., 2020).

13.3.1.1 Root Uptake (Soil or Hydroponic Route)

When CDs are introduced to the rhizosphere, either via irrigation, hydroponic solutions, or soil amendments, they are readily absorbed by the root system (Su et al., 2018). Due to their nanoscale dimensions (typically <10 nm), they move through apoplastic pathways, traveling between cell walls and intercellular spaces (Sattelmacher, 2001), and symplastic pathways, moving cell-to-cell via

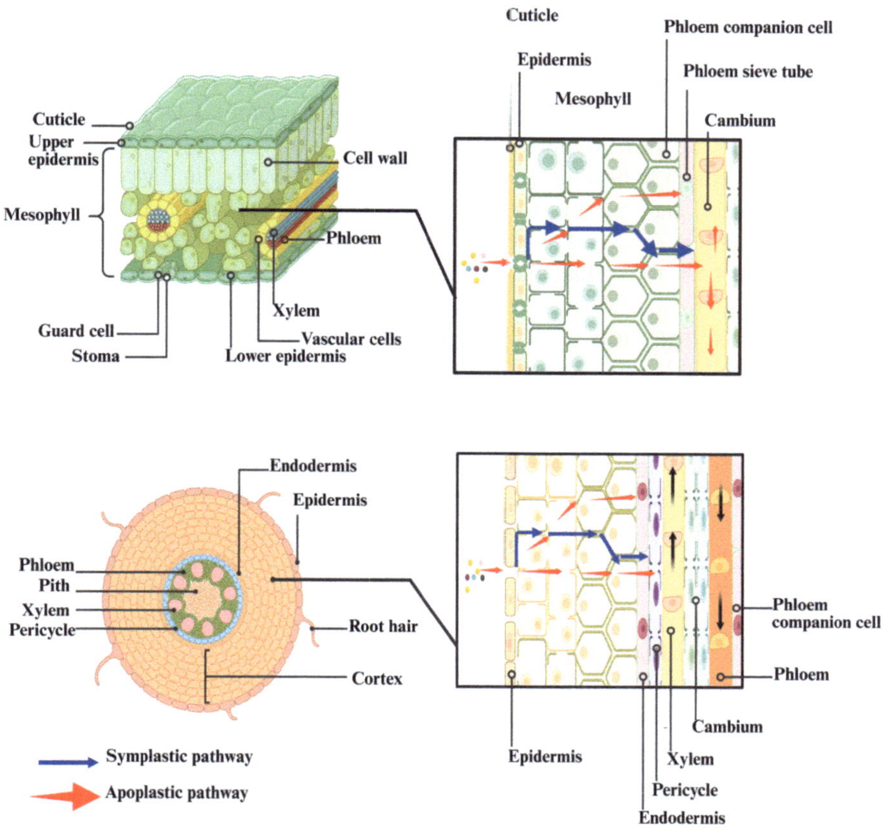

Fig. 13.1 Uptake pathway of nanoparticles in plants includes absorption through leaves (**a**) and roots (**b**) followed by transport through xylem and phloem (**c**, **d**) via apoplastic or symplastic pathways. [Created in https://BioRender.com (Reproduced with permission from *Advanced Biology*, Vol. 9, Issue 4, 2025. © John Wiley & Sons. Licensed for reuse under RightsLink License No. 6070890213548)]

plasmodesmata (Faulkner, 2018). Entry is often facilitated through root hairs, epidermal gaps, or damaged cell junctions. Once inside, CDs migrate via the xylem vessels, enabling systemic translocation toward aerial plant parts, including stems and leaves.

13.3.1.2 Foliar Uptake (Leaf-Surface Application)

Spraying CDs onto leaf surfaces enables their absorption through stomatal openings, which naturally allow gas exchange. Cuticular pores or waxes, particularly when CDs are modified with surfactants or hydrophobic carriers, and wounds or trichomes, where entry is easier due to broken or thin barriers (Afzal et al., 2024), also facilitate absorption. Once internalized, CDs move through the phloem and reach distant plant organs, making foliar delivery suitable for targeting photosynthetic or reproductive tissues (Kou et al., 2025; Pan et al., 2024).

13.3.1.3 Seed Priming and Coating

Seed treatment with carbon nanodots represents an effective pre-germination delivery strategy. In this method, seeds are soaked or coated with CDs before planting (Maholiya et al., 2023); during *imbibition* (water uptake phase), CDs penetrate seed coats and enter embryonic tissues (Zhang et al., 2023), and allows early exposure of the growing seedling to beneficial nanomaterials, often resulting in improved germination rate, vigor, and initial root–shoot development (Khan et al., 2022a, b). This method is noninvasive, scalable, and environmentally friendly, making it a promising approach for sustainable agriculture.

13.3.1.4 Stem Injection

A less common but highly targeted method involves direct injection of CDs into the plant stem using syringes or microneedles. Although invasive, this approach offers precision delivery to vascular bundles (xylem and phloem), minimal material loss compared to surface applications, and rapid and uniform systemic transport, particularly for experimental studies. Stem injection is particularly useful in controlled studies aimed at tracking fluorescence-tagged CDs within internal tissues or delivering high-value cargo, such as DNA, RNA, or hormones (Shivashakarappa et al., 2025).

13.3.1.5 In Vitro Plant Culture Exposure

In tissue culture systems, plantlets or calluses are grown on nutrient media. In this system, CDs are directly added to the growth medium. This allows for direct interaction of CDs with regenerating tissues, efficient uptake by meristematic cells, which are highly permeable and metabolically active, and their use in transformation studies, where CDs serve as carriers for genes, drugs, or growth regulators (Li et al., 2020a, b). In vitro approaches help dissect the mechanistic details of CD uptake under sterile and highly controlled conditions (Table 13.1).

These diverse entry pathways provide researchers and agronomists with multiple strategies to harness CDs for enhancing plant health, delivering nutrients or biostimulants, and achieving precision nano-farming with minimal environmental impact (Abinaya et al., 2024).

Table 13.1 Pathways of CD entry into plants

Entry method	Target site	Key features
Root uptake	Roots, xylem	Passive and active uptake; common in soil/hydroponics
Foliar uptake	Leaves, stomata, cuticle	Useful for aerial delivery; potential for systemic movement
Seed priming/coating	Embryo, radicle	Noninvasive, enhances early growth
Stem injection	Vascular system (xylem)	Precise but invasive; rapid translocation
In vitro culture medium	Callus, shoots, roots	Ideal for mechanistic studies, transformation

13.3.2 Internal Translocation and Cellular Localization

Following uptake, CDs exhibit dynamic mobility and accumulate within various plant tissues and organelles. Their nanoscale dimensions and surface functionalities enable them to penetrate plasma membranes and subsequently translocate to intracellular compartments (Hazra et al., 2024). Research has shown that CDs can localize in chloroplasts, influencing photosynthetic efficiency (Guo et al., 2022); in mitochondria, modulating respiration and energy metabolism (Law et al., 2023); in nuclei, potentially interacting with genetic material (Lin et al., 2023); and in the cytosol, where they may participate in redox regulation or enzymatic interactions (Arumugam & Lee, 2024). The inherent photoluminescent nature of CDs enables real-time tracking and visualization of their distribution within plant systems, making them suitable for imaging-based studies.

13.3.3 Interactions with Plant Cellular Components

CDs engage in multifaceted interactions at the cellular and molecular levels. These interactions are primarily governed by their surface functional groups, charge, and hydrophilicity.

Cell Wall and Membrane Binding CDs adsorb onto the cell wall matrix or integrate into the lipid bilayer, potentially influencing membrane fluidity and nutrient transport (Barceló & Laura, 2009).

Signal Transduction Certain CDs initiate or modulate signaling cascades that influence growth, stress responses, or gene expression.

Reactive Oxygen Species (ROS) Regulation CDs possess antioxidant properties and can either scavenge excess ROS or modulate the activity of ROS-associated enzymes such as superoxide dismutase (SOD), catalase (CAT), and peroxidases (PODs). This contributes significantly to stress tolerance under abiotic and biotic challenges (Das & Roychoudhury, 2014).

Nutrient and Agrochemical Carriers Due to their functional surfaces, CDs can act as nanocarriers for micronutrients (e.g., Zn, Fe, Mn) or plant growth regulators, thereby enhancing their bioavailability and facilitating transport to target sites (Chen et al., 2025a, b).

13.3.4 Influence on Gene Expression and Molecular Pathways

CDs influence gene expression patterns related to plant growth, development, and defense. Their interaction with nuclear components or regulatory molecules may modulate genes involved in cell division and elongation (Wang et al., 2020; Yan et al., 2021), pathways associated with photosynthetic apparatus biogenesis (Wang et al., 2021), expression of stress-responsive transcription factors (Cheng et al., 2023), and hormonal signaling networks such as those governed by auxins, cytokinins, and gibberellins (Yan et al., 2021). These molecular-level modulations suggest that CDs do not merely serve as passive agents but actively participate in shaping plant physiology.

13.3.5 Factors Influencing CD–Plant Interactions

The extent and nature of CD interaction with plant systems depend on several interrelated factors:

Particle Size Smaller particles (<10 nm) exhibit higher cellular penetration efficiency (Wang et al., 2023).

Surface Charge Positively charged CDs tend to bind more effectively to negatively charged plant cell membranes, enhancing uptake (Que et al., 2024).

Concentration and Dosage While low concentrations often promote growth and metabolic enhancement, higher doses may trigger toxicity and oxidative stress (Li et al., 2023).

Functional Groups Surface groups such as hydroxyl (–OH), carboxyl (–COOH), and amine (–NH_2) significantly influence the biocompatibility, solubility, and reactivity of CDs within plant environments.

13.4 Impacts of Carbon Nanodots on Plant Growth and Development

The application of carbon nanodots (CDs) has been shown to positively influence various aspects of plant growth and development. Due to their nanoscale size, high water dispersibility, and functional surface groups, CDs interact readily with plant cells, triggering a cascade of beneficial physiological responses. Their effects are especially pronounced during critical stages such as seed germination, root and shoot elongation, photosynthesis, nutrient assimilation, and overall biomass accumulation. Importantly, these effects are highly dependent on the dosage, surface chemistry, and plant species involved.

13.4.1 Enhancement of Seed Germination and Early Growth

CDs demonstrate the ability to enhance seed germination rates and improve seedling vigor. Seeds treated with an optimized concentration of CDs demonstrate faster radical emergence and plumule elongation. This is attributed to the improved water uptake, enhanced enzymatic activity, and better cellular metabolism during the initial stages of seed imbibition and growth (Zhang et al., 2023). Additionally, CDs may upregulate the activity of enzymes such as α-amylase, which helps convert stored starches into sugars, thereby supplying energy for germination. The early boost in growth sets the stage for healthier, more resilient seedlings that can withstand environmental fluctuations during their juvenile phase (Liang et al., 2023).

13.4.2 Root and Shoot Development

After the uptake of CDs by plants, they are translocated to growing tissues, influencing cell division and elongation. In roots, this often translates into increased root length, density of lateral roots, and a more developed root hair system. This improvement in root architecture enhances nutrient absorption and water uptake, contributing to robust plant establishment (Tripathi & Sarkar, 2015). Shoots benefit from CD exposure through better cell expansion and chlorophyll biosynthesis. CDs may mimic or modulate plant hormone signaling, such as auxins or gibberellins, leading to elongated internodes, increased leaf area, and overall improved shoot vigor (Li et al., 2020a, b).

13.4.3 Photosynthesis and Chlorophyll Content

CDs have a significant impact on the photosynthetic efficiency of plants. CDs penetrate mesophyll cells and localize within chloroplasts, where they interact with photosystems and light-harvesting complexes. Their inherent photoluminescent properties help harvest more light energy, especially under suboptimal lighting, thereby boosting the photosynthetic rate (Hu et al., 2022). Furthermore, CDs increase chlorophyll a and b content, resulting in greener leaves and improved energy capture. The cumulative effect is an enhanced net photosynthetic rate, resulting in increased biomass and yield potential (Li et al., 2020a, b).

13.4.4 Nutrient Uptake and Assimilation

The application of CDs in plants enhances nutrient uptake, particularly for micronutrients like Fe, Zn, Cu, and Mn, due to their high surface area and functional groups. CDs may chelate and carry essential nutrients, facilitating their transport across root membranes and subsequent movement to metabolic centers. CDs act as nano-chelating agents, enhancing nutrient solubility and availability in the rhizosphere, upregulating the expression of nutrient transporter genes, and facilitating the efficient internal redistribution of essential elements within the plant body (Hu et al., 2022).

13.4.5 Biomass Accumulation and Yield

The cumulative effects of improved germination, enhanced nutrient uptake, and increased photosynthesis ultimately result in higher dry and fresh biomass. Plants treated with CDs tend to exhibit a greater leaf area index, a higher root-to-shoot ratio, and increased total plant weight compared to untreated controls (Singh 2024). Some experimental trials have also reported higher yields of fruits or grains, especially in cereals, legumes, and leafy vegetables. This demonstrates the practical utility of CDs in yield enhancement, which is particularly relevant for resource-poor or climate-stressed agricultural systems (Özalkan et al., 2010).

13.4.6 Influence on Plant Hormonal Activity

Evidence indicates that CDs may influence endogenous hormone levels or mimic hormonal signals. For instance, their structural properties and surface functional groups allow them to participate in auxin-like signaling, thereby promoting cell elongation and differentiation. Some reports indicate elevated activity of indole acetic acid (IAA) and gibberellic acid (GA_3) in CD-treated plants, resulting in stimulated growth responses (González-Lamothe et al., 2012). Moreover, the antioxidative properties of CDs may suppress ethylene production, particularly under stress conditions, enabling plants to sustain growth when exposed to salinity or drought (Wang et al., 2022a, b).

13.4.7 Dose Dependency and Phytotoxicity Threshold

While the majority of studies highlight the growth-promoting effects of CDs, it is crucial to note that their impact is concentration-dependent. Low-to-moderate concentrations (typically <100 mg/L) tend to promote growth, whereas high doses may lead to oxidative stress, chlorosis, or inhibited root elongation (Li et al., 2023). Hence, precise dose optimization is critical before field-scale applications (Table 13.2).

The multifunctional nature of carbon nanodots makes them a powerful tool in agricultural biotechnology. Their ability to impact plant physiology at multiple stages, from germination to harvest, suggests immense potential in enhancing crop productivity, particularly under the constraints of modern agriculture.

13.5 Carbon Nanodots as Smart Delivery Vehicles

Carbon nanodots are gaining considerable attention as nanocarriers in the field of plant nanobiotechnology due to their ultrasmall size, tunable surface chemistry, and high biocompatibility. Their ability to interact with plant cellular components, coupled with their aqueous dispersibility and rich functional groups, enables CDs

Table 13.2 Overview of carbon nanodots' impacts on plant growth parameters

Growth parameter	Observed effects	Proposed mechanism	Reference
Seed germination	Faster germination, better vigor	Improved water uptake, enzyme activity (α-amylase)	Chen et al. (2017)
Root architecture	Longer roots, more lateral roots, and root hairs	Enhanced cell division and auxin-like signaling	Chen et al. (2021)
Shoot growth	Taller plants, larger leaves	Chlorophyll biosynthesis and hormone modulation	Badra et al. (2006)
Photosynthesis	Increased rate, higher chlorophyll content	Light-harvesting, photoluminescence enhancement	Guirguis et al. (2023)
Nutrient uptake	Improved micronutrient assimilation (Fe, Zn, cu)	Chelation and transport facilitation, gene upregulation	Gautam et al. (2025)
Biomass and yield	Higher dry and fresh weight, improved crop yield	Combined effects of photosynthesis, nutrient uptake, and root health	
Hormonal effects	Increased IAA, GA3, reduced ethylene under stress	Hormone-like surface interactions and ROS scavenging	Wang et al. (2022a, b)

to serve as intelligent delivery systems for a variety of agriculturally important molecules (Kohli & Parab, 2025). This includes micronutrients, growth regulators, pesticides, and even genetic material. The development of smart nanocarrier systems is crucial for modern precision agriculture, where minimizing inputs while maximizing effects is the primary goal. CDs offer advantages over traditional delivery systems through controlled release, target specificity, and reduced toxicity (Liu et al., 2020).

13.5.1 Mechanism of Delivery and Loading

Their surface functionality largely determines the delivery capability of CDs. The presence of –OH, –COOH, and —NH_2 groups enables CDs to bind and carry molecules through physical adsorption, covalent conjugation, electrostatic interaction, and hydrogen bonding (Kaurav et al., 2023). These mechanisms enable CDs to effectively encapsulate or attach essential nutrients, such as Fe^{2+}, Zn^{2+}, or Cu^{2+}, as well as phytohormones like auxins or cytokinins, agrochemicals like pesticides and herbicides, and biological molecules, including DNA, RNA, or proteins. Once loaded, CDs protect the cargo molecules from degradation (especially under UV or enzymatic stress), thereby enhancing their bioavailability and ensuring sustained release when delivered into plant tissues (Han et al., 2022).

13.5.2 Nutrient Delivery Systems

One of the most promising applications of CDs is in delivering essential plant micronutrients. Micronutrients often have limited mobility in soil or suffer from poor bioavailability due to precipitation or complexation with other elements. CDs can bind micronutrients through chelation, enhance root uptake and vascular translocation, and facilitate targeted release at deficient sites (Alikhani et al., 2023). For instance, iron-loaded CDs have shown improved uptake efficiency and translocation to chloroplasts, correcting chlorosis in iron-deficient plants more effectively than conventional Fe chelates (Lu et al., 2025).

13.5.3 Agrochemical and Pesticide Delivery

Traditional pesticide applications often suffer from nonspecific targeting, rapid degradation, and runoff losses, leading to environmental contamination (Chen et al., 2025a, b). CDs offer a more controlled approach, such as agrochemicals can be loaded or conjugated onto CDs. CDs' photoluminescent tracking ability allows real-time monitoring of pesticide delivery, and their biodegradability ensures minimal residual toxicity (Kumar et al., 2015). Moreover, CDs enhance pesticide solubility and stability, reduce the required dosage, and minimize nontarget effects, thereby contributing to safer and more sustainable crop protection (Ali et al., 2023).

13.5.4 Hormone and Plant Growth Regulator Delivery

Plant hormones play a crucial role in regulating growth and responses to environmental stimuli. However, their effectiveness often depends on timing, location, and concentration. CDs may deliver hormones such as *Indole-3-acetic acid (IAA)* for root development (Xie et al., 2020), g*ibberellic acid* (GA_3) for seed germination and elongation (Balaguera-López et al., 2008), and a*bscisic acid (ABA)* for stress response management (Hewage et al., 2020). The controlled release of these hormones via CDs enables precise temporal and spatial regulation, resulting in optimized growth and adaptive plant behavior.

13.5.5 Gene and Biomolecule Delivery (Experimental Application)

Although still in the early stages of agricultural use, CDs have shown potential in gene delivery systems due to their ability to penetrate plant cell walls and membranes, avoid degradation by nucleases, and facilitate both transient and stable gene expression. Studies in plant cell cultures and model species have utilized CDs to deliver DNA plasmids, siRNA, and proteins, providing a novel approach for gene editing, trait enhancement, or functional genomics research. CDs also exhibit low cytotoxicity, making them safer than viral vectors or synthetic polymers (Mohammadinejad et al., 2019).

13.5.6 Advantages of CDs as Delivery Vehicles

The use of carbon nanodots in delivery systems offers numerous benefits over conventional carriers (Table 13.3):

- *Biocompatibility:* CDs are naturally derived and exhibit low toxicity, ensuring compatibility with plant tissues (Kumar et al., 2023).
- *Environmental safety:* Derived from biomass and biodegradable, CDs minimize ecological risks (Karahan et al., 2020).
- *Fluorescent tracking:* CDs emit fluorescence, enabling the real-time imaging and tracking of delivery within plant tissues (Wang et al., 2017; Zhang et al., 2022).
- *Stimuli-responsive release:* CDs can be engineered to release their cargo in response to pH, light, or enzymatic cues (Karimi et al., 2017).
- *Surface modifiability:* Their functional groups enable custom modification for target-specific delivery (Jha et al., 2020).

The integration of carbon nanodots as smart delivery vehicles marks a significant advancement in nano-enabled agriculture. Their ability to combine carrier function with biological activity opens up a multidimensional approach to plant treatment, merging nutrition, growth promotion, protection, and precision management into a single platform.

Table 13.3 Applications of carbon nanodots in smart delivery systems

Target molecule	Purpose	Mode of binding	Benefits	Reference
Micronutrients (Fe, Zn)	Correct nutrient deficiency	Chelation, electrostatic	Enhanced uptake and mobility, better yield	Wijerathna (2023)
Pesticides/ herbicides	Crop protection	Surface adsorption or conjugation	Reduced dosage, less environmental pollution	Banerjee et al. (2022)
Phytohormones (IAA, GA_3)	Growth regulation	Hydrogen bonding, covalent	Targeted release, better growth response	Zhang et al. (2025)
DNA/RNA/ proteins	Gene delivery, trait enhancement	Electrostatic interaction	Controlled gene expression, low toxicity	Liu et al. (2012)
Antioxidants	Stress mitigation	Physical encapsulation	ROS scavenging, enhanced stress resistance	Chen et al. (2022)

13.6 Role of Carbon Nanodots in Stress Mitigation and Plant Defense

Plants are continuously exposed to a variety of environmental stresses, both abiotic (such as drought, salinity, heat, cold, and heavy metals) and biotic (including pathogens and insect attacks). These stressors disrupt normal physiological functions, suppress growth, and ultimately reduce agricultural productivity. The integration of carbon nanodots (CDs) into plant systems offers a promising strategy to enhance stress tolerance and activate plant defence mechanisms. Their multifunctional nature, which combines antioxidant activity, molecular signaling modulation, and biochemical reprogramming, makes CDs powerful tools in climate-resilient agriculture.

13.6.1 Abiotic Stress Mitigation

CDs significantly alleviate abiotic stress by reducing oxidative damage. Environmental stresses typically induce an overproduction of reactive oxygen species (ROS) like superoxide radicals, hydrogen peroxide, and hydroxyl radicals. While small amounts of ROS are natural signaling molecules, excessive accumulation leads to oxidative stress, damaging lipids, proteins, and nucleic acids in plant cells (Ji et al., 2023). Carbon nanodots act as potent ROS scavengers, either directly neutralizing reactive species or stimulating the plant's antioxidant defense system. CD exposures may upregulate the activity of key antioxidant enzymes such as superoxide dismutase (SOD), catalase (CAT), and peroxidases (POD). This enzymatic enhancement leads to a more favorable redox balance and increased protection of cellular structures during adverse conditions, such as drought, salinity, and temperature extremes (Xu et al., 2024). Additionally, CDs help maintain membrane integrity and osmotic balance, especially under salt stress. The regulation of ion channels and the enhancement of selective nutrient uptake by CDs help reduce Na^+ accumulation in plant tissues, thereby preventing ionic toxicity and improving water retention (Chen et al., 2020a, b).

13.6.2 Heat and Light Stress Response

High temperature and excessive light intensity impair photosynthesis by damaging photosystem complexes, especially Photosystem II (PSII). CDs, with their photoluminescent and photoprotective properties, may absorb excess light and reemit it in safer wavelengths, thus shielding chloroplasts from photoinhibition. Moreover, CDs modulate the expression of heat shock proteins (HSPs), playing a crucial role in protecting cellular proteins from denaturation during heat stress. These protective responses enhance thermotolerance, enabling plants to maintain metabolic activities under elevated temperatures (Li et al., 2021).

13.6.3 Heavy Metal Stress Alleviation

In environments contaminated with heavy metals such as cadmium (Cd), lead (Pb), and arsenic (As), plants often suffer from stunted growth and disrupted metabolic functions. CDs aid in heavy metal detoxification by binding metal ions, reducing their bioavailability, and forming metal–CD complexes that are either sequestered in vacuoles or excluded from sensitive tissues. Additionally, CDs stimulate the phytochelatin and metallothionein pathways in plants, which are naturally involved in the sequestration of heavy metals. Reducing heavy metal uptake or toxicity by CDs helps plants maintain growth in marginal or contaminated soils (Chen et al., 2020a, b; Wang et al., 2021).

13.6.4 Biotic Stress Resistance

Although the research into CDs in biotic stress management is still emerging, preliminary studies suggest that CDs may activate plant immune responses against pathogens and pests. CDs can induce systemic acquired resistance (SAR) or induced systemic resistance (ISR), two key pathways in plant immunity. CDs also exhibit direct antimicrobial activity against bacterial or fungal pathogens, likely due to their ability to disrupt microbial membranes or interfere with microbial metabolism. When used as carriers for biopesticides or defense priming agents, CDs enhance the bioavailability and targeted action of these molecules, leading to more effective disease suppression (Baruah & Sahu, 2022; Zia et al., 2024).

Moreover, carbon dots (CDs) are emerging as highly promising nanomaterials for biosensing applications due to their unique properties, such as small size, excellent water solubility, tunable photoluminescence, and biocompatibility. The optical properties, mainly strong fluorescence emission under UV or visible light, enable their use as fluorescent probes in detecting pathogens in plants. When introduced into plant tissues or onto infected surfaces, CDs interact with specific pathogen-derived molecules, resulting in detectable changes in their fluorescence intensity or emission wavelength. This property is beneficial for the early-stage, noninvasive detection of infections in crops, enabling farmers to take timely measures to mitigate yield losses (Zhou et al., 2021).

The potential of CDs in plant disease diagnostics is realized when integrated into nanosensor platforms. Functionalized CDs can be conjugated with biomolecular recognition elements such as antibodies, aptamers, or oligonucleotide probes, enabling the detection of pathogen-specific proteins, DNA/RNA sequences, or toxins. These nanosensors operate on the principle of fluorescence quenching or enhancement upon target binding, offering a rapid, label-free, and real-time approach to pathogen detection. For example, nitrogen-doped CDs have demonstrated enhanced sensitivity in detecting fungal and bacterial pathogens, such as *Xanthomonas oryzae* and *Fusarium oxysporum*, in rice and tomato plants, respectively (Li et al., 2022). Furthermore, the integration of microfluidic chips and smartphone-based readout systems with CD-based nanosensors presents opportunities for developing portable, field-deployable diagnostic tools for pathogen detection in crops.

Additionally, CD-based nanosensors offer several advantages over conventional diagnostic techniques, such as polymerase chain reaction (PCR) and enzyme-linked immunosorbent assay (ELISA), which are labor-intensive, time-consuming, and require specialized laboratory infrastructure. The high specificity, low detection limits, environmental friendliness, and rapid response time of CDs make them ideal for on-site monitoring of plant health. Furthermore, their nontoxic nature ensures safety for plants and the environment, which is crucial for sustainable agricultural practices. CDs, which enable early and precise detection of pathogens, may play a transformative role in precision agriculture, disease forecasting, and integrated pest management strategies. These strategies may reduce chemical pesticide use and promote plant health.

13.6.5 Hormonal and Genetic Regulation Under Stress

Stress mitigation by CDs is also closely linked to the modulation of phytohormones such as abscisic acid (ABA), salicylic acid (SA), and jasmonic acid (JA). For example, under drought stress, CDs promote the synthesis of ABA, playing a crucial role in stomatal closure, reducing water loss. Additionally, CDs have been reported to alter the expression of stress-responsive genes, including those encoding heat shock proteins, transcription factors (e.g., WRKY, DREB), and proteins involved in osmolyte synthesis (Wang et al., 2022a, b). This gene-level modulation strengthens the plant's capacity to adapt to environmental challenges.

13.6.6 Synergy with Other Nanomaterials and Compounds

CDs are being used in combination with other nanomaterials (like TiO_2, ZnO, or silica NPs) or organic molecules to enhance their stress-mitigating effects. These hybrid nanocomposites often provide multifunctional protection by integrating ROS scavenging, nutrient supplementation, and pathogen resistance in a single platform. Such synergistic strategies are especially beneficial in integrated crop management systems where plants are exposed to multiple stressors simultaneously. The incorporation of carbon nanodots into agricultural systems provides a multilayered defense mechanism against various stress conditions. Their unique ability to operate at both biochemical and genetic levels gives them an edge over traditional methods of stress mitigation (Jangra et al., 2024). With further research, particularly on field-level validation and biosafety, CDs have the potential to become integral components of climate-resilient and sustainable farming practices.

13.7 Toxic Impacts of CDs

13.7.1 Plant Toxicity and Concentration-Dependent Effects

In plant systems, CDs have generally been found to promote growth and development at low concentrations (typically below 100 mg/L), but they may cause toxicity at higher doses or prolonged exposure. The effects of excessive CD accumulation in plants include the following:

- Inhibited root elongation and lateral root formation.
- Altered leaf morphology, including curling and chlorosis.
- Disrupted photosynthetic pigment synthesis, resulting in reduced chlorophyll content.
- Impaired nutrient uptake, leading to nutritional deficiencies (Kumar et al., 2019).

In wheat seedlings, concentrations of CDs synthesized from glucose above 250 mg/L resulted in suppressed shoot height and reduced root biomass, likely due to metabolic overload or oxidative damage induced by ROS (Xiao et al., 2019). Similar effects have been observed in *Arabidopsis thaliana*, where high doses of CDs led to cellular membrane disruption and decreased mitotic activity in root meristem zones (Yan et al., 2021). These findings underscore the dose-sensitive nature of CDs, emphasizing the importance of optimizing concentrations specific to each plant species and growth stage. Phytotoxicity thresholds should be identified through both short-term and long-term experiments across diverse plant models.

13.7.2 Cytotoxicity, Genotoxicity, and Cellular Effects

The cellular toxicity of CDs arises from their potential to interact with and disrupt biological macromolecules. Studies using plant cell cultures, protoplasts, and callus tissues have shown that CDs may cause mitochondrial dysfunction, resulting in impaired ATP synthesis, DNA fragmentation, and chromosomal aberrations, possibly through oxidative stress or direct binding, membrane permeability alterations, leading to electrolyte leakage, and disturbed mitotic index, affecting cell cycle regulation (McCabe et al., 1997).

CDs derived from citric acid have been reported to induce nuclear condensation and apoptotic-like features in onion root tip cells at higher concentrations (Ren et al., 2020). Furthermore, doped CDs (e.g., nitrogen, sulfur, or metal-doped) may pose a risk due to increased reactivity or the presence of trace metals. Although such effects are more frequently observed in in vitro systems or under artificial exposure conditions, this provides critical insight into potential genetic and molecular toxicity that might occur under extreme or long-term field use (Qi et al., 2024). Hence, cytological and genotoxic studies should be integrated into any environmental safety evaluation.

13.7.3 Impact on Soil Microbiota and Rhizosphere Ecology

The soil microbiome, comprising bacteria, fungi, archaea, and protozoa, plays a crucial role in maintaining soil fertility, facilitating nutrient cycling, suppressing disease, and promoting plant–microbe interactions. CDs released into the soil, either through seed coatings, foliar wash-off, or root exudation, may accumulate in the rhizosphere, potentially impacting microbial diversity and function (Galhardi et al., 2022).

Key observations from recent studies include the following:

- CDs at moderate concentrations may enhance microbial respiration and enzyme activity by acting as a carbon source.
- At higher concentrations or with prolonged exposure, CDs inhibit enzymatic activities such as urease, dehydrogenase, and phosphatase, leading to impaired nitrogen and phosphorus cycling.
- Some microbial strains show inhibited growth due to membrane destabilization or oxidative stress triggered by the reactive surface of CDs.

For instance, *Rhizobium* species involved in nitrogen fixation in legumes showed reduced nodulation efficiency in the presence of high levels of sulfur-doped CDs (Abd-Alla et al., 2023). This highlights the selective toxicity of CDs toward beneficial microbes, which could have cascading effects on plant nutrition and soil health. Notably, the impact of CDs on soil systems depends on their concentration, surface charge, size, and the presence of stabilizing agents or dopants. Therefore, ecotoxicological studies using realistic soil matrices are needed to predict long-term behavior and risks.

13.7.4 Bioaccumulation and Food Chain Concerns

The accumulation of CDs in edible plant tissues and the trophic transfer through food chains significantly impact the ecosystem. While carbon nanodots are primarily composed of carbon and often biodegradable, their nanoscale nature and strong affinity for biological molecules raise concerns about their translocation to fruits, seeds, and storage tissues, interaction with edible cellular compartments, including chloroplasts and vacuoles, and metabolic persistence within plant systems (Liu et al., 2025).

In root vegetables such as radish and carrot, the systemic movement of CDs from the root to the shoot has also been observed, indicating the possibility of uptake into edible tissues. When consumed by herbivores or humans, these nanoparticles can interact with gut microbiota, epithelial cells, or immune components (Du et al., 2024; Xing et al., 2025). Although short-term studies reported that most CDs are excreted or metabolized, chronic exposure and bioaccumulation studies across multiple trophic levels are lacking. Hence, the food safety of CD-treated crops must be evaluated thoroughly, especially before integrating CDs into large-scale agricultural production or food-related industries.

13.7.5 Environmental Fate, Stability, and Degradation

Carbon nanodots synthesized via green methods, using plant waste, natural sugars, or amino acids, are generally considered biodegradable. They break down into carbon dioxide, water, and small organic molecules under exposure to sunlight, microbial action, or enzymatic degradation. However, several factors influence their environmental behavior:

- CDs produced through chemical routes or doped with metals may be more resistant to degradation.
- The adsorption of CDs onto soil particles or organic matter may limit their mobility but increase persistence.
- CDs in aquatic environments may undergo aggregation, photolytic cleavage, or form nanocomplexes with other contaminants, altering their reactivity (Chen et al., 2016).

Studies simulating environmental conditions have shown that the photodegradation half-lives of CDs may vary from a few hours to several days, depending on the light intensity, pH, and ionic strength (MacDonald, 2020; Xiang & Tan, 2022). Thus, the environmental risk assessment of CDs must consider not only their chemistry but also ecosystem-specific parameters, such as climate, soil texture, and microbial diversity. It is therefore stated that a balance must be struck between innovation and precaution, where benefits are maximized without compromising environmental and food safety.

13.8 Challenges and Future Perspectives of Carbon Nanodots in Agriculture

While the utility of carbon nanodots (CDs) in agriculture has been widely demonstrated through experimental studies, their broader adoption and commercialization face multiple challenges. These hurdles span technical, environmental, economic, and regulatory domains. Addressing these issues is crucial for integrating CDs into mainstream farming systems, especially within the frameworks of sustainable development and precision agriculture.

13.8.1 Challenges in Synthesis and Scalability

The major challenge lies in the scalable and cost-effective synthesis of high-quality CDs with consistent size, shape, and surface functionality. Although green synthesis methods using plant extracts and agro-waste offer an eco-friendly alternative, maintaining batch-to-batch uniformity and photoluminescent properties remains a concern (Semwal et al., 2025). Additionally, many laboratory-scale protocols involve long reaction times, energy-intensive steps, or low yields, factors that limit industrial-scale production. There is a pressing need to develop standardized, high-throughput synthesis routes that are not only environmentally sustainable but also economically viable for agricultural deployment, especially in resource-limited regions.

13.8.2 Understanding Mechanisms of Action

The exact biochemical and molecular mechanisms by which CDs exert these effects remain only partially understood. For instance, how CDs interact with plant hormonal signaling, gene expression networks, or cell metabolism at a systems level is still under active investigation. This lack of mechanistic clarity makes it difficult to predict long-term impacts, optimize dosages, or design CDs tailored to specific crop or environmental needs. More omics-based studies, such as transcriptomics, proteomics, and metabolomics, as well as real-time imaging techniques, are needed to unravel these interactions at both cellular and molecular resolutions.

13.8.3 Environmental and Ecotoxicological Concerns

Although CDs are considered safer than metal-based nanoparticles, their long-term fate in soil, water, and plant systems remains unclear. The issues related to bioaccumulation, soil microbial disruption, leaching into groundwater, and interaction with native flora and fauna persist. The majority of the findings are confined to controlled laboratory environments; however, field-level investigations on CDs' degradation, persistence, and ecological impact are severely lacking. Hence, comprehensive life cycle assessments (LCA) and risk assessments must be undertaken to move forward in CD synthesis and application. Furthermore, biodegradable and "self-degrading" CDs should be prioritized in future designs.

13.8.4 Regulatory and Policy Barriers

Presently, no specific regulations exist for carbon nanodots in agricultural applications. The regulatory framework for nanomaterials remains general, often grouping CDs with other engineered nanostructures, regardless of their unique properties. This regulatory ambiguity creates uncertainty for researchers, start-ups, and companies looking to commercialize CD-based agro-products. Policymakers, therefore, must work closely with scientists to establish material-specific guidelines, focusing on biosafety, permissible exposure levels, environmental monitoring, and application protocols. Additionally, globally harmonized standards are necessary to promote safe international trade and the application of CD-based technologies.

13.8.5 Economic and Adoption Challenges

Introducing nanotechnology into agriculture, particularly in developing countries, faces resistance due to high initial costs, limited awareness, and a lack of training and infrastructure. Farmers may hesitate to adopt CD-based products unless they are affordable, proven to work under field conditions, and easy to apply using conventional tools. Hence, collaborative efforts among research institutions, extension services, and local governments are necessary to promote education, pilot trials, and training programs for end-users, thereby overcoming this issue.

13.8.6 Future Directions and Research Opportunities

Despite the challenges, the future of CDs in agriculture remains highly promising. Several key directions for future research and innovation include the following:

- Designing multifunctional CDs that combine imaging, delivery, and bio-stimulant functions in a single material.
- Exploring synergistic combinations with other nanomaterials, microbial inoculants, or biofertilizers.
- Developing stimuli-responsive CDs that release cargo in response to pH, temperature, or hormonal cues within plant tissues.
- Expanding the use of CDs in gene editing, epigenetic modulation, and synthetic biology, primarily through CRISPR-Cas delivery.
- Integrating CDs into smart sensors and IoT-enabled devices for real-time crop monitoring and nutrient management.
- Conducting long-term field experiments across diverse crops and agroclimatic zones to validate their performance under real-world farming conditions.

13.9 Conclusion

It is concluded that carbon nanodots (CDs) represent a transformative innovation in sustainable and precision agriculture systems, offering multifunctional benefits that range from promoting plant growth to enhancing stress tolerance and enabling smart nutrient delivery, thereby improving productivity and yield. The unique physicochemical and biological attributes of CDs combined with eco-friendly synthesis possibilities, make them highly attractive for next-generation agro-technological applications. CDs have the potential to revolutionize plant monitoring, nutrient management, and stress resilience, contributing to climate-smart and resource-efficient agricultural practices. However, to fully harness their capabilities, it is necessary to address critical challenges, including scalability, environmental safety, regulatory frameworks, and field-level efficacy. Future research should focus on bridging these gaps, ensuring the safe deployment of CDs, and integrating them into holistic agricultural models. With targeted innovation and interdisciplinary collaboration, carbon nanodots may play a pivotal role in advancing sustainable food systems and securing global agrarian resilience in the era of changing climatic and environmental conditions.

References

Abd-Alla, M. H., Al-Amri, S. M., & El-Enany, A.-W. E. (2023). Enhancing Rhizobium–legume symbiosis and reducing nitrogen fertilizer use are potential options for mitigating climate change. *Agriculture, 13*(11). https://doi.org/10.3390/agriculture13112092

Abid, N., Khan, A. M., Shujait, S., Chaudhary, K., Ikram, M., Imran, M., Haider, J., et al. (2022). Synthesis of nanomaterials using various top-down and bottom-up approaches, influencing factors, advantages, and disadvantages: A review. *Advances in Colloid and Interface Science, 300*, 102597. https://doi.org/10.1016/j.cis.2021.102597

Abinaya, K., Raja, K., Raja, K., Moorthy, P. S., Senthil, A., & Chandrakumar, K. (2024). Impact of green carbon dot nanoparticles on seedling emergence, crop growth and seed yield in Blackgram (Vigna Mungo L. Hepper). *Scientific Reports, 14*(1), 23783. https://doi.org/10.1038/s41598-024-75366-5

Abuzeid, H. M., Julien, C. M., Zhu, L., & Hashem, A. M. (2023). Green synthesis of nanoparticles and their energy storage, environmental, and biomedical applications. *Crystals, 13*(11), 1576.

Acharya, P., Jayaprakasha, G. K., Crosby, K. M., Jifon, J. L., & Patil, B. S. (2020). Nanoparticle-mediated seed priming improves germination, growth, yield, and quality of watermelons (Citrullus Lanatus) at multi-locations in Texas. *Scientific Reports, 10*(1), 5037. https://doi.org/10.1038/s41598-020-61696-7

Afzal, S., Singh, N. K., Lal, A. F., Sohrab, S., Singh, N., Gupta, P. S., Mishra, S. K., Adeel, M., & Faizan, M. (2024). Nanostructure and plant uptake: Assessing the ecological footprint and root-to-leaf dynamics. *Plant Nano Biology, 10*, 100122. https://doi.org/10.1016/j.plana.2024.100122

Agrawal, S., Kumar, V., Kumar, S., & Shahi, S. K. (2022). Plant development and crop protection using phytonanotechnology: A new window for sustainable agriculture. *Chemosphere, 299*, 134465. https://doi.org/10.1016/j.chemosphere.2022.134465

Akash, M. S. H., & Rehman, K. (2020). Ultraviolet-Visible (UV-VIS) spectroscopy. In M. S. H. Akash & K. Rehman (Eds.), *Essentials of pharmaceutical analysis* (pp. 29–56). Springer Nature Singapore. https://doi.org/10.1007/978-981-15-1547-7_3

Ali, S., Ahmad, N., Dar, M. A., Manan, S., Rani, A., Alghanem, S. M. S., Khan, K. A., et al. (2023). Nano-agrochemicals as substitutes for pesticides: Prospects and risks. *Plants, 13*(1), 109.

Alikhani, M., Mirbolook, A., Sadeghi, J., & Lakzian, A. (2023). Improving wheat growth and nutrient uptake in calcareous soil: A novel approach with carbon dots as a slow-release zinc fertilizer. *Journal of Soil Science and Plant Nutrition, 23*(4), 6232–6245. https://doi.org/10.1007/s42729-023-01480-4

Anadebe, V. C., Thakur, A., Okoye, C. C., Ezemagu, I. G., Guo, L., & Ebenso, E. E. (2024). Carbon dots in catalysis: CO_2 conversion, H_2 evolution, and organic reactions. In *Carbon dots: Recent developments and future perspectives* (pp. 277–310). ACS Publications.

Arumugam, S. S., & Lee, N. Y. (2024). Comprehensive review on fluorescent carbon dots and their applications in nucleic acid detection, nucleolus targeted imaging and gene delivery. *The Analyst., 149*.

Badra, A., Parent, L.-É., Allard, G., Tremblay, N., Desjardins, Y., & Morin, N. (2006). Effect of leaf nitrogen concentration versus CND nutritional balance on shoot density and foliage colour of an established Kentucky Bluegrass (*Poa pratensis* L.) Turf. *Canadian Journal of Plant Science, 86*(4), 1107–1118.

Balaguera-López, H. E., Cárdenas-Hernández, J. F., & Álvarez-Herrera, J. G. (2008). Effect of Gibberellic Acid (GA3) on seed germination and growth of tomato (*Solanum lycopersicum* L.). *Acta Horticulturae, 821*(821), 141–148.

Banerjee, S., Mazumder, S., Chatterjee, D., Bose, S., & Majee, S. B. (2022). Nanotechnology for cargo delivery with a special emphasis on pesticide, herbicide, and fertilizer. In *Nano-enabled agrochemicals in agriculture* (pp. 105–144). Elsevier.

Banger, A., Gautam, S., Jadoun, S., Jangid, N. K., Srivastava, A., Pulidindi, I. N., Dwivedi, J., & Srivastava, M. (2023). Synthetic methods and applications of carbon nanodots. *Catalysts, 13*(5). https://doi.org/10.3390/catal13050858

Barceló, A. R., & Laura, V. G. R. (2009). Reactive oxygen species in plant cell walls. In L. A. Rio & A. Puppo (Eds.), *Reactive oxygen species in plant signaling* (pp. 73–93). Springer Berlin Heidelberg. https://doi.org/10.1007/978-3-642-00390-5_5

Baruah, J., & Sahu, D. (2022). Comparative studies on carbon dots applications in plant systems. In *Carbon dots in agricultural systems* (pp. 199–224). Elsevier.

Chaudhary, M., Singh, P., Singh, G. P., & Rathi, B. (2024). Structural features of carbon dots and their agricultural potential. *ACS Omega, 9*(4), 4165–4185. https://doi.org/10.1021/acsomega.3c04638

Chen, J., Dou, R., Yang, Z., Wang, X., Mao, C., Gao, X., & Wang, L. (2016). The effect and fate of water-soluble carbon nanodots in Maize (*Zea mays* L.). *Nanotoxicology, 10*(6), 818–828.

Chen, L. T., Sun, A. Q., Yang, M., Chen, L. L., Ma, X. L., Li, M. L., & Yin, Y. P. (2017). Relationships of wheat seed vigor with enzyme activities and gene expression related to seed germination under stress conditions. *Ying Yong Sheng tai xue bao= The Journal of Applied Ecology, 28*(2), 609–619.

Chen, Q., Chen, L., Nie, X., Man, H., Guo, Z., Wang, X., Jiangping, T., Jin, G., & Ci, L. (2020a). Impacts of surface chemistry of functional carbon nanodots on the plant growth. *Ecotoxicology and Environmental Safety, 206*, 111220.

Chen, Q., Liu, B., Man, H., Chen, L., Wang, X., Jiangping, T., Guo, Z., et al. (2020b). Enhanced bioaccumulation efficiency and tolerance for Cd (II) in arabidopsis thaliana by amphoteric nitrogen-doped carbon dots. *Ecotoxicology and Environmental Safety, 190*, 110108.

Chen, Q., Ren, X., Li, Y., Liu, B., Wang, X., Jiangping, T., Guo, Z., et al. (2021). Promotion effect of nitrogen-doped functional carbon nanodots on the early growth stage of plants. *Oxford Open Materials Science, 1*(1), itab002.

Chen, Q., Cao, X., Nie, X., Li, Y., Liang, T., & Ci, L. (2022). Alleviation role of functional carbon nanodots for tomato growth and soil environment under drought stress. *Journal of Hazardous Materials, 423*, 127260.

Chen, C., Teng, G., Shen, W., Yijun, L., Jin, Y., Yuan, X., Chen, K., et al. (2025a). Green carbon Dots/CaCO3/Abamectin colloidal pesticide formulation for safer and more effective pest management. *ACS Nano, 19*(1), 1007–1025. https://doi.org/10.1021/acsnano.4c12707

Chen, F., Shen, Z., Shi, R., Zhang, X., Zhang, H., Li, W., & Lei, B. (2025b). Carbon dots-mediated plant adaptive responses to abiotic stress. *Materials Research Bulletin, 182*, 113137. https://doi.org/10.1016/j.materresbull.2024.113137

Cheng, B., Ding, Z., Yue, L., Chen, F., Cao, X., Li, J., Wang, C., & Wang, Z. (2023). Carbon dots enhanced cold tolerance of lettuce (*Lactuca sativa* L.): Scavenging reactive oxygen species, modulating hormones and up-regulating gene expression. *Environmental Science: Nano, 10*(10), 2849–2860. https://doi.org/10.1039/D3EN00257H

Das, K., & Roychoudhury, A. (2014). Reactive Oxygen Species (ROS) and response of antioxidants as ROS-scavengers during environmental stress in plants. *Frontiers in Environmental Science, 2*, 53.

Du, J., Ni, Z.-J., Wang, W., Thakur, K., Ma, R.-H , Ma, W.-P., & Wei, Z.-J. (2024). Carbon dot-mediated photodynamic treatment improves the quality attributes of post-harvest Goji Berries (Lycium Barbarum L.) via regulating the antioxidant system. *Foods, 13*(6), 955.

Dubale, A. A., Su, W. N., Tamirat, A. G., Pan, C. J., Aragaw, B. A., Chen, H. M., & Hwang, B. J. (2014). The synergetic effect of graphene on Cu_2O nanowire arrays as a highly efficient hydrogen evolution photocathode in water splitting. Journal of Materials Chemistry A, 2(43), 18383–18397.

Faulkner, C. (2018). Plasmodesmata and the symplast. *Current Biology, 28*(24), R1374–R1378.

Galhardi, J. A., de Oliveira, J. L., Ghoshal, S., & Fraceto, L. F. (2022). Soil enzyme responses to polymeric nanopesticides: An ecological risk analysis approach to promote sustainable agriculture. *ACS Agricultural Science & Technology, 2*(3), 443–452. https://doi.org/10.1021/acsagscitech.1c00234

Gautam, K., Singh, H., & Sinha, A. K. (2025). Nanotechnology in plant nanobionics: Mechanisms, applications, and future perspectives. *Advanced Biology, 9*(4), 2400589.

González-Lamothe, R., El Oirdi, M., Brisson, N., & Bouarab, K. (2012). The conjugated auxin indole-3-acetic acid–aspartic acid promotes plant disease development. *The Plant Cell, 24*(2), 762–777.

Guan, W., Wei, G., Ye, L., Chenyang Guo, S. S., Pinxiang, X., & Xue, M. (2014). Microwave-assisted polyol synthesis of carbon nitride dots from folic acid for cell imaging. *International Journal of Nanomedicine*, 5071–5078.

Guerra, F. D., Attia, M. F., Whitehead, D. C., & Alexis, F. (2018). Nanotechnology for environmental remediation: Materials and applications. *Molecules, 23*(7), 1760.

Guirguis, A., Yang, W., Conlan, X. A., Kong, L., Cahill, D. M., & Wang, Y. (2023). Boosting plant photosynthesis with carbon dots: A critical review of performance and prospects. *Small, 19*(43), 2300671.

Guo, B., Liu, G., Li, W., Chaofan, H., Lei, B., Zhuang, J., Zheng, M., & Liu, Y. (2022). The role of carbon dots in the life cycle of crops. *Industrial Crops and Products, 187*, 115427. https://doi.org/10.1016/j.indcrop.2022.115427

Hallaji, Z., Bagheri, Z., Tavassoli, Z., & Ranjbar, B. (2022). Fluorescent carbon dot as an optical amplifier in modern agriculture. *Sustainable Materials and Technologies, 34*, e00493.

Han, H. S., Koo, S. Y., & Choi, K. Y. (2022). Emerging nanoformulation strategies for phytocompounds and applications from drug delivery to phototherapy to imaging. *Bioactive Materials, 14*, 182–205. https://doi.org/10.1016/j.bioactmat.2021.11.027

Hazra, N., Ray, R., & Banerjee, A. (2024). Rapid targeting and imaging of mitochondria via carbon dots using an amino acid-based Amphiphile as a carrier. *Nanoscale, 16*(20), 9827–9835.

Hewage, K. A. H., Yang, J.-F., Di Wang, G.-F. H., Yang, G.-F., & Zhu, J.-K. (2020). Chemical manipulation of abscisic acid signaling: A new approach to abiotic and biotic stress management in agriculture. *Advanced Science, 7*(18), 2001265. https://doi.org/10.1002/advs.202001265

Hu, J., Jia, W., Xinyi, W., Zhang, H., Wang, Y., Junfeng Liu, Y., Yang, S. T., & Wang, X. (2022). Carbon dots can strongly promote photosynthesis in lettuce (*Lactuca sativa* L.). *Environmental Science: Nano, 9*(4), 1530–1540. https://doi.org/10.1039/D1EN00948F

Jadhav, V., Roy, A., Kaur, K., Roy, A., Sharma, K., Verma, R., Rustagi, S., & Malik, S. (2024). Current advancements in functional nanomaterials for drug delivery systems. *Nano-Structures & Nano-Objects, 38*, 101177.

Jangra, A., Kumar, K., Maikhuri, S., Bhandari, M. S., Pandey, S., Singh, H., & Barthwal, S. (2024). Unveiling stress-adapted endophytic bacteria: characterizing plant growth-promoting traits and assessing cross-inoculation effects on Populus deltoides under abiotic stress. Plant Physiology and Biochemistry, 210, 108610.

Jha, R., Singh, A., Sharma, P. K., & Fuloria, N. K. (2020). Smart carbon nanotubes for drug delivery system: A comprehensive study. *Journal of Drug Delivery Science and Technology, 58*, 101811.

Ji, Y., Yue, L., Cao, X., Chen, F., Li, J., Zhang, J., Wang, C., Wang, Z., & Xing, B. (2023). Carbon dots promoted soybean photosynthesis and amino acid biosynthesis under drought stress: Reactive oxygen species scavenging and nitrogen metabolism. *Science of the Total Environment, 856*, 159125.

Kadian, S., Sethi, S. K., & Manik, G. (2021). Recent advancements in synthesis and property control of graphene quantum dots for biomedical and optoelectronic applications. *Materials Chemistry Frontiers, 5*(2), 627–658.

Karahan, H. E., Ji, M., Pinilla, J. L., Han, X., Mohamed, A., Wang, L., Wang, Y., et al. (2020). Biomass-derived nanocarbon materials for biological applications: Challenges and prospects. *Journal of Materials Chemistry B, 8*(42), 9668–9678.

Karimi, M., Zangabad, P. S., Baghaee-Ravari, S., Ghazadeh, M., Mirshekari, H., & Hamblin, M. R. (2017). Smart nanostructures for cargo delivery: Uncaging and activating by light. *Journal of the American Chemical Society, 139*(13), 4584–4610.

Kaurav, H., Verma, D., Bansal, A., Kapoor, D. N., & Sheth, S. (2023). Progress in drug delivery and diagnostic applications of carbon dots: A systematic review. *Frontiers in Chemistry,* 11–2023. https://www.frontiersin.org/journals/chemistry/articles/10.3389/fchem.2023.1227843

Khan, F., Pandey, P., & Upadhyay, T. K. (2022a) Applications of nanotechnology-based agrochemicals in food security and sustainable agriculture: An overview. *Agriculture, 12*(10), 1672.

Khan, I., Awan, S. A., Rizwan, M., Hassan, Z. U., Akram, M. A., Tariq, R., Brestic, M., & Xie, W. (2022b). Nanoparticle's uptake and translocation mechanisms in plants via seed priming, foliar treatment, and root exposure: A review. *Environmental Science and Pollution Research, 29*(60), 89823–89833. https://doi.org/10.1007/s11356-022-23945-2

Kohli, H. K., & Parab, D. (2025). Green synthesis of carbon quantum dots and applications: An insight. *Next Materials, 8*, 100527. https://doi.org/10.1016/j.nxmate.2025.100527

Kou, E., Luo, Z., Jingyi Ye, X., Chen, D. L., Landry, M. P., Zhang, H., & Zhang, H. (2025). Sunlight-sensitive carbon dots for plant immunity priming and pathogen defence. *Plant Biotechnology Journal., 23*, 2150.

Kumar, S., Bhanjana, G., Sharma, A., Sidhu, M. C., & Dilbaghi, N. (2015). Herbicide loaded carboxymethyl cellulose nanocapsules as potential carrier in agrinanotechnology. *Science of Advanced Materials, 7*(6), 1143–1148.

Kumar, V., Lakkaboyana, S. K., Sharma, N., Abdelaal, A. S., Maitra, S. S., & Pant, D. (2019). Chapter Four – Engineered nanomaterials uptake, bioaccumulation and toxicity mechanisms in plants. In S. K. Verma & A. K. Das (Eds.), *Comprehensive analytical chemistry* (pp. 111–131). Elsevier. https://doi.org/10.1016/bs.coac.2019.09.005

Kumar, A., Kumar, D., & Saikia, M. (2023). A review on plant derived carbon quantum dots for bio-imaging. *Materials Advances, 4*(18), 3951–3966.

Law, S. S., Yin, T. M., & Numata, K. (2023). Organelle-targeted gene delivery in plants by nanomaterials. *Chemical Communications, 59*(47), 7166–7181.

Li, Y., Gao, J., Xiaokai, X., Ying, W., Zhuang, J., Zhang, X., Zhang, H., et al. (2020a). Carbon dots as a protective agent alleviating abiotic stress on rice (*Oryza sativa* L.) through promoting nutrition assimilation and the defense system. *ACS Applied Materials & Interfaces, 12*(30), 33575–33585.

Li, Y., Xiaokai, X., Ying, W., Zhuang, J., Zhang, X., Zhang, H., Lei, B., Chaofan, H., & Liu, Y. (2020b). A review on the effects of carbon dots in plant systems. *Materials Chemistry Frontiers, 4*(2), 437–448.

Li, Y., Pan, X., Xiaokai, X., Ying, W., Zhuang, J., Zhang, X., Zhang, H., et al. (2021). Carbon dots as light converter for plant photosynthesis: Augmenting light coverage and quantum yield effect. *Journal of Hazardous Materials, 410*, 124534. https://doi.org/10.1016/j.jhazmat.2020.124534

Li, H., Yan, X., Kong, D., & Jin, R. (2022). Recent advances in carbon dot-based biosensors for plant pathogen detection. *Analytical and Bioanalytical Chemistry, 414*, 6557–6572. https://doi.org/10.1007/s00216-022-04291-z

Li, G., Jiwei, X., & Ke, X. (2023). Physiological functions of carbon dots and their applications in agriculture: A review. *Nanomaterials (Basel), 13*(19), 2684.

Li, L., Wang, T., Zhong, Y., Li, R., Deng, W., Xiao, X., Yuanyuan, X., et al. (2024). A review of nanomaterials for biosensing applications. *Journal of Materials Chemistry B, 12*(5), 1168–1193.

Liang, L., Wong, S. C., & Lisak, G. (2023). Effects of plastic-derived carbon dots on germination and growth of pea (Pisum Sativum) via seed nano-priming. *Chemosphere, 316*, 137868. https://doi.org/10.1016/j.chemosphere.2023.137868

Lin, Z., Ali, M. M., Yi, X., Zhang, L., Wang, S., & Chen, F. (2023). Unlocking the potential of carbon quantum dots for cell imaging, intracellular localization, and gene expression control in Arabidopsis Thaliana (L.) Heynh. *International Journal of Molecular Sciences, 24*(21), 15700.

Liu, C., Zhang, P., Zhai, X., Tian, F., Li, W., Yang, J., Liu, Y., et al. (2012). Nano-Carrier for gene delivery and bioimaging based on carbon dots with PEI-passivation enhanced fluorescence. *Biomaterials, 33*(13), 3604–3613.

Liu, H., Zhao, X., Wang, F., Wang, Y., Guo, L., Mei, J., Tian, C., Yang, X., & Zhao, D. (2017). High-efficient excitation-independent blue luminescent carbon dots. *Nanoscale Research Letters, 12*, 1–7.

Liu, J., Li, R., & Yang, B. (2020). Carbon dots: A new type of carbon-based nanomaterial with wide applications. *ACS Central Science, 6*(12), 2179–2195. https://doi.org/10.1021/acscentsci.0c01306

Liu, Z., Yang, F., Zhang, G., Liao, Y., Cui, M., Li, H., Liangbin, H., & Mo, H. (2025). Carbon quantum dots: A green and efficient material for fruits and vegetables preservation. *Food Reviews International*, 1–26.

Lu, Z., Li, Y., Chen, K., Chai, S., Gehong, S., Chun, W., Sun, M., et al. (2025). Multi-activity ferruginated carbon quantum dots nanozyme improves wheat seedling growth and Cd tolerance. *The Crop Journal, 13*(2), 510–523. https://doi.org/10.1016/j.cj.2025.01.016

MacDonald, R. T. (2020). The fate and transport of carbon-based nanomaterials in the environment.

Maholiya, A., Ranjan, P., Khan, R., Murali, S., Nainwal, R. C., Chauhan, P. S., Sathish, N., Chaurasia, J. P., & Srivastava, A. K. (2023). An insight into the role of carbon dots in the agriculture system: A review. *Environmental Science: Nano, 10*(4), 959–995.

Mali, S. C., Raj, S., & Trivedi, R. (2020). Nanotechnology a novel approach to enhance crop productivity. *Biochemistry and Biophysics Reports, 24*, 100821.

McCabe, P. F., Levine, A., Meijer, P.-J., Tapon, N. A., & Pennell, R. I. (1997). A programmed cell death pathway activated in carrot cells cultured at low cell density. *The Plant Journal, 12*(2), 267–280.

Midekessa, G., Godakumara, K., Ord, J., Viil, J., Lättekivi, F., Dissanayake, K., Kopanchuk, S., et al. (2020). Zeta potential of extracellular vesicles: Toward understanding the attributes that determine colloidal stability. *ACS Omega, 5*(27), 16701–16710. https://doi.org/10.1021/acsomega.0c01582

Minhans, K., Sharma, S., Sheikh, I., Alhewairini, S. S., & Sayyed, R. (2025). Artificial intelligence and plant disease management: An agro-innovative approach. *Journal of Phytopathology, 173*(3), e70084.

Mohammadinejad, R., Dadashzadeh, A., Moghassemi, S., Ashrafizadeh, M., Dehshahri, A., Pardakhty, A., Sassan, H., Sohrevardi, S.-M., & Mandegary, A. (2019). Shedding light on gene therapy: Carbon dots for the minimally invasive image-guided delivery of plasmids and noncoding RNAs – A review. *Journal of Advanced Research, 18*, 81–93. https://doi.org/10.1016/j.jare.2019.01.004

Mu, Q., Jiang, G., Chen, L., Zhou, H., Fourches, D., Tropsha, A., & Yan, B. (2014). Chemical basis of interactions between engineered nanoparticles and biological systems. *Chemical Reviews, 114*(15), 7740–7781.

Neme, K., Nafady, A., Uddin, S., & Tola, Y. B. (2021). Application of nanotechnology in agriculture, postharvest loss reduction and food processing: Food security implication and challenges. *Heliyon, 7*(12).

Njuguna, J., Ansari, F., Sachse, S., Rodriguez, V. M., Siqqique, S., & Zhu, H. (2021). Nanomaterials, nanofillers, and nanocomposites: Types and properties. In *Health and environmental safety of nanomaterials* (pp. 3–37). Elsevier.

Özalkan, Ç., Sepetoğlu, H., Daur, I., & Şen, O. F. (2010). Relationship between some plant growth parameters and grain yield of chickpea (*Cicer arietinum* L.) during different growth stages. *Turkish Journal of Field Crops, 15*(1), 79–83.

Pan, Z., Zang, H., Li, Y., Wang, X., Xia, N., Liu, C., Li, Z., et al. (2024). Foliar application of carbon dots enhances nitrogen uptake and assimilation through CEPD1-dependent signaling in plants. *Plant Physiology and Biochemistry, 217*, 109229. https://doi.org/10.1016/j.plaphy.2024.109229

Patel, J. K., Patel, A., & Bhatia, D. (2021). Introduction to nanomaterials and nanotechnology. In *Emerging technologies for nanoparticle manufacturing* (pp. 3–23). Springer.

Prokisch, J., Nguyen, D. H. H., Muthu, A., Ferroudj, A., Singh, A., Agrawal, S., Rajput, V. D., et al. (2024). Carbon nanodot–microbe–plant nexus in agroecosystem and antimicrobial applications. *Nanomaterials, 14*(15), 1249.

Qi, J., Zhang, P., Zhang, T., Zhang, R., Zhang, Q., Wang, J., Zong, M., et al. (2024). Metal-doped carbon dots for biomedical applications: From design to implementation. *Heliyon, 10*(11), e32133. https://doi.org/10.1016/j.heliyon.2024.e32133

Que, X., Fangmei, F., Yang, G., Lei, B., Ali, S., & Wang, X. (2024). A positively charged carbon dot complex improves the bioactivity of *Isaria fumosorosea* against *Plutella xylostella* (Linnaeus). *Chemical and Biological Technologies in Agriculture, 11*(1), 192.

Qureshi, Z. A., Dabash, H., Ponnamma, D., & Abbas, M. K. G. (2024). Carbon dots as versatile nanomaterials in sensing and imaging: Efficiency and beyond. *Heliyon, 10*(11), e31634. https://doi.org/10.1016/j.heliyon.2024.e31634

Ren, J., Malfatti, L., & Innocenzi, P. (2020). Citric acid derived carbon dots, the challenge of understanding the synthesis-structure relationship. *C, 7*(1), 2.

Sah, M. K., Thakuri, B. S., Pant, J., Gardas, R. L., & Bhattarai, A. (2024). The multifaceted perspective on the role of green synthesis of nanoparticles in promoting a sustainable green economy. *Sustainable Chemistry, 5*(2), 40–59.

Sajid, M. (2022). Nanomaterials: Types, properties, recent advances, and toxicity concerns. *Current Opinion in Environmental Science & Health, 25*, 100319.

Saraswat, S. K., Mustafa, M. A., Ghadir, G. K., Kaur, M., Lozada, D. F. G., Al-Ani, A. M., Alshahrani, M. Y., et al. (2024). Carbon quantum dots: A comprehensive review of green synthesis, characterization and investigation their applications in bioimaging. *Inorganic Chemistry Communications, 112279*, 162.

Sattelmacher, B. (2001). The Apoplast and its significance for plant mineral nutrition. *New Phytologist, 149*(2), 167.

Semwal, A., Gautam, K., Bhatt, M., Yadav, R., Barthwal, S., & Singh, H. (2025). Green synthesis and application of metal nanoparticles (MNPs): a sustainable strategy for protecting wood from fungal and termite decay. Next Materials, 9, 100962.

Shahid, M., Dumat, C., Khalid, S., Schreck, E., Xiong, T., & Niazi, N. K. (2017). Foliar heavy metal uptake, toxicity and detoxification in plants: A comparison of foliar and root metal uptake. *Journal of Hazardous Materials, 325*, 36–58.

Shivashakarappa, K., Marriboina, S., Dumenyo, K., Taheri, A., & Yadegari, Z. (2025). Nanoparticle-mediated gene delivery techniques in plant systems. *Frontiers in Nanotechnology, 7*, 1516180.

Singh, H. (2024). Exploring adaptive modulation in plant functional traits and its impact on the productivity of *Acacia auriculiformis* under CO_2-enriched environment. Industrial Crops and Products, 210, 118186.

Solanki, P., Bhargava, A., Chhipa, H., Jain, N., & Panwar, J. (2015). Nano-fertilizers and their smart delivery system. In *Nanotechnologies in food and agriculture* (pp. 81–101). Springer.

Su, L.-X., Ma, X.-L., Zhao, K.-K., Shen, C.-L., Lou, Q., Yin, D.-M., & Shan, C.-X. (2018). Carbon nanodots for enhancing the stress resistance of peanut plants. *ACS Omega, 3*(12), 17770–17777. https://doi.org/10.1021/acsomega.8b02604

Tang, Y., Zhao, W., Zhu, G., Tan, Z., Huang, L., Zhang, P., Gao, L., & Rui, Y. (2023). Nano-pesticides and fertilizers: Solutions for global food security. *Nanomaterials (Basel), 14*(1), 90.

Tripathi, S., & Sarkar, S. (2015). Influence of water soluble carbon dots on the growth of wheat plant. *Applied Nanoscience, 5*(5), 609–616. https://doi.org/10.1007/s13204-014-0355-9

Usman, M., & Cheng, S. (2024). Recent trends and advancements in green synthesis of biomass-derived carbon dots. *Eng, 5*(3), 2223–2263.

Varadharajan, S., Vasanthan, K. S., Mathur, V., Hariperumal, N., & Mazumder, N. (2024). Green synthesis and multifaceted applications: Challenges and innovations in carbon dot nanocomposites. *Discover Nano, 19*(1), 205. https://doi.org/10.1186/s11671-024-04124-3

Wahab, A., Muhammad, M., Ullah, S., Abdi, G., Shah, G. M., Zaman, W., & Ayaz, A. (2024). Agriculture and environmental management through nanotechnology: Eco-friendly nanomaterial synthesis for soil-plant systems, food safety, and sustainability. *Science of the Total Environment, 926*, 171862. https://doi.org/10.1016/j.scitotenv.2024.171862

Wang, B., & Lu, S. (2022). The light of carbon dots: From mechanism to applications. *Matter, 5*(1), 110–149.

Wang, F., Wang, S., Sun, Z., & Zhu, H. (2015). Study on ultrasonic single-step synthesis and optical properties of nitrogen-doped carbon fluorescent quantum dots. *Fullerenes, Nanotubes and Carbon Nanostructures, 23*(9), 769–776.

Wang, L., Li, B., Feng, X., Li, Y., Zheheng, X., Wei, D., Feng, Y., et al. (2017). Visual in vivo degradation of injectable hydrogel by real-time and non-invasive tracking using carbon nanodots as fluorescent indicator. *Biomaterials, 145*, 192–206.

Wang, B., Huang, J., Zhang, M., Wang, Y., Wang, H., Ma, Y., Zhao, X., et al. (2020). Carbon dots enable efficient delivery of functional DNA in plants. *ACS Applied Bio Materials, 3*(12), 8857–8864.

Wang, H., Kang, Y., Li, H., Huang, S., Li, W., Zheng, M., Huang, R., Lei, B., & Yang, X. (2021). Salvia Miltiorrhiza derived carbon dots and their heat stress tolerance of italian lettuce by promoting growth and enhancing antioxidant enzyme activity. *ACS Omega, 6*(47), 32262–32269. https://doi.org/10.1021/acsomega.1c05074

Wang, C., Yang, H., Yue, L., Sun, W., Chen, F., Cao, X., Zhao, X., et al. (2022a). Physiological and molecular level understanding of advanced carbon dots to enhance maize drought tolerance: Modulation of photosynthesis and signaling molecules. *Environmental Science: Nano, 9*(10), 3821–3832.

Wang, Z., Tang, J., Zhu, L., Feng, Y., Yue, L., Wang, C., Xiao, Z., & Chen, F. (2022b). Nanomaterial-induced modulation of hormonal pathways enhances plant cell growth. *Environmental Science: Nano, 9*(5), 1578–1590.

Wang, X., Xie, H., Wang, P., & Yin, H. (2023). Nanoparticles in plants: Uptake, transport and physiological activity in leaf and root. *Materials, 16*(8). https://doi.org/10.3390/ma16083097

Wang, Y., Wang, Q., Liu, W., Xin, X., & Zhao, B. (2025). A review on the synthesis of carbon dots and their applications in environmental analysis. *Crystals, 15*(5), 384.

Wijerathna, P. K. C. B. (2023). Zinc biofortification of rice using carbon nanodots (Doctoral dissertation, University of Southern Queensland).

Xia, C., Zhu, S., Feng, T., Yang, M., & Yang, B. (2019). Evolution and synthesis of carbon dots: From carbon dots to carbonized polymer dots. *Advanced Science, 6*(23), 1901316.

Xiang, S., & Tan, M. (2022). Carbon dots derived from natural sources and their biological and environmental impacts. *Environmental Science: Nano, 9*(9), 3206–3225.

Xiao, L., Guo, H., Wang, S., Li, J., Wang, Y., & Xing, B. (2019). Carbon dots alleviate the toxicity of cadmium ions (Cd 2+) toward wheat seedlings. *Environmental Science: Nano, 6*(5), 1493–1506.

Xie, L., Chen, F., Hewei, D., Zhang, X., Wang, X., Yao, G., & Benbo, X. (2020). Graphene oxide and indole-3-acetic acid cotreatment regulates the root growth of *Brassica napus* L. via multiple phytohormone pathways. *BMC Plant Biology, 20*, 1–12.

Xing, J., Bi, X., You, M., Zhang, Z., Hong, X., Chen, F., Zhou, Z., et al. (2025). Multifaceted applications of carbon dots in the postharvest treatment of horticultural products. *ACS Applied Materials & Interfaces, 17*, 37362.

Xu, Z. P. (2022). Material nanotechnology is sustaining modern agriculture. *ACS Agricultural Science & Technology, 2*(2), 232–239. https://doi.org/10.1021/acsagscitech.1c00204

Xu, Y., Liang, L., & Lisak, G. (2024). Blue-emissive antioxidant carbon dots enhance drought resistance of pea (*Pisum Sativum* L.). *ACS Applied Materials & Interfaces, 16*(30), 39090–39103.

Yadav, S., & Daniel, S. (2024). Green synthesis of zero-dimensional carbon nanostructures in energy storage applications—A review. *Energy Storage, 6*(1), e500.

Yan, F., Jiang, Y., Sun, X., Bai, Z., Zhang, Y., & Zhou, X. (2018). Surface modification and chemical functionalization of carbon dots: A review. *Microchimica Acta, 185*, 1–34.

Yan, X., Qiang, X., Li, D., Wang, J., & Han, R. (2021). Carbon dots inhibit root growth by disrupting auxin biosynthesis and transport in arabidopsis. *Ecotoxicology and Environmental Safety, 216*, 112168.

Zhang, W., Ji, Z., Zeng, Z., Jayapalan, A., Bagra, B., Sheardy, A., He, P., LaJeunesse, D. R., & Wei, J. (2022). Dark-field microscopic study of cellular uptake of carbon nanodots: Nuclear penetrability. *Molecules, 27*(8), 2437.

Zhang, Y., Zhang, A., Jing, J., Li, Z., Ahmed, A., Shi, Y., Yang, Y., Huang, J., & Zhang, W. (2023). Transcriptomic analyses reveal carbon dots-based seed priming in the regulation of root growth in rice. *Journal of Plant Growth Regulation, 42*(12), 7614–7623. https://doi.org/10.1007/s00344-023-11037-5

Zhang, H., Wang, Y., Zheng, T., Li, T., Gao, R., Liu, W., & Chi, Q. (2025). The intersection of nanotechnology and urban agriculture: Applications of carbon dots. *Environmental Science: Nano, 12*(1), 48–66.

Zhou, J., Sheng, Z., Han, H., Zou, M., & Li, C. (2021). Carbon dots for rapid and real-time detection of biological and chemical agents. *Biosensors and Bioelectronics, 171*, 112721. https://doi.org/10.1016/j.bios.2020.112721

Zia, M. U., Sambasivam, P. T., Chen, D., Bhuiyan, S. A., Ford, R., & Li, Q. (2024). A carbon dot toolbox for managing biotic and abiotic stresses in crop production systems. *EcoMat, 6*(5), e12451.

Carbon Nanodots as Fertilizers and Nutrient Carriers

14

Sudhir S. Shende, Graciela Dolores Avila-Quezada, Duyen H. H. Nguyen, Aniket K. Gade, José Milton Elias de Matos, Patrycja Golińska, József Prokisch, and Mahendra Rai

Abstract

Carbon nanodots (CNDs) have emerged as promising nanomaterials with potential applications in agriculture, particularly as fertilizers and nutrient carriers. Due to their unique properties, such as high surface area, biocompatibility, porosity, and the ability to functionalize with various chemical groups, CNDs can enhance the efficiency of nutrient delivery to plants. This capability allows for the controlled release of essential macronutrients and micronutrients, improving plant growth and productivity. Carbon nanodots can also facilitate the uptake of nutrients by plant roots, increasing nutrient availability and reducing the loss of fertilizers to the environment. Furthermore, their ability to protect nutrients from degradation or leaching offers a more sustainable and efficient approach to

S. S. Shende (✉) · P. Golińska
Department of Microbiology, Faculty of Biological and Veterinary Sciences, Nicolaus Copernicus University in Toruń, Lwowska 1, 87-100 Toruń, Poland
e-mail: sudhir.shende@umk.pl; golinska@umk.pl

G. D. Avila-Quezada
Facultad de Ciencias Agrotecnológicas, Universidad Autónoma de Chihuahua, Chihuahua, Chihuahua, Mexico
e-mail: gavilaq@gmail.com

D. H. H. Nguyen
Institute of Animal Science, Faculty of Agricultural and Food Sciences and Environmental Management, Biotechnology and Nature Conservation, University of Debrecen, Debrecen, Hungary

Institute of Life Sciences, Vietnam Academy of Science and Technology, Ho Chi Minh City, Vietnam

Doctoral School of Nutrition and Food Science, University of Debrecen, Debrecen, Hungary
e-mail: nguyen.huu.huong.duyen@agri.unideb.hu

M. Rai et al. (eds.), *Carbon Nanodots*, https://doi.org/10.1007/978-981-95-4705-0_14

fertilization. In this context, the potential of CNDs as advanced fertilizers and nutrient carriers, highlighting their benefits, such as seed germination promotion, enhancement in photosynthesis of crop plants, improved water retention and nutrient utilization capacities, enhanced plant growth and crop quality, improved abiotic stress tolerance and disease resistance, and reduced environmental impact, has been focused on. Besides this, the environmental impact of CNDs has also been discussed. However, challenges related to large-scale production, optimal application methods, and long-term environmental safety remain, requiring further research for practical adoption in agriculture.

Keywords

Agriculture · Carbon nanodots · Fertilizer · Nutrient · Sustainable

14.1 Introduction

The challenge of enhancing agricultural productivity while maintaining environmental sustainability has led to the exploration of innovative solutions for efficient nutrient delivery. One approach is the incorporation of carbon nanodots (CNDs), a class of zero-dimensional carbon nanomaterials, into agricultural practices (Rai et al., 2024). Carbon nanodots, characterized by their photoluminescent properties,

A. K. Gade
Department of Microbiology, Faculty of Biological and Veterinary Sciences, Nicolaus Copernicus University in Toruń, Lwowska 1, 87-100 Toruń, Poland

Department of Biological Sciences and Biotechnology, Institute of Chemical Technology, Mumbai, Maharashtra, India
e-mail: ak.gade@ictmumbai.edu.in

J. M. E. de Matos
Department of Chemistry, Federal University of Piaui (UFPI), Teresina, Brazil
e-mail: jmematos@ufpi.edu.br

J. Prokisch
Institute of Animal Science, Faculty of Agricultural and Food Sciences and Environmental Management, Biotechnology and Nature Conservation, University of Debrecen, Debrecen, Hungary
e-mail: joe.prokisch@gmail.com

M. Rai (✉)
Institute of Animal Science, Faculty of Agricultural and Food Sciences and Environmental Management, Biotechnology and Nature Conservation, University of Debrecen, Debrecen, Hungary

Department of Chemistry, Federal University of Piaui (UFPI), Teresina, Brazil

Nanotechnology Laboratory, Department of Biotechnology, Sant Gadge Baba Amravati University, Amravati, Maharashtra, India
e-mail: mahendrarai@sgbau.ac.in

small size, and high surface area, have demonstrated significant potential in a variety of applications, including drug delivery (Gong et al., 2022). However, their use in agriculture as fertilizers and nutrient carriers is an emerging strategy aimed at addressing critical issues such as nutrient inefficiency, environmental pollution, and rising global food demand.

Carbon nanodots, facilitate nutrient uptake by interacting with plant cell membranes and modulating molecular pathways through various mechanisms. These include non-protoplast transport, gene upregulation, and endocytosis. Studies on crops such as tomato, corn, and pakchoi have shown that CNDs with hydrophilic functional groups can traverse cell barriers via non-protoplast pathways, resulting in improved seed germination (up to +25.5%) and enhanced chlorophyll content (1.36–1.55× increase) (Chen et al., 2021).

In soybean, exposure to CNDs upregulated genes like GmNRT, GmAMT, and GmAQP, leading to an increase in nitrogen content, amino acids in shoots (up to 257.5%), and a 21.5% yield boost, along with improvements in nitrogenase activity and transporter genes (Ji et al., 2023). In wheat, *Salvia miltiorrhiza*-derived CNDs have been shown to mobilize Fe^{3+} and enter plant cells via endocytosis, enhancing Fe levels by up to 44%, while increasing Mn (+30%) and Zn (+19%) levels and reducing Cd accumulation by 20% (Zhu et al., 2023). These studies highlight the importance of the physicochemical properties of CNDs, such as their small size (<10 nm), high carbon content (>60%), and hydrophilic or doped surface groups—on molecular uptake and nutrient transport across various plant species (Blankenship & Prince, 1985).

Due to their physicochemical characteristics, CNDs offer an effective means of enhancing fertilizer efficiency by improving nutrient uptake and delivery to plants. These nanomaterials are highly reactive, functionalizable, and biocompatible, which enables them to interact with essential plant nutrients, ensuring their controlled release and targeted delivery. This is particularly beneficial compared to traditional fertilizers, whose excessive use often leads to nutrient leaching, soil degradation, and water pollution (Avila-Quezada et al., 2022). By integrating CNDs into fertilizer formulations, it is possible to mitigate these environmental impacts while enhancing plant nutrient uptake efficiency.

In addition to enhancing nutrient uptake, CNDs hold promise as carriers for bioactive compounds, such as growth regulators, pesticides, and micronutrients (Otchere et al., 2023; Prokisch et al., 2024a). This capability makes CNDs ideal candidates for the development of smart, nanotechnology-based fertilizer systems.

The mechanism by which CNDs enhance nutrient absorption and translocation involves modifications to root system architecture and alterations in cellular gene expression (Lei et al., 2024). Several studies have demonstrated that under drought stress, CNDs can increase nitrogen content in soybean tissues, stimulate root secretion, and upregulate relevant genes (Ji et al., 2023). Furthermore, studies such as those by Tripathi and Sarkar (2015) have shown that water-soluble carbon nanodots (wsCNDs) improve wheat growth by enhancing the transport of nutrients and water.

This chapter aims to explore the potential of CNDs as fertilizers and nutrient carriers, highlighting their synthesis, properties, and mechanisms of action in

Table 14.1 Some studies on CNDs and their effects on plants

Plant studied	Conditions and effects	Reference
Tomato, corn, pakchoi (Chinese cabbage)	Experimental. The nanoscale structure and abundant hydrophilic functional groups of N-FCNs enable them to regulate seed germination and plant growth by enhancing the uptake and transportation of water and nutrients.	Chen et al. (2021)
Cucumber, cotton	Experimental. Root application of negatively charged C-CNDs resulted in higher leaf fluorescence than positively charged P-CNDs in cucumber (8.09 ± 0.99 vs. 3.75 ± 0.23) and cotton (7.27 ± 1.06 vs. 3.23 ± 0.22), indicating better translocation efficiency. CNDs did not affect root cell activity, ROS levels, or photosynthesis, showing good biocompatibility. CNDs were translocated via both symplastic and apoplastic pathways, with positively charged CNDs mainly following the apoplastic route.	Chen et al. (2024)
Wheat	Under drought stress/nitrogen, Ce, CNDs, and CNDs:Ce. The root number, root length, leaf length, and plant height increased by 45%, 57%, 28%, and 46%, respectively.	Gong and Dong (2021)
Soybean	Under drought stress/nitrogen. The N content in soybean shoots and roots increased by 13.2% and 30.5%, respectively.	Ji et al. (2023)
Rice	Normal and stress conditions (NaCl, $CdCl_2$, 2,4-dichlorophenoxyacetic acid (2,4-D)/calcium, magnesium, manganese, phosphorus, potassium. The root vitality, cytoplasm content, and concentrations of ca, Mg, and Mn ions increased by 115.09%, 39.52%, 9.37%, 13.07%, and 19.29% in CNDs-pre-sprayed rice seedlings, respectively.	Li et al. (2024)
Tobacco	CNP treatments increased the K^+ concentration in tobacco plants. At concentrations of 10 $mg.L^{-1}$ and 20 $mg.L^{-1}$, the K^+ levels in the roots of tobacco seedlings were 36.61% and 37.71% higher, respectively, compared to the untreated roots.	Zhao et al. (2024)
Soybean	Drought stress/nitrogen. The application of CNDs under drought stress enhanced soybean nitrogenase activity by 8.6% and increased N content in soybean shoots and roots by 18.5% and 14.8%, respectively. Soil application of CNDs enhanced the N-fixing ability of nodules, regulated rhizosphere processes, and ultimately enhanced N and water uptake in soybeans under drought stress.	Wang et al. (2022)
Corn	Amending the soil with 200 mg CNPs per kg significantly enhanced corn growth, as evidenced by improved plant height, biomass yield, nutrient uptake, and nutrient use efficiency. These improvements were attributed to the increased availability of phosphorus and nitrogen in the amended soils.	Zhao et al. (2021)
Wheat	*Salvia miltiorrhiza*-derived CNDs are effective in regulating the accumulation of iron (Fe), manganese (Mn), zinc (Zn), and cadmium (Cd) in plants.	Zhu et al. (2023)

promoting plant growth. We also examine the opportunities associated with their application in agriculture, considering the environmental impact and long-term effectiveness.

14.2 Application of Carbon Nanodots in Crop Production

14.2.1 Carbon Nanodots as Fertilizers

14.2.1.1 Effects of Carbon Nanodots on Nutrient Uptake

Enhanced Absorption of Nitrogen and Phosphorus

Carbon nanodots may enhance nitrogen absorption across various plant species (Chen et al., 2021). Ji et al. (2023) reported that the application of CNDs to soybeans under drought stress increased nitrogen content in roots. Similarly, Wang et al. (2022) observed an 8.6% increase in nitrogenase activity and increased nitrogen content in soybean shoots and roots when treated with CNDs under drought conditions (Table 14.1).

The mechanism behind this enhanced nitrogen absorption appears to be multifaceted. Ji et al. (2023) noted that CNDs stimulated root secretion and recruited beneficial microorganisms, facilitating nitrogen activation in the soil. Additionally, the upregulation of nitrogen transport-related genes (GmNRT, GmAMT) was observed, indicating enhanced nitrogen uptake mechanisms at the molecular level.

According to the study of Zhao et al. (2021), without fertilizer, the CNPs' application did not improve P uptake by corn plants for both soils ($p > 0.05$), while the P uptake of FB200 in Spodosol soil was significantly increased compared to FB0—an increase of 267% under fertilization ($p < 0.05$).

Increased Potassium Uptake

Potassium uptake was significantly influenced by CNDs and nanoparticles. Zhao et al. (2024) demonstrated that carbon nanoparticle (CNP) treatments increased K^+ concentration in tobacco roots (Table 14.1). In shoots, K^+ concentration increased by 56.43%, 39.61%, and 41.79% at 10 $mg{\cdot}L^{-1}$. 20 $mg{\cdot}L^{-1}$, and 40 $mg{\cdot}L^{-1}$ CNPs, respectively.

The study by Zhao et al. (2024) provided insights into the mechanisms of enhanced potassium uptake. CNPs were found to alter the direction of K^+ flow in roots from efflux to influx, indicating a change in cellular activity related to ion transport. This was accompanied by changes in the expression of genes related to potassium channels and transporters, suggesting a molecular basis for the observed effects.

Enhanced Uptake of Calcium (Ca), Iron (Fe), Manganese (Mn), Magnesium (Mg), and Zinc (Zn)

Irrigation with CNDs at a concentration of 1 mg/mL, applied weekly starting at the jointing stage for 12 weeks, resulted in an 18% increase in Fe content while reducing

Cd accumulation by 20% in wheat grains. In addition, in the same study, foliar spraying of CNDs led to significant improvements in the content of Fe (44%), Mn (30%), and Zn (19%), without affecting Cd accumulation in the wheat grains (Zhu et al., 2023).

Furthermore, in the study of Li et al. (2024), root vitality, cytoplasm content, and the concentrations of Ca, Mg, and Mn ions in CND-pre-sprayed rice seedlings were significantly increased by 115.09%, 39.52%, 9.37%, 13.07%, and 19.29%, respectively.

14.2.1.2 Molecular and Physiological Mechanisms

Root Architecture Modifications

Carbon nanodots have been observed to induce significant changes in root architecture, which likely contribute to enhanced nutrient absorption. Gong and Dong (2021) reported that cerium-doped carbon quantum dots (CNDs:Ce) increased root number by 45% and root length by 57% in wheat at a concentration of 0.025 mg mL^{-1}. Similarly, Li et al. (2024) observed a 53.04% increase in root fresh weight of rice seedlings treated with 100 μg mL^{-1} CNDs.

These modifications in root architecture provide plants with a larger surface area for nutrient absorption and improved access to soil resources. The enhanced root development may also contribute to better anchorage and increased resilience to environmental stresses.

Gene Expression and Signalling Pathways

The application of CNDs modulates the expression of genes related to nutrient transport and metabolism. Ji et al. (2023) reported the upregulation of genes related to nitrogen transport (GmNRT, GmAMT) and water uptake (GmAQP) in soybeans treated with CNDs. Zhao et al. (2024) observed changes in the expression of potassium channel and transporter genes in tobacco plants treated with carbon nanoparticles. These changes in gene expression suggest that CNDs can influence nutrient uptake at the molecular level, potentially leading to long-term adaptations in plant nutrient acquisition strategies.

14.2.2 Carbon Nanodots as Nutrient Carriers and Stress Relievers

Carbon nanodots and related carbon-based nanomaterials (CB-NMs) are effective nutrient carriers in agriculture, promoting plant growth and productivity. Ashfaq et al. (2025) highlight the potential of carbon nanofibers (CNFs) to deliver essential nutrients like Fe, Cu, Zn, and Mo, which improve crop yield and nutritional quality. Li et al. (2023) mention that carbon dots have multiple roles, including positive effects on nutrient absorption and resistance to abiotic stress.

In addition, Ashfaq et al. (2017) demonstrate the successful use of Cu-loaded CNFs for controlled nutrient release, which enhances water uptake and plant growth. Jampílek and Kráľová (2017) emphasized the growing use of nanomaterials for

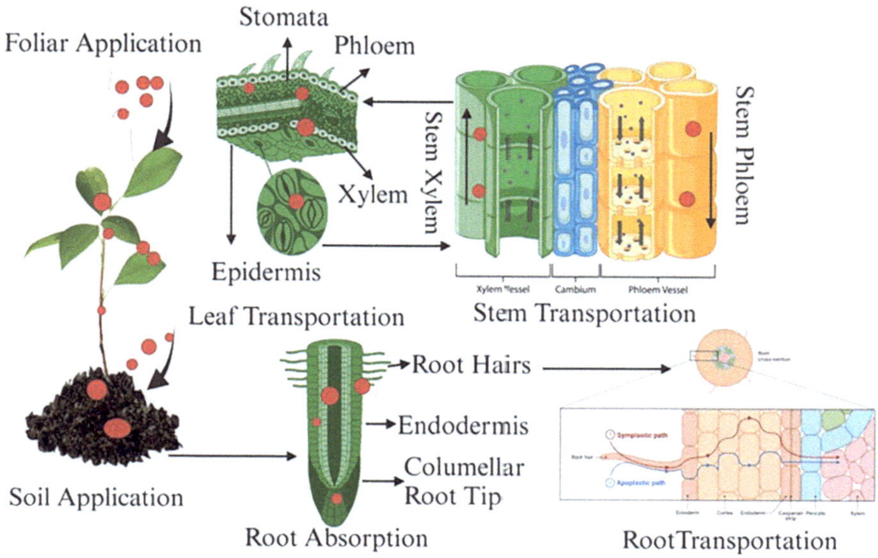

Fig. 14.1 Schematic representation of possible pathways of nanoparticle-like carbon nanodots uptake, transport, and distribution in plants: from root absorption to foliar application. (Adapted from Cao et al., 2025)

delivering macronutrients and growth-promoting compounds, showing their efficacy in optimizing fertilizer use in agriculture. Furthermore, it has been shown that carbon dots can enhance nitrogen bioavailability, leading to improved growth and nutritional quality of soybeans, under drought stress (Wang et al., 2022).

Moreover, nitrogen-doped carbon dots have been found to enhance lettuce yield and quality by promoting biomass accumulation and nutrient content (Tan et al., 2023). Carbon dots can also be useful for managing biotic and abiotic stresses in crops, potentially enhancing plant health and growth (Zia et al., 2024).

14.3 Effect of Carbon Nanodots on Plant Growth and Development

14.3.1 Carbon Nanodots' Uptake, Accumulation, and Transport in Plants

Nanotechnology has seen extensive application across various fields, and one of the emerging areas is the application of nanoparticles in agriculture. Carbon nanodots are a class of carbon-based NPs with unique physicochemical properties, including fluorescence, biocompatibility, and environmental safety. These properties make CNDs promising candidates to be used in agricultural crop plants, particularly in the areas of plant growth monitoring, stress detection, and targeted delivery systems. Understanding the uptake, accumulation, and transport of CNDs in plants is critical for their potential integration into agricultural applications (Ozfidan-Konakci et al., 2024). The possible pathways of uptake, transport, and distribution of nanodots in plants from root absorption to foliar application are schematically presented in Fig. 14.1 (Cao et al., 2025). The following sections explore these three key processes, discussing the current research findings on how CNDs interact with plant systems.

14.3.1.1 Uptake of Carbon Nanodots in Plants

The uptake of CNDs into plants is a complex process that depends on the size, surface charge, and chemical modifications of the nanodots. Carbon nanodots can be absorbed through the plant roots or foliar surfaces, and their ability to penetrate the plant's biological barriers is influenced by various factors, including the plant species, the method of application, and the physicochemical properties of the nanodots.

Root Uptake

The most common pathway for CNDs uptake in plants is via the roots. Uptake of NPs by roots is typically facilitated by root hairs, which increase the surface area for absorption. Several studies have demonstrated that CNDs, particularly those that are smaller in size (>10 nm in diameter), can penetrate the root epidermis and move into the vascular system. The surface charge and functional groups of the CNDs also play a significant role in the absorption process. Positively charged CNDs tend to have better uptake due to electrostatic interactions with negatively charged components in the plant cell wall and plasma membrane (Cheng et al., 2023).

Foliar Uptake

In addition to root absorption, CNDs can also be taken up through the leaves. The cuticle and epidermis of leaves serve as the primary barriers to NPs' entry, and modifications to the CND surface, such as coating them with surfactants or other materials, can enhance their permeability through these barriers. Some studies have indicated that CNDs with certain surface modifications can be more readily absorbed by the stomata or directly through the epidermal cells (Hong et al., 2021).

14.3.1.2 Accumulation of Carbon Nanodots in Plants

Once inside the plant, CNDs accumulate in various plant tissues depending on the extent of their uptake and the transport mechanisms. The accumulation of CNDs can vary across plant species, but they generally concentrate in the root system, vascular tissues, and leaves.

Root Accumulation

After uptake, CNDs often accumulate in the root tissue, where they can be detected in both the cortical and vascular tissues. The root cell walls and vacuoles have a high affinity for NPs, which leads to their initial accumulation in these areas. The accumulation in roots has been observed to be higher in plants with a more porous root structure or a greater number of root hairs (Cao et al., 2025).

Leaf Accumulation

In the leaves, CNDs tend to accumulate in the epidermal layers, particularly when foliar application is used. Studies have shown that surface-modified CNDs can penetrate leaf cells and accumulate in vacuoles and chloroplasts, though the levels of accumulation vary across species (Lin et al., 2009; Li et al., 2023). The localization of CNDs in plant cells can also be influenced by the age of the plant, with older leaves often showing higher accumulation rates than younger ones due to the increased surface area and the slower transpiration rates in mature tissues.

Translocation to Other Tissues

The movement of CNDs from the roots to the rest of the plant is a process that involves the xylem and phloem. Through the xylem, CNDs can be transported to the stems and leaves. The extent of this translocation depends on the physicochemical characteristics of the CNDs, such as their size and surface charge. Smaller NPs are more likely to be translocated efficiently through the xylem, while larger CNDs may be retained in the root system (Prokisch et al., 2024b). Moreover, the surface functionalization of CNDs can influence their interaction with the plant vascular tissues, affecting their movement through the plant body.

14.3.1.3 Transport of Carbon Nanodots in Plants

The transport of CNDs within plants involves both passive and active mechanisms. Once inside the plant, the movement of NPs from one tissue to another occurs mainly through the xylem and phloem vessels. The movement through the xylem is primarily driven by transpiration and capillary action, while the movement through the phloem is influenced by plant metabolic processes.

Xylem Transport

The xylem plays a critical role in the upward transport of water and nutrients, and CNDs follow this route to reach the aerial parts of the plant. Carbon nanodots that are small enough can pass through the vascular bundles and be carried by the transpiration stream, allowing for rapid transport from roots to leaves. The movement through the xylem is passive, relying on the plant's water uptake and the properties of the CNDs.

Phloem Transport

Phloem transport, which is responsible for the movement of photosynthates, i.e., the resulting products from photosynthesis such as sugars, and other solutes, can also facilitate the transport of CNDs, especially when they are functionalized with biomolecules that enable them to interact with the phloem loading and unloading mechanisms. However, the efficiency of CNDs' transport through the phloem is less well understood and may be influenced by the developmental stages and physiological conditions of plants.

14.3.1.4 Impact of Carbon Nanodots Properties on Transport

The size and morphology, surface charge, and surface chemistry of CNDs significantly influence their transport within the plant. For instance, CNDs with hydrophobic surfaces may be retained in plant tissues for longer periods, while hydrophilic CNDs are more easily translocated through vascular tissues. Furthermore, functionalization with biomolecules, such as peptides or sugars, can enhance the affinity of CNDs for specific receptors in the plant tissues, improving their transport and targeting (Verma et al., 2019; Mahor et al., 2021).

Thus, the processes of uptake, accumulation, and transport of CNDs within plants are still subjects of ongoing research. Understanding these mechanisms is crucial for harnessing the full potential of CNDs in agricultural applications. While current studies provide valuable insights, further research is needed to explore the long-term effects of CND accumulation in plants and the potential ecological impacts of their widespread use.

14.3.2 Enhancement in Photosynthesis of Crop Plants

The process by which plants convert sunlight into chemical energy is photosynthesis, which is a critical function for plant growth and food production. It is a complex process involving light absorption, electron transport, and the formation of sugars. Improving photosynthesis in crop plants is one of the most promising strategies to address global food security challenges, particularly as the world population continues to grow and climate change impacts agricultural productivity. Recent research has identified carbon-based nanomaterials, specifically CNDs, as potential enhancers of photosynthetic efficiency in plants. Carbon nanodots are nanoscale carbon particles that have unique optical, electrical, and chemical properties, making them valuable for a wide range of applications, including plant biology, in which their potential as photosynthetic enhancers is being explored, due to their ability to influence plant metabolic processes at the molecular level. The role of CNDs in enhancing the photosynthesis of crop plants, their mechanism of action, potential benefits, and challenges associated with their use have been discussed in the following sections.

14.3.2.1 Mechanism of Action: How CNDs Enhance Photosynthesis

Carbon nanodots enhance photosynthesis in plants through several mechanisms, which can be categorized into direct and indirect effects on the cellular functions of plants. The mechanisms by which they affect the photosynthetic process have been detailed in the following sections.

Improved Light Absorption

Carbon nanodots possess photoluminescent properties, meaning they can absorb and emit light. When applied to plants, CNDs may help improve light absorption efficiency by acting as light-harvesting molecules. Some studies have shown that CNDs, when applied to plant surfaces or incorporated into plant tissues, can absorb light energy more efficiently, which can aid in the enhancement of photosynthetic rates, especially under suboptimal light conditions (Li et al., 2021a, 2021b; Ji et al., 2023).

Enhanced Chloroplast Function

Carbon nanodots have been observed to interact with chloroplasts, the organelles responsible for photosynthesis in plant cells. Carbon nanodots may help stabilize chloroplast membranes and promote better functioning of the photosystem complexes, which are essential for capturing light energy and converting it into chemical energy (Tan et al., 2021). This interaction leads to more efficient photosynthetic processes and higher biomass accumulation.

Electron Transport and Charge Separation

Carbon nanodots have been shown to promote efficient electron transfer within plant cells. Photosynthesis involves the transfer of electrons from chlorophyll molecules to the electron transport chain, which ultimately leads to the production of ATP and NADPH, vital energy carriers. CNDs can act as electron shuttles, facilitating the movement of electrons and thus improving the efficiency of the photosynthetic machinery (Li et al., 2021b). By enhancing charge separation, CNDs increase the production of these energy molecules, improving plant growth and productivity.

Reduction of Photodamage

Carbon nanodots have antioxidant properties, which help protect plant cells from oxidative stress caused by excess light absorption. This stress can damage plant tissues and reduce photosynthetic efficiency, particularly in intense sunlight. The antioxidant properties of CNDs can neutralize free radicals generated during photosynthesis, minimizing photodamage and improving the overall efficiency of the process (Cheng et al., 2023).

14.3.2.2 Benefits of CNDs in Crop Photosynthesis Enhancement

Increased Crop Yield

Several studies have demonstrated that the application of CNDs to crop plants can lead to increased photosynthetic efficiency, which directly translates to higher crop yields. For example, applications of CNDs to crops like wheat, rice, and soybeans have shown improvements in growth rate, chlorophyll content, and overall biomass accumulation, leading to enhanced productivity (Ji et al., 2023).

Stress Tolerance

Carbon nanodots can help plants tolerate environmental stressors such as drought, high salinity, heavy metals, and extreme or low temperatures. Under stressful conditions, plants often experience reduced photosynthetic rates. The use of CNDs has been shown to mitigate these effects by improving cellular defence mechanisms and maintaining photosynthesis under adverse conditions (Wang et al., 2022). This makes CNDs particularly valuable in the face of climate change, where environmental stressors are becoming more frequent and severe.

Sustainability

Unlike conventional fertilizers or chemical enhancers, CNDs are environmentally friendly and can be produced from renewable resources. Their use in agriculture reduces the need for synthetic chemicals, which can have harmful effects on the environment. Carbon nanodots can also be biodegradable, making them a sustainable alternative to other nanomaterials that may pose long-term ecological risks (Tan et al., 2021).

14.3.3 Improvement in Crop Quality

The quality of crop plants is a critical factor in determining agricultural productivity, nutritional value, and market competitiveness. Traditionally, improvements in crop quality have been pursued through conventional breeding techniques, genetic modifications, and the use of fertilizers and pesticides. However, these methods have limitations and can sometimes be detrimental to the environment and human health. In recent years, CNDs have gained significant attention due to their unique properties such as small size, biocompatibility, stability, and versatility. The following section explores the potential of carbon dots in improving crop quality, focusing on their effects on plant growth, nutritional content, disease resistance, and post-harvest quality.

14.3.3.1 Mechanisms of Action: How CNDs Improve Crop Quality

Carbon dots can be synthesized using different methods, including hydrothermal, microwave, or chemical oxidation techniques. These methods allow for the production of CNDs with customizable properties, such as surface charge, size, and functional groups, enabling their tailored use in enhancing crop quality. The mechanisms by which they improve crop quality are given in the following sections.

Enhanced Plant Growth and Biomass Accumulation

One of the primary benefits of CNDs in agriculture is their ability to promote plant growth and increase biomass accumulation. Several studies have shown that when CNDs are applied to plants, they can improve seed germination rates, root development, and overall plant growth. Carbon nanodots enhance the uptake of essential nutrients such as nitrogen, phosphorus, and potassium, thereby promoting better growth conditions for crops (Guo et al., 2022; Abinaya et al., 2024). This results in healthier plants with increased biomass and improved quality. The improved growth can directly contribute to higher yields and more robust plants, thus improving the quality of the harvested produce.

Improved Nutritional Content

The application of CNDs has been associated with improvements in the nutritional content of crops. Carbon nanodots can influence the biosynthesis of secondary metabolites, which include vitamins, antioxidants, and flavonoids that enhance the nutritional quality of fruits and vegetables. For example, research on tomatoes has shown that CND treatment leads to higher concentrations of antioxidants such as lycopene and vitamin C, both of which are essential for human health (Hu et al., 2022; Tripathi & Sarkar, 2022). In leafy vegetables, such as spinach and lettuce, CNDs have been reported to increase the levels of essential nutrients like chlorophyll, amino acids, and vitamins, improving the overall nutritional value of these crops.

Stress Resistance and Improved Quality Under Adverse Conditions

Crop quality is often compromised under environmental stressors such as drought, high salinity, and extreme temperatures. One of the remarkable properties of CNDs is their ability to enhance plant tolerance to abiotic stress. Carbon nanodots have been shown to mitigate the negative effects of environmental stress by promoting antioxidant activity within plant cells. This helps reduce oxidative damage caused by stress conditions, maintaining cellular function and preserving the quality of the crop (Su et al., 2018; Wang et al., 2022). Additionally, CNDs can enhance the synthesis of stress-related proteins and enzymes, such as heat shock proteins and peroxidases, that help plants manage adverse conditions more effectively. As a result, CNDs can improve the resilience of crops, ensuring better quality even under challenging environmental conditions.

Disease Resistance and Pest Management

The use of CNDs in agriculture also holds potential for improving disease resistance and pest management. CNDs can be used to enhance the natural defence mechanisms of plants by boosting the production of antimicrobial compounds, which help protect crops from pathogens. For example, studies have shown that CNDs can induce the expression of pathogenesis-related proteins in plants, improving resistance to fungal and bacterial infections (Kou et al., 2025). In addition, the application of CNDs has been shown to reduce the severity of pest infestations, as they can interfere with the feeding behaviours of insects, potentially decreasing the damage to crops and improving quality (Li et al., 2018a, b).

Post-Harvest Quality and Shelf Life

The post-harvest quality of crops, including their shelf life and resistance to spoilage, is another area where CNDs show promise. CNDs can enhance the storage life of fruits and vegetables by reducing the rate of spoilage and degradation. They achieve this by modulating the activity of enzymes responsible for ripening and senescence, such as polygalacturonase and ethylene biosynthesis enzymes (Deepika et al., 2023; Shan et al., 2023; Chaudhary et al., 2025). By slowing down the ripening process, CNDs can extend the shelf life of produce, reducing losses during transportation and storage. This is particularly important in minimizing food waste and ensuring the availability of high-quality crops to consumers.

While the potential benefits of CNDs in enhancing photosynthesis and improving crop quality are clear, several challenges must be addressed before their widespread adoption in agriculture. One of the primary concerns is the optimal method of applying CNDs to plants. The most effective concentration, form, and delivery method are still under investigation. Additionally, the long-term effects of CNDs on soil health and biodiversity are not yet fully understood, necessitating further research into their environmental impact. Moreover, the scalability of CND production remains a challenge. Although CNDs can be synthesized through various methods, producing them on a large scale for agricultural applications is currently expensive and inefficient, and also, CNDs are generally considered non-toxic, so their accumulation in soil and water systems could have unforeseen consequences on ecosystems. Advances in synthesis techniques and cost-reduction strategies will be necessary to make CND-based solutions for commercialization in the agricultural industry. Even though challenges related to application methods, environmental impact, and production costs remain, ongoing research into the use of CNDs in crop management suggests that they have the potential to revolutionize agricultural practices. With further advancements in synthesis techniques and a deeper understanding of their ecological effects, carbon nanodots present a promising avenue for enhancing plant growth, improving disease resistance, and facilitating the delivery of nutrients or pesticides and could play a pivotal role in enhancing crop quality and ensuring sustainable food production for the future.

14.3.4 Seed Germination Promotion and Enhancing Water Retention and Nutrient Absorption Capacity

Carbon nanodots are emerging nanomaterials with significant applications in agriculture due to their unique properties, including intrinsic photoluminescence, negligible toxicity, biocompatibility, high water solubility, good conductivity, chemical inertness, high quantum yield, and resistance to photobleaching (Chaudhary et al., 2024). These properties make CNDs effective in promoting seed germination, enhancing water retention, and improving nutrient absorption capacity in plants.

Seed germination is a fundamental process that determines plant establishment and productivity. Abinaya et al. (2024) reported the synthesis of CNDs from agro-waste peanut shells through the pyrolysis method. The synthesized CNDs were used as a priming and foliar agent for improving seed quality and crop yield in black gram (*Vigna mungo* L. Hepper). They observed that the seed priming for 3 h at 200 ppm can improve water absorption, thereby improving seed germination in black gram. Biogenically synthesized CNDs from *Annona squamosa* (L.) peels using a carbonization technique through microwave-assisted hydrothermal digestion were reported by Pete et al. (2023). They reported the germination, growth, and development of *Vigna radiata* (L.) seeds when treated with CNDs. There are reports on CNDs' effectiveness in promoting seed germination, enhancing water retention, and improving nutrient absorption capacity in plants like black gram (Abinaya et al., 2024), green gram (Pete et al., 2023), wheat (Tripathi & Sarkar, 2015), and rice (Li et al., 2018b). It is the ability of CNDs to penetrate the hard seed coat, allowing water infiltration and facilitating water absorption, which makes them an ideal candidate to promote seed germination and enhance seed vitality and viability (Maholiya et al., 2023).

Carbon nanodots can significantly enhance seed germination, enhancing water retention and improving nutrient absorption capacity in plants through multiple mechanisms. Carbon nanodots can modulate phytohormone activity, particularly gibberellins (GAs) and auxins, which play a vital role in breaking seed dormancy and promoting germination; the enhanced expression of genes involved in GA biosynthesis leads to faster seedling emergence (Khan, 2025). Carbon nanodots improve seed hydration by enhancing water absorption through increased cell permeability and improve water uptake by facilitating enzymatic activation (Maholiya et al., 2023), which is crucial for breaking dormancy and initiating metabolic processes. Carbon nanodots can enhance amylase activity, which breaks down starch into sugars necessary for energy supply during germination. Increased respiration rates in germinating seeds due to CNDs contribute to improved energy metabolism as well. Seedlings grown from CD-treated seeds exhibit enhanced root and shoot development, contributing to better establishment and stress resilience. Increased chlorophyll content and improved photosynthetic efficiency result in higher growth rates and biomass accumulation (Zhang et al., 2025). Furthermore, CNDs' role in increasing water retention and nutrient uptake presents a sustainable approach to improving agricultural productivity. Future research should focus on optimizing CND formulations, assessing their long-term ecological impact, and conducting large-scale field trials to validate their practical applications.

14.3.5 Improving the Tolerance to Abiotic Stresses and Disease Resistance

The increase in salt content in the soil, drought conditions, cold, extreme temperatures, metal toxicity, and excessive use of pesticides can lead to an increase in reactive oxygen species (ROS), thereby causing abiotic stress on the plants (Chaudhary et al., 2024). The ROS, due to their strong oxidation ability, can damage the DNA and cell membrane (phospholipid bilayer) and cause denaturation of proteins. Carbon nanodots are a novel type of carbon nanomaterial with unique physicochemical properties, including enhanced water solubility, biocompatibility, antioxidant ability, radical scavenging ability, and tunable fluorescence. These attributes make them promising candidates for enhancing plant tolerance to abiotic stresses and improving disease resistance.

The multifunctional CNDs synthesized using *Salvia miltiorrhiza* via a simple one-step hydrothermal method exhibited superior antioxidant activity to effectively alleviate the oxidative damage of Italian lettuce under salt stress (Li et al., 2021a). A heteroatom-free CND used in a study on rice (*Oryza sativa* L.), revealing antioxidant and radical scavenging abilities by investigating the contents of a biomarker of oxidative cleavage by ROS in plant cells (Li et al., 2020). The antioxidant ability and radical scavenging ability of CNDs can not only reduce abiotic stress but also increase plant biomass. Moreover, the CNDs can also protect the plant against abiotic stress caused by toxic metal cadmium ions against the wheat (Xiao et al., 2019) and grapefruit (Li et al., 2019), apart from reducing the abiotic stress caused

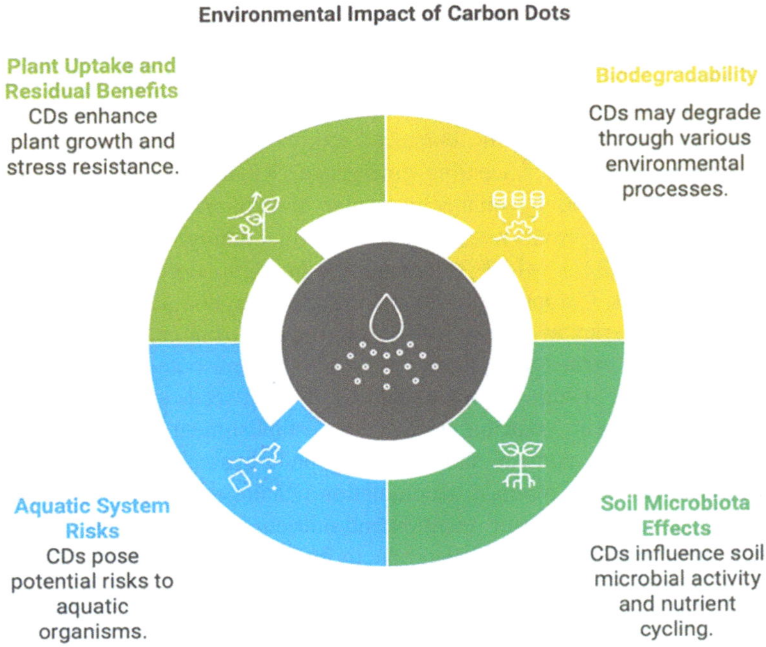

Fig. 14.2 Illustration of the environmental impact of carbon nanodots (CNDs)

by the antioxidant and radical scavenging activity. Chandrakar et al. (2020) reported tolerance in *Cicer arietinum* against arsenic by regulating the uptake of arsenic, detoxifying ROS, and interfering with defence-related gene expression. Increasing the expression of antioxidant enzymes like superoxide dismutase and catalase reduces cellular damage and promotes healthy seedling development. The presence of hydrophilic functional groups of the CNDs, which have the ability to retain water and nutrients for a longer time, is able to provide an antidrought effect to the plants, thereby increasing the yield of the plant. This was evident from the study of the drought impact on peanut plants in the presence of CNDs (Su et al., 2018).

The pathogenic infection is the main cause behind the development of biotic stress on crop plants. Nano-pesticides in general and CNDs in particular, due to their small size, reduced run-off, bioavailability, and easy uptake, are considered an excellent candidate demonstrating broad-spectrum antibacterial (Li et al., 2018a) and antifungal agents, which can be applied for controlling and managing plant diseases. Carbon nanodots synthesized using metronidazole via a hydrothermal synthesis procedure exhibited selective antibacterial activity against the obligate anaerobe *Porphyromonas gingivalis* (Liu et al., 2017). Considering the different roles of CNDs in agriculture, it can be further explored in plant disease management. Further research is necessary to optimize their application methods, understand long-term effects, and assess their environmental impact to ensure their safe and effective use in crop protection and productivity enhancement.

14.4 Environmental Impact of Carbon Dots (CNDs)

While CNDs offer various benefits as fertilizers and nutrient carriers, it is crucial to address their environmental impacts to ensure sustainable agricultural applications by examining (1) the biodegradability of CNDs, (2) the impact on soil microbiota and nutrient cycling, (3) potential risks to water systems, and (4) plant uptake and residual effects, as illustrated in Fig. 14.2.

14.4.1 Fate and Biodegradability of CNDs

The primary environmental concern regarding CNDs is their biodegradability. Depending on the synthesis route, CNDs may degrade through processes such as photodegradation, microbial activity, or oxidation, which reduces their environmental persistence. Photochemical transformations play a crucial role in the degradation of CNDs, with natural sunlight rapidly photobleaching CNDs into optically inactive nanoparticles (Frank et al., 2020). Photodegradation of CNDs can produce cytotoxic molecules that affect both normal and cancerous human cells (Y.-Y. Liu et al., 2021).

Carbon nanodots are emerging carbon-based nanomaterials notable for their unique properties, including high water solubility and low toxicity, making them less likely to accumulate in soil and water (Beker et al., 2022). However, numerous factors influence their transport and retention in aqueous environments. Water chemistry impacts the surface charge and aggregation behaviour of CNDs, particularly in relation to ionic strength and pH (Bayati et al., 2018). Higher ionic strength and lower pH generally enhance CD retention in porous media (Kamrani et al., 2018). Additionally, the presence of natural organic matter promotes CD transport (Jahan et al., 2019). While CNDs hold promise for environmental applications such as water remediation (Beker et al., 2022), their stability in suspension raises concerns about the potential long-distance transport and exposure to aquatic organisms (Bayati et al., 2018).

14.4.2 Impact on Soil Microbiota and Nutrient Cycling

Carbon nanodots affect soil microbial communities differently based on their concentration and type. At low concentrations, CNDs can boost microbial activity and growth rates (W. Liu et al., 2015). Additionally, CNDs can enhance nitrogen availability, support beneficial microbial communities, and promote plant growth under stressful conditions (Wang et al., 2022). Carbon nanodots can also impact heavy metal speciation and dissimilatory iron reduction processes in soil (Huang et al., 2024). While CNDs can potentially increase agricultural production, their use should be carefully managed to minimize long-term impacts on soil ecosystems. Huang et al. (2024) noted that CNDs influence soil acidity, microbial communities, and heavy metal speciation by encouraging iron reduction, immobilizing lead and zinc, increasing copper mobility, and affecting microbial diversity based on type and incubation period. Further research is needed to fully understand the complex interactions between CNDs and soil microbial communities.

14.4.3 Potential Risks to Aquatic Systems

Unlike other meta-nanoparticles, CNDs exhibit high water solubility and low toxicity (Beker et al., 2022). Although generally regarded as less harmful than many nanomaterials, CNDs can impact aquatic organisms across various trophic levels (Yao et al., 2018). Their high solubility allows CNDs to remain in water, potentially for trophic transfer (Y. Liu et al., 2022). The toxicity levels of CNDs vary depending on the organism, exposure time, and preparation methods (Freixa et al., 2018). Studies show that phytoplankton are most sensitive to CND exposure, followed by zooplankton, with fish showing minimal toxicity (Yao et al., 2018; Y. Liu et al., 2022). Carbon nanodots can aggregate, leading to oxidative stress, cellular structure disruption, and adversely affecting algae and zooplankton growth and behaviour (Y. Liu et al., 2022). Furthermore, CNDs may interact with other pollutants, modifying their toxicity levels (Freixa et al., 2018; Beker et al., 2022).

14.4.4 Plant Uptake and Residual Effects

Carbon nanodots can accumulate in plant tissues, promote photosynthesis, and participate in metabolic processes without significant soil impacts (Lei et al., 2024). Plants can absorb CNDs through roots and leaves, influencing growth and metabolism (Li et al., 2016). Studies have shown that CNDs can enhance seed germination, nutrient uptake, and photosynthesis, while improving crop stress resistance (Li et al., 2023). At low concentrations, CNDs have demonstrated positive effects on plant growth, with increased root and stem lengths observed in mung beans (Li et al., 2016). However, their impact on plant physiology is dose-dependent, and higher concentrations may induce stress responses. Carbon nanodots have shown promise in managing both biotic and abiotic stresses, potentially improving crop tolerance to heat, drought, toxic chemicals, and pathogens (Zia et al., 2024).

14.5 Conclusions and Future Perspectives

The potential role of CNDs as nanofertilizers and nutrient carriers depends on their ability to bind, transport, and release essential nutrients in a controlled and targeted manner, thereby improving nutrient use efficiency and reducing environmental losses. Carbon nanodots can also stimulate plant growth directly by influencing hormonal pathways, enhancing photosynthesis, and promoting seed germination and root development.

However, despite these promising qualities, some serious challenges must be addressed before CNDs can be widely commercialized as agro-inputs. Foremost among them is the concern regarding long-term toxicity and environmental impact. Although CNDs are often labelled as "eco-friendly" due to their carbon-based nature, their interactions with soil microbiota, earthworms, beneficial fungi, and other non-target organisms are not properly understood. Therefore, thorough eco-toxicological studies are needed to assess their biodegradability, accumulation, and potential impacts on macro and microfauna in diverse agroecosystems.

Scalability and cost-effective production represent another bottleneck in the transfer of CNDs from the laboratory to the field. Current synthesis methods, especially those involving complex precursors or high-energy processes, are not practicable for mass production. Green synthesis approaches using agricultural waste or plant extracts offer a sustainable alternative, but these also need optimization for yield, consistency, and reproducibility.

Another area that demands attention is targeted delivery and functionalization. Functionalized CNDs capable of recognizing plant-specific receptors or stress-induced signals could allow precise nutrient delivery or activation under specific physiological conditions. This targeted delivery could significantly minimize wastage and enhance crop productivity under both normal and stress conditions.

The future may also lie in synergistic combinations of CNDs with biofertilizers, plant growth-promoting rhizobacteria (PGPRs), and plant hormones. Such combinations may offer multifaceted benefits including enhancing nutrient solubilization, improving soil fertility, and promoting resilience against abiotic stress. Preliminary studies suggest that CNDs can modulate microbial communities beneficially and can even act as carriers for microbial inoculants.

In summary, while CNDs hold transformative potential in nutrient management and crop productivity, multidisciplinary research integrating material science, soil ecology, agronomy, and toxicology is vital. The roadmap to commercialization hinges on developing safe, cost-effective, and scalable production systems along with rigorous field evaluations to ensure environmental safety and regulatory compliance. With these advancements, CNDs could truly become the next-generation fertilizers and nutrient carriers for a climate-smart agricultural future.

Acknowledgements SSS and PG acknowledges the YUFE4Postdocs (Y4P) programme, financially supported by European Union funds as a part of the Marie Sklodowska-Curie Actions Cofund programme (Agreement No. 101081327), Ministry of Science and Higher Education under the "International Co-financed Projects" programme (Agreement No. 5678/HE/2023/2024/2). AKG acknowledges the National Science Centre and the European Union Framework Programme (Project No UMO-2022/45/P/NZ9/01571), co-funded by Research and Innovation Horizon 2020 under the Marie Skłodowska-Curie grant agreement No. 945339. MR thankfully acknowledges the Federal University of Piauí (UFPI), Teresina, Brazil, for the opportunity to serve as a Senior Visiting Professor and also to Debrecen University, Debrecen, Hungary for research visit.

Authors' Contribution Statement *SSS*: Conceptualization, Investigation, Data curation, Writing—Original draft preparation, review, and editing; *GDA-Q*: Data curation, Writing—Original draft preparation, Validation; *DHHN*: Data curation, Writing—Original draft preparation; *AKG*: Conceptualization, Methodology, Software, Supervision, Validation, Reviewing, and Editing; *JMEdeM*: Data curation, Writing—Original draft preparation, Validation; *PG*: Data curation, Writing—Original draft preparation, review, and editing; *JP*: Reviewing and Editing, Validation *MR*: Supervision, Writing—Reviewing and Editing, Validation.

References

Abinaya, K., Raja, K., Raja, K., Moorthy, P. S., Senthil, A., & Chandrakumar, K. (2024). Impact of green carbon dot nanoparticles on seedling emergence, crop growth and seed yield in blackgram (*Vigna mungo* L. Hepper). *Scientific Reports, 14*, 23783. https://doi.org/10.1038/s41598-024-75366-5

Ashfaq, M., Verma, N., & Khan, S. (2017). Carbon nanofibers as a micronutrient carrier in plants: Efficient translocation and controlled release of cu nanoparticles. *Environmental Science: Nano, 4*(1), 138–148. https://doi.org/10.1039/C6EN00385K

Ashfaq, M., Gupta, G., & Verma, N. (2025). Carbon-based nanocarriers for plant growth promotion: Fuelling when needed. *Nanoscale, 17*(2), 616–634. https://doi.org/10.1039/D4NR03268C

Avila-Quezada, G. D., Ingle, A. P., Golińska, P., & Rai, M. (2022). Strategic applications of nanofertilizers for sustainable agriculture: Benefits and bottlenecks. *Nanotechnology Reviews, 11*(1), 2123–2140. https://doi.org/10.1515/ntrev-2022-0126

Bayati, M., Dai, J., Zambrana, A., Rees, C., & Fidalgo De Cortalezzi, M. (2018). Effect of water chemistry on the aggregation and photoluminescence behavior of carbon dots. *Journal of Environmental Sciences, 65*, 223–235. https://doi.org/10.1016/j.jes.2017.03.009

Beker, S. A., Cole, I., & Ball, A. S. (2022). A review on the catalytic remediation of dyes by tailored carbon dots. *Water, 14*(9), 1456. https://doi.org/10.3390/w14091456

Blankenship, R. E., & Prince, R. C. (1985). Excited-state redox potentials and the Z scheme of photosynthesis. *Trends in Biochemical Sciences, 10*(10), 382–383.

Cao, Y., Turk, K., Bibi, N., Ghafoor, A., Ahmed, N., Azmat, M., Ahmed, R., Ghani, M. I., & Ahanger, M. A. (2025). Nanoparticles as catalysts of agricultural revolution: Enhancing crop tolerance to abiotic stress: A review. *Frontiers in Plant Science, 15*, 1510482. https://doi.org/10.3389/fpls.2024.1510482

Chandrakar, V., Yadu, B., Korram, J., Satnami, M. L., Dubey, A., Kumar, M., & Keshavkant, S. (2020). Carbon dot induces tolerance to arsenic by regulating arsenic uptake, reactive oxygen species detoxification and defense-related gene expression in *Cicer arietinum* L. *Plant Physiology and Biochemistry, 156*, 78–86. https://doi.org/10.1016/j.plaphy.2020.09.003

Chaudhary, M., Singh, P., Singh, G. P., & Rathi, B. (2024). Structural features of carbon dots and their agricultural potential. *ACS Omega, 9*(4), 4166–4185. https://doi.org/10.1021/acsomega.3c04638

Chaudhary, A., Siddiqui, M. W., Gaikwad, K. K., Ahmad, M. S., Homa, F., Singh, D. R., et al. (2025). Novel applications of carbon dots (CDs) for enhanced preservation of postharvest fruit and vegetables. *Food Control, 172*, 111182. https://doi.org/10.1016/j.foodcont.2025.111182

Chen, Q., Ren, X., Li, Y., Liu, B., Wang, X., Tu, J., et al. (2021). Promotion effect of nitrogen-doped functional carbon nanodots on the early growth stage of plants. *Oxford Open Materials Science, 1*(1), itab002. https://doi.org/10.1093/oxfmat/itab002

Chen, L., Zhu, L., Cheng, H., Xu, W., Li, G., Zhang, Y., et al. (2024). Negatively charged carbon dots employed symplastic and apoplastic pathways to enable better plant delivery than positively charged carbon dots. *ACS Nano, 18*(34), 23154–23167. https://doi.org/10.1021/acsnano.4c05362

Cheng, B., Yang, Z., Chen, F., Yue, L., Cao, X., Li, J., Qian, H.-L., Yan, X.-P., Wang, C., & Wang, Z. (2023). Biomass-derived carbon dots with light conversion and nutrient provisioning capabilities facilitate plant photosynthesis. *Science of the Total Environment, 901*, 165973. https://doi.org/10.1016/j.scitotenv.2023.165973

Deepika, Kumar, L., & Gaikwad, K. K. (2023). Carbon dots for food packaging applications. *Sustainable Food Technology, 1*(2), 185–199. https://doi.org/10.1039/D2FB00020B

Frank, B. P., Sigmon, L. R., Deline, A. R., Lankone, R. S., Gallagher, M. J., Zhi, B., Haynes, C. L., & Fairbrother, D. H. (2020). Photochemical transformations of carbon dots in aqueous environments. *Environmental Science & Technology, 54*(7), 4160–4170. https://doi.org/10.1021/acs.est.9b07437

Freixa, A., Acuña, V., Sanchís, J., Farré, M., Barceló, D., & Sabater, S. (2018). Ecotoxicological effects of carbon based nanomaterials in aquatic organisms. *Science of the Total Environment, 619–620*, 328–337. https://doi.org/10.1016/j.scitotenv.2017.11.095

Gong, Y., & Dong, Z. (2021). Transfer, transportation, and accumulation of cerium-doped carbon quantum dots: Promoting growth and development in wheat. *Ecotoxicology and Environmental Safety, 226*, 112852. https://doi.org/10.1016/j.ecoenv.2021.112852

Gong, X., Wang, Z., Zhang, L., Dong, W., Wang, R., Liu, Y., et al. (2022). A novel carbon-nanodots-based theranostic nano-drug delivery system for mitochondria-targeted imaging and glutathione-activated delivering camptothecin. *Colloids and Surfaces B: Biointerfaces, 218*, 112712. https://doi.org/10.1016/j.colsurfb.2022.112712

Guo, B., Liu, G., Li, W., Hu, C., Lei, B., Zhuang, J., Zheng, M., & Liu, Y. (2022). The role of carbon dots in the life cycle of crops. *Industrial Crops and Products, 187*(Part A), 115427. https://doi.org/10.1016/j.indcrop.2022.115427

Hong, J., Wang, C., Wagner, D. C., Gardea-Torresdey, J. L., He, F., & Rico, C. M. (2021). Foliar application of nanoparticles: Mechanisms of absorption, transfer, and multiple impacts. *Environmental Science: Nano, 8*(5), 1196–1210. https://doi.org/10.1039/D0EN01129K

Hu, J., Wenyi, J., Xuefeng, Y., Chuanhao, Y., White, J. C., Liu, J., Shen, G., Tao, S., & Wang, X. (2022). Carbon dots improve the nutritional quality of coriander (*Coriandrum sativum* L.) by promoting photosynthesis and nutrient uptake. *Environmental Science: Nano, 9*(5), 1651–1661. https://doi.org/10.1039/D1EN01079D

Huang, X., Liu, S., Liu, X., Tan, X., Guo, S., Dai, M., Chen, Q., Chen, G., & Feng, C. (2024). Doped carbon dots affect heavy metal speciation in mining soil: Changes of dissimilated iron reduction processes and microbial communities. *Environmental Science: Nano, 11*(4), 1724–1739. https://doi.org/10.1039/D3EN00823A

Jahan, S., Alias, Y. B., Bakar, A. F. B. A., & Yusoff, I. B. (2019). Transport and retention behavior of carbonaceous colloids in natural aqueous medium: Impact of water chemistry. *Chemosphere, 217*, 213–222. https://doi.org/10.1016/j.chemosphere.2018.11.015

Jampílek, J., & Kráľová, K. (2017). Nanomaterials for delivery of nutrients and growth-promoting compounds to plants. In *Nanotechnology: An agricultural paradigm* (pp. 177–226). https://doi.org/10.1007/978-981-10-4573-8_9

Ji, Y., Yue, L., Cao, X., Chen, F., Li, J., Zhang, J., et al. (2023). Carbon dots promoted soybean photosynthesis and amino acid biosynthesis under drought stress: Reactive oxygen species scavenging and nitrogen metabolism. *Science of the Total Environment, 856*, 159125. https://doi.org/10.1016/j.scitotenv.2022.159125

Kamrani, S., Rezaei, M., Kord, M., & Baalousha, M. (2018). Transport and retention of carbon dots (CDs) in saturated and unsaturated porous media: Role of ionic strength, pH, and collector grain size. *Water Research, 133*, 338–347. https://doi.org/10.1016/j.watres.2017.08.045

Khan, N. (2025). Decoding phytohormone signaling in plant stress physiology: Insights, challenges, and future directions. *Environmental and Experimental Botany, 231*, 106099. https://doi.org/10.1016/j.envexpbot.2025.106099

Kou, E., Luo, Z., Ye, J., Chen, X., Lu, D., Landry, M. P., Zhang, H., & Zhang, H. (2025). Sunlight-sensitive carbon dots for plant immunity priming and pathogen defence. *Plant Biotechnology Journal, 2025*, 1–12. https://doi.org/10.1111/pbi.70050

Lei, C., Ding, Z., Tao, M., Lu, Y., Xu, L., Cheng, B., Wang, C., & Wang, Z. (2024). Unraveling the distribution, metabolization, and catabolism of foliar sprayed carbon dots in maize and effect on soil environment. *Journal of Agricultural and Food Chemistry, 72*(36), 19710–19720. https://doi.org/10.1021/acs.jafc.4c06672

Li, W., Zheng, Y., Zhang, H., Liu, Z., Su, W., Chen, S., Liu, Y., Zhuang, J., & Lei, B. (2016). Phytotoxicity, uptake, and translocation of fluorescent carbon dots in mung bean plants. *ACS Applied Materials & Interfaces, 8*(31), 19939–19945. https://doi.org/10.1021/acsami.6b07268

Li, H., Huang, J., Song, Y., Zhang, M., Wang, H., Lu, F., et al. (2018a). Degradable carbon dots with broad-spectrum antibacterial activity. *ACS Applied Materials & Interfaces, 10*(32), 26936–26946. https://doi.org/10.1021/acsami.8b08832

Li, H., Huang, J., Lu, F., Liu, Y., Song, Y., Sun, Y., et al. (2018b). Impacts of carbon dots on rice plants: Boosting the growth and improving the disease resistance. *ACS Applied Bio Materials, 1*, 663–672. https://doi.org/10.1021/acsabm.8b00345

Li, J., Xiao, L., Cheng, Y., Cheng, Y., Wang, Y., Wang, X., & Ding, L. (2019). Applications of carbon quantum dots to alleviate Cd^{2+} phytotoxicity in *Citrus maxima* seedlings. *Chemosphere, 236*, 124385. https://doi.org/10.1016/j.chemosphere.2019.124385

Li, Y., Gao, J., Xu, X., Wu, Y., Zhuang, J., Zhang, X., et al. (2020). Carbon dots as a protective agent alleviating abiotic stress on rice (*Oryza sativa* L.) through promoting nutrition assimilation and the defense system. *ACS Applied Materials & Interfaces, 12*(30), 33575–33585. https://doi.org/10.1021/acsami.0c11724

Li, Y., Li, W., Yang, X., Kang, Y., Zhang, H., Liu, Y., & Lei, B. (2021a). *Salvia miltiorrhiza*-derived carbon dots as scavengers of reactive oxygen species for reducing oxidative damage of plants. *ACS Applied Nano Materials, 4*(1), 113–120. https://doi.org/10.1021/acsanm.0c02419

Li, Y., Pan, X., Xu, X., Wu, Y., Zhuang, J., Zhang, X., Zhang, H., Lei, B., Hu, C., & Liu, Y. (2021b). Carbon dots as light converter for plant photosynthesis: Augmenting light coverage and quantum yield effect. *Journal of Hazardous Materials, 410*, 124534. https://doi.org/10.1016/j.jhazmat.2020.124534

Li, G., Xu, J., & Xu, K. (2023). Physiological functions of carbon dots and their applications in agriculture: A review. *Nanomaterials, 13*(19), 2684. https://doi.org/10.3390/nano13192684

Li, Y., Xu, R., Qi, J., Lei, S., Han, Q., Ma, C., et al. (2024). Transcriptional and metabolic mechanism of carbon dots enhancing rice growth and resistance by promoting root. bioRxiv, 2024–08. https://doi.org/10.1101/2024.08.01.606230.

Lin, S., Reppert, J., Hu, Q., Hudson, J. S., Reid, M. L., Ratnikova, T. A., Rao, A. M., Luo, H., & Ke, P. C. (2009). Uptake, translocation, and transmission of carbon nanomaterials in rice plants. *Small, 5*(10), 1128–1132. https://doi.org/10.1002/smll.200801556

Liu, W., Yao, J., Chai, H., Zhao, Z., Zhang, C., Jin, J., & Choi, M. M. F. (2015). Concentration-dependent effect of photoluminescent carbon dots on the microbial activity of the soil studied by combination methods. *Environmental Toxicology and Pharmacology, 39*(2), 857–863. https://doi.org/10.1016/j.etap.2015.02.009

Liu, J., Lu, S., Tang, Q., Zhang, K., Yu, W., Sun, H., & Yang, B. (2017). One-step hydrothermal synthesis of photoluminescent carbon nanodots with selective antibacterial activity against *Porphyromonas gingivalis*. *Nanoscale, 9*(21), 7135–7142. https://doi.org/10.1039/C7NR02128C

Liu, Y.-Y., Yu, N.-Y., Fang, W.-D., Tan, Q.-G., Ji, R., Yang, L.-Y., Wei, S., Zhang, X.-W., & Miao, A.-J. (2021). Photodegradation of carbon dots cause cytotoxicity. *Nature Communications, 12*(1), 812. https://doi.org/10.1038/s41467-021-21080-z

Liu, Y., Ma, Y., Chen, M., Zhou, T., Ji, R., Guo, R., & Chen, J. (2022). Trophic transfer and environmental safety of carbon dots from microalgae to *Daphnia*. *Science of the Total Environment, 844*, 157201. https://doi.org/10.1016/j.scitotenv.2022.157201

Maholiya, A., Ranjan, P., Khan, R., Murali, S., Nainwal, R. C., Chauhan, P. S., Sathish, N., Chaurasia, J. P., & Srivastava, A. K. (2023). An insight into the role of carbon dots in the agriculture system: A review. *Environmental Science: Nano, 10*, 959–995. https://doi.org/10.1039/D2EN00954D

Mahor, A., Singh, P. P., Bharadwaj, P., Sharma, N., Yadav, S., Rosenholm, J. M., & Bansal, K. K. (2021). Carbon-based nanomaterials for delivery of biologicals and therapeutics: A cutting-edge technology. *C, 7*(1), 19. https://doi.org/10.3390/c7010019

Otchere, E., McKay, B. M., English, M. M., & Aryee, A. N. (2023). Current trends in nano-delivery systems for functional foods: A systematic review. *PeerJ, 11*, e14980. https://doi.org/10.7717/peerj.14980

Ozfidan-Konakci, C., Yildiztugay, E., Arikan-Abdulveli, B., Alp-Turgut, F. N., Baslak, C., & Yıldırım, M. (2024). The characterization of plant derived-carbon dots and its responses on chlorophyll a fluorescence kinetics, radical accumulation in guard cells, cellular redox state and antioxidant system in chromium stressed-*Lactuca sativa. Chemosphere, 356*, 141937. https://doi.org/10.1016/j.chemosphere.2024.141937

Pete, A. M., Ingle, P. U., Raut, R. W., Shende, S. S., Rai, M., Minkina, T. M., Rajput, V. D., Kalinitchenko, V. P., & Gade, A. K. (2023). Biogenic synthesis of fluorescent carbon dots (CDs) and their application in bioimaging of agricultural crops. *Nanomaterials, 13*(1), 209.

Prokisch, J., Törős, G., Nguyen, D. H., Neji, C., Ferroudj, A., Sári, D., et al. (2024a). Nano-food farming: Toward sustainable applications of proteins, mushrooms, nano-nutrients, and nanofibers. *Agronomy, 14*(3), 606. https://doi.org/10.3390/agronomy14030606

Prokisch, J., Nguyen, D. H. H., Muthu, A., Ferroudj, A., Singh, A., Agrawal, S., et al. (2024b). Carbon nanodot–microbe–plant nexus in agroecosystem and antimicrobial applications. *Nanomaterials (Basel), 14*(15), 1249. https://doi.org/10.3390/nano14151249

Rai, M., Ingle, A. P., Törős, G., & Prokisch, J. (2024). Assessing the efficacy of carbon nanodots derived from curcumin on infectious diseases. *Expert Review of Anti-Infective Therapy, 22*(12), 1107–1121. https://doi.org/10.1080/14787210.2024.2409401

Shan, Y., Li, T., Qu, H., Duan, X., Farag, M. A., Xiao, J., Gao, H., & Jiang, Y. (2023). Nano-preservation: An emerging postharvest technology for quality maintenance and shelf life extension of fresh fruit and vegetable. *Food Frontiers, 4*, 100–130. https://doi.org/10.1002/fft2.201

Su, L.-X., Ma, X.-L., Zhao, K.-K., Shen, C.-L., Lou, Q., Yin, D.-M., & Shan, C.-X. (2018). Carbon nanodots for enhancing the stress resistance of peanut plants. *ACS Omega, 3*(12), 17770–17777. https://doi.org/10.1021/acsomega.8b02604

Tan, T. L., Zulkifli, N. A., Zaman, A. S. K., Jusoh, M., Yaapar, M. N., & Rashid, S. A. (2021). Impact of photoluminescent carbon quantum dots on photosynthesis efficiency of rice and corn crops. *Plant Physiology and Biochemistry, 162*, 737–751. https://doi.org/10.1016/j.plaphy.2021.03.031

Tan, J., Zhao, S., Chen, J., Pan, X., Li, C., Liu, Y., et al. (2023). Preparation of nitrogen-doped carbon dots and their enhancement on lettuce yield and quality. *Journal of Materials Chemistry B, 11*(14), 3113–3123. https://doi.org/10.1039/D2TB02817D

Tripathi, S., & Sarkar, S. (2015). Influence of water soluble carbon dots on the growth of wheat plant. *Applied Nanoscience, 5*, 609–616. https://doi.org/10.1007/s13204-014-0355-9

Tripathi, S., & Sarkar, S. (2022). Carbon dots in agricultural system, in book: Carbon dots in agricultural systems, editor(s): Khan, R., Murali, S., Gogoi, S., Academic Press, pp. 175-197. https://doi.org/10.1016/B978-0-323-90260-1.00004-8.

Verma, S. K., Das, A. K., Gantait, S., Kumar, V., & Gurel, E. (2019). Applications of carbon nanomaterials in the plant system: A perspective view on the pros and cons. *Science of the Total Environment, 667*, 485–499. https://doi.org/10.1016/j.scitotenv.2019.02.409

Wang, C., Ji, Y., Cao, X., Yue, L., Chen, F., Li, J., Yang, H., Wang, Z., & Xing, B. (2022). Carbon dots improve nitrogen bioavailability to promote the growth and nutritional quality of soybeans under drought stress. *ACS Nano, 16*(8), 12415–12424. https://doi.org/10.1021/acsnano.2c03591

Xiao, L., Guo, H., Wang, S., Li, J., Wang, Y., & Xing, B. (2019). Carbon dots alleviate the toxicity of cadmium ions (Cd^{2+}) toward wheat seedlings. *Environmental Science: Nano, 6*(5), 1493–1506. https://doi.org/10.1039/C9EN00235A

Yao, K., Lv, X., Zheng, G., Chen, Z., Jiang, Y., Zhu, X., Wang, Z., & Cai, Z. (2018). Effects of carbon quantum dots on aquatic environments: Comparison of toxicity to organisms at different trophic levels. *Environmental Science & Technology, 52*(24), 14445–14451. https://doi.org/10.1021/acs.est.8b04235

Zhao, F., Xin, X., Cao, Y., Su, D., Ji, P., Zhu, Z., & He, Z. (2021). Use of carbon nanoparticles to improve soil fertility, crop growth and nutrient uptake by corn (*Zea mays* L.). *Nanomaterials, 11*(10), 2717. https://doi.org/10.3390/nano11102717

Zhao, Z., Dai, H., Wang, G., Peng, Y., Liao, F., Wu, J., & Liang, T. (2024). Carbon nanoparticles promoted the absorption of potassium ions by tobacco roots via regulation of K flux and ion channel gene expression. *Current Nanoscience, 20*, 390–398. https://doi.org/10.2174/1573413719666230418110534

Zhang, Z., Wu, C., Hu, J., Li, C., Liu, Y., Lei, B., & Zheng, M. (2025). Recent advances of carbon dots: Synthesis, plants applications, prospects, and challenges. *ACS Applied Bio Materials, 8*(2), 935–961. https://doi.org/10.1021/acsabm.4c01785

Zhu, Y., Zhang, Q., Li, Y., Pan, Z., Liu, C., Lin, D., et al. (2023). Role of soil and foliar-applied carbon dots in plant iron biofortification and cadmium mitigation by triggering opposite iron signaling in roots. *Small, 19*(35), 2301137. https://doi.org/10.1002/smll.202301137

Zia, M. U., Sambasivam, P. T., Chen, D., Bhuiyan, S. A., Ford, R., & Li, Q. (2024). A carbon dot toolbox for managing biotic and abiotic stresses in crop production systems. *EcoMat, 6*(5), e12451. https://doi.org/10.1002/eom2.12451

Part IV

Challenges and Toxicity

Challenges and Opportunities in Scaling Up of Carbon Dots

15

Ravi Parashar, Rajashri R. Naik, Ashok K. Shakya,
Maria Letizia Manca, and Preeti K. Suresh

Abstract

Carbon dots (CDs) are multifunctional fluorescent nanomaterials with significant potential in biomedical, energy, and environmental fields, attributed to their biocompatibility, adjustable photoluminescence, and environmentally sustainable manufacturing methods. However, their increasing production encounters several obstacles such as variable synthesis repeatability, elevated purifying costs, energy-demanding processes, and unresolved regulatory and toxicity issues. Progress in continuous flow synthesis, AI-assisted parameter optimization, and waste-derived eco-friendly precursors offers avenues for scalable, economical production. The commercialization of CDs is driven by their applications in bioimaging, drug delivery, catalysis, and pollution sensing, with hybrid systems, such as CD-metal/MOF composites, enhancing their capabilities. Sustainable methods must reconcile high yields with ecological efficiency via life cycle assessments. Interdisciplinary collaboration is essential for standardizing processes, validating safety, and aligning advances with international regulatory frameworks. Resolving synthesis variability, ensuring renewable precursor compatibility, and assessing long-term environmental implications will dictate the

R. Parashar · P. K. Suresh (✉)
University Institute of Pharmacy, Pt. Ravishankar Shukla University, Raipur, Chhattisgarh, India

R. R. Naik
Faculty of Allied Medical Sciences, Al-Ahliyya Amman University, Amman, Jordan

A. K. Shakya
Michael Sayegh Faculty of Pharmacy, Aqaba University of Technology, Aqaba, Jordan

M. L. Manca
Department of Life and Environmental Sciences, University of Cagliari, University Campus, Monserrato, CA, Italy

M. Rai et al. (eds.), *Carbon Nanodots*, https://doi.org/10.1007/978-981-95-4705-0_15

transfer of CDs from laboratory innovations to widely adopted technologies, promoting equitable progress in nanotechnology.

Keywords

Carbon dots · Scaling challenges · Green synthesis · Biomedical applications · Sustainability · Regulatory standards

15.1 Introduction

Carbon dots (CDs) are innovative nanoparticles characterized by biocompatibility, photoluminescence, and versatile surface engineering properties (Mishra et al., 2018). These characteristics render CDs appealing for numerous biological applications, including cellular imaging, biosensing, and drug delivery (Koutsogiannis et al., 2020). CDs can be synthesized through both top-down and bottom-up methodologies, with a recent focus on environmentally friendly synthesis techniques (Aryamol et al., 2024). Their applications include biology, environmental science, and technical domains, including supercapacitors and nanocarriers (Duran et al., 2016). The photoluminescence properties of CDs, spanning from ultraviolet to near-infrared, render them appropriate for photodynamic and photothermal therapy (Aryamol et al., 2024). Despite its potential, CD applications encounter challenges related to enhanced synthesis processes and functionalization methods (Duran et al., 2016).

The diverse prospective uses of carbon nanodots across several sectors render them highly attractive due to their distinctive features (Xiao & Sun, 2018; Adeola et al., 2023). Recently developed green synthesis methods have enhanced efficiency in environmentally friendly microwave and ultrasonic hydrothermal procedures (Usman & Cheng, 2024; Varadharajan et al., 2024). Nonetheless, obstacles to augmenting the scalability of its production, including the improvement of synthesis and purification dependability, property control, and process reproducibility, persist (Adeola et al., 2023; Usman & Cheng, 2024). The development of a circular economy framework, utilizing waste products as precursors, is addressing the issue of economic feasibility (Varadharajan et al., 2024; Usman & Cheng, 2024). Surmounting these diverse obstacles is essential for the future commercial application of CDs. Notwithstanding these challenges, CDs exhibit potential in diverse applications, including bioimaging, drug delivery, sensing, and environmental remediation (Fig. 15.1) (Xiao & Sun, 2018; Varadharajan et al., 2024).

15.2 Fundamentals of Carbon Dots

CDs are a possible class of carbon nanostructures with distinctive features that have evolved since their discovery in 2004 (He et al., 2022). Their synthesis has progressed from numerous approaches to simpler, less costly ones (Karthik et al.,

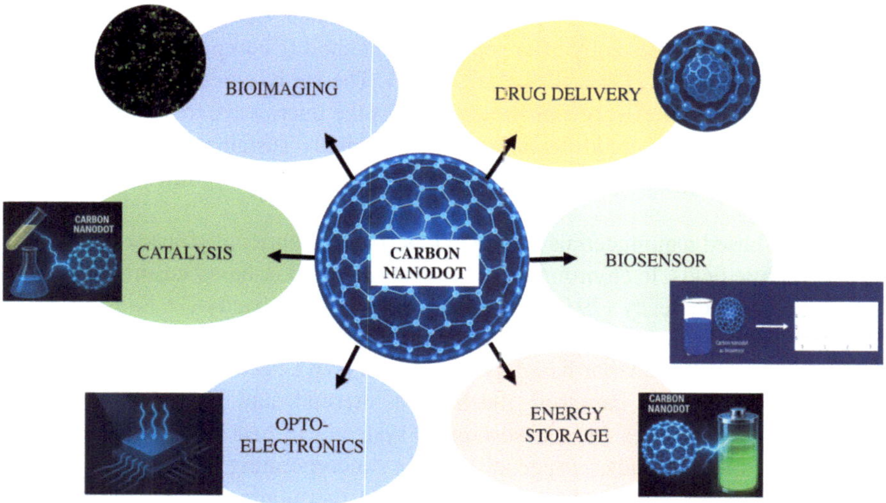

Fig. 15.1 Key properties of carbon dots

2015; He et al., 2022). CDs contain a complex photoluminescence mechanism combining the combination of surface functionalization and core structure (Sciortino et al., 2017). CDs are more biocompatible and less harmful than quantum dots (Karthik et al., 2015). Recent research has tried to understand their basic photophysics better and increase their applications in a number of industries, including medicine, energy, and optics (Liu et al., 2020).

The method employed to produce CDs influences their properties. Hydrothermal, microwave, and calcination methods produce carbon dots with comparable sizes and excitation-dependent blue-to-green luminescence. Nevertheless, calcination and microwave techniques produce superior nitrogen doping and fluorescence quantum yields compared to hydrothermal synthesis (Crista et al., 2020). Generally, CDs possess disklike or quasi-spherical geometries and measure less than 10 nm. They serve multiple functions, including bioimaging and drug delivery, owing to their biocompatibility, photostability, and tunable optical characteristics (Jorns & Pappas, 2021; Vibhute et al., 2022). Various precursors and functionalization agents may be employed to alter the characteristics of CDs (Nallayagari et al., 2022). Microwave synthesis provides superior yields and sustainability, but hydrothermal approaches provide carbon dots with the highest fluorescence quantum yield (Crista et al., 2020; Nallayagari et al., 2022).

A recent study has examined different synthesis methods for CDs, comparing batch-scale and large-scale processes regarding efficiency, yield, and quality. Microwave-assisted heating has demonstrated the capability to generate high-quality carbon dots at a 10-g scale, with quantum yields of up to 73% in less than 4 min (Li et al., 2024a). Solvothermal methods yielded 70.90% product yields and 62.98% quantum yields (Zheng et al., 2020). A kilogram-scale synthesis utilizing aldol condensation produced 1.083 kg of CDs in 2 hours at ambient temperatures (Li et al.,

2021). Nonetheless, a life cycle assessment revealed that high-yield synthesis methods may not consistently offer superior sustainability compared to conventional bottom-up techniques (Fernandes et al., 2022). The study revealed that certain old methods with comparatively high yields, such as the microwave treatment of carbon precursors with urea, could be more environmentally sustainable than high-yield technologies, suggesting that greater yields may entail significant ecological repercussions.

Carbon-based nanomaterials have demonstrated exceptional efficacy across multiple scientific fields. In environmental applications, carbon-based adsorbents exhibit good removal efficiency (79.9–93.7%) and adsorption capacity (76.8–303.6 mg/g) for dyes, with graphene-based materials showing the highest capacity (Azari et al., 2020). Carbon nanomaterial-based nanogenerators have been engineered for energy harvesting, capable of capturing mechanical, thermal, and fluid energies, hence presenting opportunities for self-sustaining systems. In heat transfer applications, carbon-based nanofluids, especially those incorporating carbon nanotubes, graphene, and nanodiamonds, exhibit improved thermophysical properties; however, dispersion stability poses a significant issue (Ali et al., 2021). Carbon nanotube-based surfaces, particularly when modified or coupled with polymers, demonstrate significant antibacterial and anti-adhesive capabilities in medical applications; nevertheless, additional study on their toxicity profile is required (Teixeira-Santos et al., 2020).

15.3 Challenges in Scaling Up Carbon Dots

15.3.1 Synthesis Challenges

The repeatability of carbon nanodot production among laboratories is impeded by many variables. Environmental factors like humidity and barometric pressure can profoundly influence growth results (Oliver et al., 2013). Essential reaction parameters, including seed particle growth duration, are important for attaining reproducible outcomes (Adams et al., 2017). To mitigate these issues, researchers advocate for the use of consistent growth protocols, regulation of environmental variables, and meticulous maintenance of provenance data (Oliver et al., 2013). Enhanced purification methods, such as dialysis or size-exclusion chromatography, can elevate the quality and uniformity of nanomaterials (Đorđević et al., 2019). By implementing these methodologies and emphasizing superior synthesis techniques, researchers may diminish run-to-run variability by over 50% and generate carbon nanodots with enhanced reproducibility across various laboratories (Oliver et al., 2013; Đorđević et al., 2019).

Recent research has underscored difficulties in replicating carbon nanodot production methodologies. Diverse bottom-up techniques, such as hydrothermal, microwave-assisted, and calcination, can yield carbon dots with distinct characteristics, even when utilizing identical precursors. The methods of synthesis influence nitrogen doping efficiency, fluorescence quantum yields, and the origins of photoluminescence (Crista et al., 2020). The selection of precursors and their functionalization affect the features of carbon dots, as evidenced by citric acid-based CDs exhibiting excitation-dependent emission, while the incorporation of urea markedly enhances quantum yields (Nallayagari et al., 2022; Otten et al., 2022). Purification techniques, including dialysis and chromatography, are essential for the isolation of high-quality carbon dots and molecular fluorophores (Đorđević et al., 2019; Otten et al., 2022). The intricate heterogeneity in synthesis methods and precursors has rendered systematic characterization of CD characteristics difficult (Crista et al., 2020). To mitigate reproducibility challenges, researchers underscore the necessity of defined methods and comprehensive characterization techniques (Đorđević et al., 2019).

Enhancing reaction parameters for carbon dot synthesis poses numerous obstacles. Temperature, precursor ratio, and reaction duration markedly affect the size, structure, and characteristics of carbon dots (Ren et al., 2020; Varvařovská et al., 2024). The hydrothermal approach is favored for its simplicity and economic efficiency (Lesani et al., 2021). Reaction time influences crystallinity, luminosity, and formation mechanisms (Papaioannou et al., 2019). Citric acid serves as a prevalent precursor, with synthesis factors influencing luminescence and quantum yields (Ren et al., 2020). Adjusting these settings can yield carbon dots with particular functional groups and dimensions (Varvařovská et al., 2024). The selection of solvent, duration of heating, and amount of dopant are critical factors in regulating the characteristics of carbon dots and their biological interactions (Lesani et al., 2021). Comprehending the synthesis-structure relationship is complex; nonetheless, refining these parameters can produce carbon dots designed for particular applications, including bioimaging (Ren et al., 2020; Lesani et al., 2021).

15.3.2 Economic and Industrial Barriers

CDs are multifunctional nanostructures with potential uses across diverse domains; yet, their manufacture encounters considerable economic and technical obstacles. The elevated expense of raw ingredients and intricate purifying procedures constrain large-scale production and commercialization (He et al., 2022). Purification is essential for obtaining dependable findings and promoting CDs as next-generation optical materials (Ullal et al., 2024). Standard purification techniques encompass dialysis, centrifugation, filtration, solvent extraction, chromatography, and electrophoresis, each possessing distinct advantages and limits (Ullal et al., 2024; Hu et al., 2024). The lack of standardized synthesis and purifying techniques, along with the existence of contaminants, obstructs the comprehension of CD production and optical properties (Adeola et al., 2023; Hu et al., 2024). To address these problems, researchers are investigating economical precursors such as coal-derived materials and formulating environmentally friendly, streamlined production methods (He et al., 2022). Interdisciplinary collaboration is advised to tackle significant problems and promote CD commercialization (Adeola et al., 2023).

Diverse synthesis techniques and precursors have been investigated to generate carbon dots with distinct properties. Pyrolysis and solvothermal synthesis with citric acid and urea as precursors have been employed to produce carbon dots with excitation-dependent emission and elevated quantum yields (Otten et al., 2022). The selection of precursor markedly influences the photocatalytic efficacy of carbon dot-based materials, with citric acid-derived carbon dots exhibiting enhanced pollutant degradation efficiency relative to ammonium citrate-derived carbon dots (Shen et al., 2022). Purification techniques, including precipitation, dialysis, and gel permeation chromatography, are essential for isolating various CD fractions and examining their characteristics (Zhou et al., 2021). Sustainable biomass resources provide a cost-effective and eco-friendly way for large-scale carbon dioxide production, employing diverse synthesis techniques such as hydrothermal treatment, pyrolysis, and microwave-induced treatment.

Coal-derived carbon dots present an economical alternative with distinct characteristics (He et al., 2022). Green synthesis techniques utilizing renewable resources are becoming increasingly popular, while maintaining consistent characteristics poses a difficulty (Chahal et al., 2021). Future research must prioritize the development of sustainable, cost-efficient production techniques and address the knowledge disparity between CDs synthesized from renewable and nonrenewable precursors to enhance their applicability (Chahal et al., 2021; He et al., 2022).

15.3.3 Environmental and Sustainability Issues

Although CDs can be synthesized from biomass and present benefits compared to conventional quantum dots, apprehensions persist over the environmental ramifications of their manufacture. The employment of non-sustainable solvents and energy-intensive methods in CD synthesis presents sustainability concerns (Ren et al., 2024). Recent research has concentrated on evaluating the environmental consequences of various CD synthesis processes through Life Cycle Assessment (LCA) (Fig. 15.2). Fernandes et al. (2022) observed that high-yield synthesis does not inherently lead to reduced environmental impacts when compared to conventional bottom-up methods. Sendão et al. (2020) and Christé et al. (2020) both reported that the microwave-assisted synthesis of carbon dots produced from urea and citric acid is predominantly more environmentally sustainable.

Research indicates that the selection of solvent and synthesis method markedly affects CD characteristics, with water and DMSO being prevalent solvents (Gyulai et al., 2019). The kind of precursor influences the CD size, surface functionality, and optical properties, with citric acid, urea, and different amines commonly employed. Christé et al. (2020) emphasized that the use of ethylenediamine as a nitrogen source in hydrothermal synthesis enhances N-doping efficiency and yield. Omar et al. (2022) examined diverse green and chemical precursors for carbon dot synthesis, highlighting the prospective applications of CDs in environmental monitoring, including the detection of heavy metals, phenols, pesticides, and nitroaromatic explosives. Notwithstanding its potential, additional investigation is required to

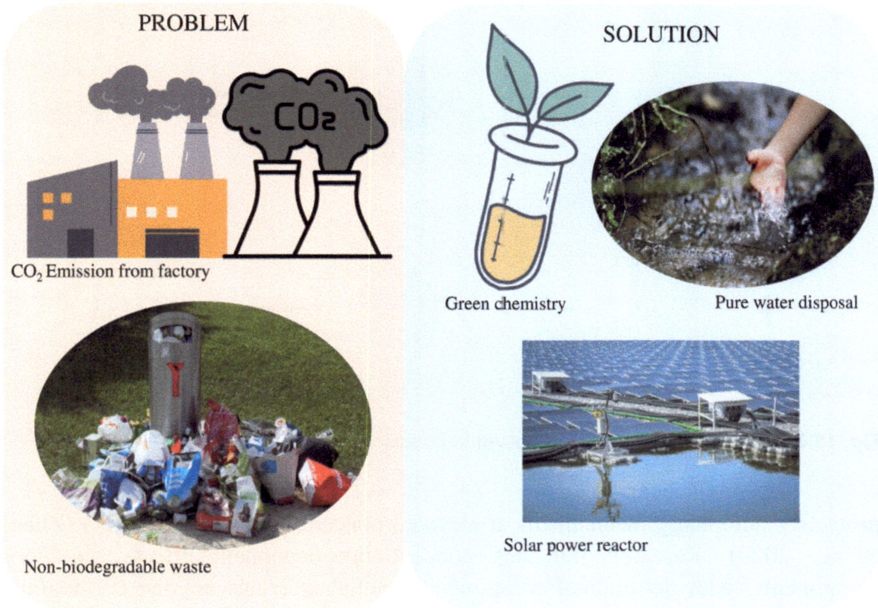

Fig. 15.2 Environmental issues and sustainability considerations in carbon dot production

mitigate the environmental hazards linked to CD synthesis, especially with solvent utilization, energy efficiency, and waste management (Ren et al., 2024).

15.3.4 Safety Concerns

Recent studies have concentrated on creating CDs for multifunctional theranostic systems that integrate imaging and therapeutic functions (Boakye-Yiadom et al., 2019). Nonetheless, obstacles persist in the translation of CD-based technologies from the laboratory to clinical applications (Ostadhossein & Pan, 2017). As research advances, there is an increasing necessity to consider safety and bioregulation aspects to promote the development of CD-based therapeutic and diagnostic materials for practical applications in biomedicine (Singh et al., 2022). Carbon dots exhibit significant potential for biomedical applications; however, their toxicological effects and biocompatibility in mammalian cells are dose-dependent (Fig. 15.3). At low doses ($\leq$0.1 mg/mL), CDs have little detrimental effects on blood components. However, elevated doses may compromise blood cell integrity and functionality, stimulate complement systems, and disrupt coagulation (Li et al., 2015). In immune cells, CDs provoke inflammatory reactions at doses of 250 μg/mL or higher (Lategan et al., 2018). A dosage of around 200 μg/mL is deemed ideal for bioimaging applications in epithelial kidney cells (Emam et al., 2017). In hepatocyte-like cells, CDs exhibit dose- and time-dependent cytotoxicity, inducing

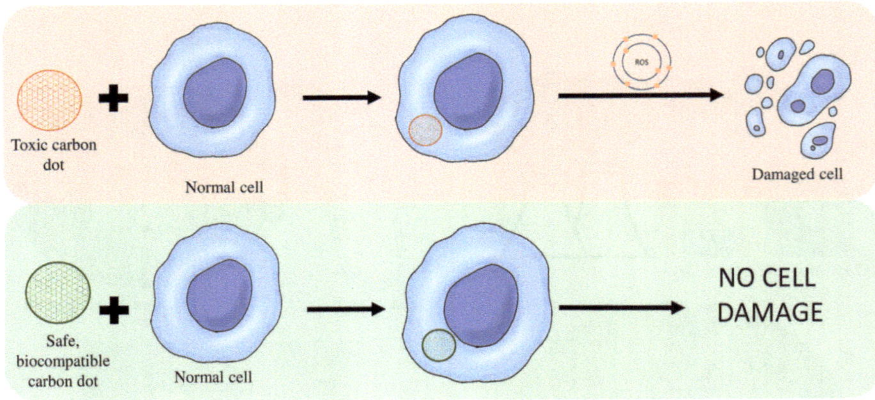

Fig. 15.3 Safety concerns of carbon dots in large-scale applications

protective autophagy mechanisms at elevated concentrations (400 μg/mL) (Zhang et al., 2020). Research indicates concentration-dependent effects on blood components, with detrimental consequences noted at levels beyond 0.1 mg/mL, including red blood cell disintegration and coagulation anomalies (Li et al., 2015). In vivo investigations demonstrated LD50 values ranging from 357 to 391 mg/kg in mice, accompanied by transient hematological alterations following repeated administration (Zheng et al., 2015). The toxicological profile of carbon dots is influenced by aspects such as synthesis process, surface functionalization, and dosage, underscoring the necessity for meticulous evaluation in their clinical development (Mishra et al., 2018; Wang et al., 2022). These findings underscore the necessity of meticulous dose evaluation when employing CDs in biological systems to reconcile their potential advantages with potentially detrimental consequences.

15.4 Opportunities and Future Prospects in Scaling Up Carbon Dots

15.4.1 Advancements in Synthesis Techniques

Carbon dots are intriguing nanomaterials with several applications; nevertheless, the scalability of their synthesis for commercial purposes poses significant challenges. Figure 15.4 presents the comparison of key attributes between batch-scale and large-scale production of carbon dots.

Continuous flow synthesis (CFS) provides substantial benefits compared to conventional batch techniques for the large-scale fabrication of nanomaterials, such as carbon dots. CFS facilitates the exact regulation of reaction parameters, leading to enhanced product quality, reduced size distributions, and increased quantum yields (Campalani & Rigo, 2023). This method exhibits exceptional efficiency, achieving unparalleled space–time yields for materials such as UiO-66

	BATCH SYNTHESIS	LARGE-SCALE PRODUCTION
Cost	HIGH	LOW
Reproducibility	INCONSISTENT	MORE STABLE
Purity	MORE IMPURITIES	HIGHER PURITY
Reaction time	LONG	SHORT
Scalability	LIMITED	HIGH

Fig. 15.4 Comparison of key attributes between batch-scale and large-scale production of carbon dots

and MIL-53(Al) (Taddei et al., 2016). Continuous flow synthesis, especially continuous flow hydrothermal synthesis (CFHS), can generate high-quality nanoparticles at rates of 100–250 kg/h, demonstrating competitive advantages over commercial technologies for environmental effect and economic feasibility (Stieberova et al., 2019). Furthermore, CFS provides reduced reaction times, enhanced repeatability, and a more economical utilization of energy and resources in comparison to conventional hydrothermal batch techniques. These benefits establish CFS as a viable method for the economical, large-scale synthesis of carbon nanodots and other sophisticated materials.

Continuous flow chemistry provides meticulous regulation of reaction parameters, potentially facilitating large-scale manufacturing with enhanced characteristics and efficiency (Campalani & Rigo, 2023). Multiple techniques for scalable carbon dioxide production have been investigated, including hydrothermal, microwave-assisted, and microfluidic methods (Chen et al., 2022). A study revealed the swift and scalable production of highly photoluminescent carbon dots utilizing glycerol as a solvent, attaining a quantum yield of 49.9% (Mai et al., 2020).

Recent progress in the eco-friendly synthesis of carbon dots from biomass and biowaste indicates potential for sustainable nanomaterial production. Diverse techniques, such as hydrothermal, microwave-assisted, and ultrasonic processes, are utilized to manufacture carbon dots with adjustable properties (Jing et al., 2023; Khan et al., 2024). These green-sourced carbon dots demonstrate distinctive optical, electronic, and sensing characteristics, rendering them appropriate for use in bioimaging, biosensing, and renewable energy applications (Sharma et al., 2024; Jing et al., 2023). The synthesis parameters, including precursor selection, reaction conditions, and surface modifications, are critical in influencing the photophysical characteristics of carbon dots (Khan et al., 2024). Recent advancements encompass hybrid methodologies and the application of eco-friendly solvents, facilitating expedited production and improved sustainability (Usman & Cheng, 2024). Nonetheless, obstacles persist in uniformly regulating CD characteristics throughout various synthesis methods, requiring sophisticated characterization techniques and process enhancements (Usman & Cheng, 2024).

15.4.2 Cost-Effective Strategies

The synthesis of CDs from waste materials, especially agricultural and industrial by-products, has garnered interest due to its cost-effectiveness and environmental advantages (Chao Kang et al., 2020). A straightforward hydrothermal technique utilizing waste materials, citric acid, and ethylenediamine has been established to generate carbon dots with fluorescence quantum yields of 40% (Fernandes et al., 2025). Optimal synthesis conditions emphasize the production of high-quality carbon dots with tunable dimensions, surface morphology, and photoluminescent characteristics. Methods to improve CD quality encompass surface passivation, morphological regulation, and heteroatom doping (Khan et al., 2024). Although biomass waste provides an eco-friendly and economical resource for CD synthesis, obstacles persist in enhancing conversion yield and scaling production (de Oliveira & da Silva Abreu, 2021; Khairol Anuar et al., 2021). Subsequent research must tackle these issues to actualize the complete potential of green CD synthesis for sustainable applications.

Machine learning (ML) techniques have demonstrated considerable potential in refining the parameters of carbon dot synthesis to enhance yield and quality. Numerous studies have shown that machine learning-guided methods can improve quantum yields and attain full-color photoluminescence (Han et al., 2020; Chen et al., 2022; Xu et al., 2022a, b; Guo et al., 2024). These machine learning models may efficiently associate synthesis parameters with targeted attributes, minimizing experimental cycles and exceeding conventional trial-and-error approaches (Han et al., 2020; Guo et al., 2024). Critical determinants for superior CD synthesis encompass precursor mass and the volume of alkaline catalyst (Han et al., 2020). Machine learning-guided synthesis has produced carbon dots with quantum yields above 60% across the spectrum and reaching 39.3% for green emission. Furthermore, ML-optimized CDs have demonstrated potential in applications like Fe^{3+} ion detection (Han et al., 2020), vibrant information encryption (Chen et al., 2022), and the monitoring of H_2O_2 in human teeth (Xu et al., 2022a).

Recent research underscores the efficacy of AI-driven optimization in carbon dot production relative to conventional approaches. Artificial intelligence methodologies have exhibited enhanced efficacy and sustainability in the synthesis of nanocrystals, minimizing resource utilization and ecological consequences (Tang et al., 2024). Machine learning methodologies have effectively been utilized to forecast synthesis parameters for red CDs, enhancing efficiency and decreasing expenses (Luo et al., 2022). These AI-driven methodologies provide benefits over traditional trial-and-error techniques, facilitating enhanced regulation of CD characteristics and the investigation of chemical space. AI-driven process optimization in CD synthesis demonstrates potential in improving efficiency, sustainability, and control relative to conventional manual techniques.

15.4.3 Industry Applications Driving Commercialization

Carbon dots are emerging as attractive nanoparticles for biomedical applications owing to their distinctive features, which encompass biocompatibility, photoluminescence, and low toxicity (Mishra et al., 2018; Li et al., 2024b). These nanoparticles, often smaller than 10 nm, provide diverse functionalization and economical synthesis (Xu et al., 2024). CDs exhibit promise across multiple biomedical domains, including bioimaging, biosensing, drug administration, and disease management (Li et al., 2024c; Xu et al., 2024). Their benefits compared to conventional materials encompass water solubility, chemical inertness, and simplicity of modification (Reddy & Gubbiyappa, 2021). CDs are being investigated for use in cellular imaging, targeted drug delivery, and theranostics (Mishra et al., 2018). Notwithstanding significant progress, obstacles persist in completely harnessing the promise of CDs in biomedicine. Current research aims to refine synthesis techniques, augment properties, and broaden applications to tackle prevailing technological issues in the biomedical sector (Reddy & Gubbiyappa, 2021; Xu et al., 2024).

CDs have arisen as potential catalytic materials owing to their distinctive characteristics and many uses. In photocatalysis, carbon dots augment light absorption and facilitate electron–hole pair formation when integrated with materials such as TiO_2 and g-C_3N_4, hence enhancing overall photocatalytic efficacy. CDs serve as electrocatalysts, providing benefits including affordability, enhanced conductivity, and facile modification (Yu et al., 2021). Their catalytic applications encompass selective oxidation, light-driven acid catalysis, and hydrogen bond catalysis (Kang & Liu, 2016). Catalysts derived from carbon dots exhibit promise in diverse electrochemical energy conversion processes, such as water splitting, alcohol oxidation, and oxygen reduction reactions, overcoming the constraints of noble metal-based catalysts (Jana et al., 2020). The many active sites and charge dispersion on CD surfaces boost their catalytic performance. Despite their potential, obstacles persist in transitioning CD-based systems from laboratory settings to clinical applications, requiring additional research to fully unlock their commercial viability in biomedical domains.

15.4.4 Role of Nanotechnology and Advanced Materials

Metal nanoparticle-carbon dot hybrid nanomaterials exhibit significant potential for augmenting photocatalytic and sensing functionalities. These hybrids integrate the distinctive characteristics of CDs, including programmable photoluminescence and excellent biocompatibility, with the benefits of metal nanoparticles (De & Karak, 2017). N-doped carbon dots and metal oxide-carbon dot composites have exhibited enhanced efficacy in chemical and biosensing applications (Sahu et al., 2022). Likewise, CD-metal organic framework (MOF) composites demonstrate augmented photocatalytic activity and sensing proficiency owing to enhanced charge transfer and solar energy harnessing. The amalgamation of metal nanoclusters with carbon nanomaterials produces adaptable hybrids with remarkable luminescent, electrical, and interfacial characteristics, rendering them appropriate for diverse applications such as imaging, sensing, and catalysis (Lopes et al., 2022).

Recent studies have investigated hybrid nanomaterials that integrate metal nanoparticles with carbon dots, emphasizing their synthesis, characteristics, and applications. Diverse integration techniques, such as microwave synthesis, have been employed to produce these hybrids (Arroyave et al., 2020; Fu et al., 2020). The resultant materials demonstrate superior optical and electronic characteristics, including surface plasmon resonance and augmented electrical conductivity, relative to their constituent components (Arroyave et al., 2020; Lopes et al., 2022). These hybrid architectures have demonstrated potential in environmental and energy applications, such as pollutant detection, water splitting, and catalysis (Fu et al., 2020; Dhinakaran et al., 2020). The synergistic interactions between metal nanoparticles and carbon dots enhance their distinctive features and performance. However, obstacles persist in refining synthesis techniques and comprehending the interactions among components (Arroyave et al., 2020; Lopes et al., 2022). Future research avenues encompass investigating green chemistry methodologies and creating more sophisticated hybrid nanomaterials for targeted applications (Lopes et al., 2022; Dhinakaran et al., 2020).

15.4.5 Prospects for Regulation and Standardization

Carbon dots are advantageous fluorescent nanomaterials utilized in sensing, imaging, and biological applications owing to their distinctive characteristics and biocompatibility (Adeola et al., 2023; Varadharajan et al., 2024). Nonetheless, several regulatory obstacles constrain their secure and scalable manufacturing. This encompasses the absence of standardized synthesis and purification techniques, ambiguous mechanisms of action, and procedural deficiencies in applications (Adeola et al., 2023). Challenges related to scalability, stability, and repeatability impede extensive use. In food safety applications, identifying specific targets within complicated samples continues to pose difficulties (Shi et al., 2019). Translational optical imaging has challenges due to limited penetration depth and inadequately regulated in vivo pharmacokinetics. Furthermore, the precise processes of in vivo biodistribution, cellular uptake, and long-term toxicological consequences of CDs require additional clarification (Wang et al., 2015).

The standardization of production protocols is essential for their extensive implementation. Critical determinants of CD efficacy encompass dimensions, concentration, and emission characteristics (Li et al., 2024c). Although high-yield synthesis technologies are available, they are not necessarily more sustainable than conventional techniques, underscoring the necessity for meticulous evaluation of environmental consequences (Fernandes et al., 2022). Essential standardization requirements must emphasize sustainable precursor procurement, efficient solvent and energy utilization, and effective waste management. Moreover, quantitative documentation of synthesis parameters and a clear examination of environmental factors are crucial for assessing scalability and sustainability (Ren et al., 2024).

These factors are essential for creating consistent, sustainable CD production standards across international manufacturing environments.

15.5 Conclusion and Future Prospects

The commercialization of carbon dots depends on addressing ongoing issues related to synthesis repeatability, scalability, and environmental effects. Present batch procedures exhibit variability, whereas novel techniques such as CFS and machine learning-driven optimization provide enhanced yields, consistency, and decreased expenses. Sustainable methods utilizing biomass and industrial waste correspond with circular economy objectives but necessitate enhanced conversion efficiency and defined protocols.

Future research should emphasize interdisciplinary collaboration to connect laboratory innovation with industrial implementation. Progress in hybrid nanomaterials, integrating carbon dots with metals or metal-organic frameworks, may reveal innovative capabilities in catalysis and sensing. Regulatory regimes must adapt to handle safety issues, dose-dependent toxicity, and long-term environmental impacts, ensuring that CDs comply with global standards for biological and environmental applications.

Ultimately, the incorporation of AI for predictive synthesis and life cycle assessments will be essential for reconciling scalability with sustainability. By overcoming these hurdles, CDs can evolve from promising nanomaterials to widely used solutions in healthcare, energy, and environmental remediation, propelling the next generation of innovations powered by nanotechnology.

References

Adams, L. A., & Fagbenro-Owoseni, K. A. (2017). Tunable carbon quantum dots from starch via microwave assisted Carbonization. Int. J. Nanoelectronics Mater, 10, 11–20.

Adeola, A. O., Clermont-Paquette, A., Piekny, A., & Naccache, R. (2023). Advances in the design and use of carbon dots for analytical and biomedical applications. *Nanotechnology, 35*(1), 012001.

Ali, M., Riaz, R., Anjum, A. S., Sun, K. C., Li, H., Ahn, S., Jeong, S. H., & Ko, M. J. (2021). Microwave-assisted ultrafast in-situ growth of N-doped carbon quantum dots on multiwalled carbon nanotubes as an efficient electrocatalyst for photovoltaics. *Journal of Colloid and Interface Science, 586*, 349–361.

Arroyave, M., Springer, V., & Centurión, M. E. (2020). Novel synthesis without separation and purification processes of carbon dots and silver/carbon hybrid nanoparticles. *Journal of Inorganic and Organometallic Polymers and Materials, 30*, 1352–1359.

Aryamol, K., Kanagaraj, K., Nangan, S., Haponiuk, J. T., Okhawilai, M., Pandiaraj, S., . . . Thirumalaivasan, N. (2024). Recent advances of carbon pathways for sustainable environment development. Environmental Research, 250, 118513.

Azari, M., Aslani, A., & Denecke, M. (2020). The effect of the COD: N ratio on mainstream deammonification in an integrated fixed-film activated sludge sequencing batch reactor. *Chemosphere, 259*, 127426.

Boakye-Yiadom, K. O., Kesse, S., Opoku-Damoah, Y., Filli, M. S., Aquib, M., Joelle, M. M. B., et al. (2019). Carbon dots: Applications in bioimaging and theranostics. *International Journal of Pharmaceutics, 564*, 308–317.

Campalani, C., & Rigo, D. (2023). Continuous flow synthesis and applications of carbon dots: A mini-review. *Next Sustainability, 1*, 100001.

Chahal, S., Macairan, J.-R., Yousefi, N., Tufenkji, N., & Naccache, R. (2021). Green synthesis of carbon dots and their applications. *RSC Advances, 11*(41), 25354–25363.

Chen, Z., Liu, Y., & Kang, Z. (2022). Diversity and tailorability of photoelectrochemical properties of carbon dots. *Accounts of Chemical Research, 55*(21), 3110–3124.

Christé, S., Esteves da Silva, J. C., & Pinto da Silva, L. (2020). Evaluation of the environmental impact and efficiency of N-doping strategies in the synthesis of carbon dots. *Materials, 13*(3), 504.

Crista, D. M., Esteves da Silva, J. C., & Pinto da Silva, L. (2020). Evaluation of different bottom-up routes for the fabrication of carbon dots. *Nanomaterials, 10*(7), 1316.

De, B., & Karak, N. (2017). Recent progress in carbon dot–metal based nanohybrids for photochemical and electrochemical applications. *Journal of Materials Chemistry A, 5*(5), 1826–1859.

de Oliveira, B. P., & da Silva Abreu, F. O. M. (2021). Carbon quantum dots synthesis from waste and by-products: Perspectives and challenges. *Materials Letters, 282*, 128764.

Dhinakaran, V., Stalin, B., Ravichandran, M., Balasubramanian, M., Chairma, C. A., & Marichamy, S. (2020). Recent Progress of hybrid architectures based on nanomaterials. *Paper presented at the IOP Conference Series: Materials Science and Engineering.*

Đorđević, L., Arcudi, F., & Prato, M. (2019). Preparation, functionalization and characterization of engineered carbon nanodots. *Nature Protocols, 14*(10), 2931–2953.

Duran, N., Simoes, M. B., de Moraes, A., Favaro, W. J., & Seabra, A. B. (2016). Nanobiotechnology of carbon dots: A review. *Journal of Biomedical Nanotechnology, 12*(7), 1323–1347.

Emam, A., Loutfy, S. A., Mostafa, A. A., Awad, H., & Mohamed, M. B. (2017). Cyto-toxicity, biocompatibility and cellular response of carbon dots–plasmonic based nano-hybrids for bioimaging. *RSC Advances, 7*(38), 23502–23514.

Fernandes, S., Esteves da Silva, J. C., & Pinto da Silva, L. (2022). Life cycle assessment-based comparative study between high-yield and "standard" bottom-up procedures for the fabrication of carbon dots. *Materials, 15*(10), 3446.

Fernandes, S., Algarra, M., Gil, A., Esteves da Silva, J., & Pinto da Silva, L. (2025). Development of a facile and green synthesis strategy for brightly fluorescent carbon dots from various waste materials. *ChemSusChem, 18*(2), e202401702.

Fu, Y., Zeng, G., Lai, C., Huang, D., Qin, L., Yi, H., et al. (2020). Hybrid architectures based on noble metals and carbon-based dots nanomaterials: A review of recent progress in synthesis and applications. *Chemical Engineering Journal, 399*, 125743.

Guo, H., Lu, Y., Lei, Z., Bao, H., Zhang, M., Wang, Z., et al. (2024). Machine learning-guided realization of full-color high-quantum-yield carbon quantum dots. *Nature Communications, 15*(1), 4843.

Gyulai, G., Ouanzi, F., Bertóti, I., Mohai, M., Kolonits, T., & Bősze, S. (2019). Chemical structure and in vitro cellular uptake of luminescent carbon quantum dots prepared by solvothermal and microwave assisted techniques. *Journal of Colloid and Interface Science, 549*, 150–161.

Han, Y., Tang, B., Wang, L., Bao, H., Lu, Y., Guan, C., et al. (2020). Machine-learning-driven synthesis of carbon dots with enhanced quantum yields. *ACS Nano, 14*(11), 14761–14768.

He, Y., Hu, C., Li, Z., Wu, C., Zeng, Y., & Peng, C. (2022). Multifunctional carbon nanomaterials for diagnostic applications in infectious diseases and tumors. *Materials Today Bio, 14*, 100231.

Hu, J., Lei, M., Yan, L., Chen, L., Yang, Y., Zheng, J., et al. (2024). Carbon dot-based afterglow composite with tunable room temperature phosphorescence and thermally activated delayed fluorescence and their anti-counterfeiting and encryption application. *Chemical Engineering Journal, 489*, 151245.

Jana, J., Ngo, Y.-L. T., Chung, J. S., & Hur, S. H. (2020). Contribution of carbon dot nanoparticles in electrocatalysis: Development in energy conversion process. *Journal of Electrochemical Science and Technology, 11*(3), 220–237.

Jing, H. H., Bardakci, F., Akgöl, S., Kusat, K., Adnan, M., Alam, M. J., et al. (2023). Green carbon dots: Synthesis, characterization, properties and biomedical applications. *Journal of Functional Biomaterials, 14*(1), 27.

Jorns, M., & Pappas, D. (2021). A review of fluorescent carbon dots, their synthesis, physical and chemical characteristics, and applications. *Nanomaterials, 11*(6), 1448.

Kang, Z., & Liu, Y. (2016). Catalytic applications of carbon dots. In *Carbon nanoparticles and nanostructures* (pp. 257–298). Springer.

Kang, C., Huang, Y., Yang, H., Yan, X. F., & Chen, Z. P. (2020). A review of carbon dots produced from biomass wastes. *Nanomaterials, 10*(11), 2316.

Karthik, P., Himaja, A., & Singh, S. P. (2015). One-pot synthesis of cost effective carbon dots from bio-degradable waste. *Advanced Science, Engineering and Medicine, 7*(6), 473–479.

Khairol Anuar, N. K., Tan, H. L., Lim, Y. P., So'aib, M. S., & Abu Bakar, N. F. (2021). A review on multifunctional carbon-dots synthesized from biomass waste: Design/fabrication, characterization and applications. *Frontiers in Energy Research 9*, 626549.

Khan, R., Qureshi, A., Azhar, M., Hassan, Z. U., Gul, S., & Ahmad, S. (2024). Recent progress of fluorescent carbon dots and graphene quantum dots for biosensors: Synthesis of solution methods and their medical applications. *Journal of Fluorescence*, 1–18.

Koutsogiannis, P., Thomou, E., Stamatis, H., Gournis, D., & Rudolf, P. (2020). Advances in fluorescent carbon dots for biomedical applications. *Advances in Physics: X, 5*(1), 1758592.

Lategan, K., Fowler, J., Bayati, M., Fidalgo de Cortalezzi, M., & Pool, E. (2018). The effects of carbon dots on immune system biomarkers, using the murine macrophage cell line RAW 264.7 and human whole blood cell cultures. *Nanomaterials, 8*(6), 388.

Lesani, P., Lu, Z., Singh, G., Mursi, M., Mirkhalaf, M., New, E. J., & Zreiqat, H. (2021). Influence of carbon dot synthetic parameters on photophysical and biological properties. *Nanoscale, 13*(25), 11138–11149.

Li, S., Guo, Z., Zhang, Y., Xue, W., & Liu, Z. (2015). Blood compatibility evaluations of fluorescent carbon dots. *ACS Applied Materials & Interfaces, 7*(34), 19153–19162.

Li, L., Li, Y., Ye, Y., Guo, R., Wang, A., Zou, G., et al. (2021). Kilogram-scale synthesis and functionalization of carbon dots for superior electrochemical potassium storage. *ACS Nano, 15*(4), 6872–6885.

Li, J., Lu, C., Yang, S., Xie, Q., Danzeng, Q., Liu, C., & Zhou, C.-H. (2024a). Integrating carbon dots and gold/silver core-shell nanoparticles to achieve sensitive detection of dopamine with fluorometric/colorimetric dual signal. *Analytical and Bioanalytical Chemistry, 416*(22), 4951–4960.

Li, P., Wang, Z., Yu, M., Cheng, C., Chen, J., Xu, Y., et al. (2024b). A facile injectable and self-healing carbon dot/oxidative polysaccharide hydrogel with sustained release capability and potent antibacterial activity. *Journal of Materials Science, 59*(28), 13090–13102.

Li, W., Wang, X., Lin, J., Meng, X., Wang, L., Wang, M., et al. (2024c). Controllable and large-scale synthesis of carbon quantum dots for efficient solid-state optical devices. *Nano Energy, 122*, 109289.

Liu, Y., Huang, H., Cao, W., Mao, B., Liu, Y., & Kang, Z. (2020). Advances in carbon dots: From the perspective of traditional quantum dots. *Materials Chemistry Frontiers, 4*(6), 1586–1613.

Lopes, R. C., Rocha, B. G., Maçôas, E. M., Marques, E. F., & Martinho, J. M. (2022). Combining metal nanoclusters and carbon nanomaterials: Opportunities and challenges in advanced nanohybrids. *Advances in Colloid and Interface Science, 304*, 102667.

Luo, J. B., Chen, J., Liu, H., Huang, C. Z., & Zhou, J. (2022). High-efficiency synthesis of red carbon dots using machine learning. *Chemical Communications, 58*(64), 9014–9017.

Mai, X.-D., Chi, T. T. K., Nguyen, T.-C., & Ta, V.-T. (2020). Scalable synthesis of highly photoluminescence carbon quantum dots. *Materials Letters, 268*, 127595.

Mishra, V., Patil, A., Thakur, S., & Kesharwani, P. (2018). Carbon dots: Emerging theranostic nanoarchitectures. *Drug Discovery Today, 23*(6), 1219–1232.
Nallayagari, A., Sgreccia, E., Pizzoferrato, R., Cabibbo, M., Kaciulis, S., Bolli, E., et al. (2022). Tuneable properties of carbon quantum dots by different synthetic methods. *Journal of Nanostructure in Chemistry*, 1–16.
Oliver, C. R., Polsen, E. S., Meshot, E. R., Tawfick, S., Park, S. J., Bedewy, M., & Hart, A. J. (2013). Statistical analysis of variation in laboratory growth of carbon nanotube forests and recommendations for improved consistency. *ACS Nano, 7*(4), 3565–3580.
Omar, N. A. S., Fen, Y. W., Irmawati, R., Hashim, H. S., Ramdzan, N. S. M., & Fauzi, N. I. M. (2022). A review on carbon dots: Synthesis, characterization and its application in optical sensor for environmental monitoring. *Nanomaterials, 12*(14), 2365.
Ostadhossein, F., & Pan, D. (2017). Functional carbon nanodots for multiscale imaging and therapy. *Wiley Interdisciplinary Reviews: Nanomedicine and Nanobiotechnology, 9*(3), e1436.
Otten, M., Hildebrandt, M., Kühnemuth, R., & Karg, M. (2022). Pyrolysis and solvothermal synthesis for carbon dots: Role of purification and molecular fluorophores. *Langmuir, 38*(19), 6148–6157.
Papaioannou, N., Titirici, M.-M., & Sapelkin, A. (2019). Investigating the effect of reaction time on carbon dot formation, structure, and optical properties. *ACS Omega, 4*(26), 21658–21665.
Reddy, M. R., & Gubbiyappa, K. S. (2021). Multi-functional carbon dots: A systematic overview. *International Journal of Applied Pharmaceutics, 13*(4), 1e22.
Ren, J., Malfatti, L., & Innocenzi, P. (2020). Citric acid derived carbon dots, the challenge of understanding the synthesis-structure relationship. *C, 7*(1), 2.
Ren, J., Opoku, H., Tang, S., Edman, L., & Wang, J. (2024). Carbon dots: A review with focus on sustainability. *Advanced Science, 11*(35), 2405472.
Sahu, Y., Hashmi, A., Patel, R., Singh, A. K., Susan, M. A. B. H., & Carabineiro, S. A. (2022). Potential development of N-doped carbon dots and metal-oxide carbon dot composites for chemical and biosensing. *Nanomaterials, 12*(19), 3434.
Sciortino, L., Messina, F., Buscarino, G., Agnello, S., Cannas, M., & Gelardi, F. M. (2017). Nitrogen-doped carbon dots embedded in a SiO 2 monolith for solid-state fluorescent detection of Cu 2+ ions. *Journal of Nanoparticle Research, 19*, 1–11.
Sendão, R., de Yuso, M. d. V. M., Algarra, M., da Silva, J. C. E., & da Silva, L. P. (2020). Comparative life cycle assessment of bottom-up synthesis routes for carbon dots derived from citric acid and urea. *Journal of Cleaner Production, 254*, 120080.
Sharma, S., Kumar, R., Kumar, K., & Thakur, N. (2024). Sustainable applications of biowaste-derived carbon dots in eco-friendly technological advancements: A review. *Materials Science and Engineering: B, 305*, 117414.
Shen, C.-L., Liu, H.-R., Lou, Q., Wang, F., Liu, K.-K., Dong, L., & Shan, C.-X. (2022). Recent progress of carbon dots in targeted bioimaging and cancer therapy. *Theranostics, 12*(6), 2860.
Shi, X., Wei, W., Fu, Z., Gao, W., Zhang, C., Zhao, Q., et al. (2019). Review on carbon dots in food safety applications. *Talanta, 194*, 809–821.
Singh, S. K., Srinivasan, A., Mitra, S., & Pattader, P. S. G. (2022). Carbon dots and methylene blue facilitated photometric quantification of hemoglobin. *Spectrochimica Acta, Part A: Molecular and Biomolecular Spectroscopy, 271*, 120906.
Stieberova, B., Zilka, M., Ticha, M., Freiberg, F., Caramazana-González, P., McKechnie, J., & Lester, E. (2019). Sustainability assessment of continuous-flow hydrothermal synthesis of nanomaterials in the context of other production technologies. *Journal of Cleaner Production, 241*, 118325.
Taddei, M., Steitz, D. A., van Bokhoven, J. A., & Ranocchiari, M. (2016). Continuous-flow microwave synthesis of metal–organic frameworks: A highly efficient method for large-scale production. *Chemistry–A European Journal, 22*(10), 3245–3249.
Tang, Y., Zhu, P., Xu, Q., & Wang, J. (2024). Machine learning assists the sensor array constructed by the tri-emission carbon dots to detect multiple metal ions. *Microchemical Journal, 201*, 110536.

Teixeira-Santos, R., Gomes, M., & Mergulhão, F. (2020). Carbon nanotube-based antimicrobial and antifouling surfaces. In *Engineered antimicrobial surfaces* (pp. 65–93). Springer.

Ullal, N., Mehta, R., & Sunil, D. (2024). Separation and purification of fluorescent carbon dots–an unmet challenge. *Analyst, 149*(6), 1680–1700.

Usman, M., & Cheng, S. (2024). Recent trends and advancements in green synthesis of biomass-derived carbon dots. *Eng, 5*(3), 2223–2263.

Varadharajan, S., Vasanthan, K. S., Mathur, V., Hariperumal, N., & Mazumder, N. (2024). Green synthesis and multifaceted applications: Challenges and innovations in carbon dot nanocomposites. *Discover Nano, 19*(1), 205.

Varvařovská, L., Bejan, A., Anisiei, A., & Jarošíková, T. Á. (2024). Optimization of parameters for fluorescent carbon nanodots preparation. *Paper presented at the 2024 E-Health and Bioengineering Conference (EHB).*

Vibhute, A., Nille, O., Kolekar, G., Rohiwal, S., Patil, S., Lee, S., & Tiwari, A. P. (2022). Fluorescent carbon quantum dots functionalized by poly l-lysine: Efficient material for antibacterial, bioimaging and antiangiogenesis applications. *Journal of Fluorescence, 32*(5), 1789–1800.

Wang, J., Zhang, P., Huang, C., Liu, G., Leung, K. C.-F., & Wáng, Y. X. J. (2015). High performance photoluminescent carbon dots for in vitro and in vivo bioimaging: Effect of nitrogen doping ratios. *Langmuir, 31*(29), 8063–8073.

Wang, B., Cai, H., Waterhouse, G. I., Qu, X., Yang, B., & Lu, S. (2022). Carbon dots in bioimaging, biosensing and therapeutics: A comprehensive review. *Small Science, 2*(6), 2200012.

Xiao, L., & Sun, H. (2018). Novel properties and applications of carbon nanodots. *Nanoscale Horizons, 3*(6), 565–597.

Xu, Q., Tang, Y., Zhu, P., Zhang, W., Zhang, Y., Solis, O. S., et al. (2022a). Machine learning guided microwave-assisted quantum dot synthesis and an indication of residual H 2 O 2 in human teeth. *Nanoscale, 14*(37), 13771–13778.

Xu, Z., Wang, K., Zhang, M., Wang, T., Du, X., Gao, Z., et al. (2022b). Machine learning assisted dual-emission fluorescence/colorimetric sensor array detection of multiple antibiotics under stepwise prediction strategy. *Sensors and Actuators. B, Chemical, 359*, 131590.

Xu, Y., Zhang, B., Xu, Z., Ye, W., Guo, B., Zhuang, J., et al. (2024). Preparation of carbon dots using aminoquinoline as nitrogen source as full ultraviolet bands absorber and application of LED UV leakage protection. *Dyes and Pigments, 225*, 112060.

Yu, J., Yong, X., Tang, Z., Yang, B., & Lu, S. (2021). Theoretical understanding of structure–property relationships in luminescence of carbon dots. *The Journal of Physical Chemistry Letters, 12*(32), 7671–7687.

Zhang, T., Qu, J., Yao, Y., Zhang, Y., Ma, Y., Wu, D., et al. (2020). N-doped carbon dots triggered the induction of ROS-mediated cytoprotective autophagy in Hepa1-6 cells. *Chemosphere, 251*, 126440.

Zheng, X., Shao, D., Li, J., Song, Y., Chen, Y., Pan, Y., et al. (2015). Single and repeated dose toxicity of citric acid-based carbon dots and a derivative in mice. *RSC Advances, 5*(111), 91398–91406.

Zheng, J., Xie, Y., Wei, Y., Yang, Y., Liu, X., Chen, Y., & Xu, B. (2020). An efficient synthesis and photoelectric properties of green carbon quantum dots with high fluorescent quantum yield. *Nanomaterials, 10*(1), 82.

Zhou, Y., ElMetwally, A. E., Chen, J., Shi, W., Cilingir, E. K., Walters, B., et al. (2021). Gel-like carbon dots: A high-performance future photocatalyst. *Journal of Colloid and Interface Science, 599*, 519–532.

Understanding the Environmental Toxicity of Carbon Nanodots

16

Vandita Kakkar, Parveen Kumari, and Parina Kumari

Abstract

Carbon nanodots (CNDs) are emerging nanomaterials known for their exceptional photoluminescence, high biocompatibility, and adaptable surface chemistry, enabling diverse applications in bioimaging, sensing, and energy systems. As their use grows, assessing their environmental toxicity has become critical. This chapter critically examines current research into CND ecotoxicity, particularly their effects on aquatic and terrestrial organisms. Biological outcomes vary widely based on synthesis methods, surface modifications, concentration, and test conditions. While many CNDs show low acute toxicity, sublethal effects such as oxidative stress, membrane disruption, and altered metabolism have been observed in multiple species. Additionally, environmental behaviors like aggregation, dissolution, and interactions with other pollutants further modulate their toxicity. These factors complicate risk assessment and highlight major knowledge gaps. The chapter emphasizes the urgent need for standardized testing protocols to better evaluate ecological risks and ensure the responsible development of CND-based technologies.

Keywords

Carbon nanodots (CNDs) · Environmental toxicity · Ecotoxicology · Aquatic organisms · Terrestrial organisms · Oxidative stress

V. Kakkar (✉) · P. Kumari
University Institute of Pharmaceutical Sciences, Panjab University, Chandigarh, India

P. Kumari
University Institute of Pharmaceutical Sciences, Panjab University, Chandigarh, India

MM College of Pharmacy, Maharishi Markandeshwar University (MMDU), Mullana, Haryana, India

M. Rai et al. (eds.), *Carbon Nanodots*, https://doi.org/10.1007/978-981-95-4705-0_16

16.1 Introduction

Carbon nanodots (CNDs), discovered in the mid-2000s, have rapidly gained prominence as a novel class of carbon-based nanomaterials due to their unique physicochemical and optical properties. Typically less than 10 nm in size and composed primarily of carbon, oxygen, nitrogen, and hydrogen, CNDs exhibit bright, tunable fluorescence across the visible spectrum—a characteristic that has redefined carbon's role in luminescent applications (Dhamodharan et al., 2022). These nanodots are highly attractive owing to their low-cost and facile synthesis, excellent water solubility, biocompatibility, and chemical stability. Compared to traditional fluorescent semiconductor quantum dots (QDs), CNDs offer several advantages, including lower toxicity, greater resistance to photobleaching, absence of blinking effects, and strong absorption in the UV–blue range. These properties have led to their widespread use in fields such as bioimaging, drug delivery, sensing, optoelectronics, and photocatalysis, as evidenced by a significant surge in research publications in recent years (Salvi et al., 2024).

Despite these promising attributes, growing concerns have emerged about the potential environmental risks associated with the increasing production and application of CNDs. While they are generally considered less toxic than metal-based nanoparticles, their small size, surface functionality, and reactive nature raise important questions about their fate, transport, transformation, and toxicity in natural systems. CNDs can interact with various environmental compartments—including water bodies, soil ecosystems, and the atmosphere—and may affect organisms in unintended ways. Furthermore, variations in synthesis methods and surface modifications can significantly alter their environmental behavior and toxicological profiles, complicating efforts to assess their safety comprehensively. This chapter aims to provide an in-depth analysis of the current understanding of CND environmental toxicity, explore underlying mechanisms of interaction with ecosystems, review available ecotoxicological data, and identify key challenges in risk assessment. Ultimately, it seeks to inform safer design strategies and contribute to the development of appropriate regulatory frameworks to ensure sustainable use of CNDs (Sciortino et al., 2018).

The terms "carbon nanodots" (CNDs) and "carbon quantum dots" (CQDs) are often used interchangeably in the literature, leading to some confusion. While certain researchers distinguish between the two, based on structural characteristics—suggesting that CNDs are quasi-spherical nanoparticles with predominantly amorphous carbon cores, unlike the more crystalline CQDs—this differentiation is largely artificial and not of major scientific consequence (Sameer et al., 2023). Both terms are commonly used to describe spherical carbon dots (CDs), which share similar core features, including small sizes (<10 nm), strong fluorescence, and surface functional groups that confer excellent water solubility and biocompatibility (Li et al., 2021). CDs can be synthesized via two broad methodological approaches: top-down methods, such as laser ablation and chemical oxidation of bulk carbon sources (e.g., graphite, carbon nanotubes), and bottom-up methods, including hydrothermal treatment, pyrolysis, and microwave-assisted synthesis using small organic molecules as precursors (Wang et al., 2019a). Advances in synthesis techniques have enabled the creation of CDs with high quantum yields—reaching up to 80%—and tunable emission ranges from deep UV to near-infrared (NIR), positioning them as compelling alternatives to traditional semiconductor quantum dots (Xiao & Sun, 2018).

16.1.1 Surface Functionalization and Stability

Carbon dots (CDs) often exhibit a spherical or quasi-spherical morphology, though irregular shapes (e.g., rods, cubes, tubes) can form depending on synthesis methods. Spherical CDs are widely preferred for their uniformity, high surface area, and strong photoluminescence. Many possess a core–shell structure, with a carbon-rich core surrounded by functional groups that enhance stability, solubility, and reactivity (Barhoum et al., 2023).

CDs vary in crystallinity—amorphous CDs, formed via bottom-up synthesis, are disordered and chemically diverse, while crystalline CDs, often made using top-down methods, exhibit ordered carbon domains, leading to improved stability, conductivity, and luminescent properties.

Structurally, CDs consist mainly of sp^2-hybridized carbon, sometimes with sp^3 content, and are rich in functional groups such as –OH, –COOH, $-NH_2$, and –C=O (Arcudi et al., 2019). These groups contribute to hydrophilicity, biocompatibility, and tunable optical properties. Surface defects and heteroatom doping can further enhance emission intensity and color control (Otun et al., 2022).

CDs are classified into three main types:

- *Graphene quantum dots (GQDs)*: Few-layer graphene fragments with quantum confinement effects.
- *Carbon nanodots (CNDs)*: Amorphous or graphitic particles with diverse optical properties.
- *Polymer dots (PDs)*: Carbon cores embedded in or surrounded by polymer chains, offering added functionality.

16.1.2 Applications

Due to their outstanding properties including photoluminescence, low toxicity, ease of functionalization, and strong electron transfer capabilities, carbon nanodots are being actively explored in a variety of fields. In biomedicine, they show great promise in bioimaging, drug delivery, and cancer therapy, particularly due to their ability to emit in the NIR region, which facilitates deeper tissue imaging with minimal background noise. Their surface chemistry can be tailored to enhance target specificity and minimize off-target effects, making them strong candidates for precision nanomedicine. Beyond biomedical applications, CNDs are gaining significant attention in biosensing, where their sensitivity to environmental changes enables the detection of ions, biomolecules, and pH variations—even within complex systems like microfluidic devices (Salvi et al., 2024). In the optoelectronics sector, CNDs have been incorporated into light-emitting diodes (LEDs), solar cells, and other energy-related technologies, though further improvements in device efficiency and emission stability are necessary. Additionally, the development of chiral carbon nanodots introduces exciting potential for use in enantioselective catalysis, chiral sensing, and even quantum communication technologies. As research continues to evolve, CNDs are poised to play a transformative role across multiple scientific and industrial domains (Li et al., 2012).

The main applications of CDs can be divided into the following categories:

1. Imaging: Multicolor fluorescent images of mammalian cells, plant cells, and microorganisms, and imaging in mice.
2. Photocatalysis: The degradation of organic molecules, the reduction in CO_2, and water splitting.
3. Optoelectronic devices: LEDs and solar cells (sensitizer/co-sensitizer, transport layer, electrolyte, and/or co-catalyst for counter electrodes).
4. Sensors: Food quality and safety, drug analysis, environmental pollution determination, immunoassay, and other fields, such as detecting heavy metal ions, anions, pesticides, molecules, small organic molecules, and/or nucleic acids.
5. Electrocatalysis: Mainly used in oxidation–reduction reactions, oxygen evolution reactions, hydrogen evolution reactions, and reduction reactions for carbon dioxide, and dual-function catalysts.
6. Biomedicine: Photodynamic therapy, photosensitizers for cancer cell destruction, radiotherapy, the tracing and delivery of drugs or genes, drug release, and anticancer drugs (Chan et al., 2021).

16.1.3 Expanding Applications in Biomedical, Electronic, and Environmental Fields

Carbon quantum dots (CQDs), a subclass of carbon nanodots (CDs), have gained significant attention for their potential applications in medicine and pharmacy due to their low toxicity, biocompatibility, stable chemical properties, and cost-effective synthesis from a wide range of raw materials. Recent studies have demonstrated that CQDs can be synthesized using nontoxic precursors and purified to remove unreactive residues, ensuring their safety for biological applications. For instance, nitrogen-doped, chitosan-based CDs showed a cell viability of 94% after 48 h when tested on human dermal fibroblasts. Similarly, CQDs derived from roasted duck and banana peel exhibited high cell viability in PC12 cells and low toxicity in nematode models, respectively, even at high concentrations. The use of nitrogen doping not only enhances fluorescence quantum yield but also improves the electronic properties of CDs, making them suitable for biosensors and nanogenerators. Furthermore, the adoption of green synthesis methods—using biomass, biowaste, and microorganisms—without toxic chemicals or additives has become a growing trend, aligning with sustainable development goals. These advances affirm the promise of CQDs as safe and efficient nanomaterials for future biomedical and pharmaceutical applications (Chan et al., 2021).

Another interesting application of carbon dots is their utilization as a potential agent for the bioimaging of cells and species under in vivo and in vitro conditions due to their photoluminescence, which is an important property of carbon dots (Banger et al., 2023). These particles have the ability to absorb UV light at the range of 270–320 nm, and can thereby assist in the presence of p–p* and n–p* electronic transition, which further proves the presence of conjugated systems and heteroatoms in carbon dots. In addition, these carbon dots produce interesting optical properties, viz., phosphorescence, up-conversion photoluminescence, solid-state fluorescence, piezo-chromic fluorescence, and excitation-dependent emission (Sameer et al., 2023).

Bioimaging of cells and tissues is a vital part of the diagnosis of various diseases, especially cancer.

16.2 Importance of Understanding Their Environmental Behavior and Safety Profile

(a) *Increasing Application and Environmental Release*
Carbon nanodots (CNDs) have gained significant attention due to their unique photoluminescence, biocompatibility, and ease of synthesis. They are widely used in bioimaging, sensing, drug delivery, and environmental remediation. The expanding production and use of CNDs raise concerns about their unintentional release into ecosystems (water, soil, air), necessitating a comprehensive understanding of their environmental fate and transport (Latiza et al., 2025).

(b) *Potential Ecotoxicity and Bioaccumulation*
Although CNDs are often described as less toxic compared to other nanomaterials (e.g., quantum dots, metal oxides), emerging studies suggest they may still pose risks to aquatic organisms, microbes, and plants. Without understanding how CNDs interact with organisms and ecosystems, there could be unforeseen harmful effects such as oxidative stress, membrane damage, or bioaccumulation that may disrupt food webs (Peng et al., 2020).

(c) *Environmental Transformation and Persistence*
CNDs can undergo physical, chemical, or biological transformations in the environment (e.g., photodegradation, oxidation, interaction with organic matter), altering their toxicity and mobility. Understanding these transformations is crucial to predicting long-term environmental impact and persistence, influencing regulatory guidelines.

(d) *Human Health and Safety Implications*
Since CNDs are used in biomedical and consumer products, there is a possibility of human exposure via inhalation, ingestion, or dermal contact. Evaluating the safety profile—including cytotoxicity, genotoxicity, and inflammatory responses—is essential to ensure safe use and guide risk management.

(e) *Guiding Sustainable Development and Regulation*
A thorough understanding of environmental behavior and safety informs the development of greener synthesis methods, safer-by-design nanomaterials, and effective waste management strategies.
It supports evidence-based policymaking and helps balance technological advancement with environmental and public health protection.

16.3 Need for Toxicological Evaluation

Carbon nanodots (CNDs) have emerged as promising nanomaterials due to their excellent optical properties, biocompatibility, and versatility in applications such as bioimaging, drug delivery, and environmental sensing.

Although CDs are widely regarded as biocompatible, their potential toxicity and environmental impact—especially when synthesized from nonrenewable or chemically intensive precursors—require careful assessment. Factors like surface functionalization, size, and precursor type influence toxicity, with some CDs posing risks to ecosystems and human health (Kamel et al., 2024). Sustainable development calls for the use of renewable sources, such as plant-based or agricultural waste, along with comprehensive toxicity testing (Latiza et al., 2025).

16.3.1 Requirement for Conducting Toxicological Evaluation

(a) *Uncertain safety profile*: Although often considered less toxic than heavy metal-based quantum dots, CNDs' interactions with biological systems at cellular and molecular levels are not fully understood. Toxic effects such as oxidative stress, inflammation, or genotoxicity may arise depending on size, surface chemistry, dose, and exposure route.
(b) *Increasing environmental and human exposure*: As the production and usage of CNDs grow, unintentional release into ecosystems and potential human exposure (via inhalation, ingestion, or dermal contact) are likely to increase, necessitating assessment of risks to both environmental and human health (Sameer et al., 2023).
(c) *Variation in synthesis and surface functionalization*: Different methods of synthesis and surface modifications can drastically alter the behavior, bioavailability, and toxicity of CNDs. Toxicological evaluation is critical to identify safe-by-design strategies and optimize synthesis protocols.
(d) *Regulatory and risk assessment needs*: For CNDs to be safely integrated into commercial products and medical applications, robust toxicological data are essential to inform regulatory guidelines, environmental monitoring, and risk management frameworks (Oualikene-Gonin et al., 2023).
(e) *Long-term and chronic effects*: Current studies often focus on acute toxicity, but long-term exposure effects, bioaccumulation potential, and environmental persistence need thorough investigation. Quantum dots (QDs), particularly

cadmium-based ones, have been shown to accumulate significantly and exhibit toxic effects on a variety of aquatic organisms, including microorganisms, invertebrates, and fish in both freshwater and marine environments. The extent of their ecotoxicity varies greatly depending on several factors such as their physicochemical properties, environmental conditions, concentration, exposure duration, and the specific species involved. Additionally, exposure to UV radiation tends to increase the toxic effects of QDs.
Studies have demonstrated trophic transfer of QDs within reconstructed food chains in both freshwater and seawater ecosystems. However, the biomagnification of QDs appears to be limited, with some detoxification mechanisms reducing their accumulation at higher trophic levels (Rocha et al., 2017).

There is a critical need to develop sensitive analytical methods to detect and quantify QDs in environmental samples, particularly in wastewater and estuarine/marine systems. Moreover, establishing standardized experimental protocols and identifying new biomarkers or endpoints are essential steps toward accurately assessing the environmental risks posed by QDs.

16.4 Sources and Environmental Release of CNDs

Carbon nanodots (CNDs) can originate from both natural and anthropogenic sources. Their presence in the environment is expected to increase due to the growing interest in their synthesis and application across various sectors, including biomedicine, optoelectronics, sensing, catalysis, and environmental remediation.

Natural sources of CNDs include the incomplete combustion of organic matter, wildfires, and volcanic activity, which release carbonaceous nanoparticles into the atmosphere. However, these naturally occurring carbon dots are heterogeneous in nature and differ from engineered CNDs in terms of structure, purity, and functional properties.

Anthropogenic sources are primarily associated with the intentional synthesis of CNDs in laboratories and industries. The widespread development and use of CNDs for commercial and biomedical applications can lead to their unintentional release into the environment through waste streams, improper disposal, or accidental spillage during manufacturing and usage processes. Additionally, end-of-life disposal of CND-containing products and effluents from research laboratories and medical facilities may contribute to environmental loading.

The environmental release pathways of CNDs include air emissions, wastewater discharges, and solid waste deposits. Once released, their small size and colloidal stability enable them to persist and disperse across various environmental compartments, including water bodies, soil, and air. Their interaction with natural organic matter, microorganisms, and other pollutants may alter their physicochemical behavior, affecting their transport, transformation, and potential ecological impact (Varadharajan et al., 2024).

Given the increasing production and utilization of CNDs, there is a pressing need for comprehensive studies into their environmental fate, bioaccumulation potential, and ecotoxicity. Understanding these factors is essential for developing appropriate risk assessment frameworks and ensuring the safe and sustainable use of CNDs across all applications.

By adopting sustainable synthesis practices, waste materials can be transformed into high-value nanomaterials, addressing pressing environmental challenges while simultaneously advancing the frontiers of nanotechnology. This chapter provides a comprehensive overview of the synthesis strategies for CDs, with a particular focus on the formation of versatile nanocomposites incorporating metal and metal oxide components. A substantial portion is dedicated to examining the multifaceted applications of these materials across diverse fields, including electronics, biomedicine, and environmental remediation.

Despite their immense promise, the broader application of CDs is currently constrained by challenges related to scalability, stability, and reproducibility. This chapter also highlights the most critical insights into overcoming these limitations and outlining potential pathways for future innovation. The research emphasizes the importance of sustainable nanotechnology as a solution to global issues, demonstrating how green synthesis can contribute to both technological advancement and environmental preservation. Furthermore, the study underscores the necessity for efficient sewage disposal systems designed to minimize toxin generation, ensuring safer and more responsible integration of nanomaterials into real-world applications.

16.5 Synthesis and Physicochemical Properties of CNDs

Common Methods of Synthesis: There are two main methods for the synthesis of carbon nanodots.

16.5.1 Top-Down Approaches

Top-down synthesis methods (Fig. 16.1) involve breaking down bulk carbon materials into nanoscale carbon dots. One of the earliest discoveries of carbon dots was reported by Xu et al. in 2004, who separated fluorescent nanodots from arc-discharged carbon nanotube soot using electrophoresis (Abid et al., 2022). These nanodots emitted bright light under UV excitation and contained only carbon, hydrogen, nitrogen, and oxygen—without any heavy metals. In 2006, Sun et al. introduced the term "carbon dots" after using laser ablation to irradiate a graphite target in the presence of cement, followed by nitric acid treatment and surface passivation with polymers like PEG1500N. The resulting carbon dots exhibited bright photoluminescence with a quantum yield of up to 10% (Table 16.1), and their emission was strongly dependent on the excitation wavelength (Sameer et al., 2023).

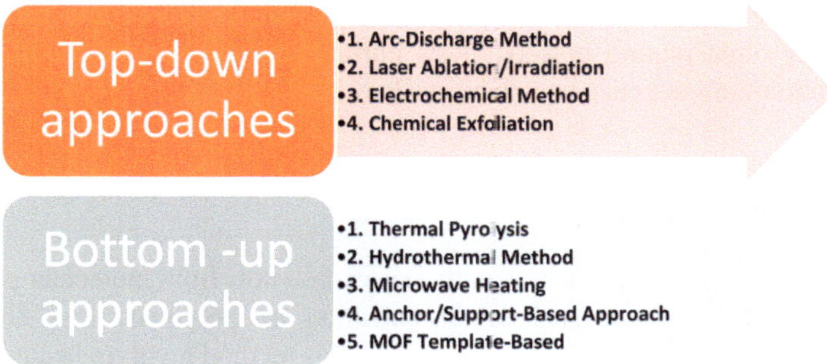

Fig. 16.1 Schematic diagram illustrating the top-down and bottom-up approaches

Table 16.1 Various features of top-down and bottom-up techniques

Sr. no.	Feature	Top-down	Bottom-up
1	Starting material	Bulk carbon (graphite, soot, etc.)	Small organics (citric acid, glucose)
2	Control over size/ morphology	Moderate–low	High
3	Surface passivation	Often post-treatment required	Often in situ during synthesis
4	Ease of method	Varies (e.g., laser ablation = complex)	Generally simpler and scalable
5	Quantum yield	Up to ~10% (initial)	Can exceed 40% + in some cases
6	Eco-friendliness	Moderate	High (esp. with natural sources)
7	Application flexibility	High (depends on treatment)	High (tunable by precursor choice)

Another significant method under the top-down category is the electrochemical approach. Ding's group used multi-walled carbon nanotubes (MWCNTs) as electrodes in an acetonitrile-based electrolyte solution, generating blue-emitting carbon dots after repeated electrochemical cycling (Banger et al., 2023). Similarly, Li et al. demonstrated that the nature of the electrolyte played a crucial role; alkaline environments favored carbon dot formation, while acidic ones did not. These electrochemically synthesized dots typically exhibited diameters under 4 nm with good water solubility (Arcudi et al., 2019).

Chemical exfoliation has also been explored for CD synthesis. Liu and colleagues treated candle soot with nitric acid to produce carbon dots, while Aja yan's group used a sulfuric and nitric acid mixture to exfoliate carbon fibers. These approaches enabled control over carbon dot size and optical properties by modifying reaction conditions, such as temperature and acid concentration. Overall, top-down techniques are effective for converting large carbon structures into carbon dots, though they often require post-synthesis purification and surface modification to enhance photoluminescence and stability (Roy et al., 2015).

16.5.2 Bottom-Up Approaches

In contrast, bottom-up approaches synthesize carbon dots from molecular precursors through chemical reactions such as carbonization and polymerization. One of the most common techniques is thermal pyrolysis. This method involves heating organic molecules, such as citric acid and ethanolamine, to induce dehydration, condensation, and carbonization. Depending on the pyrolysis temperature, the process can yield either molecular fluorophores (at low temperatures) or fully formed carbon dots (at higher temperatures). For example, pyrolysis at 180 °C primarily generates organic fluorophores, while at 230 and 300 °C, well-defined carbon dots form with decreasing quantum yield and increasing carbon content (Wang et al., 2018).

The hydrothermal method is another widely used bottom-up technique. First reported in 2011, it involves heating precursors like glucose and potassium phosphate in a sealed autoclave at high temperatures (e.g., 200 °C). This method allows for simultaneous carbon dot formation and surface passivation, and emission properties can be tuned by adjusting precursor concentrations. Natural resources such as orange juice have also been used in hydrothermal synthesis, making the method eco-friendly and cost-effective. The resulting carbon dots typically have a narrow size distribution and contain hydrophilic functional groups that enhance solubility (Nazibudin et al., 2023).

Microwave-assisted synthesis offers a rapid and energy-efficient route to carbon dots. By heating a mixture of saccharides and poly(ethylene glycol) for a few minutes, carbon dots with size-dependent emission characteristics can be produced. This technique is notable for its short reaction times, low cost, and high reproducibility. Some researchers have achieved high quantum yields, including reports of blue-emitting carbon dots with yields up to 44.9% (as shown in Table 16.1) (Wen et al., 2024).

More advanced bottom-up methods include anchor–/support-based and metal–organic framework (MOF) template-based syntheses. In the former, silica spheres or zeolites serve as substrates that promote uniform carbon dot formation and prevent aggregation during high-temperature pyrolysis. The MOF template approach utilizes the well-defined pore structure of MOFs to confine precursor molecules like glucose, allowing the synthesis of ultrasmall, uniform carbon dots with controlled size and morphology. These innovative strategies offer precise control over the carbon dots' physical and optical properties, highlighting the versatility of bottom-up synthesis routes (Roy et al., 2015).

16.6 Toxicity of CNDs

Carbon nanodots (CNDs) can be synthesized from a wide variety of raw materials, offering advantages such as low production cost, chemical stability, and the use of nontoxic precursors. In recent years, the application of CNDs in the fields of medicine and pharmacy has been extensively explored. Among carbon-based nanomaterials, carbon quantum dots (CQDs) have attracted particular attention due to their promising physicochemical and biological properties (Banger et al., 2023).

In light of their promising applications, early evaluation of the risk and toxicity of CNDs is crucial to prevent potential adverse effects on both human health and the environment (de Medeiros et al., 2019). The synthesis of CNDs typically begins with the careful selection of an appropriate carbon source and the strategic adjustment of dopants, allowing for tuning the material's optical and electronic properties. Concurrently, various biological models are employed to evaluate the toxicity and biocompatibility of the synthesized CNDs. Following exposure to these models, the biological responses are assessed to construct a comprehensive profile of the nanomaterial's potential toxicity and to explore its suitability for real-world applications (Yang et al., 2023).

Additionally, to evaluate their biocompatibility, several studies have investigated the toxicity of metal and polymer-based CNDs in both cellular and in vivo systems. For example, Janus et al. (2019) assessed the cytotoxicity of nitrogen-doped chitosan-derived CDs using human dermal fibroblasts as a model. After 48 h of exposure, cell viability remained at 94%, indicating minimal cytotoxic effects. The CDs were synthesized using nontoxic starting materials, and purification steps effectively removed unreactive residues, contributing to their favorable biocompatibility profile (Janus et al., 2019).

Carbon quantum dots (CQDs) are frequently doped with nitrogen to enhance their fluorescence quantum yield and improve their optical performance (Qu et al., 2014; Zhang & Chen, 2014). Nitrogen-doped CDs—whether derived from carbon nanotubes, graphene, or hollow spheres—exhibit distinct electronic properties. The incorporation of nitrogen facilitates electron donation into the carbon framework, altering electron transport behavior and enabling applications in areas such as sensing and nanogenerators. This has established nitrogen doping as a significant research focus in the field (Hulicova-Jurcakova et al., 2009).

In addition to conventional chemical synthesis routes, environmentally friendly ("green") approaches using natural biomass, biowaste, and microorganisms as carbon sources have gained increasing attention. These methods avoid the use of toxic chemicals and enable the production of biocompatible CQDs with reduced ecological impact (Kurian & Paul, 2021).

Several studies have reported the low toxicity of CQDs in biological systems. For example, Cong et al. investigated the cytotoxicity of CQDs synthesized from roasted duck in PC12 cells. After 36 h of exposure at a concentration of 1 mg/mL, cell viability was recorded at 91.19%, suggesting minimal cytotoxicity. Furthermore, when nematodes were exposed to CNDs at concentrations as high as 15 mg/mL, no mortality was observed after 24 h, further indicating the low acute toxicity of these nanomaterials (Cong et al., 2019).

Similarly, Atchudan et al. evaluated the toxicity of CQDs synthesized from banana peel using *Caenorhabditis elegans* as an animal model. Even at concentrations as high as 200 μg/mL, the CQDs exhibited negligible toxic effects on the nematodes. This biocompatibility was attributed to the use of nontoxic raw materials and the exclusion of passivating agents or chemical additives during synthesis (Atchudan et al., 2021). Collectively, these studies underscore the potential of green-synthesized CDs as low-toxicity nanomaterials suitable for diverse biomedical and environmental applications (Chan et al., 2021).

Chen et al. (2023) evaluated the acute toxicity of functional carbon nanodots (FCNs) and nitrogen-doped FCNs (N-FCNs) using zebrafish (*Danio rerio*) at both embryonic and adult stages. The study demonstrated that exposure to FCNs and N-FCNs induced dose- and time-dependent developmental toxicity, including growth retardation, cardiovascular defects, renal damage, and hepatotoxicity. These effects were primarily associated with oxidative stress and material biodistribution, although enhanced antioxidant responses were also observed. Notably, FCNs and N-FCNs exhibited limited permeability across embryonic barriers and were efficiently excreted via the intestine in adult fish, supporting their overall biosecurity. The LC_{50} values for zebrafish embryos at 96-h postfertilization were reported as 1610 mg/L for FCNs and 649 mg/L for N-FCNs, categorizing both materials as "practically nontoxic" under the Fish and Wildlife Service Acute Toxicity Rating Scale. These findings indicate a relatively low ecological risk for FCN-based nanomaterials in aquatic systems (Chen et al., 2023).

Acute toxicity testing using zebrafish (*Danio rerio*) is widely employed to assess the environmental and ecological hazards of chemicals and nanomaterials, owing to their high genetic similarity to humans, rapid reproductive cycle, small size, and optical transparency during early developmental stages (Xiong et al., 2022).

Huang et al. investigated the cellular toxicity of carbon dots (CDs) by culturing AD-293 cells with varying concentrations of CDs. Interestingly, the treated cells showed over 100% viability and enhanced activity, indicating low toxicity. Further in vitro toxicity assessments were conducted by exposing MCF-7 and HT-29 cells to CDs, with no significant impact on mortality, viability, or proliferation (Huang et al., 2013). For in vivo evaluation, Yang et al. exposed mice to an aqueous solution of CDs. Compared to a control group treated with 0.9% NaCl, the CD-treated group exhibited similar body weight gain and normal food intake, suggesting good biocompatibility (Havrdova et al., 2016). The basic mechanism of carbon nanodots' aquatic toxicity is illustrated in Fig. 16.2.

16.7 Ecotoxicity Studies of CNDs

Ecotoxicity studies focusing on the dispersion of contaminants in aquatic and terrestrial systems are crucial for understanding and mitigating environmental risks. These investigations delve into the fate, transport, and ultimate distribution of pollutants, considering factors such as water flow, soil composition, organismal uptake, and degradation processes. Analyzing the spatial and temporal patterns of

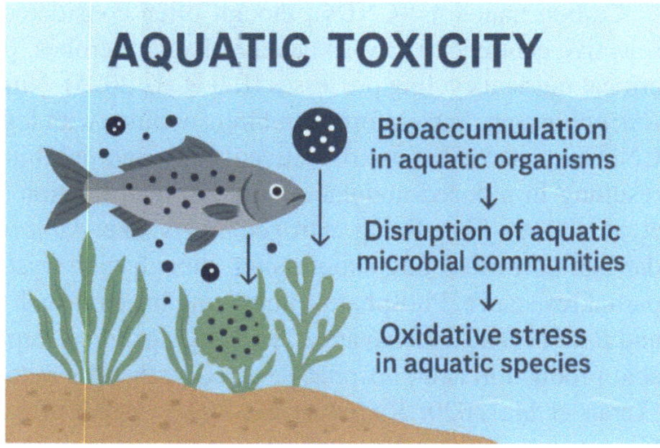

Fig. 16.2 Diagram illustrating aquatic toxicity caused by carbon nanodots

contaminant spread allows for the identification of vulnerable ecosystems and the prediction of potential ecological impacts, ranging from acute toxicity to chronic sublethal effects on individual organisms and community structures. Furthermore, understanding dispersion mechanisms informs the development of effective remediation strategies and the establishment of science-based environmental regulations to protect both aquatic and terrestrial environments from the harmful consequences of pollution.

Carbon nanodots (C-dots) are increasingly prevalent in various applications, raising concerns about their environmental dispersion and ecotoxicity in both aquatic and terrestrial systems. In aquatic environments, studies have demonstrated that C-dots can adversely affect organisms across different trophic levels. For instance, exposure to carbon quantum dots (CQDs) has been shown to inhibit the growth of microalgae such as *Scenedesmus obliquus*, with a 96-h EC_{50} value of 74.8 mg/L, indicating significant sensitivity at the base of the aquatic food web. Similarly, *Chlorella vulgaris* exhibited growth inhibition and oxidative stress responses upon exposure to degradable carbon dots, highlighting potential risks to primary producers. In terrestrial systems, while research is comparatively limited, evidence suggests that C-dots can impact soil microorganisms and plant health. For example, certain biochar-derived carbon dots have been found to be toxic to soil bacteria and affect plant growth parameters, such as root elongation and chlorophyll content. These findings underscore the necessity for comprehensive ecotoxicological assessments of C-dots to inform safe environmental practices.

However, Costa et al. (2021) reported the synthesis of carbon nanodots via hydrothermal carbonization of *Orbignya speciosa* (babassu coconut) mesocarp and assessed their ecotoxicological profile using *Daphnia magna* as a model organism. The study demonstrated that the carbon nanodots exhibited no acute toxic effects, with no immobilization observed even at concentrations as high as 1000 $mg{\cdot}L^{-1}$ following 48 h of exposure. These results suggest a low ecotoxicity potential and highlight the environmental compatibility of carbon nanodots in aquatic systems (Costa et al., 2021).

Carbon nanodots (CNDs), though often considered biocompatible, can exhibit selective ecotoxicity toward beneficial soil microbes, particularly those involved in critical nutrient cycling processes (Rai et al., 2025). Nitrogen-fixing bacteria, such as *Rhizobium* spp., which form symbiotic nodules with legumes, are highly sensitive to CND exposure due to oxidative stress and inhibition of nitrogenase activity, resulting in reduced nodulation and nitrogen fixation (Prokisch et al., 2024; Zhou et al., 2018). Free-living nitrogen fixers like *Azotobacter* also show membrane damage and metabolic suppression, though their cyst-forming ability may confer partial resistance. Phosphate-solubilizing bacteria such as *Pseudomonas fluorescens* and *Bacillus megaterium* are affected by surface adsorption of CNDs, which inhibits acid production and phosphatase activity, thereby reducing phosphorus availability (Taran et al., 2020; Rajput et al., 2021). Arbuscular mycorrhizal fungi, such as *Glomus intraradices*, are particularly vulnerable, with CNDs impeding root colonization and damaging hyphal structures. Nitrifying and denitrifying bacteria (*Nitrosomonas*, *Nitrobacter*, *Paracoccus*) also experience significant declines in activity due to the disruption of membrane-bound enzymes essential for nitrogen transformations. In contrast, decomposer bacteria like *Streptomyces* show moderate resistance, likely due to sporulation and robust cell walls (Wang et al., 2019b; Zhou & Peijnenburg, 2021). Overall, Gram-negative microbes and those with high metabolic or symbiotic activity are more susceptible to CND toxicity, which can result in disrupted nutrient cycles, impaired plant–microbe interactions, and diminished soil fertility (Singh & Singh, 2019).

It is important to recognize that the choice of precursor materials and synthesis methods can significantly influence the physicochemical characteristics of CNDs, which in turn affect their toxicological behavior in biological systems (Sivasankarapillai et al., 2020). In several studies, it was reported that functional carbon nanodots (FCNs) and nitrogen-doped FCNs (N-FCNs) were synthesized through a cost-effective and environmentally friendly electrolysis process, and the resulting nanomaterials exhibited broad biocompatibility and biosafety when tested on plants and soil microbial communities under both normal and stress conditions (Chen et al., 2020a, 2021a, b, 2022b). Given the growing interest in the application of FCNs and N-FCNs in agricultural and environmental systems, it is critical to systematically evaluate their potential nanotoxicity, with particular attention to the role of particle size, morphology, and surface chemistry in determining biological safety.

16.7.1 Mechanisms of Carbon Nanodots Toxicity

16.7.1.1 Interactions with Cells

Both nanoparticles (NPs) exhibited toxic effects on algae, and the toxicity of multiwalled carbon nanotubes (MWCNTs) was significantly stronger than that of CuO NPs, with 96-h median effective concentrations of 33.8 and 169.2 mg/L, respectively. Oxidative stress and cell membrane damage were the main reasons for the toxicity of NPs to algae, and they were concentration-dependent and the existence of CuO NPs in some groups reduced cell membrane damage caused by MWCNTs, which may be because CuO NPs formed hetero aggregation with MWCNTs, reducing the contact of nanoparticles with cell membranes, then reducing physical damage. Scanning electron microscopy (SEM) and transmission electron microscopy (TEM) results indicated cell damage, the hetero aggregation of MWCNTs-CuO NPs, and obvious nanoparticles internalization. In some groups, the presence of CuO NPs significantly reduced the reactive oxygen species (ROS) level induced by MWCNTs. However, at the highest exposure concentration, ROS generation was markedly higher than in either nanoparticlealone treatment group, which may be attributed to high MWCNT levels facilitating enhanced internalization of CuO NPs, which might be related to the high concentration of MWCNTs promoting the internalization of CuO NPs. MWCNTs and CuO NPs affected and interacted with each other, causing more complex toxic effects on aquatic organisms (Fang et al., 2022; Hong et al., 2022).

16.7.1.2 Reactive Oxygen Species (ROS) Generation

CNDs can induce ROS production, leading to oxidative stress in cells and potentially causing damage. This can happen through several pathways, including the generation of superoxide anions and singlet oxygen (Innocenzi & Stagi, 2023). For instance, Havrdova et al. (2016) reported that oxidative stress was evaluated alongside cell viability using the MTT assay after 24 h of incubation with carbon dots (CDs). A slight increase in ROS production was observed across all samples with rising CD concentrations, but a pronounced elevation occurred between 50 and 200 μg mL^{-1}. This concentration range also corresponded with a sharp decline in cell viability, suggesting that oxidative stress plays a key role in mediating CD-induced cytotoxicity from doses as low as 50 μg mL^{-1}. They reported that ROS levels declined at concentrations above 200 μg mL^{-1}, likely due to reduced viable cell numbers near the IC50 threshold, limiting further radical generation. Overall, the correlation between ROS production and reduced cell viability reinforces oxidative stress as a central mechanism of CD-induced toxicity (Havrdova et al., 2016).

16.7.1.3 Cell Damage and Apoptosis

The excess ROS generated by CNDs can disrupt cellular processes, leading to cell damage and even programmed cell death (apoptosis). Chan et al. (2021) stated that the oxidative activity of carbon dots (CDs) primarily involves photo-induced cellular damage and the disruption of biofilm matrices through the generation of reactive oxygen species (ROS), ultimately triggering cell apoptosis. Recent studies have demonstrated that CDs synthesized from organic carbon sources can selectively target and eliminate cancer cells with efficacy comparable to that of conventional chemotherapeutics, while sparing healthy cells. Notably, CDs derived from natural precursors through thermal processing tend to retain functional characteristics reflective of their source materials, which may influence their biological activity and therapeutic potential (Chan et al., 2021).

16.7.1.4 Biofilm Matrix Destruction

Some CNDs have been shown to disrupt the protective outer layer of bacteria, known as the biofilm matrix. This makes the bacteria more susceptible to other environmental stressors and can impact their survival (Chan et al., 2021).

Carbon nanodots can invade and break apart the extracellular biofilm matrix of beneficial bacteria, causing biofilm breakage, and can have an ecotoxicological impact. When establishing this mechanism, it was reported that carbon nanodots, because of their ultrasmall size and high membrane permeability, can enter biofilm matrices by either displacement or damaging key cellular components and inhibiting active pumping mechanisms—ultimately destabilizing biofilms. Another mechanism suggests that CNDs may induce ecotoxicological effects through direct interaction with microbial cells, further contributing to biofilm disruption (Ghirardello et al., 2021).

Another study also underscored that carbon-dot nanoparticles can invade biofilms in both Gram-positive and Gram-negative bacteria, effectively destroying them or preventing establishment by disorganizing the protective matrix. These studies provide in-depth explanations on the mechanistic ways carbon-dots compromise biofilm integrity and raise possibilities of ecological toxicity due to impacts on beneficial microbial communities (Kumari et al., 2024; Yu et al., 2023).

16.7.1.5 Toxicity Risk at High Concentrations

While CNDs are often considered biocompatible, studies have shown that high concentrations of CNDs can still have toxic effects, particularly in aquatic organisms. For instance, Qiang et al. (2022) investigated the cytotoxicity and underlying mechanisms of carbon quantum dots (CQDs) on *Escherichia coli* in vitro. The synthesized CQDs were ultrasmall (~2.08 nm) and exhibited strong fluorescence. While they showed good short-term biocompatibility, prolonged exposure beyond 24 h resulted in marked toxicity. This was evidenced by elevated levels of reactive oxygen species (ROS), lactate dehydrogenase release, and increased crystal violet uptake. Microscopy analyses and physicochemical assays revealed that CQDs disrupted cell membrane integrity by altering surface charge and inducing lipid peroxidation, thereby increasing cell wall permeability. At higher concentrations (200 μg/mL), CQDs also reduced extracellular osmotic pressure, contributing to cell edema and death. Additionally, comet assays confirmed CQD-induced DNA damage, indicating their potential to inhibit bacterial proliferation (Qiang et al., 2022).

Early studies reported that carbon quantum dots (CQDs) exhibited high biocompatibility, with no adverse effects observed in zebrafish embryos even at concentrations up to 5.0 mg/L. However, more recent findings have revealed contrasting outcomes in aquatic microorganisms. CQDs were shown to inhibit growth and induce oxidative stress in *Chlorella pyrenoidosa*, while treatments at 100 mg/L significantly reduced cell proliferation and chlorophyll-a content in *Staphylococcus aureus* and *Microcystis aeruginosa* (Yao et al., 2018).

Furthermore, another study also showed that repeated administration of quantum dots (2.5 mg/kg for 14 days) has been shown to cause irreversible damage to multiple organs, including the lungs, liver, spleen, kidneys, and thyroid, primarily through immunological mechanisms. While a single dose may exhibit low acute toxicity, cumulative exposure leads to delayed and permanent tissue injury, challenging earlier assumptions about their safety. The study also highlights that standard diagnostic and treatment protocols for heavy metal poisoning may be ineffective against quantum dot toxicity, underscoring the complexity and unpredictability of their biological effects (Kuznetsov et al., 2023).

16.7.1.6 Impact on Microorganisms

The ecotoxicity of CNDs extends to a wide range of microorganisms, potentially affecting their growth, metabolism, and overall population dynamics in the environment. For example, Yao et al. (2018) evaluated the ecological toxicity of carbon quantum dots (CQDs) in aquatic organisms, including zebrafish (*Danio rerio*), zooplankton (*Daphnia magna*), and phytoplankton (*Scenedesmus obliquus*). While *D. rerio* exhibited negligible toxicity and effectively depurated CQDs at concentrations up to 200 mg/L, *D. magna* showed moderate sensitivity with 48-h EC_{50} and LC_{50} values of 97.5 and 160.3 mg/L, respectively. *S. obliquus* was the most sensitive, with a 96-h EC_{50} of 74.8 mg/L, demonstrating dose- and time-dependent inhibition of photosynthesis and nutrient uptake. Mechanistic studies revealed CQD-induced oxidative stress and water acidification as key contributors to algal toxicity (Liu et al., 2021, 2022).

16.7.1.7 Surface Properties and Size

The surface properties, such as charge and functional groups, and the size of CNDs play a crucial role in their toxicity. Smaller CNDs tend to be more toxic due to their larger surface area-to-volume ratio, which allows for greater interaction with cells and potentially more efficient internalization. Surface modifications which involve interaction with cells and their susceptibility to degradation can also alter the toxicity of CDs (Chen et al., 2024). The combined mechanism of carbon nanodots is illustrated in Fig. 16.3.

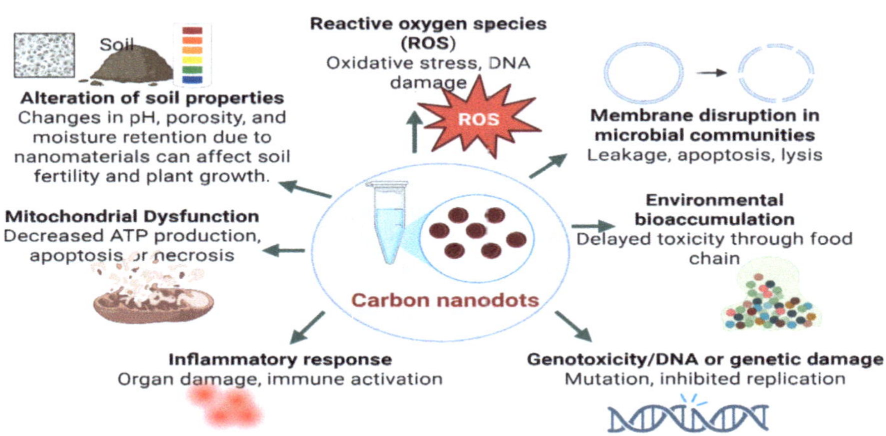

Fig. 16.3 Schematic diagram illustrating the mechanism of carbon nanodots in ecotoxicity

16.8 Regulatory Considerations and Safety Assessment

We outlined a future roadmap for the safe and sustainable adoption of carbon dots (CDs), structured around five critical steps. These include addressing variability in synthesis, enhancing scalability, and ensuring compliance with regulatory standards—each essential for advancing the commercialization of CDs. A key aspect of this roadmap is the integration of life cycle assessment (LCA) to evaluate the environmental and human health impacts of green synthesis approaches. The safe and sustainable by design (SSbD) framework plays a central role, influencing not only methodological development but also guiding the safe and responsible adoption of materials and processes throughout the entire lifecycle from design and synthesis to commercialization and end-of-life management. This approach supports the minimization of adverse environmental and human health effects, in line with the principles of green chemistry and sustainability. By adhering to this comprehensive framework, the development of green-synthesized CDs can progress toward a more sustainable and industrially viable future, contributing meaningfully to the broader implementation of green chemistry principles in the field of nanotechnology (Dhumal et al., 2024).

16.9 Strategies and Sustainable Alternatives

16.9.1 Designing Safer CNDs (Safe-by-Design Approach)

The safe-by-design (SbD) approach involves incorporating safety considerations at the earliest stages of nanomaterial development. For CNDs, this includes the selection of benign precursors, nontoxic dopants, and surface passivating agents that minimize hazardous effects on human health and the environment (Dhumal et al., 2024). Rational design strategies can also help control size, surface charge, and functional groups to reduce undesired biological interactions and improve predictability in biological systems. SbD further promotes the creation of materials with reduced persistence, lower accumulation potential, and controlled degradation profiles.

16.9.2 Biodegradable or Eco-Friendly CNDs

Developing biodegradable or environmentally benign CNDs is a vital step toward reducing long-term environmental impact. Utilizing renewable and nontoxic natural precursors, such as plant extracts, carbohydrates, or amino acids, has shown promise in synthesizing green CNDs. These eco-friendly CNDs exhibit desirable photoluminescent properties while offering improved biocompatibility and a minimal ecological footprint. Encouragingly, such green synthesis methods also contribute to circular economy models by utilizing biomass waste, thus promoting sustainable nanomaterial production (Sameer et al., 2023).

16.9.3 Waste Management and Treatment Technologies

Effective waste management and treatment technologies are essential to prevent the uncontrolled release of CNDs into the environment. This includes developing protocols for the safe disposal of CND-containing products, implementing nanowaste segregation in research and clinical settings, and applying advanced treatment techniques such as membrane filtration, adsorption, and advanced oxidation processes to remove CNDs from industrial effluents. Additionally, regulatory guidelines and risk assessment tools must evolve in parallel to address the unique challenges posed by nanomaterials, ensuring responsible handling and lifecycle management of CNDs (Atchudan et al., 2021).

16.10 Conclusions and Future Perspectives

Carbon nanodots (CNDs) have emerged as promising nanomaterials in the field of biomedicine due to their unique optical properties, excellent biocompatibility, low toxicity, and versatile surface functionality. Despite the considerable progress in their synthesis and application, several challenges and opportunities remain to be addressed to fully realize their clinical and technological potential (Sameer et al., 2023).

In the future, standardization of protocols for the purpose of synthesis will be crucial for achieving uniform size, surface chemistry, and photoluminescent properties of CNDs. This consistency is essential for ensuring reproducibility in biological systems and advancing regulatory approvals for clinical use. Furthermore, the development of green and sustainable synthesis methods using biomass and natural precursors is expected to enhance the environmental friendliness and cost-effectiveness of CND production (Nair et al., 2020).

Another important area of future research is the detailed investigation of long-term toxicity, biodistribution, and clearance mechanisms of CNDs in vivo. While current studies support their low toxicity profile, comprehensive toxicological data are required to ensure their safety for prolonged biomedical use, particularly in theranostic applications.

The integration of CNDs with advanced technologies, such as 3D-printed scaffolds, wearable biosensors, and implantable devices, holds great potential for next-generation diagnostics and targeted therapeutics. Combining CNDs with smartphone-based sensing systems may enable rapid, point-of-care diagnostics, especially in resource-limited settings (Pirsaheb et al., 2019).

In addition, the modification of CNDs with specific ligands, antibodies, or aptamers will improve their targeting capabilities, allowing for personalized medicine applications in cancer therapy, neurological disorders, and infectious diseases. Tailoring surface functionalization may also help optimize their interaction with cellular environments, enhancing uptake and therapeutic efficacy.

As interdisciplinary research continues to expand, collaborations between materials scientists, chemists, biomedical engineers, and clinicians will be vital for translating CND-based technologies from bench to bedside. With focused efforts on regulatory compliance, scalable production, and mechanistic understanding, carbon nanodots are poised to become a key component of future nanomedicine platforms, potentially transforming disease diagnosis, drug delivery, and regenerative medicine.

References

Abid, N., Khan, A. M., Shujait, S., Chaudhary, K., Ikram, M., Imran, M., Haider, J., Khan, M., Khan, Q., & Maqbool, M. (2022). Synthesis of nanomaterials using various top-down and bottom-up approaches, influencing factors, advantages, and disadvantages: A review. *Advances in Colloid and Interface Science, 300*, 102597.

Arcudi, F., Đorđević, L., & Prato, M. (2019). Design, synthesis, and functionalization strategies of tailored carbon nanodots. *Accounts of Chemical Research, 52*(8), 2070–2079.

Atchudan, R., Edison, T. N. J. I., Shanmugam, M., Perumal, S., Somanathan, T., & Lee, Y. R. (2021). Sustainable synthesis of carbon quantum dots from banana peel waste using hydrothermal process for in vivo bioimaging. *Physica E, Low-Dimensional Systems and Nanostructures, 126*, 114417.

Banger, A., Gautam, S., Jadoun, S., Jangid, N. K., Srivastava, A., Pulidindi, I. N., et al. (2023). Synthetic methods and applications of carbon nanodots. *Catalysts, 13*(5), 858.

Barhoum, A., Meftahi, A., Kashef Sabery, M. S., Momeni Heravi, M. E., & Alem, F. (2023). A review on carbon dots as innovative materials for advancing biomedical applications: Synthesis, opportunities, and challenges. *Journal of Materials Science, 58*(34), 13531–13579.

Chan, M.-H., Chen, B.-G., Ngo, L. T., Huang, W.-T., Li, C.-H., Liu, R.-S., & Hsiao, M. (2021). Natural carbon nanodots: Toxicity assessment and theranostic biological application. *Pharmaceutics, 13*(11), 1874.

Chen, Q., Chen, L., Nie, X. K., Man, H., Guo, Z. J., Wang, X. L., Tu, J. P., Jin, G., & Ci, L. J. (2020). Impacts of surface chemistry of functional carbon nanodots on the plant growth. *Ecotoxicology and Environmental Safety, 206*, 111220.

Chen, Q., Ren, X., Li, Y., Liu, B., Wang, X., Tu, J., Guo, Z., Jin, G., Min, G., & Ci, L. (2021a). Promotion effect of nitrogen-doped functional carbon nanodots on the early growth stage of plants. *Oxford Open Materials Science, 1*(1), 01–20.

Chen, Q., Cao, X., Liu, B., Nie, X., Liang, T., Suhr, J., & Ci, L. (2021b). Effects of functional carbon nanodots on water hyacinth response to Cd/Pb stress: Implication for phytoremediation. *Journal of Environmental Management, 299*, 113624.

Chen Cao, X., Li, Y., Sun, Q., Dai, L., Li, J., Guo, Z., Zhang, L., & Ci, L. (2022). Functional carbon nanodots improve soil quality and tomato tolerance in saline-alkali soils. *Science of the Total Environment, 830*, 154817.

Chen, Q., Cao, X., Yan, B., Guo, Z., Xi, Z., Li, J., et al. (2023). Ecotoxicological evaluation of functional carbon nanodots using zebrafish (Danio rerio) model at different developmental stages. *Chemosphere, 333*, 138970.

Chen, H., Luo, K., Xie, C., & Zhou, L. (2024). Nanotechnology of carbon dots with their hybrids for biomedical applications: A review. *Chemical Engineering Journal, 496*, 153915.

Cong, S., Liu, K., Qiao, F., Song, Y., & Tan, M. (2019). Biocompatible fluorescent carbon dots derived from roast duck for in vitro cellular and in vivo C. elegans bio-imaging. *Methods, 168*, 76–83.

Costa, R. S., de Castro, M. O., da Silva, G. H., Delite, F. d. S., Strauss, M., Ferreira, O. P., et al. (2021). Carbon-dots from babassu coconut (Orbignya speciosa) biomass: Synthesis, characterization, and toxicity to Daphnia magna. *Carbon Trends, 5*, 100133.

de Medeiros, T. V., Manioudakis, J., Noun, F., Macairan, J.-R., Victoria, F., & Naccache, R. (2019). Microwave-assisted synthesis of carbon dots and their applications. *Journal of Materials Chemistry C, 7*(24), 7175–7195.

Dhamodharan, D., Byun, H. S., Shree, M. V., Veeman, D., Natrayan, L., & Stalin, B. (2022). Carbon nanodots: Synthesis, mechanisms for bio-electrical applications. *Journal of Industrial and Engineering Chemistry, 110*, 68–83.

Dhumal, P., Chakraborty, S., Ibrahim, B., Kaur, M., & Valsami-Jones, E. (2024). Green-synthesised carbon nanodots: A SWOT analysis for their safe and sustainable innovation. *Journal of Cleaner Production, 480*, 144115.

Fang, R., Gong, J., Cao, W., Chen, Z., Huang, D., Ye, J., & Cai, Z. (2022). The combined toxicity and mechanism of multi-walled carbon nanotubes and nano copper oxide toward freshwater algae: Tetradesmus obliquus. *Journal of Environmental Sciences, 112*, 376–387.

Ghirardello, M., Ramos-Soriano, J., & Galan, M. C. (2021). Carbon dots as an emergent class of antimicrobial agents. *Nanomaterials, 11*(8), 1877.

Havrdova, M., Hola, K., Skopalik, J., Tomankova, K., Petr, M., Cepe, K., et al. (2016). Toxicity of carbon dots–Effect of surface functionalization on the cell viability, reactive oxygen species generation and cell cycle. *Carbon, 99*, 238–248.

Hong, M., Gong, J.-L., Cao, W.-C., Fang, R., Cai, Z., Ye, J., et al. (2022). The combined toxicity and mechanism of multi-walled carbon nanotubes and nano zinc oxide toward the cabbage. *Environmental Science and Pollution Research, 29*, 3540–3554.

Huang, X., Zhang, F., Zhu, L., Choi, K. Y., Guo, N., Guo, J., et al. (2013). Effect of injection routes on the biodistribution, clearance, and tumor uptake of carbon dots. *ACS Nano, 7*(7), 5684–5693.

Hulicova-Jurcakova, D., Kodama, M., Shiraishi, S., Hatori, H., Zhu, Z. H., & Lu, G. Q. (2009). Nitrogen-enriched nonporous carbon electrodes with extraordinary supercapacitance. *Advanced Functional Materials, 19*(11), 1800–1809.

Innocenzi, P., & Stagi, L. (2023). Carbon dots as oxidant-antioxidant nanomaterials, understanding the structure-properties relationship. A critical review. *Nano Today, 50*, 101837.

Janus, Ł., Piątkowski, M., Radwan-Pragłowska, J., Bogdał, D., & Matysek, D. (2019). Chitosan-based carbon quantum dots for biomedical applications: Synthesis and characterization. *Nanomaterials, 9*(2), 274.

Kamel, G. N., El-Shaheny, R., Shabana, R. A., & Hassan, A. H. (2024). Inherent photocatalytic activity of luminescent multi-doped carbon dots manufactured from expired medicine and its application for efficient water remediation and nanosensing. *Microchemical Journal, 201*, 110576.

Kumari, P., Beg, S., Singh, K. K., & Kakkar, V. (2024). Applicability of QbD-assisted analytical method for simultaneous detection of tetrahydrocurcumin and folic acid in developed nanostructured lipid carriers. *Current Analytical Chemistry, 20*(8), 533–548.

Kurian, M., & Paul, A. (2021). Recent trends in the use of green sources for carbon dot synthesis–a short review. *Carbon Trends, 3*, 100032.

Kuznetsov, D., Krylsky, D., Dezhurov, S., Grachev, A., Neschisliaev, V., Orlova, E., & Kuznetsova, A. (2023). Quantum dots are time bomb: Multiscale toxicological study. *Chemico-Biological Interactions, 374*, 110396.

Latiza, R. J. P., Olay, J., Eguico, C., Yan, R. J., & Rubi, R. V. (2025). Environmental applications of carbon dots: Addressing microplastics, air and water pollution. *Journal of Hazardous Materials Advances, 17*, 100591.

Li, H., Kang, Z., Liu, Y., & Lee, S. T. (2012). Carbon nanodots: Synthesis, properties and applications. *Journal of Materials Chemistry, 22*(46), 24230–24253.

Li, S., Li, L., Tu, H., Zhang, H., Silvester, D. S., Banks, C. E., Zou, G., Hou, H., & Ji, X. (2021). The development of carbon dots: From the perspective of materials chemistry. *Materials Today, 51*, 188–207.

Liu, Y.-Y., Yu, N.-Y., Fang, W.-D., Tan, Q.-G., Ji, R., Yang, L.-Y., & Miao, A.-J. (2021). Photodegradation of carbon dots cause cytotoxicity. *Nature Communications, 12*(1), 812.

Liu, Y., Ma, Y., Chen, M., Zhou, T., Ji, R., Guo, R., & Chen, J. (2022). Trophic transfer and environmental safety of carbon dots from microalgae to Daphnia. *Science of the Total Environment, 844*, 157201.

Nair, A., Haponiuk, J. T., Thomas, S., & Gopi, S. (2020). Natural carbon-based quantum dots and their applications in drug delivery: A review. *Biomedicine & Pharmacotherapy, 132*, 110834.

Nazibudin, N. A., Zainuddin, M. F., & Abdullah, C. A. C. (2023). Hydrothermal synthesis of carbon quantum dots: An updated review. *Journal of Advanced Research in Fluid Mechanics and Thermal Sciences, 101*, 192–206.

Otun, K. O., Amusat, S. O., Bello, I. T., Abdulsalam, J., Ajiboye, A. T., Adeleke, A. A., & Azeez, S. O. (2022). Recent advances in the synthesis of various analogues of MOF-based nanomaterials: A mini-review. *Inorganica Chimica Acta, 536*, 120890.

Oualikene-Gonin, W., Sautou, V., Ezan, E., Bastos, H., Bellissant, E., Belgodère, L., Maison, P., Ankri, J., & Scientific Advisory Board of the ANSM. (2023). Regulatory assessment of nano-enabled health products in public health interest. Position of the scientific advisory board of the French National Agency for the Safety of Medicines and Health Products. *Frontiers in Public Health, 11*, 1125577.

Peng, Z., Liu, X., Zhang, W., Zeng, Z., Liu, Z., Zhang, C., Liu, Y., Shao, B., Liang, Q., Tang, W., & Yuan, X. (2020). Advances in the application, toxicity and degradation of carbon nanomaterials in environment: A review. *Environment International, 134*, 105298.

Pirsaheb, M., Mohammadi, S., & Salimi, A. (2019). Current advances of carbon dots based biosensors for tumor marker detection, cancer cells analysis and bioimaging. *TrAC Trends in Analytical Chemistry, 115*, 83–99.

Prokisch, J., Nguyen, D. H., Muthu, A., Ferroudj, A., Singh, A., Agrawal, S., & Rai, M. (2024). Carbon nanodot–microbe–plant nexus in agroecosystem and antimicrobial applications. *Nanomaterials, 14*(15), 1249.

Qiang, S., Zhang, L., Li, Z., Liang, J., Li, P., Song, J., et al. (2022). New insights into the cellular toxicity of carbon quantum dots to Escherichia coli. *Antioxidants, 11*(12), 2475.

Qu, D., Zheng, M., Zhang, L., Zhao, H., Xie, Z., Jing, X., et al. (2014). Formation mechanism and optimization of highly luminescent N-doped graphene quantum dots. *Scientific Reports, 4*(1), 5294.

Rai, M., Shende, S. S., Gade, A. K., Prokisch, J., & Avila-Quezada, G. D. (2025). Carbon nanodots for crop protection and fertilizer use in agriculture. *BioNanoScience, 15*(3), 1–26.

Rajput, V. D., Singh, A., Minkina, T., Rawat, S., Mandzhieva, S., Sushkova, S., & Upadhyay, S. K. (2021). Nano-enabled products: Challenges and opportunities for sustainable agriculture. *Plants, 10*(12), 2727.

Rocha, T. L., Mestre, N. C., Sabóia-Morais, S. M. T., & Bebianno, M. J. (2017). Environmental behaviour and ecotoxicity of quantum dots at various trophic levels: A review. *Environment International, 98*, 1–17.

Roy, P., Chen, P. C., Periasamy, A. P., Chen, Y. N., & Chang, H. T. (2015). Photoluminescent carbon nanodots: Synthesis, physicochemical properties and analytical applications. *Materials Today, 18*(8), 447–458.

Salvi, A., Kharbanda, S., Thakur, P., Shandilya, M., & Thakur, A. (2024). Biomedical application of carbon quantum dots: A review. *Carbon Trends, 17*, 100407.

Sameer, M., Arif, Y., Aqil, A., Nadaf, A., Rafiya, K., Hasan, N., Kesharwani, P., & Ahmad, F. J. (2023). Carbon nanodots as a remedial nanovesicles for drug delivery. *European Polymer Journal, 200*, 112515.

Sciortino, A., Cannizzo, A., & Messina, F. (2018). Carbon nanodots: A review—from the current understanding of the fundamental photophysics to the full control of the optical response. *C, 4*(4), 67.

Singh, A. K., & Singh, H. B. (2019). Impact of engineered nanomaterials on arbuscular mycorrhizal fungi: Implications for sustainable agriculture. *Environmental Research, 172*, 69–76.

Sivasankarapillai, V. S., Kirthi, A. V., Akksadha, M., Indu, S., Dharshini, U. D., Pushpamalar, J., & Karthik, L. (2020). Recent advancements in the applications of carbon nanodots: Exploring the rising star of nanotechnology. *Nanoscale Advances, 2*(5), 1760–1773.

Taran, V., Garkusha, I., Gnidenko, Y., Krasnyj, V., Lozina, A., Taran, A., & Starikova, S. (2020). Portable ozone sterilization device with mechanical and ultrasonic cleaning units for dentistry. *Review of Scientific Instruments, 91*(8).

Varadharajan, S., Vasanthan, K. S., Mathur, V., Hariperumal, N., & Mazumder, N. (2024). Green synthesis and multifaceted applications: Challenges and innovations in carbon dot nanocomposites. *Discover Nano, 19*(1), 205.

Wang, Y., Wang, B., Shi, H., Zhang, C., Tao, C., & Li, J. (2018). Carbon nanodots in ZIF-8: Synthesis, tunable luminescence and temperature sensing. *Inorganic Chemistry Frontiers, 5*(11), 2739–2745.

Wang, X., Feng, Y., Dong, P., & Huang, J. (2019a). A mini review on carbon quantum dots: Preparation, properties, and electrocatalytic application. *Frontiers in Chemistry, 7*, 671.

Wang, H., Zhao, Y., Gao, Y., et al. (2019b). Carbon nanodots disturb nitrogen and phosphorus cycling by altering key soil microbial populations and enzymes. *Environmental Pollution, 255*, 113161.

Wen, M., Sun, N., Jiao, L., Zang, S. Q., & Jiang, H. L. (2024). Microwave-assisted rapid synthesis of MOF-based single-atom Ni catalyst for CO2 electroreduction at ampere-level current. *Angewandte Chemie International Edition, 63*(10), e202318338.

Xiao, L., & Sun, H. (2018). Novel properties and applications of carbon nanodots. *Nanoscale Horizons, 3*(6), 565–597. https://doi.org/10.1039/c8nh00106e

Xiong, Y., Chen, X., Li, F., Chen, Z., & Qin, Z. (2022). Zebrafish larvae acute toxicity test: A promising alternative to the fish acute toxicity test. *Aquatic Toxicology, 246*, 106143.

Yang, H.-L., Bai, L.-F., Geng, Z.-R., Chen, H., Xu, L.-T., Xie, Y.-C., et al. (2023). Carbon quantum dots: Preparation, optical properties, and biomedical applications. *Materials Today Advances, 18*, 100376.

Yao, K., Lv, X., Zheng, G., Chen, Z., Jiang, Y., Zhu, X., Wang, Z., & Cai, Z. (2018). Effects of carbon quantum dots on aquatic environments: Comparison of toxicity to organisms at different trophic levels. *Enviromental Science and Technology, 52*(1), 14445–14451.

Yu, M., Li, P., Huang, R., Xu, C., Zhang, S., Wang, Y., et al. (2023). Antibacterial and antibiofilm mechanisms of carbon dots: A review. *Journal of Materials Chemistry B, 11*(4), 734–754.

Zhang, R., & Chen, W. (2014). Nitrogen-doped carbon quantum dots: Facile synthesis and application as a “turn-off” fluorescent probe for detection of Hg2+ ions. *Biosensors and Bioelectronics, 55*, 83–90.

Zhou, W., & Peijnenburg, W. J. G. M. (2021). Nanomaterial-induced changes in soil microbial diversity and function: Current understanding and future perspectives. *Environmental Science: Nano, 8*(1), 11–30.

Zhou, Q., Lin, L., Xu, Y., & Gao, J. (2018). Impacts of carbon-based nanomaterials on soil microbial communities and enzyme activities: A review. *Environmental Nanotechnology, Monitoring & Management, 10*, 302–310.

MIX
Papier aus verantwortungsvollen Quellen
Paper from responsible sources
FSC® C105338

If you have any concerns about our products,
you can contact us on
ProductSafety@springernature.com

In case Publisher is established outside the EU,
the EU authorized representative is:
Springer Nature Customer Service Center GmbH
Europaplatz 3, 69115 Heidelberg, Germany

Printed by Libri Plureos GmbH
in Hamburg, Germany